C000094368

1,000,000 Books

are available to read at

www.ForgottenBooks.com

Read online
Download PDF
Purchase in print

ISBN 978-1-5278-9213-2
PIBN 10925618

This book is a reproduction of an important historical work. Forgotten Books uses state-of-the-art technology to digitally reconstruct the work, preserving the original format whilst repairing imperfections present in the aged copy. In rare cases, an imperfection in the original, such as a blemish or missing page, may be replicated in our edition. We do, however, repair the vast majority of imperfections successfully; any imperfections that remain are intentionally left to preserve the state of such historical works.

Forgotten Books is a registered trademark of FB &c Ltd.
Copyright © 2018 FB &c Ltd.
FB &c Ltd, Dalton House, 60 Windsor Avenue, London, SW19 2RR.
Company number 08720141. Registered in England and Wales.

For support please visit www.forgottenbooks.com

1 MONTH OF FREE READING

at

www.ForgottenBooks.com

By purchasing this book you are eligible for one month membership to ForgottenBooks.com, giving you unlimited access to our entire collection of over 1,000,000 titles via our web site and mobile apps.

To claim your free month visit:

www.forgottenbooks.com/free925618

* Offer is valid for 45 days from date of purchase. Terms and conditions apply.

English
Français
Deutsche
Italiano
Español
Português

www.forgottenbooks.com

Mythology Photography **Fiction** Fishing Christianity **Art** Cooking Essays Buddhism Freemasonry Medicine **Biology** Music **Ancient Egypt** Evolution Carpentry Physics Dance Geology **Mathematics** Fitness Shakespeare **Folklore** Yoga Marketing **Confidence** Immortality Biographies Poetry **Psychology** Witchcraft Electronics Chemistry History **Law** Accounting **Philosophy** Anthropology Alchemy Drama Quantum Mechanics Atheism Sexual Health **Ancient History** **Entrepreneurship** Languages Sport Paleontology Needlework Islam **Metaphysics** Investment Archaeology Parenting Statistics Criminology **Motivational**

OF

MINERALOGY,

IN WHICH

MINERALS ARE ARRANGED ACCORDING TO THE NATURAL HISTORY METHOD.

BY

ROBERT JAMESON,

REGIUS PROFESSOR OF NATURAL HISTORY, LECTURER ON MINERALOGY, AND KEEPER OF THE MUSEUM IN THE UNIVERSITY OF EDINBURGH;

FELLOW OF THE ROYAL AND ANTIQUARIAN SOCIETIES OF EDINBURGH; PRESIDENT OF THE WERNERIAN NATURAL HISTORY SOCIETY, AND MEMBER OF THE ROYAL MEDICAL AND PHYSICAL SOCIETIES OF EDINBURGH; HONORARY MEMBER OF THE ROYAL IRISH ACADEMY, AND OF THE HONOURABLE DUBLIN SOCIETY; FELLOW OF THE LINNEAN AND GEOLOGICAL SOCIETIES OF LONDON, AND OF THE ROYAL GEOLOGICAL SOCIETY OF CORNWALL; OF THE ROYAL DANISH SOCIETY OF SCIENCES; OF THE ROYAL ACADEMY OF SCIENCES OF NAPLES; OF THE IMPERIAL NATURAL HISTORY SOCIETY OF MOSCOW; OF THE SOCIETY OF NATURAL HISTORY OF WETTERAU; OF THE MINERALOGICAL SOCIETY OF JENA; OF THE MINERALOGICAL SOCIETY OF DRESDEN; HONORARY MEMBER OF THE LITERARY AND PHILOSOPHICAL SOCIETY OF NEW-YORK; OF THE NEW-YORK HISTORICAL SOCIETY, &c.

THIRD EDITION, ENLARGED AND IMPROVED.

VOL. II.

EDINBURGH:

PRINTED FOR ARCHIBALD CONSTABLE & CO. EDINBURGH:

AND HURST, ROBINSON & CO. CHEAPSIDE, LONDON.

1820.

P. Neill, Printer.

TABLE OF CONTENTS

OF

VOLUME SECOND.

SYSTEM OF MINERALOGY.

CLASS I.—EARTHY MINERALS.

Order II.—Spar,—(continued.)

	Page
Genus V. Felspar.	
1. Prismatic Felspar,	1
1st Subsp. Adularia,	2
2d ——— Glassy Felspar,	6
3d ——— Ice Spar,	8
4th ——— Common Felspar,	9
5th ——— Labrador Felspar,	18
6th ——— Compact Felspar,	22
7th ——— Clinkstone, -	25
8th ——— Earthy Common Felspar,	28
9th ——— Porcelain Earth, -	30
2. Pyramidal Felspar, or Scapolite,	35
1st Subsp. Radiated Scapolite,	ib.
2d ——— Foliated Scapolite,	38
3d ——— Compact Red Scapolite,	40
4th ——— Elaolite, - -	41
3. Prismato-Pyramidal Felpar, or Meionite,	43

4. Rhomboidal

Page

4. Rhomboidal Felspar, or Nepheline, - 46
* Chiastolite, - - 49
** Sodalite, 52

APPENDIX, including CLAY and LITHOMARGE FAMILIES.

* CLAY FAMILY.

1. Aluminite, 55
2. Common Clay,
Loam, 57
Potters-Clay, 58
3. Variegated Clay, 61
4. Slate-Clay, . 62
5. Bituminous Shale, 63
6. Clay-stone, . 66
7. Adhesive-Slate, 68
8. Polier Slate, 70
9. Tripoli, 71

** LITHOMARGE FAMILY.

1. Lithomarge.
a. Friable Lithomarge, 74
b. Indurated Lithomarge, 76
2. Mountain Soap, - 79
3. Yellow Earth. 81
4. Cimolite, 83
5. Kollyrite, 85
6. Bole, - 86
7. Sphragide, or Lemnian Earth, 89

Genus VI. SPODUMENE.
1. Prismatic Spodumene, 91

	Page
4. Prismatic Augite or Tabular Spar,	170
Genus IX. Schiller-Spar.	
1. Green Diallage,	172
2. Schiller-Spar,	174
1st Subsp. Bronzite, -	175
2d ——— Common Schiller-Spar,	177
3. Hyperstene, or Labrador Schiller-Spar,	178
4. Anthophyllite, - -	181

Order III. Mica.

	Page
Genus I. Copper Mica.	
1. Prismatic Copper Mica, - -	184
Genus II. Uranite, or Uran Mica,	186
1. Pyramidal Uranite,	187
* Uran Ochre, -	190
Genus III. Red Cobalt, or Cobalt Mica.	
1. Prismatic Red Cobalt.	
1st Subsp. Radiated Red Cobalt, -	192
2d ——— Earthy Red Cobalt,	195
3d ——— Slaggy Red Cobalt, -	197
* Cobalt Ochre, -	ib.
1. Black Cobalt Ochre.	
a. Earthy Black Cobalt Ochre,	198
b. Indurated Black Cobalt Ochre,	199
2. Brown Cobalt Ochre, -	201
3. Yellow Cobalt Ochre, -	203
Genus IV. White Antimony, or Antimony Mica.	
1. Prismatic White Antimony, -	205
* Antimony Ochre, - -	207

Genus V.

Genus V. Blue Iron. Page

1. Prismatic Blue Iron,
1st Subsp. Foliated Blue Iron, 209
2d ——— Fibrous Blue Iron, 212
3d ——— Earthy Blue Iron, 213

Genus VI. Graphite.

1. Rhomboidal Graphite.
1st Subsp. Scaly Graphite, 217
2d ——— Compact Graphite, 218

Genus VII. Mica.

1. Rhomboidal Mica, 221
1st Subsp. Mica, 222
2d ——— Pinite, 227
3d ——— Lepidolite, 230
4th ——— Chlorite.
1st Kind, Earthy Chlorite, 233
2d —— Common Chlorite, 235
3d —— Slaty Chlorite, 237
4th —— Foliated Chlorite, 239
5th ——— Green Earth, - 241
6th ——— Talc, - - 243
1st Kind, Common Talc, 245
2d —— Indurated Talc, 248
7th ——— Nacrite, - - 251
8th ——— Potstone, - 252
9th ——— Steatite, or Soapstone, 255
10th ——— Figurestone, or Agalmatolite, 261

* Clay-Slate, - 263
* Whet Slate, 271
* Drawing Slate, 273
* Alum-Slate.
1st Kind, Common Alum Slate, 276
2d —— Glossy Alum-Slate, 277

1. Native

APPENDIX. Page

1. Native Magnesia, 279
2. Magnesite, - 281
3. Meerschaum, 283
4. Nephrite.
   a. Common Nephrite, 287
   b. Axestone, - 290
5. Serpentine.
   a. Common Serpentine, - 292
   b. Precious Serpentine, - 296
      α. Splintery Precious Serpentine, 297
      β. Conchoidal Precious Serpentine, 298
6. Fullers Earth, - - 300

Order IV. MALACHITE.

Genus I. COPPER GREEN.

1. Common Copper Green.
   1st Subsp. Conchoidal Copper Green, 305
   2d ——— Earthy Ironshot Copper Green, 309
   3d ——— Slaggy Ironshot Copper Green, 310

Genus II. MALACHITE.

1. Blue Copper, or Prismatic Malachite.
   1st Subsp. Radiated Blue Copper, - 313
   2d ——— Earthy Blue Copper, 319
   * Velvet Blue Copper, - 320
2. Common or Acicular Malachite †, - 321
   1st Subsp. Fibrous Malachite, . 322
   2d ——— Compact Malachite, - 325
   * Brown Copper, - - 329

Genus III.

† Owing to some mistake, in page 321. the generic name Malachite is introduced, and the name of the Species, viz. Common or Acicular Malachite omitted.

Genus III. OLIVENITE. Page

1. Prismatic Olivenite, or Phosphat of Copper, 331

2. Di-prismatic Olivenite, or Lenticular Copper, 333

3. Acicular Olivenite.

1st Subsp. Radiated, 335
2d —— Foliated, 336
3d —— Fibrous, 338
4th —— Earthy, - 340

4. Hexahedral Olivenite, or Cube-Ore, 341

* Atacamite, or Muriate of Copper, - 343
1st Subsp. Compact Atacamite, ib.
2d —— Arenaceous Atacamite, - 345

Genus IV. EMERALD-COPPER.

1. Rhomboidal Emerald-Copper, - 347

Order V.—KERATE.

Genus I. CORNEOUS SILVER.

1. Hexahedral Corneous Silver, 350

* Earthy Corneous Silver, - 354

Genus II. CORNEOUS MERCURY. - 355

1. Pyramidal Corneous Mercury, - 356

Order VI.—BARYTE.

Genus I. LEAD-SPAR.

1. Triprismatic Lead-Spar, or Sulphate of Lead, 359
2. Pyramidal Lead-Spar, or Yellow Lead-Spar, 362

3. Prismatic Lead-Spar, or Red Lead-Spar, 366

4. Rhomboidal Lead-Spar. - -

1st Subsp. Green Lead-Spar, - 369
2d —— Brown Lead-Spar, - 374

5. Diprismatic,

5. Diprismatic Lead-Spar. [Page

1st Subsp. White Lead-Spar, 377
2d ——— Black Lead-Spar, 382
3d ——— Earthy Lead-Spar,
1st Kind, Indurated, - 384
2d —— Friable Earthy, 386

* Corneous Lead, - - 388
* Arseniate of Lead.

1st Subsp. Reniform, 390
2d ——— Filamentous, 391
3d ——— Earthy, - - 392

* Native Minium, or Native Red Oxide of Lead, - - 393

§ II. Baryte.

1. Rhomboidal Baryte or Witherite, - 394

2. Prismatic Baryte, or Heavy-Spar, - 398

1st Subsp. Earthy Heavy-Spar, - ib.
2d ——— Compact Heavy-Spar, - 400
3d ——— Granular Heavy-Spar, - 401
4th ——— Curved Lamellar Heavy-Spar, 403
5th ——— Straight Lamellar Heavy-Spar.
1st Kind, Fresh, - 405
2d —— Disintegrated, 409
3d —— Fetid, - 410
6th ——— Fibrous Heavy-Spar, - 412
7th ——— Radiated Heavy-Spar, 414
8th ——— Columnar Heavy-Spar, 416
9th ——— Prismatic Heavy-Spar, 418

3. Diprismatic Baryte, or Strontianite, 420

4. Axifrangible Baryte or Celestine.

1st Subsp. Foliated Celestine, - 423
2d ——— Prismatic Celestine, 426
3d ——— Fibrous Celestine. - 428
4th ——— Radiated Celestine, - 430
5th ——— Fine Granular Celestine, ib.

Genus III.

Genus III. TUNGSTEN OR SCHEELIUM. Page

1. Pyramidal Tungsten, - . - 432

Genus IV. CALAMINE.

1. Prismatic Calamine, 437

2. Rhomboidal Calamine, - - 440

1st Subsp. Sparry Rhomboidal Calamine, 441
2d ——— Compact Rhomboidal Calamine, 442
3d ——— Earthy Rhomboidal Calamine, 443

Genus V. RED MANGANESE.

1. Rhomboidal Red Manganese, - 445

1st Subsp. Foliated Rhom. Red Manganese, 446
2d ——— Fibrous Rhomboidal Red Manganese, - - 448
3d ——— Compact Rhomboidal Red Manganese, - - 449

Genus VI. SPARRY IRON.

1. Sparry Iron, - 451

Order VII.—HALOIDE.

Genus I. LIMESTONE.

1. Rhomb-Spar, 458

2. Dolomite, -

1st Subsp. Dolomite.

1st Kind, Granular Dolomite, 462
* White Granular Dolomite, ib.
* Brown Dolomite, 466
* Flexible Dolomite, 469
2d —— Columnar Dolomite, 470
3d —— Compact Dolomite, 471

2d Subsp. Miemite, Page
1st Kind, Granular Miemite, 473
2d —— Prismatic Miemite, 475
3d —— Brown Spar.
1st Kind, Foliated Brown Spar, 476
2d —— Columnar Brown Spar, 480
3. Limestone, - - 481
1st Subsp. Foliated Limestone.
1st Kind, Calcareous Spar, 482
2d —— Granular Foliated Limestone, - - 490
2d —— Compact Limestone.
1st Kind, Common Compact Limestone, - - 511
2d —— Blue Vesuvian Limestone, 517
3d —— Roestone, - 518
3d —— Chalk, - 521
4th —— Agaric Mineral, 526
5th —— Fibrous Limestone, 528
1st Kind, Common Fibrous Limestone, - 529
2d —— Fibrous Calc-Sinter, 530
6th —— Tuffaceous Limestone, 537
7th —— Pisiform Limestone, 539
8th —— Slate-Spar, - 541
9th —— Aphrite, - - 543
1st Kind, Scaly Aphrite, - 544
2d —— Slaty Aphrite, 546
3d —— Sparry Aphrite, 547
10th —— Lucullite.
1st Kind, Compact Lucullite, 549
a. Common Compact Lucullite, ib.
b. Stinkstone, - 553
2d Kind, Prismatic Lucullite, 556
3d —— Foliated or Sparry Lucullite, 559

11th Subsp. Marl. Page
1st Kind. Earthy Marl, 561
2d —— Compact Marl, 563
12th —— Bituminous Marl-Slate, 566
4. Prismatic Limestone or Arragonite.
1st Subsp. Common Arragonite, - 368
2d —— Coralloidal Arragonite, 572

Genus II. Apatite, - - 574
1. Rhomboidal Apatite, - 574
1st Subsp. Foliated Apatite, 575
2d —— Conchoidal Apatite, 580
3d —— Phosphorite.
1st Kind, Common Phosphorite, 583
2d —— Earthy Phosphorite, 585

Genus III. Fluor.
1. Octahedral Fluor.
1st Subsp. Compact Fluor, 587
2d —— Foliated Fluor, 589
3d —— Earthy Fluor, 597

Genus IV. Alum-Stone, - 598
1. Rhomboidal Alum-Stone, 599

Genus V. Cryolite,
1. Pyramidal Cryolite. 601

Genus VI. Gypsum, - - 604
1. Prismatic Gypsum or Anhydrite.
1st Subsp. Sparry Anhydrite, - 605
2d —— Scaly Anhydrite, 608
3d —— Fibrous Anhydrite, 609
4th —— Convoluted Anhydrite, ib.
5th —— Compact Anhydrite, 611
* Vulpinite, - 612
* Glauberite, - 613

2, Axifrangible

2. Axifrangible Gypsum. Page
1st Subsp. Sparry Gypsum or Selenite, 615
2d —— Foliated Granular Gypsum, 619
3d —— Compact Gypsum, - 624
4th —— Fibrous Gypsum, 627
5th —— Scaly Foliated Gypsum, 629
6th —— Earthy Gypsum, 630
* Montmartrite, 632

MINERAL

MINERAL SYSTEM.

CLASS I.

EARTHY MINERALS.

(CONTINUED.)

ORDER II. SPAR.—(*CONTINUED.*)

GENUS V.—FELSPAR.

THIS Genus contains four Species, viz. Prismatic Felspar, Pyramidal Felspar, Prismato-Pyramidal Felspar, and Rhomboidal Felspar.—* *Chiastolite*, ** *Sodalite.*

1. Prismatic Felspar.

Prismatischer Feldspath, *Mohs.*

THIS Species is divided into nine Subspecies, viz. 1. Adularia, 2. Glassy-Felspar, 3. Ice-Spar, 4. Common Felspar, 5. Labrador Felspar, 6. Compact Felspar, 7. Clinkstone, 8. Earthy Common Felspar; and, 9. Porcelain Earth.

First Subspecies.

Adularia.

Adular, *Werner*.

Moonstone, *Kirw.* vol. i. p. 322.—Adular, *Estner*, b. ii. s. 525. *Id. Emm.* b. i. s. 277.—Adularia, *Nap.* p. 218.—Adulaire, *La Meth.* t. ii. p. 194. *Id. Broch.* t. i. p. 371.—Feldspath nacré, *Haüy*, t. ii. p. 600.—Adular, *Reuss*, b. ii. s. 379. *Id. Lud.* b. i. s. 101.—Opalisirender Feldstein, *Bert.* s. 242.—Adular, *Suck.* 1r th. s. 389. *Id. Mohs*, b. i. s. 394.—Adularischer Feldspath, *Hab.* s. 21.—Feldspath nacré, *Lucas*, p. 50.—Opalisirender Feldspath, *Leonhard*, Tabel. s. 17.—Feldspath Adulaire, *Brong.* t. i. p. 358.—Feldspath limpide, *Brard*, p. 134. Opalisirender Feldspath, *Karsten*, Tabel. s. 34.—Adularia, *Kid*, vol. i. p. 158.—Feldspath nacré, Tabl. p. 36.—Adular, *Steffens*, b. i. s. 422. *Id. Hoff.* b. i. s. 296. *Id. Lenz*, b. i. s. 486.—Opalisirender Feldspath, *Oken*, b. i. s. 375.—Adular, *Haus.* Handb. b. ii. s. 532. *Id. Aikin*, p. 196.

External Characters.

The principal colour is greenish-white, which sometimes passes into greyish-white and milk-white, and even inclines to asparagus-green. It is frequently iridescent; and the milk-white varieties, in thin plates, when held between the eye and the light, sometimes appear pale flesh-red.

It occurs massive, and this variety is composed of granular and thick lamellar concretions; and frequently crystallized.

The primitive figure is an oblique four-sided prism, with two broad and two narrow lateral planes: the lateral edges are 120° and 60° *. The following are the most frequent secondary figures:

1. Oblique

* It is not necessary to repeat the description of the primitive form in the accounts of the other subspecies of a species, as it is understood throughout the whole work that it must be the same in all the members of the same species.

1. Oblique four-sided prism, flatly bevelled on the extremities, and the bevelling planes set on the obtuse lateral edges.

Sometimes two diagonally opposite bevelling planes become smaller than the others, and at length disappear, when there is formed

2. An oblique four-sided prism, in which the terminal planes are set on obliquely.

3. The figure N° 1. is sometimes truncated on the acute lateral edges. When these truncating planes become larger, there is at length formed

4. A broad rectangular six-sided prism, flatly bevelled on both extremities, and the bevelling planes are set on the lateral edges, which are formed by the smaller lateral planes.

Sometimes the prism becomes so broad and thin, that it may be described as a

5. Six-sided table, in which the smaller lateral planes of the preceding figure form bevelments on the terminal planes.

6. Rectangular four-sided prism, in which the terminal planes are obliquely bevelled.

Sometimes twin-crystals occur: one variety is the same as that afterwards to be described as occurring in common felspar; the other is formed by two tabular crystals of the variety 5. growing together by their broader lateral planes.

The crystals are generally middle-sized and large, sometimes very large, but seldom small. They are always superimposed, and either single or variously aggregated.

The lateral planes of the prism are longitudinally streaked.

Externally it is splendent; internally the cleavage is splendent, and the fracture shining and glistening. The lustre is intermediate between vitreous and pearly.

 The

The cleavage is threefold: two very distinct cleavages are in the direction of the terminal and smaller lateral planes of the primitive figure; and one less distinct in the direction of the broader lateral planes.

The fracture is small and imperfect conchoidal, sometimes approaching to uneven.

The fragments are indeterminate angular and sharp-edged.

It is semi-transparent, sometimes inclining to transparent, or is translucent.

The translucent varieties, when viewed in a certain direction, sometimes exhibit a silvery or pearly light *. It refracts double.

It is harder than apatite, but softer than quartz.

It is easily frangible.

Specific gravity, 2.564, *Brisson*.—2.531 & 2.560, *Hoffmann*.—2.5, *Mohs*.

Chemical Characters.

It melts before the blowpipe, without addition, into a white-coloured transparent glass.

Constituent Parts.

Silica,	64
Alumina, - -	20
Lime, -	2
Potash, -	14
	100 *Vauquelin.*

Geognostic

* This beautiful pearly light is generally seen when the specimen is viewed in the direction of the imperfect or third cleavage.

Geognostic Situation.

It occurs in cotemporaneous veins or drusy cavities in granite and gneiss: in these repositories it is associated with rock-crystal, calcareous-spar, epidote, amianthus, but principally with chlorite and common felspar.

Geographic Situation.

Europe.—It occurs in the granite of the Island of Arran; and in the granite and gneiss rocks of Norway, Switzerland, France, and Germany. The largest and most beautiful crystals are found in the mountain of Stella, a part of St Gothard.

Asia.—Rolled pieces, having a most beautiful pearly light, are collected in the Island of Ceylon.

America.—Moonstone-adularia is found in Greenland; and all the varieties in the United States.

Uses.

The variety of adularia which exhibits the bluish pearly light, is valued by jewellers, and is sold by them under the name *Moonstone.* It is cut in a low oval form, and in such a manner as to present the pearly spot in the centre of the gem. It is set in rings or brooches, with rubies and emeralds, with which it forms an agreeable contrast. Sometimes ringstones of it are set round with diamonds, and its pearly light forms a striking and agreeable contrast with the lustre and colours of that gem. The finest specimens are brought from Ceylon; but even there, perfect stones are rare.

Another variety of adularia, found in Siberia, is known to jewellers under the name *Sunstone.* It is of a yellowish-grey colour, and numberless golden spots appear distributed throughout its whole substance. These shining gol-

den

den reflections are either from minute fissures, or irregular cleavages of the mineral. The aventurine felspar of Archangel, to be afterwards mentioned, appears also to be sunstone.

Observations.

1. This mineral is known by its white colour, iridescence, pearly light, splendent external and internal lustre, conchoidal fracture, high degree of transparency, specific gravity, and considerable hardness.

2. It was first discovered by an Italian mineralogist, Professor Pini of Milan, in the mountain of Stella, belonging to the St Gothard group. He named it *Adularia Felspar*, in the belief that the mountain on which he had collected it was named *Adula*; but the truth is, the mountain of Adula does not occur near St Gothard; it is situated in the Grisons.

3. The moonstone appears to be the Hyaloides (Ὑαλοειδὴς) of Theophrastus, and the Astrios of Pliny. The Asteria of Pliny is not, as some imagine, a variety of adularia; it appears rather to belong to the Cat's-eye and Asteria-sapphire.

Second Subspecies.

Glassy Felspar.

Glasiger Feldspath, *Werner*.

Nose, Orthographische Briefe, 1. s. 128.—Nöggerath Studien. s. 27.—*Reuss*, Mineralogishe Briefe, 1. n. 2. a. a. o.—Glasiger Feldspath, *Karst.* Tabel. s. 34. *Id. Haus.* s. 88. *Id. Steffens*, b. i. s. 441. *Id. Hoff.* b. ii. s. 328. *Id. Lenz*, b. ii. s. 502. *Id. Oken*, b. i. s. 375. *Id. Haus.* Handb. b. ii. s. 532. —Glassy Felspar, *Aikin*, p. 197.

External

External Characters.

Its colour is greyish-white, sometimes passing into grey.

It occurs always crystallized, in broad rectangular four-sided prisms, bevelled on the extremities. These crystals are often very much cracked; they are generally small, seldom middle-sized, and always imbedded.

Internally it is splendent, and the lustre is vitreous.

The cleavage is the same as in adularia.

The fracture is uneven, or small and imperfect conchoidal.

It is transparent.

In all its other characters it agrees with adularia.

Specific gravity, 2.575, *Klap.*—2.518, 2.589, *Stucke.*

Chemical Characters.

Before the blowpipe, it melts without addition into a grey semi-transparent glass.

Constituent Parts.

Silica,	68.0
Alumina,	15.0
Potash,	14.5
Oxide of Iron,	0.5
	98.0

Klaproth, Beit. b. v. s. 18.

Geognostic and Geographic Situations.

It occurs imbedded in pitchstone-porphyry in Arran and Rume; in a porphyritic rock in the Siebengebirge; also in a rock composed of white felspar, and very small blackish-brown,

brown scales of mica, and fine disseminated magnetic ironstone, in the Drachenfels on the Rhine. It is an inmate of the secondary trap-rocks of the Bohemian Mittelgebirge; and has been noticed in the porphyritic pumice of Hungary. It is said also to occur in veins in Dauphiny, along with axinite and epidote; and in the lava of Solfatara.

Observations.

Glassy Felspar is distinguished from the other minerals of the felspar species, by its white colour, splendent vitreous lustre, transparency, and the frequent rents or fissures with which it is traversed.

Third Subspecies.

Ice-Spar *

Eispath, *Werner*.

Eis-spath, *Chierici*, Moll's Ephem. 5. 1. s. 126. *Id. Steffens*, b. i. s. 478. *Id. Hoff.* b. ii. s. 369. *Id. Lenz*, b. i. s. 515.

External Characters.

Its colour is greyish-white, which inclines sometimes to yellowish-white, sometimes to greenish-white.

It occurs massive, cellular, and porous; also in large granular concretions, which are composed of thin and straight lamellar concretions. It is frequently crystallized in the form of small thin longish six-sided tables, in which the shorter terminal planes are bevelled.

The

* It is named *Ice-spar*, from its icy appearance and sparry structure.

The lateral planes of the tables are longitudinally streaked.

Externally the crystals are shining, and sometimes splendent: internally shining, and the lustre is vitreous.

The cleavage is imperfect.

The fragments are indeterminate angular and sharp-edged.

The massive and other varieties are strongly translucent; the crystals are transparent.

It is as hard as common felspar, and is very easily frangible.

Geognostic and Geographic Situations.

It occurs, along with nepheline, meionite, mica, and hornblende, at Monte Somma, near Naples.

Fourth Subspecies.

Common Felspar.

Frischer Gemeiner Feldspath, *Werner.*

Spathum scintillans, *Wall.* t. i. p. 214.—Feldspath, *Wid.* s. 335. *Id. Romé de Lisle,* t. ii. p. 445.—Common Felspar, *Kirw.* vol. i. p. 316.—Blättrig Feldstein, *Estner,* b. i. s. 513.—Gemeiner Feldspath, *Emm.* b. i. s. 266.—Feldispato commune, *Nap.* p. 213.—Feldspath, *Lam.* t. ii. p. 187. *Id. Haüy,* t. ii. p. 590.—Le Feldspath commun, *Broch.* t. i. p. 362.—Gemeiner Feldspath, *Reuss,* b. ii. s. 369. *Id. Lud.* b. i. s. 100. *Id. Suck.* 1r th. s. 380.—Gemeiner Feldstein, *Bert.* s. 238.—Gemeiner Feldspath, *Mohs,* b. i. s. 407. *Id. Hab.* s. 20.—Feldspath, *Lucas,* p. 50.—Gemeiner frischer Feldspath, *Leonhard,* Tabel. s. 18.—Feldspath commun, *Brong.* t. i. p. 367.—Feldspath, *Brard,* p. 131.—Gemeiner Feldspath, *Haus.* s. 88. *Id. Karsten,* Tabel. s. 34.—Felspar, *Kid,* vol. i. p. 157.—Feldspath,

Feldspath, *Haüy,* Tabl. p. 35.—Frischer gemeiner Feldspath, *Steffens,* b. i. s. 436. *Id. Hoff.* b. ii. s. 309. *Id. Lenz,* b. i. s. 494. *Id. Oken,* b. i. s. 374. *Id. Haus.* Handb. b. ii. s. 529.—Common Felspar, *Aikin,* p. 196.

External Characters.

Its most frequent colours are white and red, seldom grey, and rarely green and blue. The white varieties are greenish-white, milk-white, yellowish-white, greyish-white, snow-white, and reddish-white; from reddish-white it passes into flesh-red, and into a colour intermediate between flesh-red and blood-red: from greenish-white it passes into apple-green, asparagus-green, grass-green, emerald-green, leek-green, mountain-green, verdigris-green; and from this latter into sky-blue: from milk-white it passes into bluish-grey, smoke-grey, and yellowish-grey. The grey varieties are generally spotted.

It occurs most frequently massive and disseminated, seldom in blunt angular rolled pieces and grains, and frequently in granular distinct concretions, from the smallest to the largest size; and sometimes crystallized, in the following figures:

1. Very oblique four-sided prism, flatly bevelled on both extremities, and the bevelling planes set on the obtuse lateral edges, fig. 90. Pl. 5. *. This may be considered as the fundamental figure.
2. The preceding crystallization, in which two diagonally opposite bevelling planes are smaller than the two others †. Sometimes the latter entirely disappear, when there is formed
3. A perfect and very oblique four-sided prism, in which the terminal planes are set on obliquely; or when

* Feldspath ditetraedre, Haüy.

† The bevelling planes mentioned above, are those that form the great-

when the prism becomes shorter, and all the planes diminish in an equal proportion, there is formed

4. An acute rhombus *, fig. 91. Pl. 5.

When the prism of the fundamental figure N° 1. becomes shorter, and the bevelling planes become much larger than the lateral planes, we can view the former as lateral planes, and the latter as bevelling planes, and thus there is formed

5. A very oblique four-sided prism, acutely bevelled on the extremities, and the bevelling planes set on the acute lateral edges. This figure sometimes passes into a kind of

6. Elongated octahedron.

7. The fundamental figure, truncated on the acute lateral edges.

8. The variety N° 3. truncated on the acute edges, in the same manner as the variety N° 7 †, fig. 92. Pl. 5.

When the truncating planes of the variety N° 7. become larger than the lateral planes, there is formed

9. A broad equiangular six-sided prism, flatly bevelled on the extremities, and the bevelling planes set on those lateral edges which are formed by the meeting of the smaller lateral planes, fig. 93. Pl. 5 ‡.

10. The preceding figure, in which the edges formed by the meeting of the larger and smaller lateral planes, are truncated. fig. 94. Pl. 5 ||.

11. The

* Feldspath binaire, Haüy.

† Feldspath prismatique, Haüy.

‡ Feldspath bibinaire, Haüy.

|| Felspath quadridecimal, Haüy.

11. The crystallization N° 9. in which the angles formed by the meeting of the smaller bevelling planes and the lateral edges on which they are set, are more or less deeply truncated, fig. 95 *. Pl. 5.
12. The preceding variety, in which the edges formed by the smaller bevelling planes and the broader lateral planes are truncated, fig. 96. Pl. 5 †.
13. The preceding variety, in which the edges formed by the meeting of the other bevelling planes with the broader lateral planes, are truncated ‡.
14. In all the preceding varieties from N° 9. the proper edge of the bevelment is sometimes truncated ‖.
15. The smaller bevelling planes in N° 11. sometimes disappear, whilst the truncating planes on the angles become larger, and form with the larger bevelling plane a new and much more acute bevelment, fig. 97. §, Pl. 5.

When two bevelling planes in variety 9. become very large, as in N° 2. whilst the prism becomes very broad and short, so that these two large bevelling planes approach near to each other, and increase in equal proportion with the broader lateral planes

* Feldspath dihexaedre, Haüy.

† Feldspath sexdecimal, Haüy.

‡ These truncating planes, along with some others, occur in Haüy's Feldspath synoptique and Feldspath decidodecaedre.

‖ This appearance is to be seen in Haüy's Feldspath apophane and Feldspath synoptique.

§ As in Haüy's Feldspath decidodecaedre.

planes with which they meet under a right angle, they form with these

16. A rectangular four-sided prism, in which the smaller lateral planes of the six-sided prism form a kind of oblique bevelment on the terminal planes, and which is variously modified by the remains of the smaller bevelling planes of the fundamental figure, and the other planes of alteration *.

17. The preceding figure truncated on the lateral edges. These truncating planes correspond with those of the 13th crystallization.

Sometimes the planes at the extremities of the figure N° 16. almost totally disappear, and there remain only the two truncating planes of the 11th crystallization, and then the figure becomes

18. A nearly perfect rectangular four-sided prism, in which the terminal planes are set on obliquely, fig. 98 †. Pl. 5.

Besides these simple crystallizations, twin-crystals also occur, of which the following are the principal varieties:

19. Twin-crystal, which we may suppose to have been formed by two prisms of Nos 9. or 15. being pushed into each other in the direction of their thickness, in such a manner that their axes are either parallel to each other, or form a more or less obtuse angle. The lateral planes, and also some of those

* Vid. Haüy, fig. 91. and 92. Romé de Lisle assumed this as the fundamental form of felspar.

† Feldspath unitaire, Haüy.

those at the extremities of the crystals, form re-entering angles.

20. Twin-crystal, which we can conceive to be formed when Nos. 16. & 17. are divided longitudinally from one extremity to the other, in the direction of the two opposite lateral planes, (the broader lateral planes of the six-sided prism), and the one-half turned completely around and applied to the other. In this way a rectangular four-sided prism is formed, in which the diagonally opposite planes of the two extremities of the single crystal will be placed together. This is the *hemitrope* crystal of Haüy *.

The crystals are generally small and middle-sized, seldom very small, large, and very large. They are generally imbedded, sometimes also superimposed, and variously aggregated, forming druses.

Internally the cleavage is shining, and sometimes splendent; the fracture is glistening, and frequently not more than feebly glistening. The lustre is intermediate between vitreous and pearly, but inclining rather more to the former than to the latter.

It has a three-fold cleavage, like that of adularia, and the folia are sometimes curved floriform.

The fracture is uneven or splintery.

The fragments are rhomboidal, and have only four splendent shining faces.

It is translucent, or only translucent on the edges.

It is hard, but in a lower degree than quartz.

It

* Romé de Lisle, t. ii. p. 478.—492. var. 10.—16. Pl. 3. fig. 94.—106.

It is very easily frangible.

Specific gravity, 2.594, *Brisson.*—2.551, 2.567, *Hoff.*

Chemical Characters.

Before the blow-pipe, it is fusible without addition into a grey semitransparent glass.

Constituent Parts.

	Siberian Green Felspar.	Flesh-red Felspar.	Felspar from Passau.
Silica,	62.83	66.75	60.25
Alumina,	17.02	17.50	22.00
Lime,	3.00	1.25	0.75
Potash,	13.00	12.00	14.00
Oxide of Iron,	1.00	0.75	a trace.
Water,	-		1.00
	96.85	98.25	98.00
	Vauquelin, Jour. des Mines, n. 49. p. 23.	*Rose*, in Scherer's Jour. der Chimie, b. 7. s. 244.	*Bucholz*, in Von Moll's Neue Jahrb der Berg und Hüttenkunde, b. 2. s. 361.

Geognostic Situation.

This is one of the most abundant minerals in nature, as it forms a principal constituent part of granite and gneiss, two of the most widely distributed rocks hitherto discovered. It occurs as an accidental mixed part in mica-slate and clay-slate. It is a constituent part of white-stone and syenite: in white-stone it is associated with garnet, mica, and hornblende: in syenite always with a subordinate portion of hornblende.

hornblende. It forms the basis of certain porphyries, and then it occurs in fine granular concretions. It appears in the form of imbedded crystals, in all the different kinds of porphyry, and very generally disseminated in quartz rock. Greenstone, a rock so abundant in Primitive country, is a compound of common felspar and hornblende, but in which the hornblende predominates. Frequently the felspar is tinged of a green colour, owing to an intermixture of hornblende, and in this state it is heavier than the pure varieties of this mineral. But it occurs not only as a constituent part, and accidentally mixed with primitive mountain rocks, but we find it also in beds alternating with these, in nests and kidneys contained in them, and in veins traversing them. The beds occurring in granite or gneiss, are sometimes entirely composed of felspar, with the addition of very little mica and quartz; or in them it is associated with hornblende, garnet, actynolite, epidote, and copper and iron ores, as in Sweden and Norway. The kidneys and nests vary from a few inches to several fathoms in extent, and are contained in granite or gneiss. The veins are of cotemporaneous formation with the granite and gneiss rocks in which they are contained: they are sometimes entirely composed of felspar: in other instances of felspar, with a little quartz and mica, or of felspar, with rock-crystal, mica, chlorite, epidote, schorl, beryl, and rutile. It is in these veins that the greater number of crystallizations of felspar occur. The most beautiful crystallizations occur in the Alps of Switzerland, in Lombardy, France, and Siberia. The green felspar analysed by Vauquelin, is said to occur in a vein in granite, in the government of Ubinsky, in the Uralian Mountains in Siberia; also in cotemporaneous masses in the granite of Onega.

Felspar is not confined to primitive rocks; it occurs abundantly in Transition mountains, and also in those of the

the Secondary class. In transition mountains, it forms an essential constituent part of granite, syenite, porphyry, greenstone, and greywacke; and occurs accidentally intermixed in other rocks of this class. In secondary rocks, it occurs in many sandstones, in porphyry, greenstone, clinkstone-porphyry, and basalt.

Geographic Situation.

As granite, gneiss, mica-slate, porphyry, syenite, greenstone, greywacke, sandstone, basalt, and other rocks in which common felspar occurs, are found in almost every great tract of country, it would be superfluous to attempt detailing the individual geographic localities of a mineral so widely distributed.

Uses.

It is one of the ingredients in the finer kinds of earthenware, and is said to be the substance used by the Chinese under the name *Petunse* or *Petunze*, in the manufacture of their porcelain *. The green varieties of felspar, which are rare, are considered as ornamental stones, and are cut and polished, and made into snuff-boxes, and other similar articles. When the green varieties are spotted with white, they are named *Aventurine Felspar*, and are prized by collectors. Other two varieties, having the same name, and much esteemed by collectors, are found in Russia: the one is a red felspar, with white spots, from the coast of the White Sea; the other a yellow felspar, with shining yellow spots, from the Island of Cedlowatoi, near Archangel. The

* Mr Clarke Abel is of opinion, that the petunse is quartz.—Travels in China, p. 218.

green felspar from South America, which is cut and polished, and sold under the name *Amazon Stone*, is found in small rolled pieces, on the banks of the river of Amazons.

Observations.

1. It is distinguished from the other subspecies of this species by its more extensive colour-suite, its want of changeability of colour, its distinct concretions, passing into fine granular, easy frangibility, and inferior translucency.

2. It has been confounded with Corundum, but it is distinguished from that mineral by its cleavage, inferior specific gravity, and inferior hardness. It is distinguished from *Chrysoberyl* by its fracture, inferior hardness, and inferior weight. The green-coloured felspar is distinguished from *Green Diallage* by its superior hardness, and its double cleavage.

3. The German name *Felspar* was given to this mineral on account of its sparry or foliated texture, and from the circumstance of its frequently occurring as a constituent part of those loose blocks of stone we observe scattered over the country, (*Feldern*). Hence it appears that the name *Felspar*, used by the English, and sometimes by French authors, is not quite correct.

Fifth Subspecies.

Labrador Felspar.

Labradorstein, *Werner*.

[illegible] Labrador, *Romé de Lisle*, t. ii. p. 497.—Feldspath, [illegible] 335.—Labradore-stone, *Kirw.* vol. i. p. 324.—La-[illegible] b. i. s. 273.—Feldspato commune, var. *Nap.*

Nap. p. 213.—Labradorite, *Lam.* t. ii. p. 197.—Feldspath opalin, *Haüy*, t. ii. p. 607.—La pierre de Labradore, *Brock.* t. i. p. 369.—Labradorstein, *Reuss*, b. ii. s. 387. *Id. Lud.* b. i. s. 100. *Id. Suck.* 1r th. s. 380.—Gemeiner Feldstein, *Bert.* s. 238.—Labradorstein, *Mohs*, b. i. s. 407.—Labradorischer Feldspath, *Hab.* s. 22.—Feldspath opalin, *Lucas*, p. 50.—Labradorischer Feldspath, *Leonhard*, Tabel. s. 18.—Feldspath opalin, *Brong.* t. i. p. 359. *Id. Brard*, p. 134.—Farbenspielender Feldspath, *Haus.* s. 88.—Labrador Feldspath, *Karst.* Tabel. s. 34.—Opaline Felspar, *Kid*, vol. i. p. 160.—Feldspath opalin, *Haüy*, Tabl. p. 36.—Labradorstein, *Steffens*, b. i. s. 432. *Id. Hoff.* b. ii. s. 304.—Labrador Felspath, *Lenz*, b. i. s. 490.—Labradorstein, *Oken*, b. i. s. 376.—Edler Feldspath, *Haus.* Handb. b. ii. s. 531.—Labrador Felspar, *Aikin*, p. 197.

External Characters.

Its most frequent colours are light and dark ash-grey, and smoke-grey, seldom yellowish-grey. When light falls on it in determinate directions, it exhibits a great variety of colours: of these the most frequent are blue and green more seldom yellow and red, and the rarest variety is pearl-grey. The blue varieties are indigo, Berlin, azure, violet, smalt, and sky blue: this latter colour passes into verdigris-green; from this variety through celandine, mountain, leek, emerald, grass, pistachio, olive, and oil, into siskin green; the siskin-green passes into sulphur-yellow, and through brass, gold, lemon, honey, and orange yellow, into yellowish and reddish-brown, copper-red, brick-red, flesh-red, brownish-red; and, lastly, into pearl-grey and bluish-grey. The same specimen exhibits different colours, which run imperceptibly into each other, and are disposed in large patches or stripes.

 It

It occurs massive, or in rolled pieces; also in large, coarse, seldom in small granular, very seldom in thick and straight lamellar concretions.

The cleavage is splendent, the fracture glistening, and the lustre is intermediate between vitreous and pearly.

The cleavage and fracture are the same as in common felspar.

It breaks into rhomboidal and sharp-edged fragments.

It is translucent, but in a low degree.

It is rather more difficultly fragible than common felspar.

Specific gravity, 2.692, *Brisson*.—2.590, *Hoffmann*.—American 2.690, *Klaproth*.—Russian 2.756, *Klaproth*.—Norwegian 2.590, *Klaproth*.

Chemical Characters.

According to Mr Kirwan, it is more infusible than common felspar.

Geognostic and Geographic Situations.

It occurs in rolled masses of syenite, in which it is associated with common hornblende, hyperstene, and magnetic ironstone, in the Island of St Paul, on the coast of Labrador, where it was first discovered, upwards of thirty years ago, by the Moravian Missionaries settled in that remote and dreary region. Some years afterwards, several varieties of it were found imbedded in a granite rock in Ingermannland; but the colours of these were neither so vivid nor numerous as in the Labrador felspar of St Paul's. In the interesting country around Laurwig in Norway, Labrador felspar occurs as a constituent part of the zircon-syenite; its colours are brighter than in the Ingermannland varieties, but not so vivid as those of St Paul. Blue is the principal colour of the

the Norwegian felspar, but it sometimes also exhibits a beautiful bluish mother-of-pearl opalescence, like that observed in adularia. A variety of this mineral is said to occur in the Hartz. Rolled pieces of it have been brought from West Greenland; and it has been found on the banks of Lake Champlain in North America *.

Uses.

On account of its beautiful colours, it is valued as an ornamental stone, and is cut into ring-stones, snuff-boxes, and other similar articles. It receives a good polish; but the streaks caused by the edges of the folia of the cleavage are frequently so prominent as to injure the appearance of the stone.

Observations.

1. This mineral is distinguished by its grey colours, and its changeability of colours.

2. The beautiful changeability of colours which Labrador felspar exhibits, appears to be caused by small rents, that run parallel with the folia of the cleavage, in this differing from the play of colour observed in the precious opal, which is owing to rents that run in every direction.

Sixth

* I have specimens in my possession, said to have been found in Aberdeenshire.

Sixth Subspecies.

Compact Felspar.

Dichter Feldspath, *Werner.*

Petrosilex æquabilis; P. semipellucidus? *Wall.* t. i. p. 268. 271.—Continuous Felspar, *Kirw.* vol. i. p. 323.—Dichter Feldstein, *Estner,* b. ii. s. 511. *Id. Emm.* b. i. s. 271.—Felspato compatto, *Nap.* p. 218.—Petrosilex agathoide, *Haüy,* Traité, t. iv. p. 385.—Le Feldspath Compacte, *Broch.* t. i. p. 367.—Dichter Feldspath, *Reuss,* b. ii. s. 366. *Id. Lud.* b. i. s. 10. *Id. Suck.* 1r th. s. 393.—Dichter Feldstein, *Bert.* s. 238.—Dichter Feldspath, *Mohs,* b. i. s. 420. *Id. Hab.* s. 19.—Feldspath compacte, *Lucas,* p. 50.—Dichter Feldspath, *Leonhard,* Tabel. s. 19.—Petrosilex, *Brong.* t. i. p. 351.—Feldspath Compacte ceroide, *Brard,* p. 133.—Dichter Feldspath, *Haus.* s. 88. *Id. Karst.* Tabel. s. 34.—Feldspath Compacte ceroide, *Haüy,* Tabl. p. 35.—Dichter Feldspath, *Steffens,* b. i. s. 442. *Id. Hoff.* b. ii. s. 334. *Id. Lenz,* b. i. s. 506. *Id. Oken,* b. i. s. 299. *Id. Haus.* Handb. b. ii. s. 534.—Compact Felspar, *Aikin,* p. 197.

External Characters.

Its colours are white, grey, green, and red: it passes from greyish-white through greenish-white, into apple-green, oil-green, inclining to olive-green, mountain-green, greenish-grey, smoke-grey, pearl-grey, flesh-red, and brick-red.

It occurs massive, disseminated, in blunt angular rolled pieces, and in small angulo-granular concretions; also crystallized in rectangular four-sided prisms.

The crystals are either middle-sized, or small, and always imbedded.

Internally

Internally it is sometimes glistening, sometimes glimmering.

The fracture is even and splintery.

It breaks into fragments which are rather sharp-edged.

It is feebly translucent, sometimes only translucent on the edges.

It is as hard as common felspar.

When pure, it is rather easily frangible.

Specific gravity, 2.609, *Kirwan.*—2.666, *La Metherie.*—2.659, *Saussure.*—2.690, *Klaproth.*

Chemical Characters.

Before the blowpipe, it melts with difficulty into a whit-i h enamel.

Constituent Parts.

Compact Felspar of Salberg in Sweden.		Compact Felspar of the Pentland Hills, near Edinburgh.			
Silica, -	68.0	Silica, -	71.17	Silica,	51.00
Alumina,	19.0	Alumina,	13.60	Alumina,	30.50
Lime, -	1.0	Lime, -	0.40	Lime,	11.25
Potash, -	5.5	Potash, -	3.19	Iron,	1.75
Oxide of Iron,	4.0	Oxide of Iron,	1.40	Natron,	4.00
Water, -	2.5	Manganese,	0.10	Water,	1.26
		Volatile Matter,	3.50		
	100				99.75
Godon de St Memin, Journal de Physique, t. 63. p. 60.			93.36	*Klaproth*, Chem. Abhandl. s. 264.	
		Loss, -	6.64		
			100		
		Mackensie, Mem. Wern. Soc. vol. i. p. 618.			

Geognostic

Geognostic Situation.

This mineral occurs in mountain-masses, beds and veins, either pure, or intermixed with other minerals, in primitive, transition, and secondary rocks. In primitive mountains, it is associated with hornblende in greenstone, and greenstone-slate; and it forms the básis of several felspar-porphyries. Beds of it in a pure state occur in gneiss, and other primitive rocks. In transition mountains, it occurs in beds, as a constituent part of porphyry and greenstone; and beds of it occur either pure, or in porphyry, or in greenstone, in secondary mountains.

Geographic Situation.

The Pentland Hills contain beds of compact felspar, associated with claystone, red sandstone, and conglomerate. It occurs in a similar situation on the hill of Tinto, described by Dr Macknight in the 2d volume of the Memoirs of the Wernerian Society. Mr Mackenzie found it along with secondary rocks in the Ochil Hills *; and Dr Fleming observed it associated with rocks of the same nature in the Island of Papa Stour, one of the Shetland group †. Beds of it, which are sometimes porphyritic, occur in the transition rocks of Dumfriesshire and Galloway; and in rocks of the same class to the north of the Frith of Forth, as in Perthshire, and the Mearns ‡. In the primitive rocks to the north of the Frith of Forth, it occurs in beds and veins, either pure, or in the state of porphyry. Examples of both occur in Perthshire, in the course of the Gary and the

* Memoirs of the Wernerian Society, vol. ii. p. 20.

† Memoirs of the Wernerian Society, vol. i. p. 170.

‡ Imrie, in Transactions of Royal Society of Edinburgh, vol. vi.

the Tilt; in the country around Castletown, in the upper part of Aberdeenshire *.

Beautiful varieties of compact felspar are found at Sala, Dannemora, Hällefors and Götheborg in Sweden: grey and green varieties occur in greenstone slate at Siebenlehn and Gersdorf, in the Saxon Erzgebirge; in green porphyry in the Hartz, and also in the same rock in Egypt.

Observations.

1. The principal characteristic distinctions of this mineral are colour, distinct concretions, lustre, fracture, translucency, hardness, and weight.

2. It has been frequently confounded with Splintery Hornstone; but is distinguished from it by colour, and distinct concretions, but principally by its lustre, inferior hardness, easier frangibility, fusibility before the blowpipe, and its being frequently intermixed with hornblende and mica.

Seventh Subspecies.

Clinkstone †.

Klingstein, *Werner.*

Phonolith, *Daubuisson.*

Hornslate, *Kirw.* vol. i. p. 307.—Porphirschiefer, *Estner,* b. ii. s. 747. *Id. Emm.* b. iii. s. 344.—Pierre sonnante, *Broch.* t. i. p. 437.

---

* Dr Macknight mentions several localities of this mineral in his elegant and interesting sketch of the scenery and mineralogy of the Highlands, in the 1st volume of the Memoirs of the Wernerian Society.

† The tabular varieties of this mineral, when struck, emit a ringing sound; hence the name *Clinkstone.*

p. 437.—Klingstein, *Klap.* Beit. b. iii. s. 229. *Id. Reuss,* b. ii. s. 340. *Id. Lud.* b. i. s. 123. *Id. Suck.* 1r th. s. 364. *Id. Bert.* s. 222. *Id. Mohs,* b. i. s. 509. *Id. Hab.* s. 16. *Id. Leonhard,* Tabel. s. 26.—Feldspath compacte sonoré, *Lucas,* p. 266.—Klingstein, *Karst.* Tabel. s. 38.—Clinkstone, *Kid,* vol. ii. App. p. 18.—Klingstein, *Steffens,* b. i. s. 338. *Id. Lenz,* b. ii. s. 613. *Id. Oken,* b. i. s. 363. *Id. Hoff.* b. ii. s. 180. *Id. Haus.* Handb. b. ii. s. 707.—Clinkstone, *Aikin,* p. 205.

External Characters.

Its most frequent colour is greenish-grey, which sometimes passes into yellowish-grey, and ash-grey; which latter passes into liver-brown, and it is occasionally mountain-green, olive-green, and oil-green.

It occurs massive; also in granular, columnar, globular, and tabular distinct concretions.

The lustre of the principal fracture is glistening and pearly; that of the cross fracture is faintly glimmering, almost dull

The principal fracture is slaty, generally thick, and often curved slaty, with a scaly foliated aspect; the cross fracture is splintery, passing into even, and flat conchoidal.

The fragments are indeterminate angular, and often slaty.

It is strongly translucent on the edges, sometimes even translucent.

It is as hard as felspar.

It is rather easily frangible.

It is brittle.

In thin plates, it emits, when struck, a ringing sound.

Specific gravity, 2.575, *Klaproth.*—2.515, *B*

Chemical Characters.

It melts before the blowpipe into a grey-coloured glass, but is more difficultly fusible than basalt.

Constituent Parts.

Silica, - -	57.25
Alumina, - -	23.50
Lime,	2.75
Natron, - -	8.10
Oxide of Iron, -	3.25
Oxide of Manganese,	0.25
Water, - -	3.00
	98.10

Klaproth, Beit. b. iii. s. 243.

Geognostic Situation.

This subspecies of felspar generally contains imbedded crystals, when it forms the rock named Clinkstone Porphyry. It is generally associated with secondary trap and porphyry rocks.

Geographic Situation.

Europe.—The Bass rock at the mouth of the Frith of Forth, North Berwick Law, Traprain Law, and the Girleton Hills, all in East Lothian, are principally composed of clinkstone, and afford many beautiful and highly characteristic varieties of this mineral. It occurs in the island of Arran, isle of Lamlash, Ochil Hills, and other parts of Scotland. The Breiddin Hills in Montgomeryshire in Wales; and Devis Mountain in the county of Antrim, afford

ford several varieties of this mineral. On the Continent of Europe, it is found in many districts where basalt abounds, as in the Bohemian Mittelgebirge; also in Bavaria, Suabia, Lusatia, Hessia, France, Italy, and Hungary.

Africa.—Along with basalt, in the island of Teneriffe.

America.—Along with trap-rocks, both in North and South America.

Observations.

1. Charpentier was the person who first directed the particular attention of mineralogists to this substance: in his Mineralogical Description of the Electorate of Saxony, he gives a very interesting account of it under the name *Hornslate*, (Hornschiefer) *. Werner afterwards examined it with more minute attention, and introduced it into the oryctognostic system as a distinct substance, under the name *Clinkstone*.

2. It has been confounded with Basalt; but is distinguished from that rock by colour, lustre, fracture, and transparency.

Eighth Subspecies.

Earthy Common Felspar.

Aufgelöster gemeiner Feldspath, *Werner*.

External Characters.

Its colours are greyish-white, yellowish-white, and reddish-white, all of which incline very much to grey.

It

* Older mineralogists were of opinion, that clinkstone was the same mineral as that described by Wallerius under [illegible] *fissilis*; hence they gave it the name Hornslate; [illegible]'s mineral appears to be hornblende-[illegible] is under their hornslate also [illegible]

It generally occurs massive, and disseminated, and sometimes in imbedded crystals, which agree in form with those of common felspar.

Internally it is sometimes glistening, sometimes glimmering, or even dull.

It has sometimes an imperfect cleavage.

The fracture is coarse and small grained uneven, which approaches to earthy.

It breaks into blunt angular pieces.

It is either translucent on the edges, or opaque.

In general, it is so soft as to yield to the nail: sometimes, however, it approaches in hardness to felspar.

It is sectile, and easily frangible.

The chemical characters and composition of this substance have not been ascertained.

Geognostic and Geographic Situations.

It occurs in granite and gneiss districts, as in Cairngorm and Arran in Scotland, and Cornwall in England. It is well known in Saxony, and other countries.

Observations.

This mineral seems in some instances to be felspar in a state of disintegration: in others, to be an unaltered substance, very nearly of the nature of common felspar. The *Growan* of Cornwall appears to contain principally the disintegrated felspar.

Ninth

Ninth Subspecies.

Porcelain Earth or Kaolin.

Porcellanerde, *Werner.*

Porcelain Clay, *Kirw.* vol. i. p. 178.—Argilla de Porcellana, *Nap.* p. 248.—La terre à Porcelaine, *Broch.* t. i. p. 320.—Feldspath argiliforme, *Haüy,* t. ii. p. 616.—Porcellanerde, *Reuss,* b. ii. s. 107. *Id. Lud.* b. i. s. 105. *Id. Suck.* 1r th. s. 492. *Id. Bert.* s. 213. *Id. Mohs,* b. i. s. 431. *Id. Hab.* s. 38. *Id. Leonhard,* Tabel. s. 21.—Argil Kaolin, *Brong.* t. i. p. 516.—Kaolin, *Haus.* s. 85. *Id. Karst.* Tabel. s. 36.—Porcelain Clay, *Kid,* vol. i. p. 165.—Feldspath decomposé, *Haüy,* Tabl. p. 36. Porcellanerde, *Steffens,* b. i. s. 445.—Kaolin, *Lenz,* b. ii. s. 546. Kieskaolin, *Oken,* b. i. s. 371.—Porzellanerde, *Hoff.* b. ii. s. 10.—Kaolin, *Haus.* Handb. b. ii. s. 450.

External Characters.

Its most frequent colour is reddish-white, of various degrees of intensity; also snow-white, and yellowish-white.

It is generally friable, and sometimes approaches to compact.

It is composed of dull dusty particles, which are feebly cohering.

It soils strongly.

It feels fine and soft, but meagre.

It adheres slightly to the tongue.

Specific gravity, 2.216, *Karsten.*

Chemical Characters.

It is infusible before the blowpipe.

Constituent

Constituent Parts.

		Porcelain Earth from Aue in Saxony.			
Silica, -	46.0	Silica,	52.00	Silica,	55.0
Alumina,	39.0	Alumina,	47.00	Alumina,	42.5
Oxide of Iron,	0.25	Iron,	0.33	Iron,	1.0
Water,	14.50			Lime,	1.0
	97.75		99.33		99.5
Klaproth, Chem. Abhandl. s. 278.		*Rose.*		*Gehlen.*	

Geognostic Situation.

It generally occurs in granite and gneiss countries, either in beds contained in the granite, or gneiss, when it appears to be an original deposite, or on the sides and bottom of granite and gneiss hills, when it is certainly formed by the decomposition of the felspar of these rocks.

Geographic Situation.

Europe.—It occurs in the different granite and gneiss districts in Scotland, and in the Shetland Isles; also in England and Ireland. One of the best known and most celebrated mines of porcelain-earth, is that of Aue in Saxony, which is used in the porcelain manufactory at Meissen. It forms a bed about three fathoms thick, which is covered with from three to six fathoms of mica-slate. It rests on fresh or unchanged granite, and is divided in the middle into two strata by a bed of disintegrated granite. There can be no doubt of this bed being an original deposite, and not felspar which has undergone a process of decomposition. A similar bed of porcelain earth occurs in granite, in the valley of Gatach, above Haussach in Wirtemberg. The Austrian porcelain is made from a fine porcelain-earth which

is

is dug near Passau. At St Yrieux la Perche, near Limoges in France, there is a bed or vein of porcelain-earth, in granite; and it has been discovered in granite near to Bayonne.

Asia.—Very valuable varieties of this mineral are found in China and Japan, where they are denominated *Kaolin.*

Uses.

This mineral forms a principal ingredient in the different kinds of porcelain. It is not used in the state in which it is found in the earth, but is previously repeatedly washed, in order to free it from impurities. After the process of washing, only fifteen parts of pure white clay remain, which is the kaolin of the Chinese. This clay, mixed in proper proportions with quartz, flint, gypsum, steatite, and other substances, forms the composition of porcelain; and this mixture is sifted several times through hair-sieves. The mixture is afterwards moistened with rain-water, in order to form a paste, which is put into covered casks. This paste is called by the workmen the *mass.* A fermentation soon takes place, which changes its smell, colour, and consistence. Sulphuretted hydrogen gas is evolved: the colour passes from white into dark-grey; and the matter becomes tougher and softer. It must be carefully moistened from time to time, to prevent it from drying. The preparation of the mixture, and the art of rightly managing the mass, are secrets in most manufactories. The second operation is to give the paste the form we wish; and this is done by first kneading it with the hands, in order to divide the mixture more completely, and then turning it on the lathe. A third operation is the baking, or firing, which is done in furnaces of a particular construction. The firing generally lasts from thirty-six to forty-eight hours; and we judge of the state of the baking by proof-pieces, as they are called, placed in convenient situations, and which we can draw out and examine from time to time. The porcelain in this state is named *biscuit porcelain* by workmen *. A fourth operation is the covering the surface of the biscuit with a varnish or enamel, which must be applied exactly over all the points of the surface, and incorporated with the paste, without cracking or flying. This enamel is composed of pure white quartz, white porcelain, and calcined crystals of gypsum, and sometimes principally of felspar: these substances are ground with the greatest care, then diffused through water, and formed into a paste. When we use it, it must be diluted in water, so as to give it considerable liquidity, and we then plunge into it the biscuit porcelain. The porcelain is now exposed to heat, sufficient to melt the enamel or covering, and then it constitutes white porcelain; and in this state it may be applied to every purpose. If the porcelain is to be painted, it must again be exposed to heat in the furnace. The colours used are all derived from metals; and many of them, though dull when applied, acquire a considerable lustre by the action of the fire. The colours are mixed with a flux, which varies in the different manufactories: in some, a mixture of glass, borax, and nitre, is employed; this mixture is melted in a crucible, and the glass is afterwards ground, and incorporated with the colour. Gum, or oil of lavender, is used as a vehicle, when we wish to lay it on the porcelain. When the painting is

 finished,

* Figures, and generally all porcelain articles which are neither to be painted nor exposed to water, have no occasion for any covering; they are [illegible] state of biscuit.

finished, the ware is exposed to a heat sufficient to melt the flux containing the colour.

The beautiful purple colours on porcelain, are from oxide of gold, called powder or *precipitate of Cassius*; the violet colours from gold precipitated by tin and silver; certain green colours by copper, precipitated from its solutions in the acids by alkalies; red colours from oxides of iron; blue from zaffre; yellow from diaphoretic antimony, mixed with glass of lead; brown and black colours from iron-filings and zaffre; and the finest green tints from oxide of chrome.

Porcelain has been manufactured in China and Japan from a very early period. The art itself was discovered in Europe by a German named Bötticher, who made his first porcelain-vessels in Dresden in the year 1706. These were of a brown and red colour. The white porcelain was not attempted until the year 1709; and the famous manufactory at Meissen, the earliest in Europe, was established in 1710.

Observations.

This mineral is distinguished from the other *Clays*, by the fineness of its particles, its soiling strongly, its fine but meagre feel, and its not becoming plastic in water.

2. Pyramidal

2. Pyramidal Felspar or Scapolite *.

Pyramidaler Feldspath, *Mohs.*

Scapolit, *Werner.*

Paranthine, *Haüy.*

THIS species is divided into four subspecies, viz. Radiated Scapolite, Foliated Scapolite, Compact Red Scapolite, and Elaolite.

First Subspecies.

Radiated Scapolite.

Strahliger & Nadelförmiger Skapolith, *Karsten,* Tabel. s. 34.—Glasartiger Scapolith, *Haus.* s. 189.—Strahliger & Glasartiger Skapolith, *Steffens,* b. i. s. 461. & 464.—Stangensteinartiger Scapolit, *Shumacher,* Verzeichniss, s. 97.—Strahliger grauer Skapolith, *Hoff.* b. ii. s. 346.—Paranthine dioctaedre, aciculaire & cylindroide, *Haüy,* Tabl. p. 46.—Strahliger Scapolite, *Haus.* Handb. b. ii. s. 514.

External Characters.

Its most frequent colour is grey, seldomer white and green: it occurs greyish-white, yellowish-white, greenish-white, yellowish-grey, greenish-grey, mountain-green, olive-green, and asparagus-green.

It occurs massive, and in distinct concretions; the concretions are radiated or fibrous, scopiform diverging, and

* *Scapolite,* from σκαπος, *a rod,* in reference to the columnar mode of aggregation of its crystals.

are collected into others which are thick and wedge-shaped. It is most frequently crystallized. The primitive figure is a pyramid of 136° 38′; 62° 56′. The secondary forms are the following:

1. Rectangular four-sided prism, flatly acuminated on the extremities with four planes, which are set on the lateral planes.
2. The preceding figure, in which the lateral edges are truncated *.

The crystals vary very much in length as well as thickness; for we meet with them from the acicular form to the thickness of a finger, and from very long to short. Sometimes the long prisms are curved, and are traversed with rents.

The crystals are frequently columnarly aggregated, or intersect one another.

The lateral planes of the crystals are deeply longitudinally streaked, and shining.

Internally it is intermediate between shining and glistening, and the lustre is intermediate between resinous and pearly.

The cleavage is double, and in the direction of the lateral planes of the prism, and also of its diagonals.

The fracture is fine-grained uneven.

The fragments are indeterminate angular, and not very sharp-edged.

It is translucent, and semitransparent in crystals.

It is as hard as apatite, and sometimes even harder; but never so hard as felspar.

It is rather easily frangible.

Specific

* Paranthine dioctaedre, Haüy.

Specific gravity, 2.740, *Laugier*.—2.691, 2.773, *Simon*.—2.857, *Schumacher*.—2.660, 2.743, *Hausmann*.—2.5, 2.8, *Mohs*.

Chemical Characters.

Green scapolite, before the blowpipe, becomes white, and melts into a white glass.

Constituent Parts.

Silica,	53.50
Alumina,	15.00
Magnesia,	7.00
Lime,	13.75
Natron,	3.50
Iron,	2.00
Manganese,	4.00
Water,	0.50
	99.24

Simon, Chem. Journ. b. iv. s. 411.

Silica,	45.0
Alumina,	33.0
Lime,	17.6
Natron,	1.5
Potash,	0.5
Iron and Manganese,	1.0
	98.6

Laugier, Annales du Museum d'Hist. Nat. cah. ix. p. 472.

Geognostic and Geographic Situations.

This mineral occurs in the neighbourhood of Arendal in Norway, where it is associated with magnetic ironstone, felspar, quartz, mica, garnet, augite, hornblende, actynolite, and calcareous-spar.

The magnetic ironstone occurs in gneiss, in the form of beds, that vary in thickness from four to sixty feet. In these beds, the scapolite and other accompanying minerals already mentioned, are either contained in cotemporaneous veins, or are irregularly disseminated throughout the beds. M. Hausmann observed it in beds of specular iron-ore or iron-glance, in the Swedish Province of Wermeland,

Wermeland, where it is associated with calcareous-spar and garnet: and the same excellent mineralogist found it at Malsjo in Wermeland, in a bed of limestone; and at Garpenberg in Dalecarlia, in beds of copper-pyrites.

Observations.

1. All the subspecies decay very readily on exposure to the weather, a circumstance which has induced Haüy to name this species *Paranthine.*

2. The *Spreustein* of Werner is said to be Fibrous Scapolite.

Second Subspecies.

Foliated Scapolite.

Micarell, *Abilgaard.*—Talkartiger Scapolit, Blättriger Scapolit, Pinitartriger Scapolit, *Schumacher,* Verzeichniss, s. 98,–100. —Wernerit, *Karsten,* Tabel. s. 34.—Arcticit, *Werner.*—Gemeiner Skapolith & Glimmeriger Skapolith, *Steffens,* b. i. s. 462. 464.—Dichter Scapolit, *Haus.* in Magaz. Natf. Freund. b. iii. s. 220.—Wernerit, *Haüy,* Tabl. p. 45.—Blättriger grauer Skapolit, *Hoff.* b. ii. s. 353.—Fuscit & Gabbronit, *Schumacher.*

External Characters.

Its principal colours are grey, green, and black. The greenish-grey passes into mountain and asparagus green. The black colours are greyish-black, and pitch-black. The colours are seldom pure, generally pale and muddy, and sometimes two colours occur in the same specimen; and greenish-grey coloured crystals sometimes appear sky-blue externally.

It

It occurs massive, disseminated, and in large, coarse, and long angulo-granular concretions; also crystallized in low eight-sided prisms, flatly acuminated with four planes, which are set on the alternate lateral planes.

The crystals are sometimes middle-sized, seldom large, or very small, and are generally superimposed, but seldom imbedded.

Externally the crystals are shining or splendent, and vitreous.

The cleavage is shining, the fracture glistening, and the lustre intermediate between resinous and pearly.

The cleavage is the same as in the radiated subspecies. The fracture is small and fine-grained uneven, or small conchoidal.

The fragments are generally indeterminate angular, and sharp-edged.

It is generally translucent, and passes sometimes into transparent, sometimes to translucent on the edges.

It yields a white streak.

It is brittle.

It is very easily frangible.

Its hardness and specific gravity are the same as in the radiated subspecies.

Geognostic and Geographic Situations.

On the north-western acclivity of the Saxon Erzgebirge, there is a considerable extent of a very compact small granular granite, named *Whitestone*, which includes cotemporaneous masses of common granite, that vary in magnitude from a few feet to some miles in extent. In these granitic masses, various minerals have been observed, as schorl, tourmaline,

tourmaline, lepidolite, and *Foliated Scapolite* *. It occurs also in Scandinavia, along with the radiated subspecies.

Third Subspecies.

Compact Red Scapolite.

Dichter Scapolite.

External Characters.

Its colour is dark brick-red, passing into pale blood-red.

It seldom occurs massive, more frequently crystallized, in long, frequently acicular, four-sided prisms, which are often curved, and are without terminal crystallizations.

Externally the crystals are rough and dull.

Internally it is very feebly glistening, almost glimmering.

The fracture is fine-grained uneven, approaching to splintery.

The fragments are indeterminate angular, and sharp-edged.

It is opaque, or very faintly translucent on the edges.

It is hard in a low degree.

It is easily frangible.

Geognostic and Geographic Situations.

It occurs along with the other subspecies, in metalliferous beds at Arendal in Norway.

Observations.

---

* 1. Prech. uber Granit.—Leonhard, Taschenbuch 1812, p. 137.

Observations.

This mineral is characterized by its red colour, low degree of lustre, compact fracture, and nearly complete opacity.

Fourth Subspecies.

Elaolite.

Eläolith, *Klaproth.*

Fettstein, *Werner.*

Dichter Wernerit, *Hausmann.*

External Characters.

The colours of this mineral are duck-blue, which inclines more or less to green, also flesh-red, which falls more or less into grey, sometimes even inclines to brown.

It occurs massive, and in very intimately aggregated granular concretions.

Internally it is shining or glistening, and the lustre is resinous.

The fracture, principally in the red variety, is flat and imperfect conchoidal. The blue variety has an imperfect double cleavage.

The fragments are indeterminate angular, and not very sharp-edged.

It is translucent in a low degree. The blue variety, when cut in a particular direction, displays a peculiar opalescence, not unlike that observed in the cat's-eye.

It has the same degree of hardness as the other subspecies.

It

It is rather easily frangible.

Specific gravity, 2.613, *Haüy*.—From 2.588 to 2.618, *Hoffmann*.

Chemical Character.

When pounded, and thrown into acids, it gelatinates. Before the blowpipe it melts into a milk white enamel.

Constituent Parts.

Silica	46.50	Silica, -	44.00
Alumina,	30.25	Alumina, -	34.00
Lime, - -	0.75	Lime, -	0.12
Potash, -	18.00	Potash and Soda,	16.50
Oxide of Iron, - -	1.00	Oxide of Iron,	4.00
Water, -	2.00		
	98.50		98.62

Klaproth, in Magazin für die Neuesten Endeckungen in der Naturkunde, &c. 3ter Jahrg, s. 45. Also *Klaproth*, Beit. b. v. s. 178.

Vauquelin, in Haüy's Tabl. Comparative, p. 178.

Geognostic and Geographic Situations.

The blue variety is found at Laurwig, and the red at Stavern and Friedrichswärn, both in the rock named *zircon syenite*.

Uses.

The pale blue variety, which has often an opalescence like that of the adularia-moonstone, is cut *en cabochon*, and used for ring-stones. When set, it is difficult to distinguish it from cat's-eye.

Observations.

Observations.

1. It is named *Elaolite* by Klaproth, and *Fettstein* by Werner, on account of its resinous lustre.

2. The *Sodaite* of Ekeberg, which is the *Natrolite* of Wollaston, appears to be a variety of Elaolite; and probably the *Lythrodes* of Karsten belongs to the same subspecies.

3. Few of the newer mineral species have had so many names given to them as Scapolite, as appears from the following enumeration:

Names given to Scapolite.

1. Paranthine; 2. Wernerite; 3. Arcticite; 4. Sodaite; 5. Natrolite; 6. Fuscite; 7. Gabbronite; 8. Elaolite; 9. Fettstein; 10. Lythrodes? 11. Spreustein? 12. Bergmannite.

3. Prismato-Pyramidal Felspar or Meionite*.

Prismato-Pyramidischer Feldspath, *Mohs.*

Meionite, *Haüy & Werner.*

Hyacinthe blanche de la Somma, *Romé de Lisle,* t. ii. p. 290.—Meionite, *Haüy,* t. ii. p. 586. *Id. Broch.* t. ii. p. 519, 520. *Id. Lucas,* p. 49. *Id. Leonhard,* Tabel. s. 17. *Id. Brong.* t. i. p. 583. *Id. Brard,* p. 130. *Id. Haus.* s. 95. *Id. Karst.* Tabel. s. 34. *Id. Haüy,* Tabl. p. 34. *Id. Steffens,* b. i. s. 458. *Id. Hoff.* b. ii. s. 361. *Id. Lenz,* b. i. s. 512. *Id. Oken,* b. i. s. 351. *Id. Haus.* Handb. b. ii. s. 549. *Id. Aikin,* p. 207.

External Characters.

Its colour is greyish-white.

The

* *Meionite,* is derived from the Greek word *μειων*, *smaller, shorter,* because the acumination of its principal crystallizations is flatter, and also lower than in similar crystallizations in other minerals.

It occurs sometimes massive, but more frequently crystallized.

The primitive figure is a pyramid, in which the angles are 136° 22′, 68° 22′.

The following are the secondary figures:

1. Rectangular four-sided prism, flatly acuminated with four planes, which are set on the lateral edges.
2. The preceding figure, truncated on the lateral edges, Fig. 99. Pl. 5. *.

   Sometimes one of the acuminating planes becomes so large that the others disappear, when there is formed
3. A four-sided prism, in which the terminal planes are set on obliquely.
4. N° 1. bevelled on the lateral edges, and the edges of the bevelment truncated; and the edges between the acuminating planes and the lateral planes also truncated, Fig. 100. Pl. 5. †.

The crystals are small, seldom middle-sized; they are superimposed, and form druses.

Externally the crystals are smooth and splendent, internally splendent and vitreous.

It has a double rectangular cleavage, in which the folia are parallel with the lateral planes of the prism.

The fragments are indeterminate angular.

It is generally transparent, or semi-transparent, seldom translucent.

It

* [illegible] dioctaèdre, Haüy.

† [illegible] soustractive, Haüy.

It is harder than common felspar, but softer than quartz.

It is easily frangible.

Specific gravity 2.5, 2.7, *Mohs*.

Chemical Characters.

It is easily fusible before the blowpipe; intumesces during fusion, and is converted into a white vesicular glass.

It has not hitherto been analysed.

Geognostic and Geographic Situations.

It occurs, along with ceylanite and nepheline, in granular limestone, at Monte Somma, near Naples. It is said also to occur in basalt, along with augite and leucite, at Capo di Bove, near Rome.

Observations.

1. This species is characterized by its white colour, simple crystallizations, splendent vitreous lustre, cleavage, transparency, hardness, low specific gravity, and the changes it experiences before the blowpipe.

2. It is distinguished from *Adularia* by its crystallizations, its cleavage, and the changes it undergoes before the blowpipe: its crystallizations, cleavage, and easy frangibility, distinguish it from *Nepheline*: it is readily distinguished from *Cross-stone*, by the flatness of its acuminations, the equality of its lateral planes, and its never occurring in twin crystals; it is further discriminated by its stronger lustre, cleavage and fusibility: It was formerly confounded with *Hyacinth* or *Zircon*, but is distinguished from that mineral by colour-suite, the flatness of its acuminations, perfect

perfect vitreous lustre, double cleavage, inferior hardness and weight, and infusibility before the blowpipe.

3. It was Romé de Lisle, who first attended to the crystallization of this mineral: it was more particularly examined by Haüy, who established it as a distinct species, under the name *Meionite*.

4. Rhomboidal Felspar, or Nepheline *.

Rhomboedrischer Feldspath, *Mohs*.

Nepheline, *Haüy* & *Werner*.

Sommite, *La Metherie*, t. ii. p. 271.—Nepheline, *Broch.* t. ii. p. 522. *Id. Haüy*, t. iii. p. 186. *Id. Lucas*, p. 72.—Sommit, *Leonhard*, Tabel. s. 16.—Nepheline, *Brong.* t. i. p. 387. *Id. Brard*, p. 176. *Id. Haus.* s. 94.—Sommit, *Karsten*, Tabel. s. 32.—Nepheline, *Haüy*, Tabl. p. 51. *Id. Steffens*, b. i. s. 476. *Id. Hoff.* b. ii. s. 365.—Sommit, *Lenz*, b. i. s. 513.—Weicher Smaragd, *Oken*, b. i. s. 319.—Nephelin, *Haus.* Handb. b. ii. s. 552.—Sommite, *Aikin*, p. 207.

External Characters.

The colours are snow-white, greyish-white, yellowish-white, and greenish-white, which latter sometimes passes into greenish-grey.

It occurs massive and crystallized.

The primitive form is a di-rhomboid of 152° 41′; 56° 15′. The secondary forms are the following.

1. Perfect equiangular six-sided prism, fig. 101 †. Pl. 5.

2. The

---

* *Nepheline*, from νεφελη, *a cloud*, because transparent pieces, when immersed in nitrous acid, become cloudy in the interior.

† Nepheline primitive of

2. The preceding figure, truncated on the terminal edges, fig. 102 *. Pl. 5.

When the prism becomes shorter, there is formed.

3. A thick six-sided table, in which the lateral edges are truncated.

The crystals are small and very small, always superimposed, and forming druses.

Externally the crystals are splendent: internally shining, and the lustre is vitreous.

A fourfold cleavage is to be observed: three of the cleavages are parallel with the lateral planes, and one with the terminal planes of the prism.

The fracture is conchoidal.

The fragments are indeterminate angular, and sharp-edged.

It is strongly translucent, passing into transparent.

It is as hard as felspar.

Specific gravity 2.6, 2.7, *Mohs.*

Chemical Characters.

It melts with difficulty before the blowpipe into a dark glass.

Constituent Parts.

Silica, - -	46	
Alumina, -	49	
Lime, -	2	
Oxide of Iron,	1	
	98	*Vauquelin.*

Geognostic

* Nepheline annulaire, Haüy.

Geognostic and Geographic Situations.

It occurs in drusy cavities in granular limestone, along with ceylanite, vesuvian, and meionite, at Monte Somma, near Naples; also in fissures of basalt at Capo di Bove, near Rome. It is mentioned also as a production of the Isle of Bourbon.

Observations.

1. This species is characterized by its white colours, which sometimes incline to green, its crystallizations, vitreous lustre, conchoidal fracture, high degree of translucency, inferior hardness, and specific gravity.

2. It is distinguished from *Meionite* by its crystallizations, fourfold cleavage, and appearance when exposed to heat: its conchoidal fracture, and superior hardness, distinguish it from *Apatite*: it is readily distinguished from *Prismatic felspar* by its crystallizations: and its colour, and inferior hardness, distinguish it from *Emerald* and *Beryl*.

3. It is described by early writers under the name *White Schorl*. La Metherie named it *Sommite*, from the place where it was first found; and Haüy denominates it *Nepheline*.

4. The small acicular crystals of this species found near Rome, are described by Fleuriau Bellevue, under the name *Pseudo-Nepheline*, and are considered as belonging to a distinct species. Judging from the accounts of this pseudo-nepheline published by authors, we are still inclined to consider it but as a variety of nepheline.

* Chiastolite.

* Chiastolite (a).

Hohlspath, *Werner*.

Robien, in Nouv. idées sur la Format. des Foss. p. 108.—Pierre de croix, *Romé de Lisle*, t. ii. p. 440.—Crucite, *Lam.* t. ii. p. 292.—Macle, *Broch.* t. ii. p. 514. *Id. Haüy*, t. iii. p. 267. —Chiastolith, *Reuss*, b. ii. s. 47. *Id. Lud.* b. i. s. 149. *Id. Suck.* 1r th. s. 476. *Id. Bert.* s. 201. *Id. Mohs*, b. i. s. 539. *Id. Hab.* s. 35.—Macle, *Lucas*, p. 85.—Chiastolith, *Leonhard*, Tabel. s. 20.—Macle, *Brong.* t. i. p. 498. *Id. Brard*, p. 200.—Chiastolith, *Haus.* s. 88. *Id. Karst.* Tabel. s. 34.—Macle, *Haüy*, Tabl. p. 56.—Chiastolith, *Steffens*, b. i. s. 447.—Hohlspath, *Hoff.* b. ii. s. 330.—Chiastolith, *Lenz*, b. i. s. 503.—Hohlspath, *Oken*, b. i. s. 324.—Chiastolith, *Haus.* Handb. b. ii. s. 540. *Id. Aikin*, p. 198.

External Characters.

Its colours are white and grey: the white colours are yellowish-white, greenish-white, greyish-white, and reddish-white: the grey colours are pearl-grey, greenish-grey, and yellowish-grey.

It occurs always crystallized.

Its primitive form appears to be an oblique four-sided prism, with lateral edges of 84° 48′, and 95° 12′ *. The following are the secondary forms.

1. Four-sided prism, in which the lateral edges are rounded †.
2. Four prisms arranged in the form of a cross ‡.

 These

(a) *Chiastolite*, from the Greek word χιαζω and λιθος, because the ends of the prisms appear marked with a figure like that of the Greek letter χ.

* Macle prismatique, Haüy. † Macle cylindroide, Haüy.

‡ Macle quaternée, Haüy.

These crystals always appear as if they had been at one time hollow, and these hollows filled up with clay-slate, the position of which varies in regard to the crystals, and gives rise to the following varieties:

a. In the centre of the crystal there is a small prism of clay-slate, the lateral planes of which are parallel with those of the crystal, and from the angles of this prism black lines run to each angle of the crystal *, fig. 87. Pl. 4.

b. In this variety there is, in addition to the central prism and black lines of the former, smaller black-coloured prisms of clay-slate, one on each angle of the crystal, and their lateral planes are parallel with those of the crystal ‡, fig. 88. Pl. 4.

c. In this variety the terminal planes of the crystal are marked with black lines, which run from each of the lateral planes, parallel with the adjacent planes, to the black diagonal lines †, fig. 89. Pl. 4.

d. A black prism, in which the lateral planes are covered with a thick or thin crust of the hollow-spar ||.

The black or clay-slate mass is often thickest in the middle, and becomes thinner towards the extremities of the crystal: in other instances it is thinnest in the middle, and becomes gradually thicker towards the extremities of the crystal; and frequently the clay-slate mass is of equal thickness throughout.

The crystals are large, middle-sized, and small; sometimes also acicular, and always imbedded.

The cleavage is double, and in the direction of the lateral planes of the prism.

The

* Macle tetragrammé, Haüy.

‡ Macle polygrammé, Haüy.

† Macle pentarhombique, Haüy.

|| Macle circonscrite, Haüy.

The lustre of the cleavage is glistening, that of the fracture glimmering.

The fracture is splintery.

It is translucent.

It is hard; it scratches glass.

It is rather difficultly frangible.

Specific gravity, 2.944, *Haüy.*—2.923, *Karsten.*

Chemical Characters.

It is infusible before the blowpipe, and becomes white and nearly opaque.

Its constituent parts have not hitherto been ascertained.

Geognostic and Geographic Situations.

It occurs in small acicular crystals in clay-slate in Wolfscrag near Keswick, and near the summit of Skiddaw in Cumberland; also at Aghavanagh, and Baltinglas-hill, in the county of Wicklow *. The largest and most beautiful crystals are found in clay-slate near to St Brieux in Brittany: smaller crystals occur in the clay-slate of St Jago di Compostella in Gallicia; the variety *d* is found in the valley of Barreges in the Pyrenees; and the variety 3. in the plain of Thourmouse, in the High Pyrenees. It has been observed in micaceous clay-slate in the Serra de Marao in Portugal; and in very small acicular crystals in clay-slate near Gefrees in Bareuth.

America.—In clay-slate near Lancaster in Massachusets; also in New Hampshire and Maine †. In emery in the Estro de las Cruces in Peru ‡.

D 2 *Observations.*

* Fitton's Mineralogy of Dublin, p. 51. & 52.

† Cleaveland's Mineralogy, p. 342.

‡ Sent to Europe by the late Mr Christian Heuland.

Observations.

Chiastolite is placed immediately after the species of the Felspar genus, on account of its supposed affinity with them; but its characters are still so imperfectly known, that it cannot be arranged in any of the present genera.

** Sodalite (a).

Sodalite, *Thomson.*

Transactions of Royal Society of Edinburgh, vol. iv. p. 390.

External Characters.

Its colour is intermediate between celandine and mountain green.

It occurs massive, and crystallized in rhomboidal or garnet dodecahedrons.

Externally it is smooth, and shining or glistening: internally the longitudinal fracture is vitreous, and the cross-fracture resinous.

It has a double cleavage.

The fracture is small conchoidal.

The fragments are indeterminate angular, and sharp-edged.

It is translucent.

It is as hard as felspar.

It is brittle, and easily frangible.

Specific gravity, 2.378.

Chemical Characters.

When heated to redness, it does not decrepitate, nor fall

to

(a) *Sodalite*, so named on account of the great quantity of soda it contains.

to powder, but becomes dark-grey; and is infusible before the blowpipe.

Constituent Parts.

Silica, - -	38.52	36.00
Alumina,	27.48	32.00
Lime, -	2.70	
Oxide of Iron, -	1.00	0.25
Soda, - -	25.50	25.00
Muriatic Acid, -	3.00	6.75
Volatile Matter, -	2.10	
Loss, - -	1.70	
	100.00	100.00
	Thomson, in Tr. R. S. of Ed. vol. vi. p. 394.	*Ekeberg*, in Tr. R. S. of Ed. vol. vi. p. 395.

Geognostic and Geographic Situations.

It was discovered at Kanerdluarsuk, a narrow tongue of land, upward of three miles in length, in lat. 61°, in West Greenland, by Sir Charles Giesecké. It is found in a bed from six to twelve feet thick, in mica-slate, and is associated with sahlite, augite, hornblende, and garnet *.

Observations.

Sodalite was first described and analysed by Dr Thomson. The description of this mineral being incomplete, its true place in the system cannot be determined. It appears nearly allied to felspar.

APPEN-

* Article *Greenland*, in Edinburgh Encyclopædia.

APPENDIX.

CLAY AND LITHOMARGE FAMILIES.

The minerals included under these titles have no regular form or cleavage, and cannot therefore be connected with any of the mineral species. We place them here on account of their affinity with some of the members of the preceding genus.

* Clay Family.

In this group or family we include the following minerals, 1. Aluminite, 2. Common Clay, 3. Variegated Clay, 4. Slate-Clay, 5. Bituminous Shale, 6. Claystone, 7. Adhesive Slate, 8. Polier Slate, 9. Tripoli.

** Lithomarge Family.

The minerals of this family have many alliances with the preceding, and hence are placed immediately after them.

1. Lithomarge, 2. Mountain Soap, 3. Yellow Earth, 4. Cimolite, 5. Kollyrite, 6. Bole, 7. Sphragide.

* Clay

* Clay Family.

1. Aluminite.

Reine Thonerde, *Werner*.

Reine Thonerde, *Wid.* s. 385. *Id.* *Wern.* Cronst. s. 176.—Native Argil, *Kirw.* vol. i. p. 175.—Argilla pura, *Nap.* p. 246. —L'Alumine pure, *Broch.* t. i. p. 318.—Reine Thonerde, *Reuss*, b. ii. s. 102. *Id. Lud.* b. i. s. 104. *Id. Suck.* b. i. s. 471. *Id. Bert.* s. 277. *Id. Mohs*, b. i. s. 434. *Id. Leonhard*, Tabel. s. 20.—Argil native, *Brong.* t. i, p. 515.—Aluminit, *Haus.* s. 85. *Id. Karsten*, Tabel. s. 48.—Alumine pure, *Haüy*, Tabl. p. 58.—Alumenit, *Steffens*, b. i. s. 194. *Id. Lenz*, b. i. s. 541. —Verwitterter Alaunstein, *Oken*, b. i. s. 368.—Reine Thonerde, *Hoff.* b. ii. s. 4.

External Characters.

Its colour is snow-white, which verges on yellowish-white.

It occurs in small reniform pieces.

It has no lustre.

The fracture is fine earthy: its consistence is intermediate between friable and solid.

It is opaque.

It soils slightly.

It affords a glistening streak.

It adheres feebly to the tongue.

It passes from very soft into friable.

It feels fine, but meagre.

Specific gravity, 1.669, *Schreber*.

Chemical Characters.

It is very difficultly fusible. It absorbs water greedily, but does not fall in pieces.

Constituent

Constituent Parts.

Alumina, -	32.50	31.0
Water, -	47.00	45.0
Sulphuric Acid,	19.25	21.5
Silica, -	0.45 }	
Lime,	0.35 }	2.0
Iron,	0.45 }	
	——	——
		99.5
	Simon, in Allgem. Journ. der Chemie, 5 Jahrg. s. 137.	*Bucholz.*

Geognostic and Geographic Situations.

It occurs, along with selenite, in calcareous loam, which rests on brown coal, in the alluvial strata around Halle in Saxony; and it is said to occur at Newhaven, near Brightelmstone in England. The white crusts sometimes observed in the clay ironstone of Scotland, appear to be aluminite.

Observations.

Steffens and Keferstein are of opinion, that this mineral, and the selenite with which it is accompanied, are formed by the decomposition of iron-pyrites: the sulphuric acid thus formed, is supposed to unite with the lime and alumina; with the lime it forms sulphate of lime or selenite, and with the alumina an alum, with a superabundance of alumina.

2. Common

2. Common Clay.

Under this head we include Loam and Potters-Clay.

Loam.

Leim, *Werner*.

Magerer Thon, *Karsten*, Tabel. s. 28.—Leimen, *Hab.* s. 42. *Id.* *Steffens*, b. i. s. 197. *Id. Lenz*, b. ii. s. 549. *Id. Oken*, b. i. s. 370.—Lehm, *Hoff.* b. ii. s. 23.

External Characters.

Its colour is yellowish-grey, sometimes inclining to greenish-grey, and is spotted yellow and brown.

It occurs massive.

It is dull, and feebly glimmering when small scales of mica are present.

The fracture is coarse and small-grained uneven in the large, and in the small earthy.

It soils slightly.

It is very easily frangible.

It is sectile, and the streak is slightly resinous.

It is intermediate between friable and soft, but inclining more to the first.

It adheres slightly to the tongue.

It feels feels rather rough, and very slightly greasy, or meagre.

It is rather heavy, bordering on light.

Geognostic and Geographic Situations.

It occurs in great beds in alluvial districts, when it sometimes contains remains of elephants, and other fossil animals; also in secondary mountains, along with wacke and basalt, and in fissures, forming veins. It appears in general

neral to be an alluvial deposite, only a comparatively small portion of it occurring in secondary rocks.

It is so very widely and generally distributed, that it is not necessary to specify any locality.

Uses.

The mud-houses we meet with in different countries are built of loam. They are generally reared on a foundation of stone and lime, to secure them from damp. It is the practice to build them in spring, and allow them to dry during the summer: they are plastered with lime in autumn, in order to protect them from rain. The loam is mixed with straw or hair, to prevent its cracking. The most advantageous practice is to form the loam into bricks, to dry these in the shade, and afterwards in the sun. The use of loam-bricks is of high antiquity; for we are told that the ancient city of Damascus, and the walls of Babylon, were built of bricks of this substance.

Observations.

It is characterized by its muddy grey colours, rough meagre feel, slight adherence to the tongue, nearly dull streak, and slight soiling. These characters distinguish it from *Potters-Clay.*

Potters-Clay.

Töpferthon, *Werner.*

There are two kinds of this clay, viz. Earthy and Slaty.

a. Earthy Potters-Clay.

Erdiger Töpferthon, *Werner.*

Erdiger Töpferthon, *Steffens,* b. i. s. 198. *Id. Lenz,* b. ii. s. 550. *Id. Hoff.* b. ii. s. 32.

External

External Characters.

Its colours are greyish and yellowish white; also yellowish, ash, pearl, smoke, greenish, and bluish grey. Very seldom mountain-green.

It occurs massive; and is friable, approaching to solid.

Internally it is dull, or feebly glimmering, from intermixed scales of mica.

The fracture in the large is coarse-grained uneven; in the small fine earthy.

It is more or less shining in the streak.

The fragments are very blunt-edged.

It is opaque.

It soils slightly.

It is very soft, passing into friable.

It is sectile.

It adheres strongly to the tongue; more strongly than loam.

It feels rather greasy.

It becomes plastic in water.

Specific gravity, 2.085, *Karsten.*—1.723. *Poole, Berger.* 1.800, 2.000, *Kirwan.*

Chemical Characters.

It is infusible.

Constituent Parts.

Silica, -	61.	63.00
Alumina, -	27.	37.00
Oxide of Iron,	1.	
Water, -	11.	
	100.	100.00
	Klaproth, Chem. Abhandl. s. 282.	*Kirwan.*

Geognostic

Geognostic Situation.

It is a frequent mineral in alluvial districts, where it sometimes occurs in beds of considerable thickness; it has also been observed in secondary or flœtz formations.

Geographic Situation.

It occurs in many districts both in England, Scotland, and Ireland.

Uses.

It is used in potteries, in the manufacture of the different kinds of earthen-ware: it is also made into bricks, tiles, crucibles, and tobacco-pipes; and is employed in improving sandy and calcareous soils.

Observations.

1. It is distinguished from *Loam* by its colour, fracture, its shining streak, and its stronger adherence to the tongue.

2. It is distinguished from *Porcelain-Earth* by greater coherence, stronger adherence to the tongue, greasy feel, and its plasticity with water.

3. The finer varieties are named *Pipe-Clay.*

b. Slaty Potters-Clay.

Schiefriger Töpferthon, *Werner.*

Schiefriger Töpferthon, *Steffens*, b. i. s. 200. *Id. Lenz*, b. ii. s. 554. *Id. Hoff.* b. ii. s. 52.

External Characters.

Its most frequent colour is dark smoke-grey, seldomer bluish and pearl grey.

It

It occurs massive.

The lustre of the principal fracture is glistening; the cross fracture dull.

The principal fracture is very imperfect slaty; the cross fracture fine earthy.

The fragments are often tabular.

It does not adhere so strongly to the tongue as the earthy kind, but becomes more shining in the streak; and it feels more greasy.

Geognostic Situation.

It occurs in considerable beds in alluvial districts, along with Earthy Potters-Clay.

3. Variegated Clay.

Bunter Thon, *Werner.*

Bunter Thon, *Steffens,* b. i. s. 200. *Id. Lenz,* b. ii. s. 554. *Id. Hoff.* b. ii. s. 54.

External Characters.

Its colours are white, grey, yellow, red and brown; the varieties are yellowish and reddish white, flesh and peach blossom red, pearl-grey, yellowish-grey, ochre-yellow, and yellowish-brown.

These colours are generally arranged in broad stripes, and often in veined and spotted delineations.

It occurs massive.

Internally it is dull.

The fracture is coarse earthy, inclining to slaty.

The fragments are blunt-edged.

It

It becomes strongly resinous in the streak, more so than the preceding kinds.

It is soft, inclining to friable.

It is sectile.

It adheres pretty strongly to the tongue.

It feels rather greasy.

Geognostic and Geographic Situations.

It occurs in alluvial deposites near Wehrau, in Upper Lusatia.

Observations.

It is closely allied to Lithomarge, and even passes into it.

4. Slate-Clay.

Schiefer Thon, *Werner.*

Slate-clay, Shale, *Kirw.* vol. i. p. 182.—L'Argile schisteuse, *Broch.* t. i. p. 327. *Id. Haüy,* t. iv. p. 446.—Schiefer Thon, *Reuss,* b. ii. s. 99. *Id. Lud.* b. i. s. 107. *Id. Suck.* 1r th. s. 490. *Id. Bert.* s. 211. *Id. Mohs,* b. i. s. 440. *Id. Hab.* s. 47. *Id. Leonhard,* Tabel. s. 22.—Argille feuilletée, *Brong.* t. i. p. 525.? Schiefriger Thon, *Karsten,* Tabel. s. 28.—Schiefer Thon, *Steffens,* b. i. s. 201. *Id. Lenz,* b. ii. s. 555. *Id. Hoff.* b. ii. s. 56.

External Characters.

Its colours are smoke and ash grey, greyish-black, and sometimes bluish and yellowish grey, and brownish-red.

It occasionally contains impressions of unknown ferns and reeds.

It

Emm. b. i. s. 289.—Schisto bituminoso, *Nap.* p. 263.—Argillite bitumineux, *Lam.* t. ii. p. 116.—Varieté de l'Argile schisteuse, *Haüy.*—Le Schiste bitumineux, *Broch.* t. i. p. 389.—Brandschiefer, *Reuss*, b. ii. s. 120. *Id. Lud.* b. i. s. 111. *Id. Suck.* 1r th. s. 504. *Id. Bert.* s. 218. *Id. Mohs*, b. i. s. 456. *Id. Leonhard*, Tabel. s. 23. *Id. Karsten*, Tabel. s. 36.—Bituminous Shale, *Kid*, vol. i. p. 189.—Brandschiefer, *Steffens*, b. i. s. 204. *Id. Lenz*, b. i. s. 573. *Id. Oken*, b. i. s. 361. *Id. Hoff.* b. ii. s. 88.

External Characters.

Its colour is light brownish-black, which sometimes passes into blackish-brown.

It occurs only massive.

Internally its lustre is feebly glimmering.

The fracture is rather thin and straight slaty.

The fragments are tabular.

It it opaque.

It becomes resinous in the streak, but the colour is not changed.

It is very soft, approaching to soft.

It is rather sectile, and easily frangible.

It feels rather greasy.

Specific gravity, 1.991, 2.049, *Kirwan.*—2.060, *Karsten.*

Constituent Parts.

Two hundred grains afforded the following parts, partly as educts, partly as products:

Carbonated

Emm. b. i. s. 289.—Schisto bituminoso, *Nap.* p. 263.—Argillite bitumineux, *Lam.* t. ii. p. 116.—Varieté de l'Argile schisteuse, *Haüy.*—Le Schiste bitumineux, *Broch.* t. i. p. 389.—Brandschiefer, *Reuss,* b. ii. s. 120. *Id. Lud.* b. i. s. 111. *Id. Suck.* 1r th. s. 504. *Id. Bert.* s. 218. *Id. Mohs,* b. i. s. 456. *Id. Leonhard,* Tabel. s. 23. *Id. Karsten,* Tabel. s. 36.—Bituminous Shale, *Kid,* vol. i. p. 189.—Brandschiefer, *Steffens,* b. i. s. 204. *Id. Lenz,* b. i. s. 573. *Id. Oken,* b. i. s. 361. *Id. Hoff.* b. ii. s. 88.

External Characters.

Its colour is light brownish-black, which sometimes passes into blackish-brown.

It occurs only massive.

Internally its lustre is feebly glimmering.

The fracture is rather thin and straight slaty.

The fragments are tabular.

It it opaque.

It becomes resinous in the streak, but the colour is not changed.

It is very soft, approaching to soft.

It is rather sectile, and easily frangible.

It feels rather greasy.

Specific gravity, 1.991, 2.049, *Kirwan.*—2.060, *Karsten.*

Constitucnt Parts.

Two hundred grains afforded the following parts, partly as educts, partly as products:

Carbonated

Carbonated Hydrogen Gas,	80 cubic inches.
Empyreumatic Oil, -	30 grains.
Thick Pitchy Oil, - -	5 do.
Ammoniacal Water, - -	4
Carbon, -	20
Silica,	87½
Alumina,	6½
Lime, -	10½
Magnesia,	1
Oxide of Iron, - - -	3

Klaproth, Beit. b. v. s. 184.

Geognostic Situation.

It occurs principally in rocks of the coal-formation, where it frequently alternates with, and passes into, slate-clay, and also into coal. It sometimes contains vegetable impressions, and also animal remains, particularly of shells. It occurs in beds of considerable magnitude in hills of iron-clay.

Geographic Situation.

It occurs in all the coal districts in this island, and also in those of Bohemia, Poland, Silesia, and other countries.

Observations.

1. From *Slate-Clay*, with which it has been confounded, it is distinguished by the streak: in Slate-Clay, the streak is always dull; whereas it is invariably shining and resinous in Bituminous-Shale.

Carbonated Hydrogen Gas,	80 cubic inches.
Empyreumatic Oil, -	30 grains.
Thick Pitchy Oil, - -	5 do.
Ammoniacal Water, - -	4
Carbon, -	20
Silica,	87½
Alumina,	6½
Lime,	10½
Magnesia,	1
Oxide of Iron, - - -	3

Klaproth, Beit. b. v. s. 184.

Geognostic Situation.

It occurs principally in rocks of the coal-formation, where it frequently alternates with, and passes into, slate-clay, and also into coal. It sometimes contains vegetable impressions, and also animal remains, particularly of shells. It occurs in beds of considerable magnitude in hills of iron-clay.

Geographic Situation.

It occurs in all the coal districts in this island, and also in those of Bohemia, Poland, Silesia, and other countries.

Observations.

1. From *Slate-Clay*, with which it has been confounded, it is distinguished by the streak: in Slate-Clay, the streak is always dull; whereas it is invariably shining and resinous in Bituminous-Shale.

Emm. b. i. s. 289.—Schisto bituminoso, *Nap.* p. 263.—Argillite bitumineux, *Lam.* t. ii. p. 116.—Varieté de l'Argile schisteuse, *Haüy.*—Le Schiste bitumineux, *Broch.* t. i. p. 389.—Brandschiefer, *Reuss,* b. ii. s. 120. *Id. Lud.* b. i. s. 111. *Id. Suck.* 1r th. s. 504. *Id. Bert.* s. 218. *Id. Mohs,* b. i. s. 456. *Id. Leonhard,* Tabel. s. 23. *Id. Karsten,* Tabel. s. 36.—Bituminous Shale, *Kid,* vol. i. p. 189.—Brandschiefer, *Steffens,* b. i. s. 204. *Id. Lenz,* b. i. s. 573. *Id. Oken,* b. i. s. 361. *Id. Hoff.* b. ii. s. 88.

External Characters.

Its colour is light brownish-black, which sometimes passes into blackish-brown.

It occurs only massive.

Internally its lustre is feebly glimmering.

The fracture is rather thin and straight slaty.

The fragments are tabular.

It it opaque.

It becomes resinous in the streak, but the colour is not changed.

It is very soft, approaching to soft.

It is rather sectile, and easily frangible.

It feels rather greasy.

Specific gravity, 1.991, 2.049, *Kirwan.*—2.060, *Karsten.*

Constituent Parts.

Two hundred grains afforded the following parts, partly as educts, partly as products:

Carbonated

7. Adhesive Slate *.

Klebschiefer, *Werner*.

Klebschiefer, *Reuss*, b. iv. s. 159.—Polierschiefer, *Leonhard*, Tabel. s. 22.—Klebschiefer, *Karsten*, Tabel. s. 26. *Id. Steffens*, b. i. s. 151. *Id. Lenz*, b. ii. s. 560. *Id. Haus.* b. ii. s. 418. *Id. Hoff.* b. ii. s. 63.

External Characters.

Its colour is very pale yellowish-grey, which passes into yellowish-white; and sometimes inclines to greenish-grey and smoke-grey.

It occurs massive.

It is dull.

The fracture is straight slaty; it is thick or thin slaty; and in the thick slaty varieties, the cross fracture is even, inclining to flat conchoidal.

The fragments are tabular.

It is feebly translucent on the edges.

It becomes shining in the streak, particularly when moist.

It is soft, passing into very soft.

It is sectile.

It splits very easily.

It exfoliates very readily in the direction of the foliated fracture, particularly when exposed in warm and dry situations.

It adheres strongly to the tongue.

It feels somewhat greasy.

Specific gravity 2.080, *Klaproth*.

Chemical

* It is so named from its adhering very strongly to

Chemical Characters.

It is infusible before the blowpipe.

Constituent Parts.

Silica, -	62.50	Silica, - -	58.0
Alumina,	00.50	Alumina, - -	5.0
Magnesia,	8.00	Magnesia, - -	6.5
Lime, -	00.25	Lime, - -	1.5
Carbon, -	00.75	Iron & Manganese,	9.0
Iron, -	4.00	Water, - -	19.9
Water, -	22.00		
	98.00		100.0
	Klaproth.		*Bucholz.*

Geognostic Situation.

It occurs in beds in secondary gypsum, and contains imbedded menilite.

Geographic Situation.

It has hitherto been found only in the gypsum formation around Paris.

Observations.

It is distinguished from *Polier Slate*, by its strong adherence to the tongue, its exfoliation, slight greasy feel, and greater hardness.

8. Polier

8. Polier or Polishing Slate.

Polierschiefer, *Werner.*

Polierschiefer, *Mohs, Karsten, Steffens, Lenz,* and *Hoffmann.*

External Characters.

Its colours are partly yellowish-white, partly yellowish-grey, which latter passes into brown and isabella-yellow. Sometimes these colours are arranged in stripes.

It occurs massive.

It is dull.

The principal fracture is straight and thin slaty, the cross fracture fine earthy.

The fragments are tabular.

It is opaque.

It soils slightly.

It is very soft, passing into friable.

It is uncommonly easily frangible.

It scarcely adheres to the tongue.

It feels fine, but meagre.

It is so light as to swim in water.

Specific gravity 0.590—0.606, *Haberle.*

Constituent Parts.

Silica, - -	79.00
Alumina, - -	1.00
Lime, - - - -	1.00
Oxide of Iron, -	4.00
Water, - -	14.00
	99.00

Bucholz, in Journ. fur d. Chemie & Physik, b. ii. s. 28.

Geognostic

Geognostic and Geographic Situations.

It forms a bed in the neighbourhood of rocks of the coal-formation, at Planitz in Saxony; also near Bilin in Bohemia; and it is said to occur at Menat, near Riom in Auvergne *. The variety found near Bilin, rests upon a mineral nearly allied to polier slate, and described by Haberle under the name *Saugschiefer*.

Uses.

It is used for polishing silver; and also for polishing marble, and other comparatively soft minerals.

Observations.

1. Werner considers it as a pseudo-volcanic production.

2. It has been described as a variety of Adhesive-slate, but it is distinguished from that mineral by its extremely thin slaty fracture, easy frangibility, meagre feel, softness, and lightness.

9. Tripoli †.

Tripel, *Werner*.

Tripela, *Wall.* t. i. p. 94.—Trippel, *Wid.* s. 353.—Tripoli, *Kirw.* vol. i. p. 202.—Tripel, *Estner*, b. ii. s. 631. *Id. Emm.* b. i. s. 307. *Id. Nap.* p. 210. *Id. La Meth.* t. ii. p. 457.—Le Tripoli, *Broch.* t. i. p. 379.—Quartz aluminifere Tripoléene, *Haüy*, t. iv. p. 467.—Tripel, *Reuss*, b. ii. s. 446. *Id. Lud.* b. i. s. 108. *Id. Suck.* 1r th. s. 428. *Id. Bert.* s. 247. *Id. Mohs*, b. i. s. 449. *Id.*

* The Auvergne mineral is described by Haüy under the name *Thermantide Tripoléene*, and is also described by Saussure.—Vid. Haüy, t. iv. p. 499, and 500.—Brongniart, Min. t. i. p. 330.

† The varieties of this mineral first used in the arts were brought from Tripoli; hence the name given to it.

Id. Hab. s. 6.—Tripoli, *Lucas,* s. 60.—Tripel, *Leonhard,* Tabel. s. 22.—Tripoli, *Brong.* t. i. p. 329.—Tripel, *Karst.* Tabel. s. 24.—Tripoli, *Kid,* Appendix, p. 31.—Tripel, *Steffens,* b. i. s. 147. *Id. Lenz,* b. i. s. 564. *Id. Oken,* b. i. s. 278. *Id. Hoff.* b. ii. s. 72. *Id. Haus.* Handb. b. ii. s. 417.

External Characters.

Its principal colour is yellowish-grey, which sometimes passes into yellowish-white, or into isabella-yellow, and ochre-yellow; it sometimes inclines to ash-grey.

It occurs massive, and in whole beds.

It is dull.

The fracture is sometimes fine, sometimes coarse earthy, and in the great inclines to slaty.

The fragments are blunt angular, or slaty.

It is opaque.

It is soft, sometimes passing into very soft.

It is not very brittle, and is rather easily frangible.

It feels meagre, and rather rough.

It does not adhere to the tongue.

Specific gravity 2.202, *Bucholz.*

Chemical Characters.

It is infusible before the blowpipe.

Constituent Parts.

				Rottenstone.	
Silica, -	81.00	Silica,	90	Silica,	4
Alumina,	1.50	Alumina,	7	Alumina,	86
Trace of Lime.		Black and Red		Carbon,	10
Black and Red		Oxide of Iron,	3		——
Oxide of Iron,	8.00		——		100
Sulphuric Acid,	3.45		100		
Water, -	4.55	[illegible]			
Loss, -	1.50				
	——				
	100	*Buchol*			

The sulphuric acid and water are considered as accidental constituent parts.

Geognostic Situation.

It occurs in beds in coal-fields; also in beds, along with secondary limestone, and alternating with clay, under basalt.

Geographic Situation.

It is found at Bakewell in Derbyshire, where it is named *Rottenstone:* also in the coal-fields of Dresden and Thuringia; in secondary trap districts in Bohemia; in Auvergne, where it is said to be associated with pseudo-volcanic rocks; in the island of Corfu; at Ronneburg and Kerms in Austria; near Burgos in Spain; and Tripoli in Barbary.

Uses.

On account of the hardness of its particles, it is used for polishing stones, metals, and glasses. If it contain any coarse quartzy particles, these must be separated by washing before it is used, because they injure the surfaces of the substances intended to be polished. When used for polishing precious stones, it is mixed with sulphur, in the proportion of two parts of tripoli to one of sulphur: these are well ground together on a marble slab, and then applied to the mineral by means of a piece of leather. When mixed with red ironstone, it is used for polishing optical glasses. It is used sometimes for moulds, in which small [illegible] or glass figures and medallions are cast.

[illegible] that a fine species of tripoli, found near Bur[illegible] is used as an ingredient in the manufacture of of porcelain. The tripoli of Corfu is reckoned the most valuable by artists.

The Rottenstone of Derbyshire, which seems to be a variety of this mineral, is used for similar purposes, and is well known in this country. Sometimes a sandy marl is dug and sold for tripoli; but its effervescence with acids, and its very rough feel, distinguish it from that substance.

Observations.

Some mineralogists are of opinion, that it is a mixture of fine sand and clay, therefore that it is a mechanical deposite: others are inclined to view it as a chemical formation,—an opinion which appears to be countenanced by its geognostic relations.

** Lithomarge Family.

1. Lithomarge (a).

Steinmark, *Werner.*

There are two kinds, viz. Friable Lithomarge, and Indurated Lithomarge.

a. Friable Lithomarge.

Zerreiblicher Steinmark, *Werner.*

Friable Lithomarge, *Kirwan,* vol. i. p. 187.—Zerreiblicher Steinmark, *Reuss,* b. ii. s. 49. *Id. Leonhard,* Tabel. s. 26. *Id. Karsten,*

(a) It occurs in veins and [illegible] in rocks, somewhat like marrow in bones; hence the nam[illegible] Marrow.

Karsten, Tabel. s. 28. *Id. Steffens*, b. i. s. 246. *Id. Lenz*, b. ii. s. 618. *Id. Hoff.* b. ii. s. 201. *Id. Haus.* Handb. b. ii. s. 455. —Lithomarge, *Aikin*, p. 240.

External Characters.

Its colours are snow-white, and yellowish-white.

It occurs massive, disseminated, and sometimes in crusts.

It consists of very fine scaly or dusty, feebly glimmering particles.

It becomes shining in the streak.

It is generally slightly cohering, seldom loose.

It soils slightly.

It feels rather greasy.

It adheres to the tongue.

It is light.

It phosphoresces in the dark.

Constituent Parts.

Earth or Lithomarge of Sinopia.

Silica, - -	32.00
Alumina, -	26.50
Iron, -	21.00
Muriate of Soda,	1.50
Water, -	17.00
	98.00

Klaproth, Beit. b. iv. s. 349.

Geognostic

Geognostic and Geographic Situations.

It generally occurs in small quantity, often associated with compact lithomarge, and for the most part in tinstone veins, where it is accompanied with tinstone, fluor-spar, quartz, and sometimes ores of silver. It is found in tinstone veins at Ehrenfriedersdorf, also at Penig; in fissures in greywacke, in the Hartz; in manganese veins with red ironstone at Walkenried; and it is said also in Nassau, Bavaria, and Transylvania.

Observations.

1. Klaproth describes a substance under the name *Earth of Sinopis*, which is found in Pontus. It is of a dark red colour, and, according to Karsten, is but a variety of the friable lithomarge. The analysis here given is of the Earth of Sinopis.

2. Colour, appearance of the particles, adherence to the tongue, fine and greasy feel, and geognostic situation, are characteristics of this mineral. Colour, aspect of the particles, and greasy feel, distinguish it from *Porcelain Earth*.

b. Indurated Lithomarge.

Verhärtetes Steinmark, *Werner.*

Terra miraculosa Saxoniæ, *Schütz*, in Nov. Act. Cæs. Nat. Curios. 3. App. p. 93.—Steinmark, *Hoffmann*, Bergm. Journ. 1788, 1. 2. s. 520.—Indurated Lithomarge, *Kirwan*, vol. i. p. 188. —La Möelle de Pierre, ou Lithomarge, *Broch.* t. i. p. 447.—Argil Lithomarge, *Haüy*, t. iv. p. 444.—Steinmark, *Reuss*, b. ii. s. 164. *Id. Leonhard*, Tabel. s. 26.—Argile Lithomarge, *Brong.*

Brong. t. i. p. 521.—Verhärtetes Steinmark, *Haus.* s. 86. *Id. Karst.* Tabel. s. 28. *Id. Steffens,* b. i. s. 248. *Id. Lenz,* b. ii. s. 619. *Id. Hoff.* b. ii. s. 202. *Id. Haus.* Handb. b. ii. s. 453.

External Characters.

Its colours are yellowish and reddish *white*, which latter passes from pearl-*grey*, through lavender-*blue*, pale plum-blue, into flesh *red*, and nearly into brick-red; the yellowish-white passes into ochre-*yellow*. The white and red varieties are generally uniform; but the others are disposed in clouded, spotted, veined, and striped delineations.

It occurs massive, disseminated, and globular or amygdaloidal *.

It is dull.

The fracture is fine earthy in the small, and large conchoidal, and sometimes even, in the great.

The fragments are indeterminate angular, and rather blunt edged.

It is opaque.

It becomes shining in the streak.

It is very soft, sectile, and easily frangible.

It adheres strongly to the tongue.

It feels fine and greasy.

Specific gravity, 2.419, *Kopp.*—2.435—2.492, *Breithaupt.*

Chemical Characters.

It is infusible before the blowpipe. Several of the varieties

* It sometimes occurs in prisms, like those of prismatic felspar. These prisms are supposed to be felspar changed into lithomarge. For accounts of these, vid. Estner, Min. b. ii. s. 771.; also Klaproth's Chem. Abhandl.

ties phosphoresce when heated; and others, when moistened with water, afford an agreeable smell, like that of nuts.

Constituent Parts.

	Red Lithomarge of Rochlitz.	Lithomarge from Flachenseifen.
Silica, -	45.25	58
Alumina, -	36.50	32
Oxide of Iron,	2.75	2
Water, -	14.0	7
Trace of Potash.		
	98.50	99
	Klaproth, Chem. Abhandl. s. 287.	Id. s. 285.

Geognostic Situation.

It occurs in veins in porphyry, gneiss, grey-wacke, and serpentine: in drusy cavities in topaz-rock; or nidular, in basalt, amygdaloid, and serpentine; and it is said also in beds, in a coal-formation.

Geographic Situation.

At Rochlitz in Saxony, it occurs in cotemporaneous veins, traversing clay-porphyry; at Ehrenfriedersdorf and Altenberg, also in Saxony, the white and red varieties occur in veins in gneiss; at Zöblitz, it traverses serpentine, in the form of veins; the yellow variety lines the drusy cavities of the topaz-rock: it is remarked of the topazes, that when the accompanying lithomarge is yellow, they also have the same colour, and when the topazes are white, the lithomarge is white; and the Saxon *Terra miraculosa*, a variety of this mineral, appears to occur in small beds, in a coal-formation near Planitz. In the Hartz, it occurs in veins that

that traverse grey-wacke; and it is described as a production of the mountains of Bavaria, Bohemia, and Norway. In secondary hills, it occurs in balls in amygdaloid.

Uses.

The Chinese are said to use it, when mixed with the root of Veratrum album, in place of snuff: in Germany it is employed for polishing serpentine; and it was formerly an article of the Materia Medica. The blue variegated variety from Planitz in Saxony, named *Terra miraculosa Saxoniæ*, used to be kept in apothecaries shops.

Observations.

1. The friable kind is characterized by its scaly particles, soiling, and low degree of coherence; the indurated by fracture, streak, softness, and sectility.

2. It is distinguished from *Potters-Clay*, by its colours, greater hardness, not soiling, rather greater specific gravity, and also by its fracture, and geognostic situation: it is nearly allied to *Variegated Clay*, but that mineral is softer, lighter, and wants the conchoidal or even fracture; and its strong adherence to the tongue, fracture, and inferior weight, distinguish it from *Steatite*.

2. Mountain Soap *.

Bergseife, *Werner.*

Bergseife, *Wid.* s. 436. *Id. Emm.* b. i. s. 360.—Le Savon de Montagne, *Broch.* t. i. p. 453.—Bergseife, *Reuss*, b. ii. s. 171. *Id.*

---

* This mineral is named *Mountain Soap*, on account of its greasiness, sectility, and softness.

Id. Lud. b. i. s. 127. *Id. Suck.* 1r th. s. 502. *Id. Bert.* s. 208. *Id. Mohs,* b. i. s. 522. *Id. Leonhard,* Tabel. s. 26. *Id. Karst.* Tabel. s. 28. *Id. Haus.* s. 86. *Id. Steffens,* b. i. s. 256. *Id. Lenz,* b. ii. s. 625. *Id. Oken,* b. i. s. 385. *Id. Hoff.* b. ii. s. 206. *Id. Haus.* Handb. b. ii. s. 456.

External Characters.

Its colour is pale brownish-black.
It occurs massive.
It is dull.
The fracture is fine earthy.
The fragments are indeterminate angular.
It is opaque.
It becomes shining in the streak.
It writes, but does not soil.
It is very soft, and perfectly sectile.
It is easily frangible.
It adheres strongly to the tongue.
It feels very greasy.
It is light, bordering on rather heavy.

Geognostic and Geographic Situations.

It occurs in trap-rocks in the Island of Skye. It was formerly found at Olkutzk in Gallicia, but is now no longer to be met with in that quarter. It is said to occur in a bed in the district of Nassau; and in a bed, immediately under the soil, along with potters-clay and loam, near Waltershaus, at the foot of the mountains of the Forest of Thuringia.

Use.

It is valued by painters as a crayon.

Observations.

Observations.

1. This mineral is characterized by its colour, fracture, streak, greasy feel, perfect sectility, adherence to the tongue, its writing without soiling, and low specific gravity.

2. Its colour and property of writing, distinguish it from *Lithomarge:* from *Bole*, it is distinguished by its dull and fine earthy fracture, its writing, greater sectility and greasiness, and its not falling into pieces in water, which is the case with bole; and its colour distinguishes it from *Fullers Earth.*

3. It is allied to Bole and Lithomarge.

4. It was first established as a distinct mineral by Werner, and particularly described by Stift, in Moll's Ephemer. 4. 1. s. 31.; and by Schlottheim, in the Magaz. Naturf. Fr. in Berlin, 1. 4. s. 406.

5. Its black colour is alleged to be owing to bitumen; but its chemical constitution is still unknown.

3. Yellow Earth.

Gelberde, *Werner.*

Argile ocreuse jaune graphique, *Haüy.*

Gelberde, *Wid.* p. 427.—Yellow Earth, *Kirw.* vol. i. p. 194.—Gelberde, *Estner*, b. i. s. 362.—La Terre jaune, *Broch.* t. i. p. 455.—Gelberde, *Reuss*, b. ii. s. 101. *Id. Lud.* b. i. s. 128. *Id. Suck.* 1r th. s. 524. *Id. Bert.* s. 302. *Id. Mohs*, b. i. s. 524. *Id. Hab.* s. 48. *Id. Leonhard*, Tabel. s. 26. *Id. Karsten*, Tabel. s. 48. *Id. Steffens*, b. i. s. 261. *Id. Lenz*, b. i. s. 626. *Id. Oken*, b. i. s. 372. *Id. Hoff.* b. ii. s. 210. *Id. Haus.* Handb. b. ii. s. 457.

External Characters.

Its colour is ochre-yellow, of different degrees of intensity.

It occurs massive.

It is dull on the cross fracture, but glimmering on the principal fracture.

The fracture in the large inclines to slaty; in the small, it is earthy.

The fragments are tabular, or indeterminate angular.

It becomes somewhat shining in the streak.

It is opaque.

It soils and writes slightly.

It is very soft, passing into friable.

It is easily frangible.

It adheres pretty strongly to the tongue.

It feels rather greasy.

Specific gravity 2.240, *Breithaupt.*

Chemical Characters.

Before the blowpipe, it is converted into a black and shining enamel.

Constituent Parts.

From Bitry in France.

Silica, - -	92
Alumina,	2
Lime, - - -	3
Iron, - - -	3
	100

Merat Guillot, Brong. Min. t. i. p. 544.

Geognostic and Geographic Situations.

It is found at Wehraw, in Upper Lusatia, where it is associated with clay, and clay ironstone; near Meissen, mixed ed with quartz sand; in the district of Berry, and at Bitry, in the department of Nievre in France.

Uses.

It may be employed as a yellow pigment; and when burnt, it is sold by the Dutch under the name of *English red.* The remains in Pompeii, show that it was used as a pigment, both in its yellow and red state, by the ancient Romans. It appears even to have been known to Theophrastus as a yellow pigment.

4. Cimolite *.

Cimolith, *Klaproth.*

Creta Cimolia, *Plin.* Hist. Nat. xxxv. 57.—Cimolith, *Klaproth,* Beit. b. i. s. 291.—La Cimolite, *Haüy,* t. iv. p. 446. *Id. Reuss,* b. ii. s. 169. *Id.* b. i. s. 150. *Id. Suck.* 1r th. s. 500. *Id. Bert.* s. 212. *Id. Leonhard,* Tabel. s. 21. *Id. Haus.* s. 86. *Id. Karsten,* Tabel. s. 28. *Id. Steffens,* b. i. s. 260.—Kimolit, *Lenz,* b. ii. s 544.—Cimolith, *Oken,* b. i. s. 372. *Id. Haus.* Handb. b. ii. s. 463.

External Characters.

Its colours are greyish-white, and pearl-grey, which become reddish by the action of the weather.

It occurs massive.

It is dull.

The fracture is earthy, sometimes inclining to slaty.

It is opaque.

It becomes shining in the streak.

* So named from Cimolia, the island where it is principally found.

It soils very slightly.
It is very soft.
It is rather easily frangible.
It adheres pretty strongly to the tongue.
Specific gravity, 2.00, *Klaproth*.—2.187, *Karsten*.

Chemical Characters.

It is infusible.

Constituent Parts.

Silica,	63.00		Silica,	54.00
Alumina,	23.00		Alumina,	26.50
Iron,	1.25		Iron,	1.50
Water,	12.00		Potash,	5.50
			Water,	12.00
	99.25			99.50
	Klaproth, Beit. b. i. s. 299.			*Klaproth*, Chem. Abhandl. s. 284.

Geognostic and Geographic Situations.

It appears to occur in beds, in the islands of Argentiera or Cimolia, and Milo, in the Mediterranean Sea.

Uses.

It was highly prized as a medicine by the ancients; they also used it for cleansing woollen and other stuffs, for which purpose it is excellently suited.

Observations.

1. This mineral is mentioned by several ancient writers, as Theophrastus, Dioscorides, Strabo, Pliny, and Ovid; and in modern times, first by Tournefort, and next by Klaproth,

Klaproth, who in the year 1794 received specimens of it from Mr Hawkins, who had collected it in the island of Cimolia.

2. It appears to be nearly allied to Fullers Earth.

5. Kollyrite *.

Kollyrit, *Karsten.*

Natürlicher Alaunerde, *Klaproth,* Beit. b. i. s. 257.—*Fichtel,* Mineralog. Aufsätze, 170.—Kollyrit, *Leonhard,* Tabel. s. 21. *Id. Karsten,* Tabel. s. 48. *Id. Haus.* s. 85. *Id. Steffens,* b. i. s. 259. *Id. Lenz,* b. ii. s. 543. *Id. Oken,* b. i. s. 370. *Id. Hoff.* b. iv. s. 161. *Id. Haus.* Handb. b. ii. s. 446.

External Characters.

Its colours are snow, greyish, reddish, and yellowish white.

It occurs massive.

Internally it is dull; but the reddish-white variety is feebly glimmering.

The fracture is fine earthy in the small, and flat conchoidal in the large.

The fragments are indeterminate angular and rather sharp edged.

The snow-white is feebly, the reddish-white is strongly translucent on the edges.

It becomes shining and slightly resinous in the streak.

It soils slightly.

It

* It is the κολλυριον of Dioscorides, which he describes as an earth having the property of adhering strongly to the tongue.

It is very soft; the snow-white variety friable, the reddish-white approaching to very soft.

It is rather brittle, and very easily frangible.

It adheres strongly to the tongue.

It feels greasy, but in a low degree.

It is light.

Chemical Characters.

It is infusible. It becomes transparent in water, and falls into pieces with a crackling noise.

Constituent Parts.

Silica, - - -	14
Alumina, - -	45
Water, - - -	42

Klaproth, Beit. b. i. s. 257.

Geognostic and Geographic Situations.

It is found in the Stephen's pit at Schemnitz in Hungary, where it forms a vein from four to five inches wide in porphyry; and it occurs in veins in sandstone, at Weissenfels in Saxony.

6. Bole *.

Bol, *Werner.*

Bolus, *Waller.* t. i. p. 51.—Bole, *Kirw.* vol. i. p. 191.—Bol, *Estner*, b. ii. s. 784. *Id. Emm.* b. i. s. 381.—Bolo, *Nap.* p. 256. —Le

* The name *Bole*, is derived from the Greek word *βωλος*, *a lump of earth.* All the clayey substances formerly us[illegible], were named *bolus.*

—Le Bol, *Broch.* t. i. p. 459.—Bol, *Reuss,* b. ii. s. 115. *Id. Lud.* b. i. s. 129. *Id. Suck.* 1r th. s. 495. *Id. Bert.* s. 207. *Id. Mohs,* b. i. s. 525. *Id. Hab.* s. 39. *Id. Leonhard,* Tabel. s. 26. —Le Bol Armenie, *Brong.* t. i. p. 543.—Bol, *Haus.* s. 86. *Id. Karsten,* Tabel. s. 28.—Bole, *Kid,* vol. i. p. 179.—Bol, *Steffens,* b. i. s. 253. *Id. Lenz,* b. ii. s. 634. *Id. Hoff.* b. ii. s. 226. *Id. Haus.* Handb. b. ii. s. 458.

External Characters.

Its colour is pale yellowish-brown, which passes on the one side into reddish-brown, and isabella-yellow, and very rarely into pale flesh-red; on the other into chesnut-brown and brownish-black. Sometimes it is spotted and dendritic.

It is massive, and disseminated.

Internally its lustre is glimmering, and very rarely dull.

The fracture is perfect conchoidal.

The fragments are indeterminate angular, and rather sharp-edged.

The red variety is feebly translucent, the yellow translucent on the edges, and the brown and the black opaque.

It is very soft, approaching to soft.

It is rather sectile, and very easily frangible.

It feels greasy.

It becomes shining and resinous in the streak.

It adheres to the tongue.

Specific gravity, 1.922, *Karsten.*—From 1.4 to 2.00, *Kirwan.*—1.977, 2.051, *Breithaupt.*

Chemical Characters.

When immersed in water, it breaks in pieces with an audible noise, with the evolution of air-bubbles, and falls into powder.

Before

Before the blowpipe, it melts into a greenish-grey coloured slag.

Constituent Parts.

Silica, - - .	47.00
Alumina, - -	19.00
Magnesia, - -	6.20
Lime,	5.40
Iron,	5.40
Water, -	7.50

Bergmann, Opusc. t. iv. p. 152.

It is still uncertain whether this analysis of Bergmann is of true bole.

Geognostic Situation.

The geognostic situation of this mineral is rather circumscribed, it having been hitherto observed only in secondary or flœtz trap-rocks, principally in trap-tuff, wacke, and basalt, in which it occurs in angular pieces, and disseminated *.

Geographic Situation.

Europe.—It is found at Strigau in Silesia; at Artern in Thuringia; in the Habichtwald in Hessia; the chesnut and reddish brown varieties are found at Sienna in Tuscany, and known under the name *Ochria di Siena.* The yellowish-brown occurs in the island of Lemnos.

Uses.

* It is said to occur in she[illegible] the neighbourhood of Waltersdorf in Gotha.—Von Schlo[illegible]aturf Freunde zu Berlin, [illegible]

Uses.

It was formerly an article of the Materia Medica, and was used as an astringent, and in some places is still employed in veterinary practice. It is said that tobacco-pipes are sometimes made of bole, and that it is an ingredient in the glaze of some kinds of earthen-ware.

Observations.

1. Formerly a number of clayey brick-red and brownish coloured clays, were preserved in collections under the name *Bole*. The bole of modern mineralogists, of which we have given a description, was first established as a distinct mineral by Werner.

2. It inclines sometimes to Lithomarge, sometimes to Clay.

7. Sphragide, or Lemnian Earth.

Sphragid, *Werner*.

Λημνια σφραγις of the Greeks.—Terra Lemnia, *Galenus*, De Simpl. Med. Facult. l. ix. the first variety.—Sphragid, *Karsten*, Tabel. s. 28.—Lemnische Erde, *Steffens*, b. i. s. 255.—Sphragid, *Lenz*, b. ii. s. 643. *Id. Oken*, b. i. s. 384. *Id. Haus.* Handb. b. ii. s. 460.

External Characters.

Its colours are yellowish-grey, and yellowish-white. On the surface, it appears frequently marbled with rust-like spots.

It is dull.

The fracture is fine earthy.

It

It is meagre to the feel.

It adheres slightly to the tongue.

When immersed in water, it falls into pieces, and numerous air-bubbles are evolved.

Constituent Parts.

Silica, - -	66.00
Alumina, - -	14.50
Magnesia, -	0.25
Lime, -	0.25
Natron, -	3.50
Oxide of Iron,	6.00
Water, -	8.50
	99.00

Klaproth, Beit. b. iv. s. 336.

Geographic Situation.

Its geognostic situation is unknown, and it has hitherto been found only in the island of Stalimene (Lemnos of the ancients) in the Mediterranean.

Uses.

In Stalimene or Lemnos, it is dug but once a-year, on the 15th of August, in the presence of the clergy and magistrates of the island, after the reading of prayers. The clay is cut into spindle-shaped pieces, of an ounce weight, and each of them is afterwards stamped with a seal, having on it the Turkish name of the mineral. Even so early as the time of Homer, this substance was used as a medicine againt poison and the plague, and was then in great repute, as it is at present, in eastern countries. In early times, it

was

was also sold, bearing on it the impression of a seal: hence it was called σφραγὶς, *sigillum*; and it was in such estimation, that none but priests durst handle it, and severe punishments were inflicted on those who presumed to dig for it at any other but the stated period. It is mentioned, that Scultetus Montanus, physician to the Emperor Rodolph, in the year 1568, ordered this earth to be kept in apothecaries shops.

Observations.

The only analysis we possess is that by M. Klaproth, who received specimens of it from Mr Hawkins.

GENUS VI. SPODUMENE *.

THIS genus contains but one species, viz. Prismatic Spodumene.

1. Prismatic Spodumene.

Prismatischer Triphan Spath, *Mohs.*

Triphane, *Haüy.*

Spodumene, *D'Andrada,* Scherer's Journ. b. iv. 19. s. 30. *Id. Reuss,* b. ii. s. 495. *Id. Lud.* b. ii. s. 162. *Id. Suck.* 1r th. s. 725.

---

* On exposure to the blowpipe, it first separates into golden-coloured scales, and then into a kind of powder or ash; hence the name *Spodumene,* from σποδόω, *I change into ash,* or σποδὸς, *ashes.*

s. 725. *Id. Bert.* s. 174.—Triphane, *Lucas*, p. 209.—Spodumene, *Leonhard*, Tabel. s. 19.—Triphane, *Brong.* t. i. p. 388. *Id. Brard*, p. 417. *Id. Haus.* s. 88.—Spodumen, *Karsten*, Tabel. s. 34—Triphane, *Haüy*, Tabl. p. 37.—Spodumene, *Steffens*, b. i. s. 474. *Id. Hoff.* b. ii. s. 341.—Triphan, *Lenz*, b. i. s. 525. *Id. Oken*, b. i. s. 372. *Id. Haus.* Handb. b. ii. s. 525.—Spodumene, *Aikin*, p 198.

External Characters.

Its colour is intermediate between greenish-white and mountain-grey, and sometimes passes into oil-green.

It occurs massive, disseminated, and in large and coarse granular concretions.

The cleavage is shining, the fracture glistening, and the lustre is pearly.

It has a distinct threefold cleavage; two of the cleavages are parallel with the lateral planes of an oblique four-sided prism of about 100°, and the third with the smaller diagonal of the basis of the same prism.

The fracture is fine-grained uneven.

It sometimes breaks into very oblique rhomboidal fragments, but more frequently into such as are tabular and indeterminately angular.

It is translucent.

It is as hard as felspar.

It is uncommonly easily frangible.

Specific gravity, 3.192, *Haüy*.—3.218, *D'Andrada*.—3.0, 3.1, *Mohs*.

Chemical Characters.

Before the blowpipe, it first separates into small gold-yellow coloured folia; and if the heat is continued, they melt into a greenish-white coloured glass.

Constituent Parts.

Silica,	63.50	64.4
Alumina, -	23.50	24.4
Lime,	1.75	3.0
Potash, -	6.00 *	5.0
Oxide of Iron,	2.50	2.2
Water, -	2.00	
Manganese, -	trace.	
	99.25	99.0
	Vogel, in Annal. of Phil. Nov. 1818.	*Vauquelin*, Haüy's Tabl. p. 168.

Geognostic and Geographic Situations.

This mineral was first discovered in the island of Utön, in Sudermanland, in Sweden, where it is associated with red felspar and quartz. It has been lately found in the vicinity of Dublin †; and in the Tyrol, on the road to Sterzing, in granite, and along with tourmaline.

GENUS VII.

* According to some analyses, it contains 8 *per cent.* of a new alkali named *lithina*.

† It was first found in Ireland by Dr Taylor.

Genus VII.—KYANITE *.

Disthene Spath, *Mohs.*

This genus contains one species, viz. Prismatic Kyanite.

1. Prismatic Kyanite.

Prismatischer Disthene Spath, *Mohs.*

Kyanite, *Werner.*

Sappare, *Saussure*, Voyages, § 1900. & Jour. de Phys. 1789, p. 213.—Cyanite, *Wid.* s. 475. *Id. Kirwan*, vol. i. p. 209.—*Id. Estner*, b. ii. s. 690. *Id. Emm.* b. i. s. 412. *Id. Nap.* p. 328. *Id. Lam.* t. ii. p. 256.—Disthene, *Haüy*, t. iii. p. 220.—La Cyanite, *Broch.* t. i. p. 501.—Cyanit, *Reuss*, b. ii. 2. s. 61. *Id. Lud.* b. i. s. 139. *Id. Suck.* 1r th. s. 463. *Id. Bert.* s. 285. *Id. Mohs*, b. i. s. 575. *Id. Hab.* s. 34.—Disthene, *Lucas*, p. 76.—Cyanit, *Leonhard*, Tabel. s. 30.—Disthene, *Brong.* t. i. p. 423. *Id. Brard*, p. 186.—Cyanit, *Karsten*, Tabel. s. 48.—Kyanit, *Haus.* s. 102.—Cyanite, *Kid*, vol. i. p. 182.—Disthene, *Haüy*, Tabl. p. 54.—Kyanit, *Steffens*, b. i. s. 299.—Cyanit, *Lenz*, b. ii. s. 696.—Talkschorl, *Oken*, b. i. s. 303.—Kyanit, *Hoff.* b. ii. s. 313. *Id. Haus.* Handb. b. ii. s. 634. *Id. Aikin*, p. 189.

External Characters.

Its principal colour is Berlin-blue, which passes on the one side into bluish-grey, and milk-white, on the other into sky-

* *Kyanite*, from the Greek word *κυανος*, *sky-blue*, a frequent colour of this mineral.

sky-blue, celandine-green, and greenish-grey. The white varieties are often marked with blue-coloured flame delineations.

It occurs massive and disseminated; also in distinct concretions, which are large and longish angulo-granular, and also wedge-shaped prismatic, which are straight or curved, and sometimes disposed in scopiform or stellular directions. It is sometimes regularly crystallized. The primitive figure is an oblique four-sided prism, in which the lateral edges meet under angles of 106° 15', and 73° 45' *. The following are the secondary forms.

1. Oblique four-sided prism, truncated on the two opposite acute lateral edges.
2. Preceding figure, in which all the lateral edges are truncated.
3. Twin-crystal: it may be considered as two flat four-sided prisms joined together by their broader lateral planes †.

The narrow lateral planes are longitudinally streaked, and glistening: the broad are smooth, or delicately transversely streaked, and splendent.

The crystals are middle-sized, small, and very small: are singly imbedded, or intersect one another.

The lustre is splendent and pearly.

It has a three-fold cleavage, in which the folia are parallel with the lateral and terminal planes of the prism. Of these cleavages or folia, those parallel with the broader lateral planes are the most distinct; the others, indeed, are very imperfect.

The

* Forme primitive, Hauy.

† Disthene perihexaedre, Hauy.

‡ Disthene double, Hauy.

The fragments are splintery, or imperfectly rhomboidal.

The massive varieties are translucent; the crystals are in general transparent.

The surface of the broader lateral planes is as hard as apatite, while that of the angles is as hard as quartz.

It is rather brittle.

It is easily frangible.

Specific gravity, 3.470, *Karsten.*—3.517, *Saussure.*—3.680, *Klaproth.*—3.5, 3.7, *Mohs.*

Physical Characters.

When pure, it is idio-electric. Some crystals, by rubbing, acquire negative electricity, even on perfectly smooth planes, others positive electricity: hence the name *Disthene* given by Haüy to this mineral, on account of its double electrical powers.

Chemical Character.

It is infusible before the blowpipe.

Constituent Parts.

Silica,	29.2	Silica,	38.50	Silica, -	43.00
Alumina,	55.0	Alumina,	55.50	Alumina,	55.50
Magnesia,	2.0	Lime,	0.50	Iron, -	0.50
Lime, -	0.25	Iron, -	2.75	Trace of Potash.	
Iron, -	6.65	Water,	0.75		
Water,	4.9				99.00
Saussure the Son, Voyages dans les Alpes, N° 1900.		*Laugier*, Annales du Mus. t. v. 25 cahier, p. 17.		*Klaproth*, Beit. b. v. s. 10.	

Geognostic

Geognostic Situation.

It has been hitherto found only in primitive mountains, where it occurs in compact granite (white-stone), mica-slate and talc-slate, accompanied with several other minerals.

Geographic Situation.

Europe.—It occurs in primitive rocks, near Banchory in Aberdeenshire, and Boharm in Banffshire; in mica-slate near Sandlodge *, and in the same rock near Hillswick † in Mainland, the largest of the Shetland Islands. At Airolo, on St Gothard, it is found in a beautiful silver-white mica-slate, associated with felspar, garnet, grenatite, and quartz; in the Saualp in Carinthia, with quartz, calcareous-spar, garnet, and common actynolite; in the Zillerthal in the Tyrol, with quartz and hornblende; and imbedded in white-stone at Waldenberg, in the Saxon Erzgebirge; Prizbram in Bohemia; also in France, Transylvania, Hungary, and Spain.

Asia.—In the Uralian Mountains; also in India.

America.—It is found in Maryland associated with grenatite, garnet, and magnetic ironstone, in mica-slate. In Pennsylvania in crystals upwards of a foot in length; in mica-slate in Connecticut; in Massachusets along with garnets and quartz; and in the district of Maine.

At Maniquarez in South America; and in Brazil.

* Jameson's Mineralogy of Shetland.

† Fine specimens of this mineral were found near Hillswick by Dr Hibbert.

Uses.

In India it is cut and polished, and sold as an inferior kind of sapphire.

Observations.

1. It is distinguished from *Actynolite* by its cleavage and infusibility; from *blue-coloured Quartz*, and *Sapphire*, by its inferior hardness: from *Mica* by its superior hardness, its infusibility, and its being common flexible, whereas mica is elastic-flexible; from *Tremolite*, by colour, figure, and infusibility.

2. It was first described as a kind of schorl, under the names *violet schorl*, *blue schorl-spar*, *pseudo-schorl*; afterwards as belonging to the mica or talc species, under the names *blue mica* and *blue talc*. Some observers arranged it with felspar, and named it *skye-blue foliated felspar*; and by others it was denominated *foliated beryl*. It was Werner who first correctly pointed out its characters.

Genus VIII. AUGITE *.

This genus contains four species, viz. 1. Oblique-edged Augite; 2. Straight-edged Augite; 3. Prismatoidal Augite; 4. Prismatic Augite.

1. Oblique

---

* *Augite* is a name applied to a particular mineral by Pliny, and is derived from the Greek word αὐγη, *lustre*, because this character is striking in several of the varieties of this species.

1. Oblique-edged Augite *.

Schiefkantiger Augit, *Mohs.*

This species contains seven subspecies, viz. Foliated Augite, Granular Augite, Conchoidal Augite, Common Augite, Coccolite, Diopside, and Sahlite.

First Subspecies.

Foliated Augite.

Blættriger Augit, *Werner.*

Schorl des volcans, *Daubenton,* Tabl. p. 11.—Pyroxene noire des terrains volcaniques, *Haüy,* Tabl. p. 41.—Blättriger Augit, *Hoff.* b. i. s. 453.

External Characters.

Its colour passes from velvet-black through greenish-black, into blackish-green, and sometimes even approaches to dark leek-green.

It has hitherto been found only crystallized, and the crystals are sharp-edged, and imbedded. The primitive form is an oblique four-sided prism, in which the lateral planes meet under angles of 92° 18′, and 87° 42′. The following are the secondary forms.

 1. Broad

* In all the varieties of this species, the edge of the bevelment on the extremity of the prism, is generally oblique; hence the specific name *Oblique-edged.* This name is retained, until one more appropriate shall be discovered.

1. Broad six-sided prism, with two opposite acuter lateral edges *, Pl. 6. fig. 112. It is generally
   - *a.* Flatly bevelled on the extremities, and the bevelling planes set on the acuter lateral edges †, Pl. 7. fig. 113. Sometimes also
   - *b.* The edge of the bevelment is slightly truncated, and, in some instances,
   - *c.* The acuter angles of the bevelment are deeply truncated.
2. Broad six-sided prism, bevelled on the extremities, and truncated on the acuter lateral edges ‡, Pl. 6. fig. 114.
3. Broad, nearly equiangular, eight-sided prism, bevelled on the extremity, in the same manner as N° 1., and one of the angles of the bevelment truncated ||, Pl. 6. fig. 115.
4. Eight-sided prism, in which the obtuse terminal edges of two and two opposite planes, that meet under acute angles, are truncated §, Pl. 6. fig. 116.
5. Twin-crystal, in which the crystals are joined together by their broader lateral planes ¶, Pl. 6. fig. 117.
6. Twin-crystal, in which the crystals intersect each other **, Pl. 6. fig. 118.

The crystals are middle-sized, and small.

They are all around crystallized, and therefore originally imbedded.

The surface is sometimes smooth, sometimes rough; when smooth it is shining, when rough glimmering.

Internally it is shining, inclining to splendent, and the lustre is resino-vitreous.

It

* Pyroxene peri-hexaedre, Haüy.
† Pyroxene bis-unitaire, Haüy.
‡ Pyroxene tri-unitaire, Haüy.
|| Pyroxene soustractif, Haüy.
§ Pyroxene dioctaedre, Haüy.
¶ Pyroxene hemitrope, Haüy.
** Pyroxene hemitrope, Hauy.

It has a distinct cleavage, in which the folia are parallel to the sides and to the diagonal of the primitive form.

The fracture is conchoidal.

The fragments are indeterminate angular, and rather sharp-edged.

It is opaque, or translucent on the edges.

It is harder than apatite, but softer than felspar.

It is rather easily frangible.

Specific gravity,—

Green variety, out of basalt,	3.471,	*Werner.*
Crystallized, from Frascati,	3.400,	*Klaproth.*
	3.350, 3.397,	*Hoffmann.*
From Lipari,	3.459,	*Hoffmann.*
From Vesuvius,	3.357,	*Haüy.*
	3.226,	*Haüy.*
	3.2, 3.5,	*Mohs.*

Chemical Characters.

Fusible with difficulty into a black enamel.

Constituent Parts.

	From Ætna.	From Frascati.
Silica,	52.00	48.00
Alumina, -	3.33	5.00
Magnesia, -	10.00	8.75
Lime, -	13.20	24.00
Oxide of Iron,	14.66	12.00
Oxide of Manganese,	2.00	1.00
		Trace of Potash.
	95.19	98.75
	Vauquelin, Journ. de Min. N. 39. p. 176.	*Klaproth*, Beit. b. 5. s. 166.

Geognostic

Geognostic Situation.

It occurs only in secondary trap-rocks, and in lava.

Geographic Situation.

It is found in basalt in different districts in Scotland. On the continent, it occurs in the basalts of Bohemia, Auvergne, and in the lavas and secondary trap-rocks of Vesuvius, Frascati, and Ætna.

Observations.

1. The most characteristic features of this mineral are, besides the colour and crystallization, the strong internal lustre, and the perfection of the cleavage.

2. *Distinctive Characters.*—*a.* Between Augite and *Basaltic Hornblende.*—Basaltic hornblende has always a velvet-black colour, augite generally a green colour; the lustre of basaltic hornblende is vitreous and splendent, that of augite is resino-vitreous, and shining; basaltic hornblende is softer than augite; in basaltic hornblende, the edge of the bevelment on the extremity of the prism is generally straight, whereas it is oblique in augite; and, lastly, basaltic hornblende is easily fusible before the blowpipe, but augite very difficultly fusible.—*b.* Between Augite and *Schorl.* In schorl the lateral planes of the crystals are deeply longitudinally streaked, whereas those of augite are smooth; the lustre of schorl is vitreous, but that of augite is resino-vitreous; it is harder than augite; it becomes electrical by heating, but augite does not; and schorl is more easily fusible than augite.—*c.* Between Augite and *Melanite.* In melanite the crystallizations are tessular, whereas those of augite are prismatic; and the specific gravity of melanite is 3.8, that of augite 3.2—3.5.

Second

Second Subspecies.

Granular Augite.

Körniger Augit, *Werner.*

Pyroxene, cristaux noirs de Norwege, *Haüy,* Tabl. p. 41.—Körniger Augit, *Hoff.* b. i. s. 449.

External Characters.

Its colour is greenish-black.

It occurs massive, and in coarse and small angulo-granular concretions. Also crystallized, and in the following figures:

1. Broad six-sided prism, with two opposite acuter edges. This prism is generally
   - *a.* Flatly bevelled on the extremities, and the bevelling planes are set obliquely, but parallelly, on the acuter lateral edges *. It is seldom
   - *b.* Acuminated on the extremities with four planes, which are set on the obtuser lateral edges. Sometimes
   - *c.* The bevelment, and also
   - *d.* The acumination, are truncated. The truncating planes are often convex, and thus there is formed
   - *e.* Prisms with convex terminal faces.
2. Six-sided prism truncated on the acuter lateral edges †.

   When these truncating planes become as large as

* Pyroxene bis-unitaire, Haüy.

† Pyroxene tri-unitaire, Haüy.

as the other planes of the prism, there is formed a

3. Broad, nearly equiangular eight-sided prism, which has the same bevelments and acuminations as N° 1.
4. Four-sided prism, which is formed when the broader lateral planes of N° 1. disappear. Both the eight and four sided prisms occur very rarely.

The crystals are seldom sharp-edged and perfect.

They are generally middle-sized, attached by one end, and forming druses.

The surface is rough and glistening. Internally it is glistening and resinous.

It has an imperfect cleavage.

The fracture is uneven.

The fragments are indeterminate angular, and rather sharp-edged.

It is opaque.

Hardness same as that of foliated augite.

Specific gravity, 3.318, 3.388, *Hoffmann.*—3.448, 3.465, *Schumacher.*—3.573, *Hausmann.*

Chemical Characters.

According to Simon does not melt before the blowpipe.

Constituent Parts.

Silica, - -	$50\frac{1}{4}$	
Alumina, -	$3\frac{1}{4}$	
Magnesia,	7	
Lime, -	$25\frac{1}{4}$	
Iron, -	$10\frac{1}{2}$	
Manganese, - -	$2\frac{1}{4}$	
Water, -	$\frac{1}{2}$	
	$99\frac{1}{2}$	*Simon.*

Geognostic and Geographic Situations.

This subspecies of augite has been hitherto found only at Arendal in Norway, in several of the iron-mines, particularly that named Ulve-Grube. It occurs in beds of magnetic ironstone in gneiss, where it is associated with common garnet, epidote, hornblende, and calcareous-spar.

Observations.

1. The essential characters of this subspecies are its constant black colour, the bluntness of the crystals, feeble lustre, fracture, and opacity.

2. It is nearly allied to *Coccolite*, but differs from it by its darker colours, resinous lustre, less perfect cleavage, closer aggregation of the distinct concretions, opacity, and rather greater hardness. It is distinguished from *Epidote* by its darker green colour, crystallization, different cleavage, and opacity; and from *Common Garnet* by its crystallization, resinous lustre, foliated fracture, and inferior hardness and weight.

Third Subspecies.

Conchoidal Augite.

Muschlicher Augit, *Werner*.

Muschlicher Augit, *Hoff.* b. i. s. 462.

External Characters.

Its colour is greenish-black, passing into blackish-green; also into a very dark olive-green, and sometimes even into liver-brown.

It

It occurs in imbedded grains.

Its lustre is splendent, and is resino-vitreous.

The fracture is imperfect, and flat conchoidal.

It is translucent on the edges, or translucent.

It agrees in its other characters with the foregoing subspecies.

Constituent Parts.

Silica,	52.00	55.00
Alumina,	5.75	5.50
Magnesia,	12.75	13.75
Lime,	14.00	12.50
Oxide of Iron,	12.25	11.00
Manganese,	0.25	Trace.
Potash,	0.25	1.00
	97.25	98.78

Klaproth, Beit. b. v. s. 159, 162.

Geognostic and Geographic Situations.

It occurs only in secondary trap-rocks, and is the rarest of the subspecies of this species. The finest specimens, from two to three inches in diameter, are found in the vesicular basalt of Fulda.

Observations.

1. This species of augite is characterised by its high lustre, conchoidal fracture, translucency, and colour.

2. It somewhat resembles Vesuvian, and also Olivine. Its granular form, conchoidal fracture, strong lustre, and inferior translucency, distinguish it from *Vesuvian*; and its colour, stronger lustre, more perfect conchoidal fracture, and rather greater hardness, distinguish it from *Olivine*.

Fourth

Fourth Subspecies.

Common Augite.

Gemeiner Augit, *Werner.*

Gemeiner Augit, *Hoff.* b. i. s. 464.

External Characters.

Its colours are blackish-green and velvet-black.

It occurs in large and small imbedded grains.

Internally its lustre is intermediate, between shining and glistening, and is resinous.

The fracture is coarse, and small-grained uneven. Sometimes inclining to imperfect conchoidal.

It is translucent on the edges, seldom translucent.

In its other characters it agrees with the foliated subspecies.

Geognostic and Geographic Situations.

It occurs principally in secondary trap-rocks, as basalt and greenstone, and also in lavas. The secondary trap-rocks of France, Germany, and Britain, and the lavas of Vesuvius and Iceland, in many cases abound with this mineral.

Observations.

The dark colour, granular form, inferior lustre, compact fracture, and low transparency, are the distinguishing characters of this subspecies.

Fifth

Fifth Subspecies.

Coccolite *.

Kokkolith, *Werner.*

Körniger Augit, *Karsten.*

Coccolith, *D'Andrada,* Scherer's Journal, b. iv. 19. s. 30. *Id. Schumacher,* Verzeichn. s. 30. *Id. Broch.* t. ii. p. 504. *Id. Haüy,* t. iv. p. 355. *Id. Reuss,* b. i. s. 86. *Id. Lud.* b. ii. s. 134. *Id. Suck.* 1r th. s. 184. *Id. Bert.* s. 159. *Id. Mohs,* b. i. s. 55. *Id. Lucas,* p. 194. *Id. Leonhard,* Tabel. s. 2.—Pyroxene Coccolithe, *Brong.* t. i. p. 447.—Pyroxene granuleux, *Brard,* p. 141.—Körniger Augit, *Karsten,* Tabel. s. 40. *Id. Haus.* s. 98.—Pyroxene granuliforme, *Haüy,* Tabl. p. 42. —Kokkolith, *Steffens,* b. i. s. 347. *Id. Hoff.* b. i. s. 443.—Körniger Augit, *Lenz,* b. i. s. 208.—Coccolith, *Oken,* b. i. s. 336. *Id. Aikin,* p. 228.

External Characters.

Its principal colour is leek-green, which passes on the one side into pistachio-green, blackish-green, even into olive-green, and oil-green, and on the other into mountain-green.

It occurs massive, also in distinct concretions, which are coarse or small, seldom fine angulo-granular, and are so loosely aggregated together, as frequently to be separable by the simple pressure of the finger. Sometimes the concretions are longish granular. It is sometimes crystallized, in the following forms:

1. Six-

* *Coccolith,* from the Greek word κοκκος, granum, and λιθος, on account of the granular concretions that characterize it.

1. Six-sided prism, with two opposite acute lateral edges, and bevelled on the extremities; the bevelling planes set on the acute lateral edges. Sometimes two additional planes occur, when the bevelment passes into a four-planed acumination; and in other varieties two of the opposite lateral planes disappear, when the planes that meet under acute angles form a
2. Four-sided prism.

The crystals are generally blunt, or rounded on the angles and edges, even appear with convex lateral faces, and hence pass into longish grains.

The crystals are generally middle-sized, seldom small, and occur either singly imbedded, as is the case with the grains, or in druses.

The surface of the distinct concretions is sometimes rough, and strongly glimmering, sometimes smooth and glistening.

Externally the crystals are sometimes smooth, sometimes rough; the first is shining and glistening, the other strongly glimmering.

Internally it is shining, sometimes approaching to glistening, and the lustre is vitreous, inclining to resinous.

It has a double oblique angular cleavage.

The fracture is uneven.

The fragments are more or less sharp-edged.

It is translucent, or translucent on the edges.

It is hard in a low degree: it scratches apatite, but not felspar.

It is brittle.

It is very easily frangible.

Specific gravity, 3.316, *D'Andrada*.—3.303, *Karsten*.—3.15, 3.06, *Schumacher*.—3.373, *Haüy*.

Chemical

Chemical Characters.

It is very difficultly fusible before the blowpipe.

Constituent Parts.

Silica, -	50.0
Lime, -	24.0
Magnesia,	10.0
Alumina,	1.5
Oxide of Iron, -	7.0
Oxide of Manganese,	3.0
Loss, - -	4.5
	100 *Vauquelin.*

Geognostic Situation.

It occurs in mineral beds subordinate to the primitive trap formation, where it is associated with granular limestone, garnet, and magnetic ironstone.

Geographic Situation.

It occurs at Arendal in Norway; in the iron mines of Hellsta and Assebro in Sudermanland; and in many places in Nericke, in Sweden. The mountain-green variety is found at Barkas in Finland. It is mentioned as occurring in the Harzeburg Forest in the Hartz, in Lower Saxony; and also in Spain.

Observations.

1. *Distinctive Characters.—a.* Between Coccolite and *Common Garnet:* The internal lustre of common garnet is resinous, that of coccolite vitreous; common garnet has no cleavage, while coccolite has a distinct cleavage; common

mon garnet scratches quartz, coccolite only apatite; common garnet has a specific gravity of 3.75, coccolite of 3.33.—*b*. Between Coccolite and *Common Augite:* The colour-suites of the two minerals are different; the lustre of common augite is resinous, that of coccolite is vitreous; in common augite the distinct concretions are grown together, whereas they are so loosely aggregated in coccolite, as frequently to be separable by the mere pressure of the fingers; coccolite is rather softer than augite; and the concretions in coccolite are frequently enveloped in an extremely delicate crust, which is not the case with common augite.

2. It was first described by D'Andrada, under its present name.

Sixth Subspecies.

Diopside.

Diopsid, *Werner*.

Cristaux gris-verdâtres, transparens; formes tres prononcés, du Depart. du Po; *Alalite* de *Bonvoisin*, Journal de Physique, Mai 1806, p. 409, &c.—Varieté du *Diopside*, Journal des Mines, n. 115. p. 65. &c.—Cristaux gris-verdâtres, ou blancs grisâtres, offrant la forme primitive peu prononcé, de Depart. du Po; *Mussite* de *Bonvoisin*, Journal de Physique, ib.—Varieté du Diopside, Journal des Mines, ib.—*Pyroxene;* also Pyroxene cylindroide, comprimé, et fibro-granulaire, *Haüy*, Tabl. p. 41, 42.—Diopsiod, *Steffens*, b. i. s. 349. *Id. Hoff.* b. ii. s. 467. *Id. Lenz*, b. i. s. 212.—Strahliger Pyroxen, *Oken*, b. i. s. 335.—Diopsid, *Haus.* Handb. b. ii. s. 494.

External

---

* *Diopside*, from δις and οψις, because when viewed as a distinct species, the cleavage appeared to point out a double series of crystallizations.

External Characters.

Its colours are greenish-white, greenish-grey, and pale mountain-green.

It occurs massive, disseminated, in lamellar concretions, which sometimes approach to prismatic; and crystallized in the following figures:

1. Low, oblique four-sided prism, sometimes equilateral, sometimes broad *.
2. The preceding figure truncated on the acute lateral edges, bevelled on the obtuse edges, and the edge of the bevelment truncated: also rather acutely acuminated by four planes, two of the large acuminating planes set on the truncating planes of the acute lateral edges of the prism, the two smaller on the truncating planes of the bevelment. The two edges of the larger truncating planes, and the truncating planes of the acute edges of the prism, and the apex of the acumination, truncated. The broader lateral planes of the prism, and the truncating planes of the acumination, belong to the primitive form †.
3. Eight-sided prism, with alternate broader and smaller lateral planes, acuminated with four planes, the acuminating planes set on the smaller lateral planes, and this acumination again acuminated with four planes, set obliquely on the planes of the lower acumination, and of which two adjacent planes are large, and two small. The summit of the second acumination, and also the angles of the truncated summit, and the lateral edges of the larger planes, are truncated. The smaller lateral planes of the prism,

* Diopside primitive, Haüy; Mussite, Bonvoisin.

† Diopside didodecaedre, Haüy; Alalite, Bonvoisin.

prism, and acuminating planes of the upper acumination, belong to the primitive form *.

The broader lateral planes are deeply longitudinally streaked; but the smaller lateral planes, and the acuminating planes, are smooth.

The crystals are middle-sized and small; occur resting on one another, intersecting one another, and collected into scopiform groups.

Externally it is shining and glistening, and pearly; internally it is shining and vitreous.

The cleavage is the same as in sahlite.

The fracture is uneven, sometimes inclining to imperfect and small conchoidal.

The fragments are splintery, or indeterminate angular.

It is translucent.

It is as hard as augite.

Specific gravity 3.310, *Haüy*.

Chemical Characters.

It melts with difficulty before the blowpipe.

Constituent Parts.

Silica, - - -	57.50
Magnesia, - - -	18.25
Lime, - - -	16.50
Iron and Manganese, -	6.00

Laugier.

* Diopside octovigesimal, Haüy.

Geognostic and Geographic Situations.

It is found in the hill of Ciarmetta in Piedmont; also in the Black Rock at Mussa, near the town of Ala, in veins, along with epidote or pistacite, and hyacinth-red garnets; and in the same district, in a vein traversing serpentine, along with prehnite, calcareous-spar, and iron-glance or specular iron-ore. It is said also to occur at St Nicolas, in the Upper Valais.

Observations.

1. This mineral was discovered by Dr Bonvoisin, who formed it into two species, named *Alalite* and *Mussite:* the white-coloured and massive varieties, and those crystallized in the form N° 2. he refers to alalite; the green, with scopiformly aggregated crystals, and radiated fracture, he considers as mussite. Haüy having ascertained that mussite and alalite have the same primitive form, ranked them in the system as one species, under the name *Diopside;* afterwards, he ascertained that the primitive form of diopside did not differ from that of augite, and consequently abolished the diopside species, and arranged it in the system as a variety of augite.

2. The Fassaite of Werner, the Pyrogom of Breithaupt, is a particular variety or subspecies of oblique-edged augite.

Seventh

Seventh Subspecies.

Sahlite *.

Sahlit, *Werner.*

Sahlit, *D'Andrada,* Scherer's Journal, b. iv. 19. s. 81. *Id. Schumacher,* Verzeichniss, s. 32. *Id. Haüy,* t. iv. p. 379. *Id. Broch.* t. ii. p. 518. *Id. Reuss,* b. ii. s. 474. *Id. Lud.* b. i. s. 158.—Malacolith, *Suck.* 1r th. s. 186. *Id. Bert.* s. 162. *Id. Mohs,* b. i. s. 488.—Malacolithe, *Lucas,* p. 201.—Sahlit, *Leonhard,* Tabel. s. 31.—Malacolithe, *Brong.* t. i. p. 445. *Id. Brard,* p. 414.—Sahlit, *Karsten,* Tabel. s. 44.—Salait, *Haus.* s. 98.—Pyroxene laminaire gris-verdâtre, *Haüy,* Tabl. p. 42.—Malacolith, *Steffens,* b. i. s. 354.—Sahlit, *Lenz,* b. ii. s. 700.—Schaliger Pyroxene, *Oken,* b. i. s. 333.—Sahlit, *Hoff.* b. ii. s. 319. *Id. Aikin,* p. 228.

External Characters.

Its colours are greenish-grey, mountain, leek, and blackish green.

It occurs massive, and in straight lamellar and coarse granular concretions; also crystallized in the following figures:

1. Broad rectangular four-sided prism, which approaches to the tabular form.
2. Preceding figure, with truncated lateral edges, and in which the terminal planes are set on obliquely.

The crystals are occasionally superimposed, and are middle-sized and small.

* So named from Sala in Sweden where it was first found.

Internally the lustre of the principal fracture is shining, splendent and vitreous; that of the cross fracture dull.

It has a fivefold cleavage: one of the cleavages is parallel with the terminal planes; two with the lateral planes; and two with the diagonals of the prism: the prism splits easily in the direction of all the planes, but most easily in the direction of the terminal planes.

The fracture is uneven.

The fragments are sometimes indeterminate angular, sometimes rhomboidal.

It is strongly translucent on the edges.

It is harder than augite.

It is rather brittle.

It is rather easily frangible.

Specific gravity, 3.223, *Haüy*.—3.265, *Breithaupt*.—3.473, *Dr Wollaston*.

Chemical Characters.

It melts with great difficulty before the blowpipe.

Constituent Parts.

Silica,	53.00
Magnesia, -	19.00
Alumina,	3.00
Lime, - -	20.00
Iron, and Manganese, -	4.00

Vauquelin, in Haüy, t. iv. p. 302.

Geognostic and Geographic Situations.

Europe.—It occurs in the island of Unst in Shetland: in granular limestone in the island of Tiree, one of the Hebrides: in limestone in Glen Tilt: in Rannoch: in the

silver

silver mines of Sala, in Westmanland in Sweden, associated with asbestous actynolite, calcareous-spar, iron-pyrites, and galena; near Arendal in Norway, along with magnetic iron-stone, common hornblende, calcareous-spar, and seldom with felspar and black mica.

Asia.—At Odon-Tschelong, near the river Amour in Syberia, along with beryl, mica, and calcareous-spar, and on the shores of the Lake Baikal *.

America.—In South Greenland †; and on the banks of Lake Champlain.

Observations.

This mineral was first described as a variety of felspar, from which, however, it is distinguished by colour, lustre, cleavages, inferior hardness, and greater specific gravity: It is more nearly allied to hornblende, but its crystallizations, cleavage, greater transparency, and hardness, show that it is a different species; and its dark green colours, granular concretions, and inferior translucency, distinguish it from *Diopside.*

2. Straight-Edged Augite ‡.

Rechtkantiger Augit, *Mohs*

THIS Species contains five Subspecies, viz. Carinthin, Hornblende, Actynolite, Tremolite, and Asbestus.

First

* The variety from Baikal has been described under the name *Baikalite.*

† Giesecké.

‡ In most of the varieties of this species, the edge formed by the meeting of the bevelling planes on the ends of the prisms, is straight; hence the name *Straight-edged Augite.*

First Subspecies.

Carinthin *.

Karinthin, *Werner.*

Keraphyllit, *Steffens,* Handb. b. i. s. 308.—Blättricher Augit, *Karsten,* in Klap. Beit. b. iv. s. 185. *Id. Hoff.* b. i. s. 459.—Blättricher Strahlstein, *Haus.* Handb. b. ii. s. 723.—Amphibole laminaire, *Haüy,* Tabl. p. 40.

External Characters.

Its colours are greenish and velvet black.

It occurs massive and disseminated; and the massive varieties in coarse granular concretions.

Internally it is splendent, and the lustre is resino-vitreous.

It has a distinct double cleavage, in which the folia meet under angles of 55° 50′, and 124° 50′.

The fracture is conchoidal.

The fragments are indeterminate angular, and very sharp-edged.

The greenish-black varieties are strongly translucent on the edges, but the velvet-black is opaque.

It is as hard as hornblende.

Specific gravity, 3.085, *Klaproth.*—3.161, 3.194, *Breithaupt.*

Chemical Characters.

It is difficultly fusible.

Constituent

* *Carinthin,* from Carinthia, the country where it occurs.

Constituent Parts.

Silica, -	52.50
Alumina,	7.25
Magnesia, -	12.50
Lime,	9.00
Potash,	0.50
Oxide of Iron, - -	16.25
	98.00

Klaproth, Beit. b. iv. s. 189.

Geognostic and Geographic Situation.

It occurs in the Saualpe in Carinthia, in a bed in primitive rock, associated with quartz, kyanite, garnet, and zoisite.

Observations.

1. This mineral is distinguished from *Hornblende* by its high degree of lustre, kind of lustre, perfect conchoidal fracture, greater hardness, and rather inferior specific gravity. Its cleavage and inferior hardness distinguish it from *Augite*.

2. It was first described as a particular species under the name Saualpite; afterwards by Werner and Karsten, as a variety of Foliated Augite; by Hausmann, as Foliated Actynolite; by Steffens as a new species, under the name *Keraphyllit*; and, lastly, again, by Werner, as a new species, under the name Carinthin.

Second

Second Subspecies.

Hornblende *.

Hornblende, *Werner.*

THIS Species is divided into three Kinds, viz. Common Hornblende, Hornblende-Slate, and Basaltic Hornblende.

First Kind.

Common Hornblende.

Gemeiner Hornblende, *Werner.*

Corneus facie spatosa, striata, *Wall.* gen. 26. spec. 171.—Hornblende, *Kirw.* vol. i. p. 213.—Gemeiner Hornblende, *Estner,* b. ii. s. 699. *Id. Emm.* b. i. s. 322. & b. iii. s. 267.—Orniblenda commune, *Nap.* p. 276.—La Hornblende commune, *Broch.* t. i. p. 415.—Amphibole laminaire, *Haüy,* t. iii. p. 63. —Gemeiner Hornblende, *Reuss,* b. ii. s. 144. *Id. Lud.* b. i. s. 118. *Id. Suck.* 1r th. s. 118. *Id. Bert.* s. 185. *Id. Mohs,* b. i. s. 492. *Id. Hab.* s. 31. *Id. Leonhard,* Tabel. s. 24.—Amphibole schorlique commun, *Brong.* t. i. p. 452.—Gemeiner Hornblende, *Karsten,* Tabel. s. 38. *Id. Haus.* s. 91.—Amphibole lamellaire, *Haüy,* Tabl. p. 40.—Gemeiner Hornblende, *Steffens,* b. i. s. 304. *Id. Lenz,* b. i. s. 317. *Id. Oken,* b. i. s. 323. *Id. Hoff.* b. ii. s. 147. *Id. Haus.* Handb. b. ii. s. 700.—Common Hornblende, *Aikin,* p. 221.

External Characters.

Its most frequent colour is greenish-black, sometimes greyish-

* This mineral is frequently found mixed with others; hence the name *blende*, from the Swedish *blandat*, mixed; and the prefix *horn*, refers to its toughness, or difficult frangibility, in which it agrees with horn.

greyish-black, seldom velvet-black; from greyish-black it passes into dark greenish-grey; and from greenish-black into blackish-green, leek-green, and dark olive-green.

It occurs massive and disseminated, and in distinct concretions, which are large, coarse, and small granular, and also wedge-shaped prismatic. It is rarely crystallized, and the principal forms are the following:

1. Broad, long, thin, very oblique four-sided prism, in which the obtuse lateral edges are sometimes rounded, thus giving a reed-like form to the crystals; or the acuter edges are truncated, and the prism bevelled on the one extremity, the bevelling planes set on the acuter lateral edges, and often the edge of the bevelment is truncated. Very frequently there are no terminal crystallizations.

When the truncating planes of the acuter lateral edges of the oblique prisms increase in magnitude, there is formed

2. A six-sided prism, with four opposite broader lateral planes, and very flatly acuminated on one extremity, with three planes, which are set on the alternate lateral edges.

The crystals are small and middle-sized; often occur intersecting one another, or scopiformly or stellularly aggregated, and are either imbedded or superimposed.

The lateral planes of the prism are deeply longitudinally streaked.

Internally the lustre is shining and pearly.

The cleavage is twofold, and oblique angular, and the folia are longitudinally streaked. The angles of the cleavage the same as in carinthin.

The fracture is coarse and small grained uneven.

The

The fragments are indeterminate angular, and rather sharp-edged.

The black coloured varieties are opaque, but the green generally translucent on the edges.

It is harder than apatite, but not so hard as felspar; it is not so hard as augite.

It yields a mountain-green, inclining to a greenish-grey coloured streak.

When breathed on or moistened, even when brought from a colder to a warmer place, it yields what is called a bitter smell.

It is rather brittle.

It is rather difficultly frangible.

Specific gravity, 3.202, 3.287, *Karsten*.—3.243, *Klaproth*.

Chemical Characters.

It melts before the blowpipe, with violent ebullition, into a greyish-black coloured glass.

Constituent Parts.

	Common Hornblende from Nora in Westmanland.
Silica, -	42.00
Alumina, -	12.00
Lime,	11.00
Magnesia, -	2.25
Oxide of Iron, - -	30.00
Ferruginous Manganese,	0.25
Water, - -	0.75
Trace of Potash.	
	98.25

Klaproth, Beit. b. v. s. 158.

Geognostic

Geognostic Situation.

It forms an essential ingredient in several mountain rocks: is sometimes accidentally intermixed with others; and it frequently occurs in beds of considerable magnitude. Thus, it forms an essential ingredient of syenite and primitive greenstone; also of transition syenite and greenstone; and of secondary greenstone. It occurs occasionally in granite, gneiss, mica-slate, clay-slate, and porphyry; and beds of it, frequently associated with ores of different kinds, as magnetic-ironstone, and iron-pyrites, appear in gneiss, mica-slate, and clay-slate.

Geographic Situations.

Europe.—It occurs very abundantly in Scotland, in greenstone and syenite; and imbedded in limestone, gneiss, and mica-slate. It is found in similar rocks in England; and plentifully in the primitive and secondary trap-rocks of Ireland. On the Continent, it occurs abundantly in Sweden; in Norway, as at Arendal, where it is associated with coccolite, felspar, quartz, granular limestone, titanite, and magnetic-ironstone: in Lower Saxony, as in the Hartz, where it forms a constituent part of transition greenstone; in Upper Saxony, as in the Erzgebirge; also in Hessia, Silesia, Franconia, Bavaria, Switzerland, Austria, Hungary, Transylvania, Italy, France, and Spain.

Asia.—It occurs very abundantly in many parts of Siberia, as Kolyvan, Irkutzk, Catharinenburg, &c.

America.—In North America, it has been observed in primitive and secondary, and also in transition rocks, from Greenland, and the shores of Hudson's Bay, to the Isthmus of Darien.

Observations.

Observations.

This mineral is characterized by its frequent dark-green colours, crystallizations, lustre and fracture, distinct concretions, weight and smell. It is distinguished from *Actynolite* by its darker colours, kind of lustre, opacity, darker streak, inferior hardness, and greater weight.

It sometimes very nearly resembles *Epidote*, from which, however, it may be distinguished by its cleavage, opacity, streak, and inferior hardness.

Second Kind.

Hornblende-Slate.

Hornblende Schiefer, *Werner.*

Corneus rigidus non nitens, apparenter lamellis parallelis; Corneus fissilis, *Wall.*—Schistose Hornblende, *Kirw.* vol. i. p. 222.—La Hornblende schisteuse, *Broch.* t. i. p. 428.—Schiefriger Hornblende, *Reuss,* b. ii. s. 151.—Hornblende-Schiefer, *Lud.* b. i. s. 120. *Id. Suck.* 1r th. s. 238. *Id. Bert.* s. 187. *Id. Hab.* s. 32. *Id. Leonhard,* Tabel. s. 25.—Amphibole hornblende shisteux, *Brong.* t. i. p. 453.—Schiefrige Hornblende, *Karst.* Tabel. s. 38.—Hornblende-Schiefer, *Steffens,* b. i. s. 310.—Schiefrige Hornblende, *Lenz,* b. i. s. 321.—Hornblende-Schiefer, *Oken,* b. i. s. 323. *Id. Hoff.* b. ii. s. 155. *Id. Haus.* Handb. b. ii. s. 700.

External Characters.

Its colour is intermediate between greenish-black and blackish-green.

It occurs massive, and in thin promiscuous prismatic concretions.

Internally it is glistening, passing into shining and pearly.

The

The fracture is straight slaty.

The fragments are thick tabular.

It is opaque.

It yields a greenish-grey coloured streak.

It is semi-hard, passing into soft.

It is rather difficultly frangible.

In other characters it agrees with the foregoing.

It is not always pure, being frequently intermixed with mica and felspar.

Geognostic Situation.

It occurs in beds, in granite, gneiss, mica-slate, quartz-rock, sometimes also in clay-slate, and frequently along with beds of primitive limestone. It occasionally accompanies metalliferous beds, that contain magnetic-ironstone, chlorite, and other minerals. It is frequently intermixed with mica; and sometimes with quartz or iron-pyrites.

Geographic Situation.

Europe.—In Scotland, it occurs in gneiss, in the districts of Braemar and Aberdeen, in Aberdeenshire; in Banffshire, as near Portsoy; in Argyleshire, as in the islands of Coll, Tiree, &c.; in Inverness-shire, as in the islands Rona, Lewis, &c.; and in many other parts in Scotland; and also in England and Ireland, as will be mentioned in the fourth volume of this work. On the Continent, it occurs in Norway, Sweden, Saxon Erzgebirge, Lusatia, Bohemia, Silesia, Franconia, Bavaria, Moravia, Switzerland, Stiria, the Tyrol, Hungary, France, and Spain.

Asia.—It occurs abundantly in many places in Siberia, as Nertschinsk, Kolywan, and Catharinenburg.

Third

Third Kind.

Basaltic Hornblende*.

Basaltische Hornblende, *Werner.*

Schorl opaque rhomboidal, *Romé de Lisle,* t. ii. p. 379.—Basaltische Hornblende, *Wid.* s. 417.—Basaltine, *Kirw.* vol. i. p. 219.—Basaltische Hornblende, *Estner,* b. ii. s. 719. *Id. Emm.* b. ii. s. 330. *Id.* b. iii. s. 269.—Orniblenda basaltica, *Nap.* p. 281.—Amphibole, *Lam.* t. ii. p. 330.—Amphibole crystallizée, *Haüy,* t. iii. p. 58.—Basaltische Hornblende, *Reuss,* b. ii. s. 159. *Id. Lud.* b. i. s. 120. *Id. Suck.* 1r th. s. 242. *Id. Bert.* s. 188. *Id. Mohs,* b. i. s. 500. *Id. Hab.* s. 32. *Id. Leonhard,* Tabel. s. 25.—Amphibole schorlique basaltique, *Brong.* t. i. p. 452.—Basaltische Hornblende, *Haus.* s. 91. *Id. Karsten,* Tabel. s. 38. *Id. Steffens,* b. i. s. 311. *Id. Lenz,* b. i. s. 322. *Id. Oken,* b. i. s. 324. *Id. Hoff.* b. ii. s. 157. *Id. Haus.* Handb. b. ii. s. 700.—Basaltic Hornblende, *Aikin,* p. 221.

External Characters.

Its colours are velvet-black or brownish-black.

It occurs crystallized, in the following figures:

1. Unequiangular six-sided prism, flatly acuminated with three planes, set on the alternate lateral edges of the prism †, Fig. 109. Pl. 6.
2. The preceding prism, flatly acuminated on one extremity by four planes, which are set on the four opposite

---

* It is named *Basaltic Hornblende,* because it occurs principally in basaltic rocks.

† Amphibole dodecaedre, Haüy.

opposite lateral planes, and on the other extremity bevelled, the bevelling planes set on the two opposite lateral edges *, Fig. 110. Pl. 6.

3. The six-sided prism flatly acuminated on one extremity by three planes, which are set on the alternate lateral edges; on the other bevelled, the bevelling planes set on the opposite lateral edges †, Fig. 110[3]. Pl. 6.
4. Six-sided prism, in which two opposite lateral planes are broader than the others, and doubly acuminated on the extremities; first, with four planes, which are set on those edges which one of the broader lateral planes always forms with an adjacent smaller one; and again acuminated with four planes, which are set on the first, under very obtuse angles ‡.

The crystals are small and middle-sized, seldom large, and are imbedded, and all around crystallized. Their surfaces are smooth.

The lustre of the cleavage is splendent and vitreous, approaching to pearly; that of the cross fracture is glistening.

It has a distinct double cleavage, in which the folia meet under angles of 55° 50′, and 124° 50′.

The fracture is small-grained uneven, approaching to conchoidal.

The fragments are indeterminate angular, and sometimes indistinctly rhomboidal.

It is always opaque.

It

* Amphibole equi-différent, Haüy.

† Amphibole ondecimal, Haüy.

‡ Amphibole surcomposé, Haüy.

It is rather harder than common hornblende.

It is rather brittle.

It is more easily frangible than the preceding subspecies.

It affords a dark greyish-white streak.

Specific gravity, 3.158, 3.199, *Karsten*.—3.158, *Breithaupt*.

Chemical Characters.

Before the blowpipe, it melts into a black glass, but is rather more refractory than common hornblende.

Constituent Parts.

	Basaltic Hornblende from Fulda.
Silica,	47.00
Alumina,	26.00
Lime,	8.00
Magnesia,	2.00
Oxide of Iron,	15.00
Water,	0.50
	98.50

Klaproth, Beit. b. v. s. 154.

Geognostic Situation.

It occurs imbedded in basalt, along with olivine and augite; also in wacke and trap-tuff; in small quantity in some kinds of porphyry, and frequently in lava.

Geographic Situation.

Europe.—It occurs in the basalt of Arthur's Seat, and other similar hills around Edinburgh; in the basalt of Fifeshire, and that of the islands of Mull, Canna, Eig, and Skye.

Skye. It is also an inmate of the basaltic rocks in England and Ireland. Upon the Continent, it occurs in the secondary trap-rocks of Fulda, the Saxon Erzgebirge, Bohemia, Spain, and other countries.

America.—It is frequent in the basaltic rocks of Mexico.

Observations.

1. It is distinguished from the other kinds of *Hornblende* by its colour, crystallization, and splendent lustre; from *Augite* by its form, splendent lustre, and more oblique cleavage. It has been confounded with Schorl, by some authors; but it is distinguished from it by colour, form, lustre, foliated fracture, and inferior hardness.

2. It decomposes more slowly than basalt: hence we frequently find unaltered crystals dispersed through the clay formed by the decomposition of basaltic rocks.

3. Beyer describes a mineral under the name *Kohlenhornblende*, (Coal Hornblende), which appears to be nearly allied to hornblende; hence it deserves to be noticed in this part of the system. He describes it in the following terms:—Its colour is velvet-black, passing into brownish-black. It occurs massive and disseminated. The principal fracture is imperfect foliated, almost slaty, sometimes straight, sometimes curved, and inclining to fibrous; the cross fracture is small-grained uneven. The lustre of the principal fracture is shining and glistening, and pearly; the cross fracture glimmering, or dull. It is opaque. It affords a dark greenish grey-coloured streak. It is soft. It emits a clayey smell when breathed on. It occurs im-

bedded in pitchstone-porphyry, between Zwickau and Planitz *.

Third Subspecies.

Actynolite †.

Strahlstein, *Werner*.

THIS subspecies is divided into three kinds, viz. Asbestous Actynolite, Common Actynolite, and Glassy Actynolite.

First Kind.

Asbestous Actynolite.

Asbestartiger Strahlstein, *Werner*.

Asbestartiger Strahlstein, *Wid.* s. 479.—Amianthinite, *Kirw.* vol. i. p. 164.—Asbestartiger Strahlstein, *Emm.* b. i. s. 416.—Asbestoid, *Lam.* t. ii. p. 371.—Actinote aciculaire, *Haüy,* t. iii. p. 75.—La Rayonnante asbestiforme, *Broch.* t. i. p. 504.—Asbestartiger Strahlstein, *Reuss,* b. ii. 1. s. 174. *Id. Lud.* b. i. s. 140. *Id. Suck.* 1r th. s. 252. *Id. Bert.* s. 156. *Id. Mohs,* b. i. s. 581. *Id. Hab.* s. 61. *Id. Leonhard,* Tabel. s. 30. Amphibole actinote aciculaire, *Brong.* t. i. p. 455.—Asbestartiger Strahlstein, *Karst.* Tabel. s. 40. *Id. Haus.* s. 99.—Asbestinite, *Kid,* vol i. p. 116.—Amphibole, *Haüy,* Tabl. p. 40.—Asbestartiger Strahlstein, *Steffens,* b. i. s. 281. *Id. Lenz,* b. ii. s. 683.—Strahlige Hornblende, *Oken,* b. i. s. 322.—Asbestartiger Strahlstein, *Hoff.* b. ii. s. 293. *Id. Haus.* b. ii. s. 727.

External

---

* Vid. Beyer, in Crell's Chem. Annal. 2. 11. 381.—Lenz, Tabel. s. 33.—[illegible] Tasch. b. i. s. 267.

† [illegible], from the Greek words *ακτιν*, *ray*, and λιθος, *stone*, on ac-[illegible] radiated or prismatic concretions.

External Characters.

Its colour is greenish-grey, which passes on the one side through mountain-green into a kind of sky-blue, on the other through olive-green into yellowish-brown, and liver-brown.

It occurs massive, in distinct concretions, which are fibrous, and sometimes collected into others which are promiscuous wedge-shaped, and granular. It rarely occurs crystallized, in delicate capillary, rigid, moss-like, superimposed crystals.

Internally the lustre is glistening and pearly.

The fracture, owing to the smallness of the concretions, is not visible.

The fragments are splintery and wedge-shaped.

It is opaque, or slightly translucent on the edges.

The fibres or concretions in groups are soft, but individually equally hard with the other varieties of actynolite.

Specific gravity, 2.809, *Karsten.*

Chemical Characters.

It melts with difficulty before the blowpipe, into a black or dark green coloured glass.

Constituent Parts.

Silica, - -	47.0
Lime, -	11.3
Magnesia,	7.3
Oxide of Iron, -	20.0
Oxide of Manganese,	10.0
Loss, - -	4.4
	100.0

Vauquelin, in Haüy, t. iv. p. 335.

 This

This is an analysis of the variety of Asbestous Actynolite, named *Byssolite* by Saussure.

Geognostic Situation.

It occurs in beds in gneiss, mica-slate, and granular limestone, along with magnetic ironstone, iron-glance, iron-pyrites, copper-pyrites, variegated copper-ore, malachite, galena, blende, common actynolite, amethyst, garnet, and asbestus.

Geographic Situation.

Europe.—In Norway, it occurs at Arendal, Kongsberg, and Röraas; in Sweden, at Sala, and other places; in the Hartz, the Saxon Erzgebirge, Bohemia, Franconia, Silesia, Switzerland, Hungary, Italy, and France.

America.—Greenland; and the metalliferous mountains of Zacatecas in Mexico.

Observations.

1. It is distinguished from the other subspecies by its colour-suite, fibrous concretions, capillary crystallizations, pearly lustre, and inferior degree of hardness. It is distinguished from *Asbestus* by its fibrous, wedge-shaped and granular concretions.

2. Those varieties of asbestous actynolite which occur in very thin scopiformly aggregated acicular elastic-flexible crystals, have been considered as forming a distinct species, and named *Byssolite* by Saussure, *Amianthoid* by Haüy, and *Asbestoid* by some French mineralogists.

Second

Second Kind.

Common Actynolite.

Gemeiner Strahlstein, *Werner.*

Basaltes radiis minimis, fibrosis, nitidis, compositus; Basaltes fibrosus, *Wall.* gen. 22. sp. 152.—Gemeiner Strahlstein, *Wid.* s. 480.—Schorlaceous Actynolite, and Common Asbestoid, *Kirw.* vol. i. p. 166. & 168.—Gemeiner Strahlstein, *Estner,* b. ii. s. 887. *Id. Emm.* b. i. s. 418.—Stralite commune, *Nap.* p. 323.—Zillerthite, *Lam.* t. ii. p. 357.—Actinote etalé, *Haüy,* t. iii. p. 75.—La Rayonnante commune, *Broch.* t. i. p. 507.—Gemeiner Strahlstein, *Reuss,* b. i. s. 176. *Id. Lud.* b. i. s. 140. *Id. Suck.* 1r th. s. 140. *Id. Bert.* s. 185. *Id. Mohs,* b. i. s. 583. *Id. Hab.* s. 59. *Id. Leonhard,* Tabel. s. 31.—Amphibole Actinote hexaedre, *Brong.* t. i. p. 454.—Gemeiner Strahlstein, *Karst.* Tabel. s. 40. *Id. Haus.* s. 99.—Actynolite, *Kid,* vol. i. p. 116.—Amphibole comprimé, *Haüy,* Tabl. p. 40.—Gemeiner Strahlstein, *Steffens,* b. ii. s. 284. *Id. Lenz,* b. i. s. 681. *Oken,* b. i. s. 322. *Id. Hoff.* b. ii. s. 296. *Id. Haus.* Handb. b. ii. s. 725.

External Characters.

Its principal colour is leek-green, which passes on the one side into blackish-green, on the other into olive-green and grass-green.

It occurs massive and disseminated; also in wedge-shaped prismatic concretions, which are thin, thick, scopiform, stellular, and promiscuous; these sometimes pass into coarse, small, and long angulo-granular concretions. Frequently the prismatic concretions are collected into large granular concretions.

Internally

Internally it is shining, inclining to glistening, and is pearly inclining to vitreous.

It has a distinct double oblique angular cleavage, in which the angles are 124° 50′ and 55° 50′.

The fracture is uneven and conchoidal.

The fragments are splintery and wedge-shaped, seldom indeterminate angular.

It is generally translucent on the edges.

It is rather harder and more brittle than hornblende.

Specific gravity, 2.994, 3.293, *Kirwan.*

Chemical Characters.

Before the blowpipe, it melts into a greenish-grey or blackish glass.

Constituent Parts.

Silica, - -	64.00
Magnesia,	20.00
Alumina,	2.70
Lime,	9.30
Iron, -	4.00
	100 *Bergmann.*

Geognostic Situation.

It occurs in beds in gneiss, mica-slate, and talc-slate, sometimes alone, sometimes accompanied with ores of different kinds, as galena, magnetic ironstone, copper-pyrites, and blende. Small and irregular veins occasionally occur in transition-trap, and minute portions in secondary or flœtz-trap rocks.

Geographic

Geographic Situation.

It occurs at Eilan Reach in Glenelg, in Inverness-shire; near Fortrose in Cromarty; in the parish of Sleat, in the isle of Skye; different places in the isle of Lewis. In Cornwall, as in the neighbourhood of Redruth *. On the Continent, it is not uncommon in Saxony, Bohemia, Silesia, Sweden, and Norway.

Observations.

1. This is the most common kind of Actynolite. It never occurs regularly crystallized; the crystallized varieties of actynolite formerly included under this kind, being now referred by Werner to the Glassy Actynolite.

2. The mineral named *Pargasite* is now arranged with common actynolite.

Third Kind.

Glassy Actynolite.

Glasartiger Strahlstein, *Werner*.

Glasartiger Strahlstein, *Wid.* s. 438.—Glassy Actynolite, *Kirw.* vol. i. p. 168.—Glasartiger Strahlstein, *Estner*, b. ii. s. 893. *Id. Emm.* b. i. s. 422.—Stralite vetrosa, *Nap.* p. 326.—La Rayonnante vitreux, *Broch.* t. i. p. 510.—Glasartiger Strahlstein, *Reuss*, b. i. s. 182. *Id. Lud.* b. i. s. 141. *Id. Bert.* s. 155. *Id. Mohs*, b. i. s. 386. *Id. Leonhard*, Tabel. s. 31.—Amphibole actinote fibreux, *Brong.* t. i. p. 455.—Glasartiger Strahlstein, & Muschlicher Strahlstein, *Karst.* Tabel. s. 40.—Glasartiger Strahlstein, *Haus.* s. 99.—Amphibole etalé et fibreux, (in part), *Haüy*, Tabl. p. 40.—Glasartiger Strahlstein, *Steffens*,

* Greenough.

fens, b. i. s. 286. *Id. Lenz*, b. ii. s. 685. *Id. Oken*, b. i. s. 322. *Id. Hoff.* b. ii. s. 298. *Id. Haus.* b. ii. s. 726.

External Characters.

Its principal colour is mountain-green, which passes into grass-green, and leek-green, also into blackish-green, and greenish-grey.

It occurs massive; also in prismatic distinct concretions, which are very thin or fibrous, and thin or radiated, and are arranged in a scopiform, and rarely in a promiscuous manner; and these are again collected into thick and wedge-shaped prismatic, or large granular concretions. It sometimes occurs crystallized, and in the following figures:

1. Very oblique four-sided prism, which is long, thin, and often acicular, and truncated on the acute edges.
2. Rather flat six-sided prism, with two opposite acute lateral edges.

If the crystals are fully crystallized, and not broken, the edges of the obtuse lateral edges, and the angles of the acute lateral edges, are generally truncated; and sometimes the terminal edges are truncated.

The lateral planes are longitudinally streaked, seldom smooth and splendent.

Internally it is shining, sometimes splendent, and intermediate between vitreous and pearly.

It has a distinct double oblique angular cleavage.

The fracture is not visible, owing to the smallness of the concretions.

The fragments are splintery and wedge-shaped.

It is translucent, or semi-transparent.

It is brittle.

It is uncommonly easily frangible.

It is traversed by numerous parallel rents.

It is as hard as hornblende, but more brittle.

Specific gravity, 3.175, *Karsten*.

Chemical

Chemical Characters.

Before the blowpipe, it melts with difficulty into an opaque green-coloured glass.

Constituent Parts.

Glassy Actynolite from Zillerthal in the Tyrol.

Silica, -	50.00
Magnesia,	19.25
Alumina,	0.75
Lime,	9.75
Potash, -	0.50
Oxide of Iron, -	11.00
Oxide of Manganese,	0.50
Oxide of Chrome, -	3.00
Carbonic Acid, and Water,	5.00
Loss, - -	0.25

Laugier, Annales du Mus. t. v. p. 79.

Geognostic and Geographic Situations.

Europe.—It occurs in primitive rocks in the isle of Skye: in veins, along with rock-crystal, axinite, and epidote, at Bourg d'Oisans in Dauphiny; in beds of indurated talc, with limestone and common talc, on St Gothard; also in a similar repository in the Zillerthal, in the Tyrol; and in Sweden.

Asia.—It appears to be associated with talc at Bialoyarsk, in the Uralian Mountains.

Observations.

1. It is distinguished from the preceding kind by its vitreous lustre, crystallizations, and parallel cross rents.

2. The fibrous varieties of glassy actynolite have been confounded with Amianthus; but they are distinguished from

from it by lustre, cross rents, and the rough feel of their powder.

Fourth Subspecies.

Tremolite *.

Tremolith, *Werner.*

THIS subspecies is divided into three kinds, viz. Asbestous Tremolite, Common Tremolite, and Glassy Tremolite.

First Kind.

Asbestous Tremolite.

Asbestartiger Tremolit, *Werner.*

Asbestartiger Tremolith, *Emm.* b. i. s. 425. *Id. Estner,* b. ii. s. 893.—Grammatite, *Haüy,* t. iii. p. 227.—La Tremolith asbestiforme, *Broch.* t. i. p. 514.—Asbestartiger Tremolith, *Reuss,* b. i. s. 136. *Id. Lud.* b. i. s. 142. *Id. Suck.* 1r th. s. 272. *Id. Bert.* s. 166. *Id. Mohs,* b. i. s. 589. *Id. Leonhard,* Tabel. s. 31.—Grammatite, *Brong.* t. i. p. 475. *Id. Lucas,* p. 77. *Id. Brard,* p. 188.—Asbestartiger Tremolith, *Karsten,* Tabel. s. 44.—Amphibole blanc et soyeux, *Haüy,* Tabl. p. 41.—Asbestartiger Tremolith, *Steffens,* b. i. s. 290. *Id. Lenz,* b. i. s. 689.—Asbestartiger Grammatite, *Oken,* b. i. s. 327.—Asbestartiger

* The name is derived from *Tremola,* a valley in the Alps, where it is said to have been first found. It would, however, appear, that it was first discovered in Transylvania by M. Von Fichtel, and described by him under the name *Saulen* and *Stern-spath:* it was afterwards, in the year 1788, found in the Valley of Tremola.

bestartiger Tremolit, *Hoff.* b. ii. s. 306. *Id. Haus.* Handb. b. ii. s. 732.

External Characters.

Its most common colour is greyish-white; it is found also yellowish-white, and greenish-white, rarely reddish-white, and pale violet blue *.

It occurs massive; also in fibrous or very thin prismatic distinct concretions; these are generally scopiform or stellular, and collected into thick and wedge-shaped prismatic, and into large granular concretions.

Internally it is shining, approaching to glistening, and is pearly †.

The fracture is not visible.

The fragments are wedge-shaped, or splintery.

It is translucent on the edges.

It is rather easily frangible.

It is soft, approaching to very soft, in the mass, but the individual concretions are as hard as the other kinds of this subspecies.

It is rather sectile.

Physical Characters.

When struck gently, or rubbed in the dark, it emits a pale reddish-coloured light; when pounded, and thrown on coals, a greenish-coloured light. It phosphoresces more than any of the other subspecies.

Chemical

* The rare violet-blue variety has been hitherto found only at St Marcel in Piedmont.

† It has a lower lustre than any of the other subspecies.

Chemical Characters.

Before the blowpipe it melts into a white opaque mass.

Geognostic Situation.

It occurs most frequently in granular foliated limestone, or in dolomite; sometimes in chlorite; and more rarely in secondary trap-rocks.

Geographic Situation.

Europe.—It occurs in foliated granular limestone in Glen Tilt, in Perthshire, and in Glen Elg in Inverness-shire; in dolomite in Aberdeenshire and Icolmkill; and in basalt in the Castle Rock of Edinburgh. In Norway, it is an inmate of foliated granular limestone; in Bohemia, it is imbedded in limestone, along with calcareous-spar, slate-spar, brown-spar, fluor-spar, and quartz; at Dognatska in Hungary, with galena, copper-pyrites, iron-pyrites, compact and foliated magnetic ironstone, and garnet; Switzerland in dolomite; in granular limestone, along with augite, on Mount Vesuvius.

America.—It is found in foliated granular limestone in many places in the United States, and in Greenland.

Asia.—Patrin found it at Kadainsk, in Siberia.

Observations.

Its fibrous concretions, pearly lustre, and softness, characterize it as a particular kind of Tremolite. It is distinguished from *Amianthus*, by its concretions, comparative brittleness, and meagre feel; and its higher lustre, inferior translucency and hardness, distinguish it from *Fibrous Zeolite.*

Second

Second Kind.

Common Tremolite.

Gemeiner Tremolit, *Werner*.

Gemeiner Tremolith, *Estner*, b. ii. s. 901. *Id. Emm.* b. i. s. 426. —Grammatite, *Haüy*, t. iii. p. 227.—La Tremolithe commune, *Broch.* t. i. p. 515.—Gemeiner Tremolith, *Reuss*, b. i. s. 188. *Id. Lud.* b. i. s. 142. *Id. Suck.* 1r th. s. 274. *Id. Bert.* s. 164. *Id. Mohs*, b. i. s. 590.—Grammatite, *Lucas*, p. 77.—Tremolith, *Hab.* s. 61. *Id. Leonhard*, Tabel. s. 31.—Grammatite, *Brong.* t. i. p. 475. *Id. Brard*, p. 188.—Gemeiner Tremolith, *Karsten*, Tabel. s. 44.—Gemeiner Grammatite, *Haus.* s. 97. —Amphibole grammatite, *Haüy*, Tabl. p. 40.—Gemeiner Tremolith, *Steffens*, b. i. s. 291. *Id. Lenz*, b. ii. s. 691.—Gemeiner Grammatite, *Oken*, b. i. s. 327.—Gemeiner Tremolit, *Hoff.* b. ii. s. 308.—Gemeiner Grammatite, *Haus.* Handb. b. ii. s. 730.

External Characters.

Its most frequent colour is white, and principally greyish and yellowish white, seldom greenish and reddish white; the greyish-white passes into smoke-grey, and the greenish-white into pale asparagus-green.

It occurs massive; also in distinct concretions, which are prismatic, and these are collected into large and coarse longish granular concretions. It is sometimes crystallized, and in the following figures:

1. Very oblique-four-sided prism, truncated on the acute lateral edges.
2. Same prism, truncated on the obtuse lateral edges.
3. Same prism, truncated on all the lateral edges.

4. Same

4. Same prism, bevelled on the obtuse lateral edges. When these bevelling planes increase so much that the original ones disappear, there is formed
5. An extremely oblique four-sided prism.
6. Very oblique four-sided prism, very flatly bevelled on the extremities, the bevelling planes set on the acute lateral edges *.
7. The preceding figure †, truncated on the acute lateral edges.
8. N° 6. truncated on all the lateral edges ‡.
9. N° 6., in which all the lateral edges are rounded off ‖.

The lateral planes are longitudinally streaked.

The crystals are middle-sized or small; sometimes singly imbedded, sometimes superimposed, or promiscuously aggregated.

The lustre is shining, and intermediate between vitreous and pearly ¶.

It has a double oblique angular cleavage §, the angles of which are 124° 50′, 55° 50′.

The

* Grammatite di-tetraedre, Haüy.

† Grammatite bis-unitaire, Haüy.

‡ Grammatite tri-unitaire, Haüy.

‖ Grammatite cylindroide, Haüy.

¶ It has a higher degree of lustre than any of the other subspecies.

§ This mineral splits easily, not only in the direction of the planes of the prism, but also in that of its diagonals, particularly the longest diagonal. When we break across one of these prisms, we observe on the fracture-surface a line in the direction of the longer diagonal, which is so strongly marked, that at first sight we are apt to consider it as pointing out these as hemitrope or twin-crystals. The name *Grammatite*, formerly given to this mineral by Haüy, is derived from the character just stated. It is also worthy of remark, that in the fracture of tremolite, even in crystals, there is

The fracture is uneven, or conchoidal.

The fragments are splintery and wedge-shaped, or indeterminate angular.

It is translucent, or semi-transparent.

It is as hard as hornblende.

It is rather brittle.

It is easily frangible.

Its powder is rough to the feel.

Specific gravity, 2.9257, 3.2, *Haüy*.—2.832, *Karsten*.—3.000, *Wid.*

Chemical Character.

Before the blowpipe, it loses its colour and transparency, melts with great difficulty, often only on the edges, and with considerable ebullition, into an opaque glass.

Constituent Parts.

Silica,	27.0	Silica,	35.5	Silica,	52.0	Silica,	50
Magnesia,	18.5	Magnesia,	16.5	Magnesia,	12.0	Magnesia,	25
Lime,	21.0	Lime,	16.5	Lime,	20.0	Lime,	18
Alumina,	6.0	Carbonic acid		Carbon. acid,	12.0	Carbonic acid	
Carbon. acid,	26.0	and Water,	23.0	A trace of iron.		and Water,	5
	Chenevix.		*Buchols.*		*Lowitz.*		*Laugier.*

Geognostic Situation.

Like the asbestous subspecies, it occurs principally in granular limestone, or dolomite, and in metalliferous beds. These

is a tendency to the fibrous structure: the stroke of a hammer, or even the simple pressure of the finger in some cases, will separate folia or radia into fibres as delicate as those of amianthus, and which are somewhat elastic-flexible.

These beds contain, besides the tremolite, quartz, calcareous-spar, garnet, blende, galena, copper-pyrites, and vitreous copper-ore, or copper-glance. It sometimes occurs in indurated talc, along with rhomb-spar; or in common talc, with calcareous-spar and rutile. It occurs rarely in serpentine and granite.

Geographic Situation.

Europe.—It occurs in Glen Tilt and Glen Elg, and in Unst, one of the Shetland Islands; also at Clicker Tor in Cornwall. On the Continent, it occurs at Kongsberg in Norway, along with ores of silver; in the island of Senjen in Nordland, in thin beds, resting on limestone, and covered with a bed of massive garnet; in Sweden, Hessia, Bohemia, Silesia, Moravia, Switzerland, the Tyrol, Carinthia, Carniola, Hungary, Transylvania, Italy, and France.

Asia.—It is found on the borders of the Lake Baiakal.

America.—It occurs in several districts of the United States.

Africa.—Egypt.

Third Kind.

Glassy Tremolite.

Glasartiger Tremolith, *Werner.*

Glasartiger Tremolith, *Estner,* b. ii. s. 907. *Id. Emm.* b. i. s. 429.—Grammatite, *Haüy,* t. iii. p. 229.—La Tremolithe vitreuse, *Broch.* t. i. p. 516.—Glasartiger Tremolith, *Reuss,* b. i. th. 1. s. 193. *Id. Lud,* b. i. s. 145. *Id. Suck.* 1r th. s. 277. *Id. Bert.* s. 165. *Id. Mohs,* b. i. s. 392.—Grammatite, *Lucas,* p. 77.—Glasartiger Tremolith, *Leonhard,* Tabel. s. 31.—Tremolith, *Hab.* s. 61.—Grammatite, *Brong.* t. i. p. 475. *Id. Brard,*

Brard, p. 188.—Glasartiger Tremolith, *Karsten*, Tabel. s. 44. *Id. Haus.* s. 97.—Amphibole Grammatite, *Haüy*, Tabl. p. 40. Glasartiger Tremolith, *Steffens*, b. i. s. 294. *Id. Lenz*, b. ii. s. 694.—Glasartiger Grammatite, *Oken*, b. i. s. 327.—Glasiger Tremolit, *Hoff*. b. ii. s. 311.—Glasartiger Grammatite, *Haus.* Handb. b. ii. s. 729.

External Characters.

Its colours are greyish, greenish, yellowish, and reddish white.

It occurs massive; also in distinct concretions, which are thin, very thin, straight and scopiform prismatic, with numerous cross-rents, and these are again grouped into thick and wedge-shaped concretions.

It is frequently crystallized in long acicular crystals.

Its lustre is shining, but in a lower degree than the preceding subspecies, and intermediate between vitreous and pearly.

The fragments are splintery and wedge-shaped.

It is translucent.

It is nearly as hard as hornblende.

It is very easily frangible, and very brittle.

Specific gravity 2.863, *Karsten*.

Physical Character.

It is phosphorescent in a low degree.

Chemical Character.

It is said to be infusible before the blowpipe.

Constituent

Constituent Parts.

		Tremolite from St Gothard.					
Silica,	65.00	Silica,	35.5	Silica,	28.4	Silica,	41.00
Magnesia,	10.33	Lime,	26.5	Lime,	30 6	Lime,	15.00
Lime,	18.00	Magnesia,	16.5	Magnesia,	18.0	Magnesia,	15.25
Iron,	0.16	Water, and		Water, and		Water, and	
Carbon. acid,		Carbon. acid,	23.0	Carbon. acid,	23.0	Carbon. acid,	23.00
and Water,	6.05					Loss,	5.75
			101.5		100.0		
							100.0
Klaproth.		*Laugier*, Annales du Museum, 34 cahier, t. vi. p. 232.					

Geognostic Situation.

It is the same as that of the preceding subspecies, occurring principally along with granular limestone.

Geographic Situation.

Europe.—In Scotland, it occurs along with the other kinds. It is found at Arendal in Norway; Sweden; in Bavaria, Salzburg, the Tyrol, Switzerland, and Hungary.

Asia.—In the island of Ceylon; in the Uralian Mountains.

America.—In the United States.

Observations.

1. On a general view, Tremolite is characterized by its white colours; Actynolite by its light green colours; and Hornblende by its dark green colours.

2. A mineral found in ironstone in Normark in Sweden, has been lately described by Werner, under the name *Calamite.* It appears to be a variety of tremolite. The following is the description of it, as given in Leonhard's Taschenbuch for 1816:—

Colour

Colour intermediate between asparagus and pistachio green, sometimes approaching to mountain-green. Occurs in reed-like prismatic crystals, in which the lateral planes are deeply longitudinally streaked. Is shining and splendent, and vitreous, inclining to metallic. Oblique double cleavage. Uneven fracture. As hard as actynolite. Easily frangible.

Another mineral, found in Pfitschthal in the Tyrol, has been described under the name *Rhatizit*, but which appears to be but a variety of Glassy Tremolite. The following description is given in Leonhard's Taschenbuch for 1816:

Colour milk-white, seldomer yellowish and greyish white, and isabella-yellow, and smoke-grey. Occurs massive. Fracture scopiform and stellular radiated. Fragments splintery. Distinct concretions large granular. Translucent on the edges. Semihard. Rather brittle. Rather easily frangible.

Fifth Subspecies.

Asbestus.

Asbest, *Werner.*

This subspecies is divided into four kinds, viz. Rock-Cork, Amianthus, Common Asbestus, and Rock-Wood.

First Kind.

Rock-Cork *.

Berg Cork, *Werner.*

Aluta montana, *Wall.* t. i. p. 414.; Suber montanum, Id. p. 415. —Berk Kork, *Wid.* s. 469.—Suber montanum, Corium montanum, *Kirw.* vol. i. p. 163.—Berg Kork, *Estner,* b. ii. s. 864. *Id. Emm.* b. i. s. 399.—Sughero montano, *Nap.* p. 319.—Varieté d'Amianthe, *Lam.* p. 367.—Le Siege de montagne, *Broch.* t. i. p. 492.—Asbeste tresse, *Haüy,* t. iii. p. 248.—Holz Asbest, *Reuss,* b. ii. 2. s. 253.—Kork Asbest, *Lud.* b. i. s. 137.—Berg Kork *Suck.* 1r th. s. 263. *Id. Bert.* s. 148. *Id. Mohs,* b. i. s. 567.—Asbeste tresse, *Lucas,* p. 81.—Berg Kork, *Leonhard,* Tabel. s. 29.—Asbeste suberiforme, *Brong.* t. i. p. 479.—Asbeste tresse, *Brard,* p. 194.—Schwimmender Asbest, *Karst.* Tabel. s. 42. *Id. Haus.* s. 99.—Compact spongy Amianthus, *Kid,* vol. i. p. 103.—Asbeste tresse, *Haüy,* Tabl. p. 55.—Berg Kork, *Steffens,* b. i. s. 278. *Id. Lenz,* b. ii. s. 670.—Korkichter Asbest, *Oken,* b. i. s. 326.—Berg Kork, *Hoff.* b. ii. s. 278.—Schwimmender Asbest, *Haus.* Handb. b. ii. s. 738.—Mountain Cork, *Aikin,* p. 232.

External Characters.

Its colours are yellowish and greyish white, also yellowish and ash grey, and pale ochre yellow.

It occurs massive, in plates that vary in thickness †, corroded,

---

* This mineral is named *Rock Cork,* on account of its resembling cork in lightness, and its receiving impressions from the nail.

† The variety in plates has received the following names: *Mountain Flesh,* Bergfleish, Caro montana, Chaire de montagne, Chair fossile; *Mountain Paper,* Papiere fossile, Bergpapier; *Mountain Leather,* Bergleder, Corium montanum, Cuir de montagne.

roded, and with impressions; and these forms are composed of delicate and promiscuous fibrous concretions.

Internally it is feebly glimmering, or dull.

The fracture is fine-grained uneven, inclining to slaty in the large.

The fragments are indeterminate angular, and blunt-edged.

It is opaque.

It is very soft.

It becomes shining in the streak.

It is sectile, almost like common cork.

It is slightly elastic flexible.

It is difficultly frangible.

It adheres slightly to the tongue.

It emits a grating sound when we handle it.

It feels meagre.

It is so light as to swim on water.

Specific gravity, 0.679, 0.991, *Brisson*.—0.991, *Haüy* *.

Chemical Characters.

It melts with great difficulty before the blowpipe into a milk-white nearly translucent glass.

Constituent Parts.

Silica,	56.2	62.0
Magnesia,	26.1	22.9
Alumina,	2.0	2.8
Lime,	12.7	10.0
Oxide of Iron,	3.0	3.2
	100	100

Bergmann, Opusc. t. iv. p. 170.

Geognostic

* This low specific gravity is owing to the loose texture of the mass.

Geognostic Situation.

It occurs in cotemporaneous veins in serpentine, and in red sandstone; also in metalliferous veins in primitive and transition rocks; and occasionally in mineral beds.

Geographic Situation.

It occurs in veins in the serpentine of Portsoy, and in the red sandstone of Kincardineshire; in plates, in the lead-veins at Lead Hills and Wanlockhead in Lanarkshire; and in small quantities at Kildrummie in Aberdeenshire. At Sala in Sweden, it occurs in a metalliferous bed, along with asbestus, steatite, calcareous-spar, rhomb-spar, and brown-spar: in veins along with ores of silver, calcareous spar, and heavy-spar, at Kongsberg in Norway; in the silver-mines of Johanngeorgenstadt in Saxony; at Valecas in Spain, in beds along with meer-schaum and talc: and in primitive rocks in Carinthia, Idria, France, Moravia, &c.

Second Kind.

Amianthus, or Flexible Asbestus *.

Amiant, *Werner.*

Αμιαντος of the Greeks.—Amiantus, *Plin.* Hist. Nat. xxxvi. 19. p. 31.—Asbestus maturus, *Wall.* t. i. p. 410.; Amianthus, Id. p. 408.—Amianth, *Wid.* s. 464.—Amianthus, *Kirw.* vol. i. p. 161.—Amianth, *Estner,* b. ii. s. 368. *Id. Emm.* b. i. s. 402. —Amiantho, *Nap.* p. 316.—L'Amianth, *Lam.* t. ii. p. 365.— *Id.*

---

* *Amianthus*, from αμιαντος, *unstained, unsoiled*, which refers to the property this substance possesses, of remaining unsoiled in the fire. It is also named Rock-flax and Rock-wool.

Id. Broch. t. i. p. 494.—Asbest flexible, *Haüy*, t. iii. p. 245. Biegsamer Asbest, *Reuss*, b. 2. ii. s. 243.—Amianth, *Lud.* b. i. s. 137.—Amianth-asbest, *Suck.* 1r th. s. 265.—Amianth, *Bert.* s. 149. *Id. Mohs*, b. i. s. 569.—Amianth-asbest, *Hab.* s. 64.—Asbest flexible, *Lucas*, p. 81.—Biegsamer Asbest, *Leonhard*, Tabel. s. 30.—Asbest amianthe, *Brong.* t. i. p. 478.—Asbest flexible, *Brard*, p. 194.—Biegsamer Asbest, *Karst.* Tabel. s. 42.—Amianth, *Haus.* s. 99.—Loosely fibrous and flexible Amianthus, *Kid*, vol. i. p. 101.—Asbest flexible, *Haüy*, Tabl. p. 55.—Amiant, *Steffens*, b. i. s. 276. *Id. Lenz*, b. ii. s. 672. Biegsamer Asbest, *Oken*, b. i. s. 325.—Amianth, *Hoff.* b. ii. s. 281. *Id. Haus.* Handb. b. ii. s. 736.—Amianthus, *Aikin*, p. 232.

External Characters.

Its most common colour is greenish-white, of different degrees of intensity, which passes into greenish-grey, and rarely into light olive-green. It is sometimes blood-red, particularly when it occurs in veins in serpentine.

It occurs massive, and in small veins, also in fibrous distinct concretions, which are parallel, generally straight, and sometimes curved.

Internally its lustre is shining and pearly, occasionally approaching to semi-metallic.

The fracture is not visible.

The fragments are generally long splintery, or thread-like.

It is translucent on the edges, or opaque.

It is very soft.

It is sectile.

It is perfectly flexible.

It splits easily.

Specific gravity, 2.444, *Muschenbröck*. According to *Brisson*, it varies considerably in specific gravity: he found the

the long silky amianthus to vary from 0.9088 to 2.3134, before it had absorbed water; from 1.5662 to 2.3803, after it had absorbed water.

Chemical Characters.

Before the blowpipe, it phosphoresces, and melts with difficulty into a whitish or greenish slag.

Constituent Parts.

	Asbestus of Swartioick, in Sweden.	Asbestus of Tarentaise, in Savoy.	Asbestus of Torias, in Spain.	
Silica, - -	64.0	64.0	72.00	59.00
Carbonate of Magnesia,	17.2	18.6	12.19	25.00
Alumina, - -	2.7	3.3	3.03	3.00
Carbonate of Lime,	13.9	6.9	10.05	9.05
Barytes, - -		6.0		
Oxide of Iron, -	2.2	1.2	2.02	2.25
	Bergmann, Opusc. t. iv.	*Bergmann*, Id.	*Bergmann*, Id.	*Chenevix*.

Geognostic Situation.

It occurs frequently along with common asbestus, in contemporaneous veins in serpentine; in similar veins in primitive and secondary greenstone, gneiss, and mica-slate; and it occasionally forms one of the constituent parts of metalliferous beds.

Geographic Situation.

Europe.—It occurs in serpentine in the islands of Mainland *, Unst and Fetlar in Shetland; and in the same rock at Portsoy; in veins in mica-slate, at Glenelg in Inverness-shire:

* Hibbert.

shire; in different parts of Aberdeenshire, and Argyleshire: in secondary greenstone in the middle division of Scotland, as in Fifeshire, particularly in Inchcolm, and other quarters. In England, it occurs in veins in serpentine, at St Kevern's in Cornwall *. On the Continent, it occurs in the Hartz, in veins in primitive greenstone; in Bohemia, in metalliferous beds, along with magnetic ironstone: in Upper Saxony, in veins in serpentine; and in a similar situation in Silesia and Switzerland. In Dauphiny, and in St Gothard, it is found in cotemporaneous veins in gneiss and mica-slate, along with felspar, earthy and common chlorite, and rock-crystal. Uncommonly beautiful white and long fibrous varieties are met with in the Val de Serre in Savoy, at Cogne in Piedmont, and in the island of Corsica †.

Asia.—It abounds in serpentine rocks in the Uralian and Altain mountains.

America.—In veins that traverse serpentine in Maryland, Delaware, New Jersey, Connecticut, and Massachusets ‡.

Uses.

This mineral, on account of its flexibility, and its resisting the action of considerable degrees of heat, was woven into those incombustible cloths in which the ancients sometimes wrapped the bodies of persons of distinction, before they were placed on the funeral-pile, that their ashes might be

* Greenough.

† It is so abundant in Corsica, that Dolomieu used it in place of hay and tow for packing his collections of minerals.

‡ Cleaveland's Mineralogy, p. 328.

be collected free from admixture *. After the body was consumed, the cloth was withdrawn from the fire, the ashes taken out of it, washed with milk and wine, and sprinkled with consecrated water, and inclosed in an urn, either with or without the fossile-cloth in which the body had been consumed †. The goodness of the amianthus for this purpose, depends on the length of its fibres, which vary from an inch to a foot in length, its whiteness and flexibility. In preparing the cloth, the amianthus is previously well washed, to free it of all impurities, then combed straight, and woven with flax. The cloth is placed on glowing coals, by which the flax and oil used in the operation of weaving are consumed, and the cloth is deprived of its stains ‡. In this manner are manufactured, not only large pieces of cloth, but also gloves, purses, belts, and napkins. All these articles have a shining appearance, and white colour; but various tints may be communicated to them by artificial means. At Nerwinski in Siberia, gloves, caps and purses are made of amianthus; and it is worked into girdles, ribbons, and other articles, in the Pyrenees. The finest girdles are made by weaving the most beautiful varieties of amianthus with silver-wire: they are much prized by the women, not only on account of their beauty, which is certainly

* Dioscorides says: "Amianthus lapis in Cypro nascitur, scissili alumini similis, quo elaborato utpote flexili, telas spectaculi gratia texunt, quæ ignibus injectæ ardent quidem sed flammis invectæ splendidiores exeunt."

† Dolomieu informs us, that he saw in the Library of the Vatican in Rome, a shroud, containing ashes and burnt bones, which had been found in a sarcophagus; and in Italy, asbestus-cloth, containing ashes, has been frequently found inclosed in urns.

‡ We are told that the Emperor Charlemagne had a table-cloth of amianthus, which he used to throw into the fire after dinner, that it might burn clean, by way of amusing his guests.

certainly very considerable, but from certain mysterious properties they are supposed to possess. When a number of fibres are placed together, we can use them as a wick for lamps; and it is remarked, that such a wick readily attracts the oil, and affords a pretty lively flame. It is said the Romans made use of this kind in the lamps placed in their temples and cemeteries; hence, it has been alleged that these lamps never required to be renewed *. It is well known, however, that the duration of amianthus wicks is not considerable; for Rozier found that they did not continue for more than twenty hours †. Paper has been made of this mineral, but it is too hard for use. It has been proposed to preserve valuable documents from fire, by writing them on paper of amianthus. Such a plan might deserve consideration, if we possessed fire-proof ink. Dolomieu informs us, that it is used by the Corsicans in the composition of a kind of pottery, which is thereby rendered very light, and less liable to be broken by sudden alternations of temperature, or even by falling, than other kinds of pottery. The Chinese pound and knead it with gum-tragacanth, and form it into a kind of furnace, which they affirm to be very durable. Ancient physicians prescribed it for different diseases. Thus, in the state of salve, it was considered as very useful in restoring vigour to enfeebled limbs: the itch was said to yield readily to its drying powers; and in affections of the stomach, it was not to be

* The incombustibility of bodies made of amianthus, gave rise among the antients to many fables. Thus, Pliny says, that the asbest (our amianthus) is obtained from an Indian plant, which grows in an arid region of the earth, never refreshed by the rain or dew of heaven, and hence it is able to resist the most violent degrees of heat.

† It is said that the natives of Greenland make use of amianthus for the wicks of their lamps.

be diregarded, as it restored the appetite when entirely lost.

Observations.

1. It is distinguished from *Common Asbestus* by its higher lustre, its fibres being more easily separated, and its flexibility.

2. It is said to be the *Lapis caristius* of Strabo.

Third Kind.

Common Asbestus *.

Gemeiner Asbest, *Werner.*

Asbestus immaturus, *Wall.* t. i. p. 411.—Gemeiner Abest, *Wid.* s. 471.—Asbestus, *Kirwan*, vol. i. p. 159.—Gemeiner Asbest, *Estner*, b. ii. s. 872. *Id. Emm.* b. i. s. 406.—Asbesto commune, *Nap.* p. 314.—Asbeste, *Lam.* t. ii. p. 369.—Asbeste dur, *Haüy*, t. iii. p. 247.—L'Asbeste commune, *Broch.* t. i. p. 497.—Gemeiner Asbest, *Reuss*, b. ii. 2. s. 248. *Id. Lud.* b. i. s. 138. *Id. Suck.* 1r th. s. 267. *Id. Bert.* s. 150. *Id. Mohs*, b. i. s. 571. *Id. Hab.* s. 63.—Asbeste dur, *Lucas*, p. 81. —Gemeiner Asbest, *Leonhard*, Tabel. s. 30.—Asbest dur, *Brong.* t. i. p. 479. *Id. Brard*, p. 194.—Gemeiner Asbest, *Karsten*, Tabel. s. 42. *Id. Haus.* s. 99.—Asbeste dur, *Haüy*, Tabl. p. 55.—Gemeiner Asbest, *Steffens*, b. i. s. 274. *Id. Lenz*, b. i. s. 679.—Steifer Asbest, *Oken*, b. i. s. 325.—Gemeiner Asbest, *Hoff.* b. ii. s. 288.—Talkartiger Asbest, *Haus.* Handb. b. ii. s. 736.—Common Asbest, *Aikin*, p. 233.

External

* The literal signification of this term is *unextinguishable*; but as the verb [illegible] is metaphorically used in the sense of *aboleo*, or *perdo*, it may be [illegible] *imperishable*; this explanation being more appropriate than the [illegible] peculiar character of this substance.—*Kid.*

External Characters.

Its colours are dark leek-green, and mountain-green; also greenish-grey and yellowish-grey.

It occurs massive; and in distinct concretions, which are parallel, slightly curved, and coarsely fibrous, and intimately aggregated together.

It is rarely crystallized in capillary crystals *.

Internally it is glistening and pearly.

The fracture is not visible.

The fragments are long splintery.

It is translucent, or only translucent on the edges.

It is soft, approaching to very soft.

It is rather brittle.

It is difficultly frangible.

It feels rather greasy.

Specific gravity, 2.000, *Karsten.*—2.542, *Kirwan.*—2.591, *Breithaupt.*

Chemical Characters.

It melts before the blowpipe into a blackish glass.

Constituent Parts.

According to Mr Chenevix, it contains nearly the same constituent parts as amianthus. Gehlen discovered chrome in the leek-green asbestus of Zöblitz, and manganese in a variety from Siberia.

Geognostic

* Count de Bournon found them to be tetrahedral rhomboidal prisms.—*Cat. Min.* p. 123.

Geognostic Situation.

Like amianthus, it occurs in veins in serpentine, and in primitive greenstone: it also occurs in metalliferous beds, along with magnetic ironstone, iron-pyrites, magnetic-pyrites, calcareous-spar, garnet, and indurated talc, and sometimes along with ores of copper, viz. copper-pyrites, copper-glance, and grey copper-ore.

Geographic Situation.

Europe.—It occurs in the serpentine of Shetland, Portsoy, Anglesey, and Cornwall; and on the Continent of Europe, it is found in all the serpentine districts, and in metalliferous beds in the Saxon Erzgebirge, Salzburg, &c.

Asia.—It is found at Sisertskoi and Sawod, and other parts in Siberia.

America.—In the United States.

Fourth Kind.

Rock-Wood or Ligneous Asbestus.

Bergholz, *Werner.*

Bergholz, *Wid.* s. 473.—Ligniform Asbestus, *Kirw.* vol. i. p. 161. —Bergholz, *Estner,* b. ii. s. 877. *Id. Emm.* b. i. s. 410.—Ligno Montano, *Nap.* p. 321.—Asbeste ligniforme, *Haüy,* t. iii. p. 240.—Le Bois de Montagne, *Broch.* t. i. p. 499.—Asbest ligniforme, *Brong.* t. i. p. 480.—Holzasbest, *Reuss,* b. ii. 2. s. 253. *Id. Leonhard,* Tabel. s. 30.—Asbeste ligniforme, *Lucas,* p. 81. *Id. Brong.* t. i. p. 48. *Id. Brard,* p. 195. —Holzasbest, *Karst.* Tabel. s. 42.—Holzartiger Asbest, *Haus.* s. 99.—Ligniform Asbestus, *Kid,* vol. i. p. 105.—Asbest ligniforme,

niforme, *Haüy*, Tabl. p. 55.—Bergholz, *Steffens*, b. i. s. 280. *Id. Lenz*, b. ii. s. 680.—Holzicher Asbest, *Oken*, b. i. s. 326. Bergholz, *Hoff.* b. ii. s. 291.—Holzasbest, *Haus.* Handb. b. ii. s. 737.—Mountain Wood, *Aikin*, p. 232.

External Characters.

Its colour is wood-brown, of various degrees of intensity.

It occurs massive, and in plates: also in delicate and promiscuous fibrous concretions.

Internally its lustre is glimmering.

The fracture is curved slaty.

The fragments are tabular.

It becomes shining in the streak.

It is soft, passing into very soft.

It is opaque.

It is sectile.

It is rather difficultly frangible.

It is slightly elastic-flexible.

It feels meagre, and adheres slightly to the tongue.

Specific gravity, before immersion, 1.534; after immersion, 2.225, *Breithaupt.*

Chemical Characters.

It is infusible before the blowpipe.

Geognostic and Geographic Situations.

It occurs at Sterzing in the Tyrol, along with many different fossils, as common asbestus, actynolite, quartz, garnet, blende, iron-pyrites, galena, and calamine; and its repository, as Mohs remarks, appears to be a bed, as it is accompanied with minerals that often occur in such situations.

tions. It is also found in Dauphiny, and in Stiria; and Steffens conjectures, from the descriptions of Georgi, that it occurs in different places in the mountains of Archangel and Olocnezk.

Observations.

It is distinguished from *Rock-Cork*, by its wood-brown colour, higher lustre, fracture, and greater specific gravity.

3. Prismatoidal Augite.

Prismatoidischer Augit, *Mohs.*

This Species contains two subspecies, viz. Epidote and Zoisite.

First Subspecies.

Epidote or Pistacite.

Epidote, *Haüy.*

Pistazit, *Werner.*

Schorl vert de Dauphiny, *Romé de Lisle*, t. ii. p. 401.—Thallite, *Daubenton*, Tabl. p. 9.—Thallite, *La Metherie*, Theor. de la Terre, 2d edit. t. ii. p. 319.—Delphinite, *Saussure*, Voyage dans les Alpes, n. 1918.—Acanticone, *Dandrada*, Journ. Chem. von Schœrer, t. iv. p. 19.—Thallite, *Karsten*, Mineral. Tabellen, p. 20.—Arendalite, *ib.* p. 34.—Thallite, *Reuss*, b. ii th. i s. 117.—Epidote, *Haüy*, t. iii. p. 102.—Thallite, Arendalite, *Lud.* b. ii. s. 136, 137.—Epidot, *Suck.* 1r th. s. 256. —Thallite, Acanticone, *Bert.* s. 196. 173.—Epidote, *Mohs*, b. i.

b. i. s. 57. *Id. Lucas*, p. 59. *Id. Brong.* t. i. p. 410. *Id. Brard*, p. 153.—Thallite, *Kid.* vol. i. p. 242.—Epidot, *Haüy*, Tabl. p. 43. *Id. Steffens*, b. i. s. 66.—Pistacit, *Hoff.* b. ii. s. 654.—Thallit, *Haus.* Handb. b. ii. s. 672.—Epidote, *Aikin*, p. 222.

External Characters.

Its principal colour is pistachio-green, which passes on the one side into blackish-green and greenish-black, on the other side into dark olive-green, oil-green, and siskin-green.

It occurs massive; also in distinct concretions, which are coarse and small granular, and stellular or scopiform fibrous, which latter are collected into wedge-shaped prismatic concretions.

It is frequently crystallized. The primitive figure is an oblique four-sided prism, in which the lateral planes meet under angles of 114° 37′, and 65° 23′. The following are the secondary figures.

1. Very oblique four-sided prism, bevelled on the extremities; the bevelling planes are set either on the obtuse or on the acute lateral edges, and in the latter case the bevelment is a little flatter than in the first *.
2. The preceding figure truncated on the acute edges, and flatly bevelled on the extremities; the bevelling planes set on the truncating planes; or we may describe it as a broad unequiangular six-sided prism, flatly bevelled on the extremities; the bevelling planes set on the opposite smaller lateral planes, fig. 119. Pl. 6.

* In the first variety, the angle of the bevelment is 110° 6′, in the latter 117° 14′.

[illegible] in the [illegible] in the [illegible] the acumina[illegible] [illegible] bevelling of two [illegible] plane [illegible]

[illegible] planes, and the [illegible] fig. 121.

[illegible] lateral edges, [illegible] by six [illegible] planes, and the [illegible] edges formed [illegible] acuminating planes [illegible] fig. 122. Pl. 5.

[illegible] in the [illegible] in the [illegible] planes, the apex of the acumi[illegible] bevelled by two [illegible] fig. 123. Pl. 6.

[illegible] occur, [illegible] bevelling, but [illegible] in the [illegible] and the in[illegible]

[illegible] and are promiscu[illegible] and are general[illegible]

[illegible] are more or less deeply [illegible] acuminating and bevelling [illegible] and the terminal planes diagonally streaked.

Externally, [illegible] from splendent to glistening.

tening, and is vitreous; internally, it is shining or glistening, and is resinous, inclining to pearly.

It has a twofold cleavage, and the cleavages are parallel with the lateral planes of the oblique four-sided prism. Of these cleavages one only in general is perfect.

The fracture is small and flat conchoidal, sometimes small-grained uneven, sometimes even or splintery.

The fragments are indeterminate angular, and sharp-edged.

It alternates from translucent to translucent on the edges, and to nearly transparent.

It is harder than felspar, but not so hard as quartz.

It is brittle, and easily frangible.

Specific gravity, 3.452, *Lucas*.—3.452, *Breithaupt*.—Variety named Acanticone, from Norway, 3.407, *Lowry*.

Chemical Characters.

Before the blowpipe it is converted into a brown-coloured scoria, which blackens by continuance of the heat.

Constituent Parts.

	Epidote from the Valais.	From Oisans.	From Arendal.
Silica,	37.0	37.0	37.0
Alumina,	26.0	27.0	21.0
Lime,	20.0	14.0	15.0
Oxide of iron,	13.0	17.0	24.0
Oxide of manganese,	0.6	1.5	1.5
Water, -	1.8	3.5	1.5
Loss,	1.0	0	0
	100.0	100.0	100.0
		Descotils.	*Vauquelin.*

Laugier, Ann. du Mus. d'Hist. Nat. t. v. p. 149.

Geognostic Situation.

It occurs in beds and veins, and sometimes as an accidental constituent part of rocks. The beds in which it occurs are primitive, and contain augite, garnet, hornblende, quartz, calcareous-spar, and magnetic ironstone, as at Arendal in Norway; or, besides the epidote, they contain calcareous-spar, copper-pyrites, and variegated copper-ore, as in the Bannat and other places. The veins of which it forms a part are small, and of very old formation, usually traverse gneiss, and contain besides the epidote, felspar, rock-crystal, axinite, chlorite, asbestus, prehnite, octahedrite, and several other minerals. The varieties that occur in veins, are distinguished from those that occur in beds, by their lighter colours, and the more needle-shaped aspect of the crystals. The rocks in which it occurs are syenite, porphyry, and undefined granitous rocks.

Geographic Situation.

Europe.—In Arran it occurs in secondary syenite, and clay-slate: in Mainland in Shetland in syenite *. In the island of Icolmkill, in a rock composed of red felspar and quartz: in the island of Rona, also one of the Hebrides, in slender veins, traversing a rock composed of felspar and quartz, and felspar and hornblende: in the syenite of Glencoe and the neighbouring districts; in similar rocks among the Malvern Hills in Worcestershire; in quartz, at Wallow Crag near Keswick in Cumberland; near Marazion in Cornwall; and in granitous rocks in the islands of Guernsey and Jersey †. Upon the Continent of Europe it occurs in

* Hibbert.

† Geological Transactions, vol. i. p. 292.

in magnificent crystals at Arendal; and in porphyry near Christiania in Norway; also in Sweden; imbedded in rolled masses of a granitous rock in Mecklenburg; Bavaria, France, Italy, and Switzerland.

Asia.—Imbedded in granular limestone in Siberia; and in India along with corundum.

Africa.—Found imbedded in common quartz on the banks of the Orange River, by Dr Somerville.

America.—Upon the banks of Lake Champlaine, along with tremolite *; and in the mountains of South Carolina.

Observations.

1. *Distinctive Characters.*—*a.* Between epidote and *Actynolite.* The colour-suite of actynolite differs from that of epidote; in actynolite, the primitive figure has an angle of 124° 34′; but the primitive figure of epidote has an angle of 114° 37′; in actynolite, both the cleavages are distinctly seen; but, in epidote, frequently only one cleavage is to be seen; the crystals of actynolite are generally imbedded, and their terminal edges and angles truncated, whereas the crystals of epidote are frequently superimposed, and their extremities are bevelled or acuminated: actynolite is softer than epidote: and actynolite, before the blowpipe, melts into a greyish-white enamel; epidote into a black scoria.—*b.* Between epidote and *Asbestus.* Asbestus, when pounded, feels soft, whereas epidote feels rough; and asbestus fuses into an enamel, but epidote into a scoria.

2. Klaproth describes, under the title *Scorza*, a substance which probably belongs to this species. Its colour

* Greenough.

lour is intermediate between pistachio and siskin green: it occurs in fine, roundish, dull, and meagre grains, that scratch glass, and have the specific gravity of 3.135. It contains silica, 43.; alumina, 21.; lime, 14.; oxide of iron, 16.5; oxide of manganese, 0.25. It is found in small nests, in a grey-coloured clayey stone, in a valley near the town of Muska, on the river Aranyos in Transylvania. The Wallachian name for this substance, viz. Scorza, has been retained by Klaproth. It might be arranged as a subspecies of epidote, under the title *Arenaceous Epidote.* Karsten names it *Arenaceous Thallite.*

3. Hausmann, according to Steffens, describes a mineral said to belong to this species, under the name *Earthy Epidote (erdiger Epidot).* It has a pale siskin-green colour; occurs disseminated, and in membranes. Internally it is dull, and the fracture earthy; it is meagre to the feel, and soils. It occurs in granite, at Trolhatta in Sweden and, I believe, in the island of Rona, and other parts of the Highlands of Scotland.

4. Hausmann describes another mineral under the title *Capillary Epidote.* It is said to have a very dark pistachio-green colour, and to occur in very delicate capillary crystals, which have a lustre intermediate between silky and vitreous, and to incrust small drusy cavities, at Hackedal in Norway.

5. Karsten divides this species into three subspecies, viz. *Common, Splintery,* and *Arenaceous.* The arenaceous is the scorza already mentioned; the splintery includes all those varieties that have been described under the names Arendalite and Acanticone.

6. Epidote was first described by Romé de L'Isle as a variety of schorl, under the name Green Schorl. La Metherie afterwards gave an account of it, and describes it as

a

a new species, which he named Thallite. Saussure, who found it in the Alps, names it Delphinite; and other varieties found in Norway, were described by Dandrada, by the names Acanticone and Arendalite. Haüy and Werner, nearly about the same time, particularly examined this mineral. Haüy named it Epidote, and Werner Pistacite.

Haüy published a description of the species, which Werner has not done; therefore the name Epidote, given to it by Haüy, has been very generally adopted.

Second Subspecies.

Zoisite *.

THIS subspecies is divided into two kinds, viz. Common Zoisite and Friable Zoisite.

First Kind.

Common Zoisite.

Zoisite, *Werner*.

Zoisit, *Karsten*, in Klaproth, Beit. b. iv. s. 180.—Epidot, *Haüy*, Journ. des Mines, n. 113. p. 465. *Id. Haüy*, Tabl. p. 44.—Zoisit, *Steffens*, b. i. s. 74. *Id. Hoff.* b. ii. s. 665. *Id. Haus.* Handb. b. ii. s. 676. *Id. Aikin*, p. 223.

External Characters.

Its colours are yellowish-grey, and light bluish-grey, which approaches to smoke-grey. Sometimes also of a colour

* *Zoisite*, in honour of the discoverer Baron von Zois, an Austrian gentleman.

lour intermediate between yellowish-brown and reddish-brown.

It occurs massive; also in large and longish granular and thin straight prismatic distinct concretions.

It occurs crystallized in very oblique four-sided prisms, in which the obtuse lateral edges are often rounded, so that the crystals have a reed-like form. Their surface is shining or glistening.

The crystals are middle-sized, and always imbedded.

Internally, it is shining on the cleavage, and glistening, on the fracture surface, and the lustre is resino-pearly.

The cleavage is double; but in general only one cleavage can be detected, which is parallel with the axis of the oblique prism.

The fracture is small-grained uneven.

The fragments are indeterminate angular, and sharp-edged.

It is feebly translucent, or only translucent on the edges.

It is as hard as epidote.

It is very easily frangible.

Specific gravity, 3.249, 3.290, *Breithaupt.*—3.315, *Klaproth.*

Chemical Characters.

Before the blowpipe it is affected nearly in the same manner as epidote.

Constituent Parts.

Silica, -	43.
Alumina, -	29.
Lime, -	21.
Oxide of Iron,	3.

Klaproth, Beit. b. iv. s. 183.

nostic

Geognostic and Geographic Situations.

It was first observed in the Saualp in Carinthia, where it occurs imbedded in a bed of quartz along with kyanite, garnet, and augite; or it takes the place of felspar, in a granular rock, composed of quartz and mica. It also occurs imbedded in a coarse granular granite from Thiersheim near Wunsiedel, in Bareuth in Franconia; and in Bavaria, Salzburg, the Tyrol, Carniola, and Switzerland. I have it from Glen Elg in Inverness-shire, and from Shetland, I believe the island of Unst.

Second Kind.

Friable Zoisite.

Murber Zoisit, *Karsten.*

Murber Zoisit, *Karsten*, Magazin de Berlin, Geselch. 2. Jahrg. 3. quart. 1808, s. 187. *Id. Steffens*, b. i. s. 76. *Id. Klaproth*, Beit. b. v. s. 41.

External Characters.

Its colour is reddish-white, which is spotted with pale peach-blossom red.

It is massive, and in very fine loosely aggregated granular concretions.

It is very feebly glimmering.

The fracture is intermediate between earthy and splintery.

The fragments are not very sharp-edged.

It

It is translucent on the edges.

It is semi-hard.

It is brittle.

It is rather heavy.

Specific gravity 3.300, *Klaproth.*

Constituent Parts.

Silica,	44.
Alumina, -	32.
Lime, - -	20.
Oxide of Iron, - -	2.50
	98.50

Klaproth, Beit. b. v. s. 43.

Geognostic and Geographic Situations.

It occurs imbedded in green talk at Radelgraben in Carinthia.

4. Prismatic Augite, or Tabular Spar.

Prismatischer Augitspath, *Mohs.*

Schaalstein, *Werner.*

Tafelspath, *Karsten.*

Tafelspath, *Reuss*, b. ii. s. 435. *Id. Lud.* b. ii. s. 144. *Id. Suck.* 1r th. s. 422.—Schaalstein, *Bert.* s. 166. *Id. Mohs*, b. ii. s. 1.–3. —Tafelspath, *Leonhard*, Tabel. s. 35. *Id. Karsten*, Tabel. s. 44.—Spath en tables, *Haüy*, Tabl. p. 66.—Schaalstein, *Lenz*, b. ii. s. 763.—Spathiger Conit, *Oken*, b. i. s. 392.— Schaalstein,

Schaalstein, *Hoff.* b. iii. s. 55.—Tafelspath, *Haus.* Handb. b. ii. s. 582.—Tabular Spar, *Aikin*, p. 183.

External Characters.

Its most common colour is greyish-white, which passes into greenish and yellowish white, and reddish-white.

It occurs massive, and coarsely disseminated; also in distinct concretions, which are coarse, long, and broad angulo-granular, and these are again composed of others which are thin and straight lamellar.

Internally the lustre varies from shining to glistening, and is pearly, inclining to vitreous.

The cleavage is double, and the folia are in the direction of an oblique prism of about 105°,—*Mohs.*

The fracture is splintery.

It is translucent.

It is harder than fluor-spar, but not so hard as apatite.

It is brittle, and easily frangible.

Specific gravity, 3.2—3.5, *Mohs.*

Constituent Parts.

Silica, - -	50
Lime,	45
Water,	5
	100

Klaproth, Beit. b. iii. s. 291.

Geognostic and Geographic Situations.

Europe.—It occurs in primitive rocks at Orawicza in the Bannat of Temeswar, where it is associated with brown garnets,

garnets, blue-coloured calcareous-spar, tremolite, actynolite, and variegated copper-ore.

Asia.—It has been lately discovered in the Island of Ceylon, associated with cinnamon-stone in gneiss.

GENUS IX.—SCHILLER-SPAR.

Schiller Spath, *Mohs.*

THIS Genus contains four Species, viz. 1. Green Diallage, 2. Schiller-Spar, 3. Hyperstene, 4. Anthophyllite.

1. Green Diallage.

Prismatische Schiller Spath, *Mohs.*

Diallage Verte, *Haüy.*

Körniger Strahlstein, *Werner.*

External Characters.

Its colours are grass-green, which sometimes inclines to emerald-green, or to mountain-green.

It occurs massive and disseminated.

Internally it is shining, glistening and pearly.

It

It has an imperfect double cleavage; one only of the cleavages is visible.

Its fragments are indeterminate angular, and rather sharp-edged.

It is translucent on the edges, sometimes passing into translucent.

Some varieties are harder than fluor-spar, and others harder than apatite, but none so hard as felspar.

It is brittle.

Specific gravity 3.0, 3.2., *Mohs.*

Chemical Characters.

It melts before the blowpipe into a grey or greenish enamel.

Constituent Parts.

Silica, - -	50.0
Alumina, - -	11.0
Magnesia,	6.0
Lime, -	13.0
Oxide of Iron,	5.3
Oxide of Copper, -	1.5
Oxide of Chrome,	7.5
	94.3

Vauquelin, An. d. Chimie, No. 88.

Geognostic and Geographic Situations.

Europe.—It occurs in the Island of Corsica, along with Saussurite; and with the same mineral in Mont Rosa in Switzerland, and at La Rivera, in the Valley of Susa in Piedmont.

Asia.—In India, along with quartz and rutile.

America.—In Labrador, associated with Saussurite.

Uses.

Uses.

The compound of green diallage and saussurite, named *Gabbro* by the Italians, *Euphotide*, by the French, and by artists *Verde di Corsica duro*, when cut and polished has a beautiful appearance, and is much prized as an ornamental stone. It is cut into snuff-boxes, ring-stones, for in-laid work, and other similar purposes.

Observations.

1. It has been confounded with Hornblende and Felspar; but it is distinguished from the first by its pearly lustre, single distinct cleavage, and inferior hardness; from the latter, by its inferior hardness, and by its cleavage; felspar having always a distinct double cleavage, whereas in this mineral there is but one distinct cleavage.

2. This mineral was first discovered by Saussure, who named it from its colour, Smaragdite, and Emeraudite, which other mineralogists changed into Emerald-Spar, and Prime d'Emeraude. It has been described as a felspar, and also as a variety of hornblende. Haüy remarks, that as the minerals with which this substance had been confounded, have at least two distinct cleavages, whereas it has but one, he chose a name which would recall this difference; hence the origin of the name Diallage, which signifies difference.

2. Schiller-Spar.

Schiefer Schiller-Spath, *Mohs.*

This Species contains two subspecies, viz. Bronzite and Common Schiller-spar.

First

First Subspecies.

Bronzite.

Blättriger Anthophyllit, *Werner.*

Diallage metalloide fibro-laminaire, *Haüy.*

Bronzit, *Leonhard,* Tabel. s. 29.—Diallage metalloide, *Brong.* t. ii. p. 443.—Bronzit, *Karsten,* Tabel. s. 40.—Diallage metalloide fibro-laminaire, *Haüy,* Tabl. p. 47.—Bronzit, *Steffens,* b. i. s. 325. *Id. Lenz,* b. ii. s. 663.—Blättriger Anthophyllit, *Hoff.* b. ii. s. 676.—Bronzit, *Haus.* Handb. b. ii. s. 717. Bronzite, *Aikin,* p. 230.

External Characters.

Its colour is intermediate between clove-brown, yellowish-brown and pinchbeck-brown; it occurs also yellowish-grey.

It occurs massive, and in coarse and small granular distinct concretions.

Internally it is shining, and the lustre is metallic-pearly.

It has a double slightly oblique cleavage: one of the cleavages is very distinct, the other indistinct: they belong to a prism, in which one of the angles appears to be about 100°. The folia are curved, and their surface streaked. Sometimes the folia appear fibrous.

The fragments are indeterminate angular and blunt-edged.

It is translucent on the edges, sometimes approaching to translucent.

It is harder than fluor-spar, but not so hard as apatite.

It affords a white streak.

It

It is difficultly frangible.

Specific gravity, 3.200, *Klaproth*.—3.213—3.281, *Blöde*. 3.271, *Breithaupt*.—3.0, 3.3, *Mohs*.

Chemical Characters.

It is infusible before the blowpipe.

Constituent Parts.

Silica, - -	60.00
Magnesia, - -	27.50
Iron, - - -	10.50
Water, - -	0.50
	98.50

Klaproth, Beit. b. v. s. 34.

Geognostic and Geographic Situations.

It occurs in greenstone in the Island of Skye: in large masses in a bed of serpentine near Kraubat in Upper Stiria; at Kupferberg in Bareuth, in small globular masses, sometimes associated with asbestus, and disseminated magnetic ironstone, in serpentine; in small masses in serpentine near Peinach, on the Pacher Alp in Lower Stiria; and in the vicinity of Hoff in Franconia.

America.—In the island of Cuba.

Observations.

1. Its dark-brown colours, metallic-pearly lustre, distinct single and curved fibro-laminar cleavage, and granular concretions, are its distinguishing characters.

2. It is distinguished from *Common Schiller-spar* by its curved and fibro-laminar cleavage, greater hardness, and brittleness.

Second

Second Subspecies.

Common Schiller-Spar.

Schillerstein, *Werner.*

Diallage metalloide laminaire, *Haüy.*

Schillerspath, ou Spath chatoyant, *Broch.* t. i. p. 421.—Schillerende Hornblende, *Reuss,* b. ii. 1. s. 153.—Schillerstein, *Lud.* b. i. s. 134. *Id. Suck.* 1r th. s. 134. *Id. Bert.* s. 532. *Id. Mohs,* b. i. s. 557. *Id. Hab.* s. 30. *Id. Leonhard,* Tabel. s. 28.—Diallage chatoyant, *Brong.* t. i. p. 442.—Smaragdit, *Karsten,* Tabel. s. 40.—Schillerende Hornblende, *Haus.* Nordeutsche Beit. b. i. s. 1.—Diallage metalloide, *Haüy,* Tabl. p. 47.—Schillerstein, *Steffens,* b. i. s. 371. *Id. Lenz,* b. ii. s. 661. *Id. Hoff.* b. ii. s. 264. *Id. Haus.* Handb. b. ii. s. 715.

External Characters.

Its colours are olive-green, which passes on the one side into mountain-green and greenish-grey, on the other into yellowish-brown and pinchbeck-brown.

It seldom occurs massive, generally disseminated, and sometimes in granular distinct concretions.

Internally it is shining and splendent, and the lustre is pearly, or metallic-pearly.

It has a distinct straight single cleavage.

The fragments are indeterminate angular or tabular.

It is faintly translucent on the edges, or is opaque.

It is softer than bronzite.

The streak is greenish-grey, and dull.

It is easily frangible, and slightly inclining to sectile.

Specific gravity, 2.882?

Geognostic and Geographic Situations.

It occurs imbedded in serpentine in Fetlar and Unst in Shetland, and at Portsoy in Banffshire; in the greenstone rocks of the island of Skye; also in the greenstone rocks of Fifeshire; in the porphyritic rock of the Calton Hill, and the trap-rocks of Craig Lockhart, near Edinburgh; in similar rocks near Dunbarton; in serpentine at Cortachie in Forfarshire; and in the same rock between Ballantrae and Girvan in Ayrshire *. In Cornwall it occurs in serpentine and hornblende-slate. At Basta in the Hartz, it is found in primitive greenstone, which rests on granite, associated with compact felspar, pinchbeck-brown mica, amianthus, mountain-cork, precious serpentine, steatite, copper-pyrites, and iron-pyrites. Also disseminated in the serpentine of Zöblitz in Saxony, of Gastein in Salzburg, and of the Pinzgau in the Tyrol.

Observations.

It is distinguished from *Bronzite* by its green colours, straight cleavage, and inferior hardness.

3. Hyperstene, or Labrador Schiller-Spar.

Labradorische Schiller-Spath, *Mohs.*

Hyperstene, *Haüy.*

Labrador Hornblende, *Kirw.* vol. i. p. 221.—Diallage metalloide, *Haüy*, t. iii. p. 127.—Labradorische Hornblende, *Leonhard*, Tabel. s. 25.—Hyperstene, *Brong.* t. i. p. 444. *Id. Karsten*, Tabel.

* Allan.

Tabel. s. 40. *Id. Haus.* s. 98. *Id. Haüy,* Tabl. p. 44. *Id. Steffens,* b. i. s. 322. *Id. Lenz,* b. ii. s. 664. *Id. Oken,* b. i. s. 328.—Paulit, *Hoff.* b. ii. s. 143.—Hypersthen, *Haus.* Handb. b. ii. s. 718. *Id. Aikin,* p. 230.

External Characters.

Its colour is intermediate between greyish and greenish black, but it is nearly copper-red on the cleavage, and brownish-black, or blackish-brown on the fracture surface.

It occurs massive, disseminated, also in thin curved lamellar concretions, which are collected into coarse granular.

On the cleavage the lustre is shining and glistening, and is metallic-pearly, but on the fracture it is glimmering and pearly.

It has a double oblique angular cleavage, the folia meeting under angles of about 100°, and 80°; but of these cleavages one only is distinct; there is a third, but indistinct cleavage, in the direction of the shorter diagonal of the terminal plane of an oblique four-sided prism; and all the cleavages are frequently more or less curved.

The fragments are indeterminate angular or rhomboidal.

It is opaque, or feebly translucent on the edges.

It is greenish-grey in the streak.

It is as hard as felspar.

It is brittle, and rather easily frangible.

Specific gravity, 3.390, *Klaproth.*—3.376, *Karsten.*—3.3, 3.4, *Mohs.*

Chemical Character.

It is infusible before the blowpipe.

Constituent Parts.

Silica,	54.25
Magnesia, -	14.00
Alumina, - -	2.25
Lime, -	1.50
Oxide of Iron,	24.50
Water, - -	1.00
Oxide of Manganese, a trace.	
	97.50

Klaproth, Beit. b. v. s. 49.

Geognostic and Geographic Situations.

It was first discovered on the coast of Labrador, where t occurs as a constituent part of a rock composed of Laorador felspar, and sometimes also of common hornblende and magnetic ironstone. Giesecké found it in granitous rocks in Greenland; MacCulloch detected it forming a constituent part of a mountain rock at Loch Scavig in the sland of Skye; and in greenstone, near Portsoy.

Uses.

When cut and polished, it has a beautiful copper-red colour, and metallic pearly lustre, and is made into ringstones and brooches.

Observations.

Observations.

1. Hyperstene was originally described as a variety of hornblende, under the name Labrador Hornblende; but it is distinguished from *Hornblende* by its metallic-pearly luster and cleavage.

2. This mineral, although nearly allied to Anthophyllite, differs from it in being harder, heavier, its cleavage less distinct, and its lustre more metallic.

4. Anthophyllite *.

Gerader Schiller-Spath, *Mohs.*

Anthophyllith, *Schumacher.*

Strahliger Anthophyllit, *Werner.*

Anthophyllith, *Schumacher*, Verzeichniss, s. 96. *Id. Leonhard*, Tabel. s. 42. *Id. Brong.* t. i. p. 444. *Id. Karst.* Tabel. s. 82. *Id. Haus.* s. 92. *Id. Haüy*, Tabl. p. 58. *Id. Steffens*, b. i. s. 324. *Id. Lenz*, b. i. s. 527.—Strahliger Anthophyllit, *Hoff.* b. i. s. 673.—Anthophyllit, *Haus.* Handb. b. ii. s. 720. *Id. Aikin*, p. 223.

External Characters.

Its colour is intermediate between dark yellowish-grey and clove-brown.

It generally occurs massive; also in narrow or broad prismatic distinct concretions, which are scopiform or promiscuous, and in which the surface is streaked.

It

* This mineral is named *Anthophyllite*, on account of the similarity of its colour with that of the anthophyllum.

It is rarely crystallized in reed-like very oblique four-sided prisms.

The surface of the crystals is longitudinally streaked.

The lustre is shining and glistening, and metallic-pearly.

It has a fourfold cleavage; two of the cleavages, and these are the most distinct, are parallel with the sides of an oblique four-sided prism, in which one of the angles is about 100°; the other two cleavages are parallel with the diagonals of the prism.

The fragments are wedge-shaped and splintery; and sometimes rhomboidal.

It is translucent on the edges, or translucent.

It is as hard as felspar.

Specific gravity, 3.3, 3.4, *Mohs.*—3.285, *Haüy.*

Chemical Characters.

It becomes dark greenish-black before the blowpipe, but is infusible.

Constituent Parts.

Silica,	56.00
Alumina, - -	13.30
Magnesia,	14.00
Lime,	3.33
Iron, -	6.00
Oxide of Manganese,	3.00
Water, -	1.43

John, Chem. Untersuchungen, 1. s. 200, 201.

Geognostic

Geognostic and Geographic Situations.

Europe.—It occurs in beds in mica-slate, at Kongsberg in Norway, along with common hornblende, mica, and asbestous-tremolite; at Modum cobalt mines, also in Norway, along with common hornblende, cobalt-glance, and copper-pyrites.

America.—In mica-slate, and along with garnets in Greenland.

Observations.

It is named *Anthophyllite*, from the resemblance of its colour to that of the Anthophyllum. This name was given to it by Schumacher, the naturalist who first described it.

ORDER *III.*

ORDER III. MICA.

THIS order contains seven genera, viz. 1. Copper-mica, 2. Uranite, 3. Red Cobalt, 4. White Antimony, 5. Blue Iron, 6. Graphite, 7. Mica.

GENUS I. COPPER-MICA.

This genus contains one species, viz. Prismatic Copper Mica.

1. Prismatic Copper-Mica.

Prismatischer Kupferglimmer, *Mohs.*

Kupferglimmer, *Werner.*

Blättriges Olivenerz, *Karsten,* Journ. de Phys. an 10. p. 348.—Arseniate of Copper in hexaedral laminæ, with inclined sides, *Bournon,* Phil. Trans. part i. 1801.—Blättriches Olivenerz, *Reuss,* b. iii. s. 504.—Kupferglimmer, *Mohs,* b. iii. s. 294.—Cuivre arseniaté lamelliforme, *Brong.* t. ii. p. 230.—Kupferglimmer, *Karsten,* Tabel. s. 64.—Cuivre arseniaté lamelliforme, *Haüy,* Tabl. p. 90.—Kupferglimmer, *Hoff.* b. iii. s. 162. *Id. Haus.* Handb. b. iii. s. 1043.—Hexahedral Arseniate of Copper, *Aikin,* p. 93.

External Characters.

Its colour is emerald-green, which in some varieties inclines to verdigris-green.

It

It occurs massive, disseminated, and in granular distinct concretions; seldom crystallized in very thin equiangular six-sided tables, in which the alternate terminal planes are set on obliquely.

Externally it is smooth and splendent.

Internally it is splendent, and the lustre is pearly.

It has a distinct single cleavage, which is parallel with the sides of the table, or with the sides of a prism, when the table is viewed as a short prism.

The fracture is small-grained uneven, inclining to conchoidal.

The fragments are indeterminate angular and tabular.

The massive varieties are translucent; the crystallized transparent.

It scratches gypsum slightly, but does not affect calcareous-spar.

Its streak is green.

It is sectile.

It is rather brittle.

Specific gravity, 2.548, *Bournon.*—2.5, 2.6, *Mohs.*

Chemical Characters.

It decrepitates before the blowpipe; and passes, first, to the state of a black spongy scoria, after which it melts into a black globule, of a slightly vitreous appearance.

Constituent Parts.

Oxide of Copper, -	39	58
Arsenic Acid,	43	21
Water, -	17	21
	99	100
	Vauquelin, Journ. des Mines, N. 55. p. 562.	*Chenevix*, Phil. Tr. for 1801, p. 201.

Geognostic

Geognostic and Geographic Situations.

It has been hitherto found only in veins in the copper-mines in Cornwall, where it is accompanied with red copper-ore, copper-pyrites, copper-glance or vitreous copper-ore, variegated copper-ore, copper-black or black oxide of copper, compact and fibrous malachite, ironshot copper-green, azure copper-ore, indurated tile-ore, oliven-ore, and brown iron-ochre.

Observations.

1. This mineral is distinguished by its colour, crystallization, cleavage, softness, and sectility.

2. It is distinguished from *Foliated Talc* by its colour, cleavage, want of flexibility, and its greasy feel. It is also nearly allied to *Mica* in external appearance, but is readily distinguished from it by colour, and want of flexibility; and its cleavage and form distinguish it from *Malachite.*

GENUS II. URANITE OR URAN-MICA.

This genus contains one species, viz. Pyramidal Uranite.

1. Pyramidal

1. Pyramidal Uranite.

Pyramidaler Uran-Glimmer, *Mohs.*

Uran-Glimmer, *Werner.*

Chalkolith, *Wern.* Pabst. b. i. s. 290.—Grün Uranerz, *Wid.* s. 990.—Micaceous uranitic Ore, *Kirw.* vol. i. p. 304.—Grün Uranerz, *Emm.* b. ii. s. 584.—Oxide d'Uranit avec Cuivre, *Lam.* t. i. p. 410.—Urane oxidé, *Haüy,* t. iv. p. 283.—L'Urane micacé, *Broch.* t. ii. p. 463.—Uranglimmer, *Reuss,* b. iv. s. 556. *Id. Lud.* b. i. s. 308. *Id. Suck.* 2ter th. s. 469. *Id. Bert.* s. 511. *Id. Mohs,* b. iii. s. 721.—Uran oxydé, *Lucas,* p. 177.—Uranglimmer, *Leonhard,* Tabel. s. 81.—Uran oxidé micacé, *Brong.* t. ii. p. 103. *Id. Brard,* p. 379.—Uranglimmer, *Karsten,* Tabel. s. 74.—Uran oxydé, *Haüy,* Tabl. p. 115.—Micaceous Uranite, *Kid,* vol. ii. p. 221.—Uran-oxyd, *Haus.* Handb. b. i. s. 327. Uranglimmer, *Hoff.* b. iv. s. 275.—Uranite, *Aikin,* p. 138.

External Characters.

Its chief colour is grass-green, which passes on the one side into apple-green and emerald-green, and on the other into siskin-green and sulphur-yellow.

It is very seldom massive, sometimes in flakes; the massive varieties are disposed in angulo-granular concretions. It is frequently crystallized.

Its primitive form is a pyramid, in which the angles are 95° 13′, and 144° 56′. The secondary forms are the following:

1. Rectangular four-sided table, or short prism. This is sometimes elongated.
2. The four-sided table bevelled on the terminal planes, and the bevelling planes set on the lateral planes.

3. The

3. The terminal edges of the table truncated, thus forming an eight-sided table. This table is sometimes elongated.
4. The terminal planes of the four-sided table bevelled; and sometimes the edges of the bevelment truncated.
5. When the bevelling planes of N° 4. increase very much in size, there is formed a very acute double four-sided pyramid, in which the apices are more or less deeply truncated.
6. Sometimes the figure N° 4. is acuminated on both extremities with four planes, which are set on the lateral planes, and the apices of the acuminations deeply truncated.

The crystals are small and very small, superimposed, and form druses.

The terminal planes of the table are streaked, but the lateral planes are smooth.

Externally it is usually shining, and sometimes splendent.

Internally it is shining, approaching to glistening, and the lustre is pearly.

Its cleavage is fourfold and rectangular; of these one only is very distinct, and is that parallel with the base of the prism, or lateral planes of the table.

It is transparent and translucent.

It scratches gypsum, but not calcareous-spar.

Its streak is green.

It is sectile.

It is not flexible.

It is easily frangible.

Specific gravity, 3.121, *Champeaux*.—3.3, *Gregor*.—3.1, 3.2, *Mohs*.

Chemical

Chemical Characters.

It decrepitates violently before the blowpipe on charcoal; loses about 33 *per cent.* by ignition, and acquires a brass-yellow colour; with borax it yields a yellowish-green glass; it dissolves in nitric acid without effervescence, and communicates to it a lemon-yellow colour..

Constituent Parts.

	Cornwall.
Oxide of Uranium, with a trace of Oxide of Lead, - -	74.4
Oxide of Copper,	8.2
Water, -	15.4
Loss,	2.
	100

Gregor, in Annals of Phil. vol. v. p. 284.

Geognostic and Geographic Situations.

It occurs in veins in primitive rocks. In Cornwall in tinstone and copper veins that traverse granite and clay-slate. In Saxony, partly in silver-veins, along with pitch-ore, and partly in red ironstone, and tin veins. It is also found in the district of Autun in France; at Welsendorf in Bavaria, with fluor-spar; and at Bodenmais, also in Bavaria, along with beryl and felspar *.

Observations.

1. This mineral is nearly allied to copper-mica; but is distinguished from it by its pyramidal crystallization, and the

* Heuland.

the same character distinguishes it from *Chlorite*, *Talc*, and *Mica*.

2. When first discovered, it was described under the name *Green* Mica; afterwards it was named *Chalcolite*, on account of its supposed cupreous nature; and some authors have named it *Uranium Spar*, and *Torberit*, in Honour of Sir Torbern Bergmann.

Werner describes a soft mineral, found along with Uranite, under the name Uran-Ochre. It does not appear to form a distinct species, nor can it be considered as a subspecies of Uranite. It is here placed immediately after Uranite.

* Uran-Ochre.

Uran-Ocker, *Werner*.

There are two kinds of this mineral, viz. Friable and Indurated.

1. Friable Uran-Ochre.

Zerreibliche Uranocker, *Werner*.

Uran oxydé pulverulent, *Haüy*, t. iv. p. 285.—Uranocher, *Reuss*, b. iv. s. 561. *Id. Leonhard*, Tabel. s. 81. *Id. Karsten*, Tabel. s. 74.—Zerreibliche Uranocker, *Hoff*. b. iv. s. 280.

External Characters.

Its colour is lemon-yellow, which passes into straw-yellow and sulphur-yellow, and also into orange yellow.

It

It occurs usually as a coating or efflorescence on pitch-ore, sometimes small reniform.

It is friable, and composed of dull, dusty, and weakly cohering particles.

It feels meagre.

Geognostic Situation.

It occurs always on pitch-ore.

2. Indurated Uran-Ochre.

Feste Uranocker, *Werner.*

Verhärtete Uranocher, *Karsten,* Tabel. s. 74.—Feste Uranocker, *Hoff.* b. iv. s. 279.

External Characters.

Its colours are straw-yellow, lemon-yellow, and orange-yellow; and this latter passes into aurora-red and hyacinth-red, and into reddish and yellowish brown.

It occurs massive, disseminated, and superimposed; and sometimes there is a tendency to fibrous concretions.

Internally it is glimmering, and glistening and resinous.

The fracture is imperfect conchoidal.

It is opaque.

It is soft and very soft.

It is rather sectile.

Specific gravity, 3.1500, *La Metherie.*—3.2438, *Haüy.*

Chemical Characters.

According to Klaproth, the yellow varieties are pure oxide of uranium, but the brownish and reddish contain also a little iron.

Geognostic

Geognostic and Geographic Situations.

It is found at Joachimsthal, and Gottesgab in Bohemia, and at Johanngeorgenstadt in Saxony.

GENUS III.—RED COBALT.

Kobalt Glimmer, *Mohs.*

This Genus contains one Species, viz. Prismatic Red Cobalt.

1. Prismatic Red Cobalt.

Prismatischer Kobalt Glimmer, *Mohs.*

THIS Species is divided into three Subspecies, viz. Radiated Red Cobalt, Earthy Red Cobalt, and Slaggy Red Cobalt.

First Subspecies.

Radiated Red Cobalt, or Cobalt-Bloom.

Kobaltblüthe, *Werner.*

Flos Cobalti, *Wall.* Syst. Min. t. ii. p. 181.—Koboldblüthe, *Wern.* Pabst. b. i. s. 206. *Id. Wid.* s. 939.—Cobaltic Germinations, Flowers of Cobalt, of some, *Kirw.* vol. ii. p. 278.—Koboldblüthe, *Emm.* b. ii. s. 507.—Le Fleurs de Cobalt, ou Cobalt terreux rayonné rouge, *Broch.* t. ii. p. 403.—Cobalt arseniaté

arseniaté aciculaire, *Haüy*, t. iv. p. 217.—Strahlicher rother Erdkobold, *Reuss*, b. iv. s. 420.—Kobaltblüthe, *Lud.* b. i. s. 288. *Id. Mohs*, b. iii. s. 672.—Strahliger rother Erdkobold, *Leonhard*, Tabel. s. 77.—Cobalt arseniaté aciculaire, *Brong.* t. ii. p. 119.—Strahlige Kobaltblüthe, *Karsten*, Tabel. s. 72. —Cobalt arseniaté aciculaire, *Haüy*, Tabl. p. 108.—Strahlige Kobaltblüthe, *Haus.* Handb. b. iii. s. 1125. *Id. Hoff.* b. iv. s. 203.—Red Cobalt, *Aikin*, p. 130.

External Characters.

Its principal colour is crimson-red, which passes on the one side into peach-blossom, on the other into columbine-red; it is rarely greenish-grey, and olive-green.

It occurs massive disseminated, often in membranes, small reniform, small botryoidal; also in stellular and scopiform radiated or fibrous concretions, which are sometimes collected into granular concretions.

It also occurs crystallized. Its primitive form is not known. The following are the only crystallizations hitherto met with.

1. Rectangular four-sided prism.
2. Compressed acute double six-sided pyramid?

The crystals are generally acicular or capillary, and are scopiformly or stellularly aggregated.

Externally it is shining, passing into splendent.

Internally it is shining and glistening, and the lustre is pearly.

A single cleavage is observable, and in the direction of the axis of the prism.

The fragments are splintery and wedge-shaped.

It is more or less translucent; sometimes translucent on the edges.

Its colour is not changed in the streak.

It is harder than gypsum, but softer than calcareous spar.

It is rather sectile.

It is easily frangible.

Specific gravity 4.0, 4.3, *Mohs*.

Chemical Characters.

Before the blowpipe it becomes grey, and emits an arsenical odour, and tinges borax glass blue.

Constituent Parts.

Cobalt, -	39
Arsenic Acid, - -	38
Water,	23
	100

Buckholz. in J. d. Min. t. 25. p. 158.

Geognostic Situation.

It occurs in veins, in primitive, transition, and secondary rocks, along with silver-white cobalt, tin-white cobalt, grey cobalt, and other cobaltic minerals; also with copper-nickel, nickel-ochre, copper-pyrites, grey copper-ore, blue copper-ore, iron-shot copper-green, native bismuth, brown ironstone, galena or lead-glance, and blende; the vein-stones are heavy-spar, calcareous-spar, brown-spar, ironstone, and quartz.

Geographic Situation.

It occurs in veins in secondary rocks at Alva, in Stirlingshire; in limestone of the coal formation in Linlithgowshire; formerly in small veins in sandstone of the coal formation,

mation, along with galena and blende, at Broughton, in Edinburgh; in the Clifton lead-mines near Tyndrum, already described; and at Dolcoath in Cornwall. On the Continent, it is met with at Modum in Norway, Riegelsdorf in Hessia; Schneeberg, Annaberg, and Saalfeldt in Saxony; Kupferberg in Silesia; Wittichen in Furstemberg, and Alpersbach in Wurtemberg; Allemont in France; and in Salzburg and Hungary.

Observations.

A mixture of red cobalt, black cobalt, with ochre of nickel and native silver, occurs in the mines of Allemont, and of Schemnitz in Hungary, and is known to the miners by the name of *Goose-Dung Ore*. It is the *minera argenti mollior diversicolor*, Wall. t. 2. p. 346.; *Mine d'argent merde d'oie*, Delisle, t. 3. p. 150.; *Cobalt merde d'oie*, Brong.; *Cobalt arseniaté terreux argentifère*, Lucas; and the *Ganskothigsilber* and *Ganiskothigererz* of the Germans. Some other mixtures of silver-ores have received the same name. This is the case with a mixture of native arsenic, red silver-ore, and earthy silver-glance; and slaggy yellow orpiment, is named *Goose-Dung Ore* in the Hartz.

Second Subspecies.

Earthy Red Cobalt, or Cobalt-Crust.

Koboldbeschlag, *Werner.*

Ochra Cobalti rubra, *Wall.* Syst. Min. t. ii. p. 181.—Koboldbeschlag, *Wid.* s. 938.—Cobalt Incrustations, *Kirw.* vol. ii. p. 279.—Koboldbeschlag, *Emm.* b. ii. s. 509.—Le Cobalt terreux rouge, pulverulent, *Broch.* t. ii. p. 405.—Cobalt arseniaté

niaté pulverulent, *Haüy*, t. iv. p. 218.—Erdiger rother Erdkobold, *Reuss*, b. iv. s. 419.—Koboldbeschlag, *Lud.* b. i. s. 287. *Id. Mohs*, b. iii. s. 671.—Cobalt arseniaté, *Lucas*, p. 161.—Erdiger rother Erdkobold, *Leonhard*, Tabel. s. 77.—Cobalt arseniaté pulverulent, *Brong.* t. ii. p. 119. *Id. Brard*, p. 357. Gemeine Koboldblüthe, *Karsten*, Tabel. s. 72.—Arseniate of Cobalt, *Kid*, vol. ii. p. 211.—Cobalt arseniaté pulverulent, *Haüy*, Tabl. p. 108.—Erdige Kobaltblüthe, *Haus.* Handb. b. iii. s. 1125.—Koboldbeschlag, *Hoff.* b. iv. s. 201.

External Characters.

Its colour is peach-blossom red, of different degrees of intensity, which sometimes inclines to crimson-red, sometimes verges on cochineal-red, and also passes into reddish-white.

It seldom occurs massive or disseminated, generally in velvety crusts, and also small reniform and botryoidal.

It is generally friable, and is composed of scaly and dusty particles, which are feebly glimmering or dull.

The massive varieties have a fine earthy fracture.

The fragments are indeterminate angular, and blunt-edged.

It is very easily frangible.

It is very soft, or friable.

It is sectile.

The streak is shining.

It does not soil.

Third

Third Subspecies.

Slaggy Red Cobalt.

Schlackige Kobaltblüthe, *Hausmann.*

Id. Haus. Syst. d. Unorgan. Natk. s. 140. *Id. Haus.* Handb. b. iii. s. 1126.

External Characters.

Its colours are muddy crimson-red, and dark hyacinth-red, which passes into chesnut-brown.

It occurs in thin crusts, and sometimes reniform.

Externally it is smooth.

The lustre is shining and resinous.

The fracture is conchoidal.

It is translucent.

It is soft, and brittle.

Geognostic and Geographic Situations.

It occurs in veins along with other cobaltic minerals, in the mine of Sophia at Wittichen in Furstemberg.

COBALT-OCHRE.

The Black, Brown and Yellow Cobalt Ochres, and other similar minerals, ought to be arranged together, and form a particular Order by themselves. In the mean time, I place them beside the Red Cobalt, on account of their being often associated with that mineral.

1. Black

*

1. Black Cobalt Ochre.

Schwarz Erdkobold, *Werner*.

It is distinguished into Earthy Black Cobalt-ochre, and Indurated Black Cobalt-ochre.

a. Earthy Black Cobalt-Ochre.

Schwarzer Kobold Mulm, *Werner*.

Cobalt oxide noire terreux, *Haüy*.

Wern. Pabst. b. i. s. 205.—Zerreiblicher schwarzer Erdkobold, *Wid.* s. 933.—Loose Black Cobalt-ochre, *Kirw.* vol. ii. p. 275.—Schwarzer Kobold Mulm, *Emm.* b. ii. s. 498.—Le Cobalt terreux noire friable, *Broch.* t. ii. p. 397.—Cobalt oxydé noire terreux, *Haüy*, t. iv. p. 215.—Zerreiblicher schwarzer Erdkobold, *Reuss*, b. iv. s. 411.—Schwarzer Kobaltmulm, *Lud.* b. i. s. 285.—Zerreiblicher schwarzer Erdkobold, *Leonhard*, Tabel. s. 76.—Cobalt oxydé terreux, *Brong.* t. ii. p. 118. —Lockere Kobaltschwärze, *Haus.* Handb. b. i. s. 332.—Koboldmulm, *Hoff.* b. iv. s. 192.

External Characters.

Its colour is intermediate between brownish and blackish brown.

It is friable, and composed of dull coarse particles, which soil very little.

The streak is shining.

It is meagre to the feel.

It is light.

Chemical

Chemical Characters.

Before the blowpipe, it yields a white arsenical vapour; and it colours borax blue.

b. Indurated Black Cobalt-Ochre.

Fester Schwarz Erdkobold, *Werner.*

Minera Cobalti scoriformis, *Wall.* Syst. Min. t. ii. p. 180.—Verhärteter schwarzer Erdkobold, *Wid.* s. 933.—Indurated Black Cobalt-ochre, *Kirw.* vol. ii. p. 275.—Verhärteter schwarzer Erdkobold, *Emm.* b. ii. s. 499.—Le Cobalt terreux noire endurci, *Broch.* t. ii. p. 397.—Cobalt oxydé noire, var. 1.–3. *Haüy,* t. iv. p. 215.—Verhärteter schwarzer Erd-cobalt, *Reuss,* b. iv. s. 413. *Id. Lud.* b. i. s. 286. *Id. Mohs,* b. iii. s. 665. *Id. Leonhard,* Tabel. s. 76.—Cobalt oxidé vitreux, *Brong.* t. ii. p. 118.—Feste Kobaltschwärze, *Haus.* Handb. b. i. s. 333. *Id. Hoff.* b. iv. s. 193.

External Characters.

Its colour is distinct bluish-black.

It occurs massive, disseminated, in crusts, small botryoidal, small reniform, fruticose, moss-like, stalactitic, corroded, specular, and with pyramidal impressions. Sometimes it occurs in thin and curved lamellar concretions.

The surface is feebly glimmering.

Internally it is dull, or very feebly glimmering.

The fracture is fine earthy, sometimes passing into conchoidal.

The fragments are indeterminate angular, and blunt-edged.

It is opaque.

The

The fracture is fine earthy, approaching to conchoidal in the large.

The fragments are indeterminate angular, and blunt-edged.

It is opaque.

The streak is shining and resinous.

It is very soft.

It is sectile.

It is very easily frangible.

It is light.

Chemical Characters.

Before the blowpipe it emits an arsenical odour, and communicates a blue colour to borax.

Constituent Parts.

It is considered to be a compound of Brown Ochre of Cobalt, Arsenic, and Oxide of Iron.

Geognostic Situation.

It appears to occur principally in secondary mountains, and is generally accompanied with red and black cobalt-ochre, ochry-brown ironstone, and lamellar heavy-spar.

Geographic Situation.

It is found at Kamsdorf and Saalfeld in Saxony; Alpirsbach in Wurtemberg; and in the valley of Gistain in Spain.

Use.

It is used for making smalt, but is not so valuable as the black cobalt.

Observations.

Observations.

It is distinguished from *Umber*, *Bole*, and other minerals of the same description, by its streak and softness.

3. Yellow Cobalt-Ochre.

Gelber Erdkobold, *Werner*.

Cobalt arseniaté terreux argentifere (?) *Haüy*.

Ochra Cobalti lutea et alba, *Wall.* t. ii. p. 183.—Gelber Erdkobold, *Wid.* s. 936.—Yellow Cobalt-ochre, *Kirw.* vol. ii. p. 277. Gelber Erdkobold, *Emm.* b. ii. s. 504.—Le Cobalt terreux jaune, *Broch.* t. ii. p. 401.—Gelber Erdkobold, *Reuss*, b. iv. s. 417. *Id. Lud.* b. i. s. 287. *Id. Suck.* 2ter th. s. 407. *Id. Bert.* s. 488. *Id. Leonhard*, Tabel. s. 76.—Cobalt oxidé jaune, *Brong.* t. ii. p. 118.—Erdkobalt, *Haus.* Handb. b. i. s. 334. (in part.)—Gelber Erdkobold, *Hoff.* b. iv. s. 199.

External Characters.

The colour is muddy straw-yellow, which in some varieties passes through light yellowish-grey into yellowish-white.

It occurs massive, disseminated, corroded, and incrusting.

It frequently appears rent in different directions.

Internally it is dull.

The fracture is fine earthy, in the small; conchoidal in the large.

The fragments are indeterminate angular, and blunt-edged.

The streak is shining.

It

It is soft, passing into friable.

It is sectile.

It is very easily frangible.

Specific gravity, 2.677, *Kirwan*, after having absorbed water.

Chemical Characters.

It emits an arsenical odour before the blowpipe, and colours borax blue. It appears to be the purest of the cobalt ochres. It generally contains a portion of silver.

Geognostic Situation.

It occurs in the same geognostic situation as the preceding, and is almost always associated with earthy red cobalt, and sometimes with radiated red cobalt, nickel-ochre, iron-shot copper-green, and azure copper-ore.

Geographic Situation.

It occurs at Saalfeld in Thuringia; Kupferberg in Silesia; Wittichen in Furstenberg, and Alpirsbach in Wurtemberg in Swabia; and Allemont in France.

Use.

It affords a better smalt than the preceding, and, owing to the silver it contains, in the countries where it occurs, is also valued as an ore of silver.

GENUS IV.

Genus IV. WHITE ANTIMONY.

Spiessglass-glimmer, *Mohs.*

This genus contains one species, viz. Prismatic White Antimony.

1. Prismatic White Antimony.

Prismatischer Spiessglass-glimmer, *Mohs.*

Weiss-spiesglaserz. *Werner.*

Id. Wern. Pabst. b. ii. s. 203. *Id. Wid.* s. 920.—Muriated Antimony, *Kirw.* vol. ii. p. 251.—Muriate d'Antimoine, *De Born,* t. ii. p. 147.—Weiss-spiesglaserz, *Emm.* b. ii. s. 480.—Antimoine muriatique, *Lam.* t. i. p. 348.—Antimoine oxydé, *Haüy,* t. iv. p. 273.—Antimoine blanc, *Broch.* t. ii. p. 381.—Weiss-spiesglanzerz, *Reuss,* b. iv. s. 382. *Id. Lud.* b. i. s. 281. *Id. Suck.* 2ter th. s. 392. *Id. Bert.* s. 470. *Id. Mohs,* b. iii. s. 710.—Antimoine oxydé, *Lucas,* p. 173.—Weiss-spiesglanzerz, *Leonhard,* Tabel. s. 70.—Antimoine oxydé, *Brong.* t. ii. p. 128. *Id. Brard,* p. 374.—Weiss-spiesglanz, *Karsten,* Tabel. s. 72.—Antimoine oxydé, *Haüy,* Tabl. p. 113.—Spiessglanzweiss, *Haus.* Handb. b. i. s. 341.—Weiss-spiessglanzerz, *Hoff.* b. iv. s. 119.—White Antimony, *Aikin,* p. 125.

External Characters.

Its colours are snow-white, greyish-white, light ash-grey, and yellowish-white, which latter colour is the most com-

It

It seldom occurs massive, more frequently disseminated, and in membranes; also in distinct concretions, which are coarse and small granular, and scopiform and stellular radiated. Is often crystallized. Its primitive form is a prism, but its dimensions are unknown. The following figures have been observed.

1. Rectangular four-sided prism, bevelled on the extremities.
2. Oblique four-sided prism.
3. Rectangular four-sided table.
4. Six-sided prism.
5. Acicular and capillary crystals.

The tables are small and very small, usually adhering by their lateral planes, and sometimes, although seldom, manipulárly aggregated, and often intersecting each other, in such a manner as to form cellular groups.

The crystals are sometimes smooth, sometimes feebly longitudinally streaked, and splendent.

Internally it is shining, and the lustre is intermediate between pearly and adamantine.

It has a cleavage in the direction of the lateral planes of the prism.

The fragments are indeterminate angular, or wedge-shaped.

It is translucent.

Its hardness is indeterminate, between that of talc and gypsum.

It is rather sectile.

Specific gravity, 5.0, 5.6, *Mohs.*

Chemical Characters.

Before the blowpipe it melts very easily, and is volatilised in the form of a white vapour.

Constituent

Constituent Parts.

	Allemont.
Oxide of Antimony, -	86
Oxides of Antimony and Iron,	3
Silica, - -	8
	98

Vauquelin, Haüy, t. iv. p. 274.

Geognostic and Geographic Situations.

It occurs in veins in primitive rocks, and is usually accompanied with the other ores of antimony.

At Prizbram in Bohemia, it occurs along with crystallized galena or lead-glance; and at Allemont, with native antimony, and grey and red antimony. It has also been found in Malaxa in Hungary.

Observations.

1. It is distinguished from *Calamine* by its inferior hardness; from *White Lead-spar* by its inferior hardness and crsytallization; from *Strontianite* and *Arragonite* by inferior hardness, and superior weight.

2. The foliated varieties are found only at Prizbram, the radiated chiefly in Hungary.

* Antimony-Ochre.

Spiesglanzocker, *Werner.*

Spiesglanzocher, *Reuss*, b. iv. s. 388. *Id. Lud.* b. i. s. 282. *Id. Suck.* 2ter th. s. 394. *Id. Bert.* s. 478. *Id. Mohs*, b. iii. s. 713. *Id. Leonhard*, Tabel. s. 79. *Id. Karsten*, Tabel. s. 72.—Antimoine

timoine oxydé terreux, *Haüy*, Tabl. p. 113.—Spiessglanzocher, *Haus.* Handb. b. i. s. 339. *Id. Hoff.* b. iv. s. 124.—Antimonial Ochre, *Aikin*, p. 125.

External Characters.

Its colour is straw-yellow, of different degrees of intensity, which inclines on the one side into yellowish-grey, on the other into yellowish-brown.

It scarcely occurs massive, and disseminated, generally incrusting crystals of grey antimony.

It is dull.

The fracture is earthy, and sometimes inclines to radiated.

It is opaque.

It is soft, passing into very soft.

It is brittle, and easily frangible.

Chemical Characters.

Before the blowpipe, on charcoal, it becomes white, and evaporates without melting. With borax, it intumesces, and is partly reduced to the metallic state.

Geognostic and Geographic Situations.

It occurs always in veins, and accompanied with grey antimony, and sometimes with red antimony.

It is found at Huel Boys in Endellion in Cornwall. At Dublowitz, near Saltschaw in Bohemia; Telkebanya in Hungary; Toplitz in Transylvania; Braunsdorf, in the kingdom of Saxony; on the Sonnenberg, near Mittersill in Salzburg; and in Siberia.

GENUS V.

GENUS V. BLUE IRON.

Eisen Glimmer, *Mohs.*

THIS genus contains one species, viz. Prismatic Blue Iron.

1. Prismatic Blue Iron.

This species is divided into three subspecies, viz. Foliated Blue Iron, Fibrous Blue Iron, and Earthy Blue Iron.

First Subspecies.

Foliated Blue Iron.

Blättriches Eisenblau, *Hausmann.*

Blättriches Eisenblau, *Uttinger,* Moll's Eph. b. iv. s. 71.—Fer phosphaté crystallisé, *Haüy,* Tabl. p. 99.—Fer phosphaté crystallisé ou laminaire, *Lucas,* t. ii. p. 413.—Blättriches Eisenblau, *Haus.* Hand. b. iii. s. 1075.—Kristallisirte Blaueisenerde, *Hoff.* b. iv. s. 144.

External Characters.

Its colour is dark indigo-blue, and sometimes bluish-grey; also leek-green, and inclining to sky-blue.

Its primitive form is an oblique four-sided prism, the dimensions of which are not known. The secondary forms are the following:

1. Broad rectangular four-sided prism, in which the lateral edges are truncated, (the truncating planes are set obliquely on the smaller lateral planes, and are the original

original planes of the oblique four-sided prism), flatly bevelled on the extremities; the bevelling planes set obliquely on the broader lateral planes.

2. Eight-sided prism, acuminated with four planes.

The crystals are sometimes acicular, and deeply longitudinally streaked. They are small, or middle-sized, and superimposed.

Externally it is shining or splendent.

Internally shining, passing into splendent, and pearly inclining to adamantine.

It has a perfect and straight single cleavage, which is parallel with the broader lateral planes of the prism.

The fragments are long tabular, or splintery.

It is translucent on the edges, or strongly translucent.

It is as hard as gypsum.

The colour is paler blue in the streak.

It is sectile, and easily frangible.

It is flexible in thin pieces.

Specific gravity, 2.70, *Breithaupt.*—2.80, 3.0, *Mohs.*

Constituent Parts.

	From the Isle of France.
Oxide of Iron, -	41.25
Phosphoric Acid, -	19.25
Water, -	31.25
Ironshot Silica,	1.25
Alumina, - -	5.00
	98

Fourcroy and *Laugier*, in Ann. du Mus. t. iii. p. 405.

Geognostic

Geognostic and Geographic Situations.

Europe.—It occurs in Whealkind Mine, in St Agnes's in Cornwall; along with iron-pyrites, and magnetic-pyrites, in gneiss, in the Silberberg, at Bodenmais, in Bavaria; and in the department of Allier in France.

Africa.—In the Isle of France.

America.—It is said to occur in drusy cavities in bog-iron-ore in New Jersey, United States; and it is mentioned as a Brazilian mineral.

Observations.

1. This mineral is described by Reuss as Kyanite *; and by Brauner as Foliated Gypsum †. Its true nature was first ascertained by Uttinger of Sonthofen, in a paper in Von Moll's Ephemeriden, already quoted.

2. In the fourth volume of Hoffman's Mineralogy, there is a description by Werner of a new mineral, under the name *Vivianite*, in compliment to Mr Vivian of Cornwall, and which appears to be but a variety of this species.

* Reuss, Lehrbuch der Mineralogie.

† Annalen der Berg und Hüttenkunde, b. iii. lif. 2. s. 296.

Second Subspecies.

Fibrous Blue Iron.

Fasriges Eisenblau, *Hausmann.*

Fasriges Eisenblau, *Haus.* Handb. b. iii. s. 1076.

External Characters.

Its colour is indigo-blue.

It occurs massive, and sometimes intimately connected with hornblende, and in roundish blunt angular pieces; also in delicate fibrous concretions, which are scopiform or promiscuous.

Internally it is glimmering and silky.

It is opaque.

It is soft.

Geognostic and Geographic Situations.

Europe.—It occurs in transition syenite at Stavern in Norway *.

America.—In West Greenland †.

Third

* Hausmann's Reise durch Scandinavien, b. ii. s. 109.

† Schumacher, Verz. s. 139.

Third Subspecies.

Earthy Blue Iron.

Blau Eisenerde, *Werner.*

Erdiges Eisenblau, *Hausmann.*

Coeruleum berolinense naturale, *Wall.* t. ii. p. 260.—Ocre martiale bleu; Bleu de Prusse natif, *Romé de Lisle,* t. iii. p. 295. —Prussiate de Fer natif, *De Born,* t. ii. p. 275.—Blaue Eisenerde, *Werner,* Pabst. b. i. s. 169. *Id. Wid.* s. 835.—Blue Martial Earth, *Kirw.* vol. ii. p. 185.—Blaue Eisenerde, *Emm.* b. ii. s. 359.—Prussiate de Fer natif, *Lam.* t. i. p. 247.—Fer azure, *Haüy,* t. iv. p. 119,–122.—Le Fer terreux bleu, *Broch.* t. ii. p. 288.—Blaue Eisenerde, *Reuss,* b. iv. s. 146. *Id. Lud.* b. i. s. 257. *Id. Mohs,* b. iii. s. 433. *Id. Leonhard,* Tabel. s. 68.—Fer phosphaté azure, *Brong.* t. ii. p. 179.—Blau Eisenerde, *Karsten,* Tabel. s. 66.—Erdiges Eisenblau, *Haus.* s. 138.—Phosphate of Iron; Native Prussian Blue, *Kid,* vol. ii. p. 189.—Fer phosphaté terreux, *Haüy,* Tabl. p. 99. —Blaue-eisenerde, *Hoff.* b. iv. s. 302.—Erdiges Eisenblau, *Haus.* Handb. b. iii. s. 1077.—Earthy Blue Iron-ore, *Aikin,* p. 105.

External Characters.

In its original repository it is said to be white, but afterwards becomes indigo-blue, of different degrees of intensity, which sometimes passes into smalt-blue.

It is usually friable, sometimes loose, and sometimes cohering.

It occurs massive, disseminated, and thinly coating.

Its particles are dull and dusty.

It soils slightly.

It

It feels fine and meagre.

It is rather light.

Chemical Characters.

Before the blowpipe, it immediately loses its blue colour, and becomes reddish-brown, and, lastly, melts into a brownish-black coloured slag, attractable by the magnet.

It communicates to glass of borax a brown colour, which at length becomes dark yellow. It dissolves rapidly in acids.

Constituent Parts.

		From Eckartsberg.
Oxide of Iron,	-	47.50
Phosphoric Acid,	-	32.00
Water,	-	20.00
		99.50

Klaproth, Beit. b. iv. s. 122.

Geognostic Situation.

It occurs in nests and beds in clay-beds, also disseminated in bog iron-ore, or incrusting turf and peat.

Geographic Situation.

Europe.—On the surface of peat-mosses in several of the Shetland Islands; and in river-mud at Toxteth, near Liverpool; Iceland; Helsingor on the Island of Seeland; Schonen in Sweden; Russia; Maschen in Hanover; Steinbach, Oberlichtenau, and Weissig in Upper Lusatia; Silesia; Suabia; Upper Palatinate; Bavaria; Carniola *; France.

Asia.

---

* [illegible], upon the subject of this mineral, in a letter to Dr Bruce, [illegible]—"That it occurs [illegible] of the Cimmerian Bosphorus, now called

Asia.—Borders of the Lake Baikal in Siberia.

America.—Along with bog iron-ore in alluvial soil in New Jersey *.

Uses.

It is sometimes used as a pigment. It is principally employed in water-colours, because, when mixed with oil, the colour is said to change into black †. Beautiful green and olive colours have been formed, by mixing it with other colours. It would appear that this mineral was known to the ancients; for a substance answering to blue iron-earth is mentioned by Pliny, as being collected in the marshes of Egypt, and ground and washed, and used as a pigment.

GENUS VI.

---

called the Straits of Taman, between the Sea of Azoph and the Black Sea. It lies there associated with extraneous fossil remains of animals, whose decomposition, it is conjectured, afforded phosphoric acid to the metal."—Bruce's Journal, p. 123.

* Cutbush, in Bruce's American Mineralogical Journal, p. 86.

† Mr Cutbush was informed, that a piece of this mineral, by grinding with oil, afforded a beautiful blue colour, which shows that the American variety is different from that used by painters in Europe.—Vid. Cutbush, in Bruce's Journal, p. 87, 88.

GENUS VI. GRAPHITE *.

Kohlen Glimmer, *Mohs.*

THIS genus contains one species, viz. Rhomboidal Graphite.

1. Rhomboidal Graphite.

Graphit, *Werner.*

Ferrum molybdena, *Wall.* t. ii. p. 249.—Plombagine, *Romé de Lisle,* t. ii. p. 500. *Id. De Born,* t. ii. p. 295.—Graphites plumbago, *Lin.* Syst. Nat. edit. 13. cura Jo. Frid. Gmelin, t. iii. p. 284.—Plumbago, *Kirw.* vol. ii. p. 58.—Graphit, *Emm.* b. ii. s. 97. *Id. Wid.* s. 651.—Graphite, *Broch.* t. ii. p. 76.—Fer carburé, *Haüy,* t. iv. p. 98.—Graphite, *Reuss,* b. iii. 3. s. 176. *Id. Lud.* b. i. s. 196. *Id. Suck.* 2ter th. s. 73. *Id. Bert.* s. 335: *Id. Mohs,* b. ii. s. 327. *Id. Leonhard,* Tabel. s. 50.—Plumbago, *Kid,* vol. ii. p. 58.—Plumbago, *Aikin,* p. 59.

This species is divided into two subspecies, viz. Scaly Graphite, and Compact Graphite.

First

* *Graphite,* from γραφω, *I write,* on account of its writing quality.

First Subspecies.

Scaly Graphite.

Schuppiger Graphit, *Werner.*

Graphite lamellaire, *Brong.* t. ii. p. 53.—Schuppiger Graphit, *Karst.* Tabel. s. 58. *Id. Haus.* s. 115.—Graphite granulaire, *Haüy*, Tabl. p. 70.—Schuppiger Graphit, *Lenz*, b. ii. s. 1084.—Blättriger Graphit, *Haus.* Handb. b. i. s. 67.—Schuppiger Graphit, *Hoff.* b. iii. s. 312.

External Characters.

Its colour is dark steel-grey, which approaches to light iron-black.

It occurs massive, disseminated; in coarse, small, and fine granular concretions; and crystallized.

Its primitive form appears to be a rhomboid, the dimensions of which are unknown. The only secondary form hitherto met with, is the equiangular six-sided table.

Internally it is shining, passing into splendent, and the lustre is metallic.

It has a distinct single cleavage, which is parallel with the terminal planes of the prism.

The fracture is scaly foliated.

The fragments are indeterminate angular, and blunt-edged.

The streak is shining, even splendent, and its lustre is metallic.

Its hardness is sometimes equal to that of gypsum.

It is perfectly sectile.

It is rather difficultly frangible.

It

It writes and soils.

Its streak is black.

It feels very greasy.

Specific gravity, (but uncertain whether of scaly or compact graphite),

Kirwan,	1.987	2.267
Brisson,	2.1500	2.456
Haüy,	2.0891	2.245
Mohs,	1.9	2.1

Second Subspecies.

Compact Graphite.

Dichter Graphit, *Werner.*

Graphite granuleux, *Brong.* t. ii. p. 54.—Dichter Graphit, *Karsten,* Tabel. s. 58. *Id. Haus.* s. 115. *Id. Lenz,* b. ii. s. 1085.

External Characters.

The colour is nearly the same with the preceding, only rather blacker.

It occurs massive, and disseminated; also in columnar concretions.

Internally it is glimmering, sometimes glistening, and the lustre is metallic.

The fracture is small and fine-grained uneven, which passes into even, and also into large and flat conchoidal; in the large it is sometimes slaty longitudinal.

The fragments are indeterminate angular, and blunt-edged, and sometimes also tabular.

In other characters it agrees with the preceding subspecies.

Chemical

Chemical Characters.

When heated in a furnace, it burns without flame or smoke, and during combustion emits carbonic acid, and leaves a residuum of red oxide of iron.

Constituent Parts.

				Graphite of Pluffier.	
Carbon,	90.9	Carbon,	81	Carbon,	23
Iron, -	9.1	Oxygen,	9	Iron, -	2
	——	Iron,	10	Alumina,	37
	100.00		——	Silica,	38
	Berthollet.		100		——
			Scheele.		100
					Journal des Mines, N. 12. p. 16.

According to John, it sometimes contains Chrome, Nickel, and Manganese; and Schrader mentions Oxide of Titanium as one of its ingredients.

Geognostic Situation.

It occurs usually in beds, sometimes disseminated, and in imbedded masses, in granite, gneiss, mica-slate, clay-slate, foliated granular limestone, coal and trap formations.

Geographic Situation.

Europe.—It occurs in imbedded masses, and disseminated in gneiss in Glen Strath Farrar in Inverness-shire; in the coal formation near Cumnock in Ayrshire, where it is imbedded in greenstone, and in columnar glance-coal *. At Borrodale in Cumberland, it occurs in a

* Jameson's Mineralogical Description of Dumfriesshire, p. 161.

a bed or beds of very varying thickness, included in a bed of trap, which is subordinate to clay-slate. This trap varies in its nature, being sometimes greenstone, or trap-tuff, in other instances amygdaloid, which is occasionally slaty, and contains agates. On the Continent it is met with in the granite of Langsdorf in Bavaria. At Geffrees in Bareuth, imbedded in foliated granular limestone; at Arendal and Friedrschwärn, in Norway; in mica-slate, near Monte-Rosso in Calabria; in gneiss in Piedmont; in serpentine in the mountain of Mora, near to Marbella in Andalusia: and in Iceland, in trap, along with green-earth and zeolite. It is also enumerated amongst the mineral productions of France, Savoy, Bohemia, Austria, Stiria, Salzburg, Hungary and Transylvania.

America.—In the United States it is found at Sparta in New Jersey, imbedded in foliated granular limestone; in syenite near New York; in marble in the county of Ulster; at Freeport in Maine in granite; at Bath in granite; in transition rocks in Rhode Island; and in several other places in foliated granular limestone *. Also in granite in Greenland.

Asia.—At Thutskoi Noss.

Africa.—It is said to occur in rocks near the Cape of Good Hope.

Uses.

The finer kinds are first boiled in oil, and then cut into tables or pencils: the coarser parts, and the refuse of the sawings, are melted with sulphur, and then cast into coarse pencils for carpenters; they are easily distinguished by their sulphureous

---

* Cleaveland's Mineralogy, and Bruce's Mineralogical Journal.

sulphureous smell. It is also used for brightening and preserving grates and ovens from rust; and, on account of its greasy quality, for diminishing the friction in machines. Crucibles are made with it, which resist great degrees of heat, and have more tenacity and expansibility than those manufactured with the usual clay mixtures.

Genus VII. MICA.

Talk-glimmer, *Mohs.*

This genus contains one species, viz. Rhomboidal Mica.

1. Rhomboidal Mica.

Rhomboedrischer Talk-glimmer, *Mohs.*

This species is subdivided into ten subspecies, viz. 1. Mica, 2. Pinite, 3. Lepidolite, 4. Chlorite, 5. Green Earth, 6. Talc, 7. Nacrite, 8. Potstone, 9. Steatite, 10. Figure-Stone. * Clay-slate, Whet-Slate, Black Chalk, Alum-Slate.

First

First Subspecies.

Mica *.

Glimmer, *Werner.*

Mica, *Wall.* t. i. p. 383.—Glimmer, *Wid.* s. 403.—Mica, *Kirw.* vol. i. p. 210.—Glimmer, *Estner,* b. ii. s. 673. *Id. Emm.* b. i. s. 31.—Mica, *Lam.* t. ii. p. 337. *Id. Nap.* p. 272. *Id. Broch.* t. i. p. 402. *Id. Haüy,* t. iii. p. 208.—Glimmer, *Reuss,* b. ii. s. 72. *Id. Lud.* b. i. s. 114. *Id. Suck.* 1r th. s. 474. *Id. Bert.* s. 202. *Id. Hab.* s. 41.—Mica, *Lucas,* p. 75.—Glimmer, *Leonhard,* Tabel. s. 23.—Mica, *Brong.* t. i. p. 508. *Id. Brard,* p. 182. —Glimmer, *Haus.* s. 89. *Id. Karsten,* Tabel. s. 30.—Mica, *Kid,* vol. i. p. 183. *Id. Haüy,* Tabl. p. 53.—Glimmer, *Steffens,* b. i. s. 215. *Id. Lenz,* b. ii. s. 585. *Id. Oken,* b. i. s. 387. *Id. Hoff.* b. ii. s. 115. *Id. Haus.* Handb. b. ii. s. 487.—Mica, *Aikin,* p. 199.

External Characters.

Its most common colours are yellowish and greenish grey, seldomer smoke and ash grey. The yellowish-grey passes into pinchbeck-brown, and brownish-black, and also into yellowish and silver white. The greenish-grey passes through leek-green and blackish-green into greenish-black, and the ash-grey into velvet-black. It is very rarely peach-blossom red.

It occurs massive, and disseminated; also in distinct concretions, which are large, coarse, and small granular, and wedge-shaped prismatic. Sometimes regularly crystallized. Its primitive figure is a rhomboid, the dimensions of which are

* *Mica,* from the Latin word *mico,* to shine, given to it on account of its lustre.

are not accurately known; and the following are the secondary forms:

1. Equiangular six-sided prism, fig. 104. Pl. 5.
2. Equiangular six-sided table, fig. 105. Pl. 5.
3. Equiangular six-sided table, truncated on four of the terminal edges, fig. 106. Pl. 5.
4. Equiangular six-sided table, bevelled on the terminal planes, and the edges of the bevelment truncated, fig. 107. Pl. 5.
5. Rectangular four-sided table.
6. Rectangular four-sided prism.
7. Six-sided pyramid, with alternate broader and narrower lateral planes, fig. 108. Pl. 5.

The crystals are middle-sized and small, seldom large.

The tables generally adhere by their terminal planes, seldom by their lateral planes, and form druses. They are sometimes arranged in rows, rarely in the rose-form, and seldom intersecting each other.

The lateral planes of the tables, and the terminal planes of the prism, are smooth and splendent: the terminal planes of the table are longitudinally streaked, and the lateral planes of the prism are transversely streaked.

Internally it is generally splendent, seldom shining, generally pearly, sometimes semimetallic, and in the silver-white variety passing into metallic.

It has a perfect single cleavage, which is parallel with the terminal planes of the prism, or with the lateral planes of the table. The folia of the cleavage are sometimes spherical * and undulating curved, or are floriform.

The fracture is not discernible.

The fragments are tabular and splintery.

* This is the Mica hæmispherica of Wallerius and Linnæus, which occurs at Skogboll in Sweden.

It is translucent or transparent in thin plates, but rarely in crystals of considerable thickness or length †.

It is sectile.

It affords a grey-coloured dull streak.

It is harder than gypsum, but not so hard as calcareous spar.

It feels fine and meagre, or smooth.

It is elastic-flexible.

Specific gravity, 2.654, 2.034, *Haüy*.—2.726, *Karstèn*.

Chemical Characters.

Before the blowpipe, it melts into a greyish-white enamel.

Constituent Parts.

Common Mica of Zinnwald.		Large foliated Mica from Siberia.		Black Mica from Siberia.	
Silica,	47.00	Silica, -	48.00	Silica,	42.50
Alumina,	22.00	Alumina,	34.25	Alumina,	11.50
Oxide of Iron,	15.50	Oxide of Iron,	4.50	Oxide of Iron,	22.00
Oxide of Manganese,	1.75	Oxide of Manganese,	0.50	Oxide of Manganese,	2.00
Potash,	14.50	Potash,	8.75	Potash,	10.00
		Loss by heating,	1.25	Magnesia,	9.00
				Loss by heating,	1.00
	98.75		97.25		98.00
Klaproth, Beit. b. v. s. 69.		*Klaproth*, Ib. s. 73.		*Klaproth*, Ib. s. 78.	

Geognostic

* Count de Bournon mentions crystals of mica in his valuable collection, of considerable thickness, which are transparent in the direction of their axes. He also notices particularly the difference of colour observed as we look in the direction of the axis or across the crystal: thus, he observed in a transparent crystal from Pegu, that the colour in the direction of the axis was yellowish-green; but at right angles to the axis, was beautiful *vert d'herbe*. In other crystals, the colour in the line of the axis was of a beautiful green, whilst in the opposite direction it was orange; and in some other crystals, the colour parallel with the axis was white; but perpendicular to it flesh-red.

Geognostic Situation.

This mineral occurs as an essential constituent part of several primitive rocks, and accidentally intermixed with others, both of the primitive, transition, secondary or flœtz, and alluvial classes. Thus, along with felspar and quartz, it forms granite and gneiss, and with quartz mica-slate: it is occasionally intermixed with clay-slate, quartz-rock, primitive limestone, sienite, porphyry, greenstone, hornblende-slate and hornblende-rock, whitestone, greywacke, greywacke-slate, sandstone, wacke, amygdaloid, basalt, and various alluvial deposites. It sometimes forms short beds in granite, and other primitive rocks; or it appears in globular, oval, tuberose, or irregular-shaped cotemporaneous masses, in granite or gneiss. It also occurs in veins, as in those formed of granite or quartz, or in such as contain ores of different kinds, as tinstone and copper-pyrites *.

Geographic Situation.

The rocks in which mica occurs, are so universally distributed, that it is not necessary to enter into any detail of localities: we may merely mention, that most of the mica of commerce is brought from Siberia, and the borders of the Caspian Sea, where it occurs in large plates or crystals, in granite.

Uses.

In some countries, as in Siberia, mica is an article of commerce, and is regularly mined. In Siberia, the principal mica mines are those on the banks of the Wettin, the Aldan,

* At Zinnwald in Bohemia, it occurs in veins, in a variety of granite, which contains little or no mica, and is known under the name *Greisen*.

Aldan, and other rivers that fall into the Lena. It occurs in nests, often of considerable magnitude, imbedded in granite *. The mica is extracted by means of hammers and chisels, is then washed of the adhering earth, and assorted into different kinds, according to goodness, purity, and size. The plates or tables intended for sale, must be clear, well coloured, and as free as possible from spots. The greenish-coloured and imperfectly transparent, or the spotted varieties, are laid aside, and sold at a low rate. It is exported in considerable quantity from Russia. In 1781, 200 puds were sent from St Petersburgh to Lubec, and a very considerable quantity to England and Ireland.

In Siberia, where window-glass is scarce, it is used for windows; also for a similar purpose in Peru, and, I believe, also in New Spain, as it appears that the mineral named *Teculi* by Ulloa, and which is used for that purpose, is a variety of mica. It is also used in lanterns, in place of glass, as it resists the alternations of heat and cold better than that substance. In Russia, it is employed in different kinds of inlaid work. It is sometimes intermixed with the glaze in particular kinds of earthen-ware: the heat which melts the glaze has no effect on the mica; hence it appears dispersed throughout the glaze, like plates or scales of silver or gold, and thus gives to the surface of the ware a very agreeable appearance. Some artists use it in the making of artificial aventurines.

Observations.

1 Mica is distinguished from *Talc* by lustre, elastic [illegible], superior hardness, and different colours; from [illegible] *Chlorite*, by colour, streak, superior hardness, elastic

* [illegible] sometimes three or four feet square.

tic flexibility, and crystallizations; and from *Common Hornblende*, by form, single distinct cleavage, streak, and inferior hardness.

2. The following works contain information in regard to the situation, mode of mining, and uses of Russian mica.

1. Georgi's Geograph. Physikal. u. Naturhist, od Beschreib. Russischen. Reichs. b. 3. s. 236.
2. Gmelin's Reise, b. 2. s. 322.
3. Nov. Comment. Acad. Petrop. 1766, p. 549.
4. Hanover'sches Magazin. 9. s. 79.
5. Neue Nordische Beiträge, b. ii. s. 356.
6. Beckmann's Vorbereitung zur Waarenkunde, b. ii. s. 233.

3. The yellow and brown varieties are by the vulgar named *Cat-gold*, the white varieties *Cat-Silver*. The Russian *Sliuda*, is mica, and their *Frauen-glass* or Muscovy glass is the same mineral.

Second Subspecies.

Pinite *.

Pinit, *Werner*.

Micarelle, *Kirwan*.

Pinit, *Reuss*, b. ii. s. 69. *Id. Lud.* b. ii. s. 149. *Id. Suck.* 1r th. 469. *Id. Bert.* s. 298. *Id. Mohs*, b. i. s. 480. *Id. Lucas*, p. 280. *Id. Leonhard*, Tabel. s. 24. *Id. Brong.* t. i. p. 507. *Id. Brard*, p. 185. *Id. Karsten*, Tabel. s. 48. *Id. Haüy*, Tabl. p. 53. *Id. Steffens*, b. i. s. 219. *Id. Lenz*, b. ii. s. 592. *Id.* Oken,

* It is named *Pinite*, from the Pini mine gallery, where it was first found.

Oken, b. i. s. 389. *Id. Hoff.* b. ii. s. 127. *Id. Haus.* Handb. b. ii. s. 507. *Id. Aikin*, p. 190.

External Characters.

Its colour is blackish-green, altered on the surface by brown or red iron-ochre into brownish-red. It is sometimes iron-shot.

It occurs massive, also in distinct concretions, which are thick and thin lamellar, collected into large and coarse granular, and crystallized in the following figures:

1. Equiangular six-sided prism.
2. The preceding figure truncated or bevelled on all the lateral edges. Owing to the number of planes, figures of this description have a cylindrical form. The terminal angles are sometimes truncated.
3. Rectangular four-sided prism.

The crystals are seldom middle-sized, generally small. They are imbedded, and frequently intersect each other.

The cleavage is shining; the fracture is glistening and glimmering, and the lustre is resinous.

The cleavage is imperfect and single, the folia being parallel with the terminal planes of the prism.

The fracture is small-grained uneven.

The fragments are blunt-angular, seldom tabular.

It is opaque, or faintly translucent on the edges.

It is soft, passing into very soft.

It is sectile, and easily frangible.

It is not flexible.

It feels somewhat greasy.

Specific gravity, 2.914, *Haüy*.—2.974, *Kirwan*.

Chemical Character.

[illegible] before the blowpipe.

Constituent

Constituent Parts.

Silica,	29.50	Silica,	46.0
Alumina, -	63.75	Alumina, -	42.0
Oxide of Iron,	6.75	Oxide of Iron,	2.5
		Loss by calcination,	7.0
	100.00	Loss, -	2.5
			100.0
	Klaproth, Jour. des Mines, N. 100. p. 311.		*Drappier*, Jour. des Mines, N. 100. p. 311.

Geognostic and Geographic Situations.

It is found imbedded in the granite of St Michael's Mount in Cornwall: in porphyry in Ben Gloe and Blair-Gowrie; in granite at Schneeberg in Saxony, and in the porcelain-earth of Aue, also in Saxony; in a greyish porous felspar-porphyry in the Puy de Dome, in Auvergne; in Dauphiny, along with epidote, axinite, rock-crystal, chlorite, and iron-ochre; in Regenbei near Bodenmais in Bavaria, and in the Lisenz-Alp in the Tyrol, where it is very rare *.

Observations.

1. It is distinguished from *Mica*, with which it has been confounded, by its circumscribed series of colour, its peculiar truncations, its never inclining to the tabular form, and its lustre, fracture, and want of flexibility.

2. It was first established as a distinct mineral by Werner,

* Heuland.

ner, and named *Pinite*, from the Pini Gallery in the mines of Schneeberg, where it was first found.

3. According to Bernhardi, in Von Moll's Ephemerid. b. iii. st. 1. pinite is nearly allied to Schorl; and Hausmann is of opinion, that it is a mixture of mica and andalusite. In opposition to these conjectures, and in confirmation of its close affinity to mica, we have its form, cleavage, softness and specific gravity.

Third Subspecies.

Lepidolite *.

Lepidolith, *Werner*.

Lepidolith, *Wid.* s. 378. *Id. Kirw.* vol. i. p. 208. *Id. Emm.* b. iii. s. 324. *Id. Estner*, b. ii. s. 228. *Id. Nap.* p. 167. *Id. Lam.* t. ii. p. 315. *Id. Broch.* t. i. p. 399. *Id. Haüy*, t. iv. p. 375. *Id. Reuss*, b. ii. s. 402. *Id. Lud.* b. i. s. 114. *Id. Suck.* 1r th. s. 397. *Id. Bert.* s. 17. *Id. Mohs*, b. i. s. 465. *Id. Hab.* s. 40. *Id. Lucas*, p. 199. *Id. Leonhard*, Tabel. s. 23. *Id. Brong.* t. i. p. 506. *Id. Brard*, p. 411. *Id. Haus.* s. 91. *Id. Karsten*, Tabel. s. 30. *Id. Kid*, vol. ii. p. 246. *Id. Haüy*, Tabl. p. 64. *Id. Steffens*, b. i. s. 213. *Id. Lenz*, b. ii. s. 582. *Id. Oken*, b. i. s. 390. *Id. Hoff.* b. ii. s. 111. *Id. Haus.* Handb. b. ii. s. 500. —Mica, *Aikin*, p. 200.

External Characters.

Its colour is peach-blossom-red, inclining sometimes to rose-red, sometimes to lilac-blue; it also passes into pearl-grey, yellowish-grey, and greenish-grey.

It

* *Lepidolite*, from the Greek word λεπις, *a scale*, given it on account of its foliated appearance.

It occurs massive, and in small granular distinct concretions.

Internally its lustre is glistening, passing into shining, and pearly.

It has a distinct single cleavage.

The fracture is coarse splintery.

The fragments are indeterminate angular and blunt-edged.

It is feebly translucent.

It is soft.

It is rather sectile.

It is rather easily frangible.

Specific gravity, 2.816, *Klaproth*.—2.58, *Karsten*.

Chemical Characters.

Before the blowpipe it intumesces, and melts very easily into a milk-white nearly translucent globule.

Constituent Parts.

Silica, - -	54.50	Silica, -	54.00
Alumina, -	38.25	Alumina,	20.00
Potash, -	4.00	Potash, -	18.00
Manganese & Iron,	0.75	Fluat of Lime,	4.00
Loss, partly Water,	2.50	Manganese,	3.00
		Iron, -	1.00
	100		100
	Klaproth, Beit. b. ii. s. 195.		*Vauquelin*, Jour. de Min. t. ix. p. 235.

Geognostic

Geognostic and Geographic Situations.

It occurs disseminated, in foliated and granular limestone, at Dalmally; in primitive limestone in a quarry on the north side of Loch Fyne, opposite the Inn of Cairndow, situated on the south side; and in primitive limestone, from a quarry on the east side of Loch Leven, nearly opposite to the Inn at Balachulish, situated on the west side *.

It occurs imbedded in granite, in the mountain of Hradisko, near to Rosena in Moravia; in quartz in granite, in the Riesengebirge in Silesia; also in beds of ironstone subordinate to gneiss, along with apophyllite, in Utön in Sweden; in Norway; in the vicinity of Limoges in France; and in the Isle of Elba †; and associated with rubellite and common felspar, at Perm, in the government of Catharineburg, in Siberia.

Uses.

It is sometimes cut into snuff-boxes, which are admired for their colour; but, owing to the softness of the mineral, they have rather a dull greasy-like surface.

Observations.

1. It is nearly allied to Mica, from which, however, it is distinguished by colour and fracture.

2. The first account published of this mineral, was by M.

* The above localities are on the authority of Mr Holme of Peterhouse, Cambridge, who found this mineral in the places above mentioned.

† Haüy and Tondi received, through Mr Schultz, a mineral from Bavaria, which they consider as a variety of lepidolite.—Von Moll's Neue Jahrbücher d. Berg & Hüttenkunde, 2. B. 1. Lif. s. 111.

M. Von Born, in the Chem. Annalen for 1791, who considered it as zeolite.

3. The grey variety from Utön, has been described as a distinct species, under the name *Petalite*.

Fourth Subspecies.

Chlorite *.

Chlorit, *Werner*.

This subspecies is divided into four kinds, viz. Earthy Chlorite, Common Chlorite, Slaty Chlorite, and Foliated Chlorite.

First Kind.

Earthy Chlorite.

Erdiger Chlorit, *Karsten*.

Chlorite in a loose form; Peach of the Cornish Miners, *Kirw.* vol. i. p. 147.—Erdiger Chlorite, *Reuss*, b. ii. s. 81. *Id. Lud.* b. i. s. 116. *Id. Suck.* 1r th. s. 479. *Id. Bert.* s. 426. *Id. Mohs*, b. i. s. 484. *Id. Leonhard,* Tabel. s. 24. *Id. Haus.* s. 90. *Id. Karst.* Tabel. s. 12.—Talc Chlorit terreux, *Haüy*, Tabl. p. 56.—Erdiger Chlorit, *Steffens*, b. i. s. 221. *Id. Lenz*, b. ii. s. 600. *Id. Oken*, b. i. s. 382. *Id. Hoff.* b. ii. s. 134.—Schuppiger Chlorit, *Haus.* Handb. b. ii. s. 491.

External

* *Chlorite*, from the Greek word χλωρος, green, on account of its green colour.

External Characters.

Its colours are dark mountain and leek green, and sometimes olive-green.

It occurs massive, disseminated, in crusts, and moss-like, inclosed in adularia and rock-crystal.

It is glimmering or glistening, and the lustre is pearly.

It consists of fine scaly particles, which are more or less cohering, and feels rather greasy.

It does not soil.

Its streak is of a mountain-green colour.

Specific gravity 2.612, 2.699.

Chemical Characters.

It melts before the blowpipe into a blackish slag.

Constituent Parts.

Silica, -	26.00
Alumina, -	18.50
Magnesia, - -	8.00
Muriate of Soda, or Potash,	2.00
Oxide of Iron, -	43.00
Loss, -	2.50
	99.00

Vauquelin, Journ. des Mines, N. 39. p. 167.

Geognostic and Geographic Situations.

It occurs in veins along with common chlorite at Forneth Cottage in Perthshire. In felspar and Adularia veins in St Gothard; also in Dauphiny, where it encrusts rock-crystal, axinite, and sphene; and in veins intersecting serpentine at Waldheim near Freyberg.

Observations.

Observations.

1. It is characterized by its green colour, scaly glimmering particles, slightly greasy feel, and its not soiling.

2. The great quantity of iron it contains, is by Karsten considered more as an accidental than as a regular constituent part.

3. The scaly parts, according to Haüy, when viewed by the microscope, appear to be regular six-sided tables.

Second Kind.

Common Chlorite.

Gemeiner Chlorit, *Werner.*

Indurated Chlorite, *Kirwan*, vol. i. p. 148.—Gemeiner Chlorit, *Reuss*, b. ii. s. 84. *Id. Lud.* b. i. s. 117. *Id. Suck.* 1r th. s. 488. *Id. Bert.* s. 426. *Id. Mohs*, b. i. s. 485. *Id. Hab.* s. 59. *Id. Leonhard*, Tabel. s. 24.—Chlorite commune, *Brong.* t. i. p. 500. —Blättricher Chlorit, *Haus.* s. 90. *Id. Karsten*, Tabel. s. 42. *Id. Steffens*, b. i. s. 222. *Id. Lenz*, b. ii. s. 60. *Id. Oken*, b. i. s. 382.—Gemeiner Chlorit, *Hoff.* b. ii. s. 137. *Id. Haus.* Handb. b. ii. s. 492.

External Characters.

Its colour is intermediate between dark blackish-green and leek-green.

It occurs massive and disseminated.

Its lustre is glimmering, or glistening, and is pearly, inclining to resinous.

The fracture is fine earthy, and fine scaly foliated.

The

The fragments are blunt-edged.

It is opaque.

It becomes light mountain-green in the streak, with a feeble lustre.

It is soft, passing into very soft.

It is sectile.

It does not adhere to the tongue.

It feels somewhat greasy.

Specific gravity 2.832, *Wid.*

Geognostic and Geographic Situations.

It occurs not only disseminated through rocks of different kinds, as granite and mica-slate, but also in beds and veins. The granite of Mont Blanc contains common chlorite in veins, or disseminated through it: in Saxony, Salzburg, and other countries, it occurs in beds, which contain magnetic ironstone, copper-pyrites, iron-pyrites, arsenical pyrites, hornblende, actynolite, and calcareous-spar. In the Island of Arran, it occurs in quartz veins that traverse clay-slate; in similar repositories in the Island of Bute, and in several other districts in Scotland, in granite and other rocks. In England, it occurs in the Wherry Mine, Penzance, and other places in Cornwall *.

Third

* Greenough.

Third Kind.

Slaty Chlorite or Chlorite-Slate.

Chlorit-Schiefer, *Werner.*

Schiefriger Chlorit, *Karsten.*

Chlorit Schiefer, *Reuss*, b. ii. s. 88. *Id. Lud.* b. i. s. 117. *Id. Suck.* 1r th. s. 484. *Id. Bert.* s. 427. *Id. Mohs*, b. i. s. 487. *Id. Hab.* s. 59. *Id. Leonhard*, Tabel. s. 24.—Chlorit schisteuse, *Brong.* t. i. p. 501.—Schiefriger Chlorit, *Haus.* s. 90. *Id. Karst.* Tabel. s. 42.—Talc Chlorite fissile, *Haüy*, Tabl. p. 56.—Schiefriger Chlorit, *Steffens*, b. i. s. 223.—Chlorit Schiefer, *Lenz*, b. ii. s. 605. *Id. Oken*, b. i. s. 383. *Id. Hoff.* b. ii. s. 139. *Id. Haus.* Handb. b. ii. s. 493.

External Characters.

Its colour is intermediate between dark mountain and leek green, and sometimes passes into blackish-green and greenish-black.

It occurs massive, and in whole beds.

The lustre is glistening, sometimes inclining to shining, and is intermediate between pearly and resinous.

The fracture is more or less perfect slaty, seldom straight, generally waved slaty, and sometimes scaly foliated.

The fragments are tabular.

It is opaque.

It affords a pale mountain-green streak.

It is soft.

It is sectile; and rather easily frangible.

It does not adhere to the tongue.

It feels slightly greasy.

Specific

Specific gravity, 2.905, *Saussure.*—2.822, *Karsten.*—2.794, *Grüner.*

Constituent Parts.

Silica, - - -	29.50
Alumina, - - -	15.62
Magnesia, - - -	21.39
Lime, - -	1.50
Iron, - - -	23.3
Water, - - -	7.38
	Grüner.

Geognostic Situation.

It occurs principally in beds, subordinate to clay-slate, and is occasionally associated with potstone and talc-slate. It also occurs in beds in gneiss, mica-slate, and quartz rock. It frequently contains octahedral crystals of magnetic ironstone; also garnets, schorl, and rhomb-spar.

Geographic Situation.

It occurs in beds, in the clay-slate districts of the Grampians, and other parts of Scotland. On the Continent, it is found in Norway, Sweden, Saxony, Switzerland, Corsica, and other countries.

Observations.

It passes into Clay-slate, and also into Potstone.

Fourth

Fourth Kind.

Foliated Chlorite.

Blättriger Chlorit, *Werner.*

Blättriger Chlorit, *Reuss*, b. ii. s. 86. *Id. Lud.* b. i. s. 118. *Id. Suck.* 1r th. s. 481. *Id. Mohs*, b. i. s. 486. *Id. Leonhard*, Tabel. s. 24. *Id. Haus.* s. 90. *Id. Karsten*, Tabel. s. 62.—Talc Chlorit, *Haüy*, Tabl. p. 56.—Blättriger Chlorit, *Steffens*, b. i. s. 224. *Id. Lenz*, b. ii. s. 603. *Id. Oken*, b. i. s. 383. *Id. Hoff.* b. ii. s. 140. *Id. Haus.* Handb. b. ii. s. 490.

External Characters.

Its colour is dark blackish-green, which in some rare varieties is dark olive-green.

It occurs massive, disseminated, in granular concretions, and crystallized in four-sided prisms, and in six-sided tables. These tables are aggregated together, in such a manner as to form the two following figures:

A. Cylinder terminated by two cones.

B. Two truncated cones, joined base to base.

If we suppose the six-sided table N° 1. to revolve around an axis which passes through its two opposite angles, the figure A will be formed; but if it revolves around an axis which passes through two opposite sides, the figure B will be formed. The streaking on the surfaces shows the mode of aggregation of the tables.

The crystals are generally longitudinally streaked, and are small or middle-sized.

Externally

Externally it is glistening, approaching to shining, and is resinous; internally it is shining and pearly.

It has a single imperfect cleavage, in which the folia are often curved.

The fragments are indeterminate angular, or tabular.

It is opaque, or translucent on the edges.

It is soft, passing into very soft.

It is sectile, and rather difficultly frangible.

It feels rather greasy.

Its colour is lighter in the streak.

Specific gravity 2.823, *Karsten.*

Constituent Parts.

Silica, - - -	35.00
Alumina, - - -	18.00
Magnesia, - - -	29.90
Iron, - -	9.70
Water, - - -	2.70

Lampadius, Handbuch zur Chem. Analyse der Mineral Körper, s. 229.

Geognostic and Geographic Situations.

Europe.—It occurs in the island of Jura, one of the Hebrides, in quartz rock. On the Continent of Europe, it is found in St Gothard, where it is associated with adularia, rock-crystal, and rutile; also in the valley of Fusch in Salzburg, where it occurs along with amethyst and adularia, and seldom with prehnite; and in Sweden, Saxony, and Corsica.

Asia.—It occurs in Siberia, along with slaty chlorite.

OBSERVATIONS

OBSERVATIONS ON CHLORITE.

1. It was Saussure the Father, who first directed the attention of mineralogists to this mineral; and Werner was the first who ascertained its oryctognostic relations.

2. It is nearly allied to Talc and Mica, and also to Potstone. The foliated kind approaches the nearest to Mica, the common and slaty to Potstone.

3. Hausmann, in his "Entwurf eines Systems der Unorganisirten Naturkörper," describes a substance under the name *Conchoidal Chlorite*, which deserves to be more particularly examined. The following is his account of it:

"*Conchoidal Chlorite.* Colour leek-green; internally dull; but shining and resinous on the surface of the fissures. Fracture flat conchoidal, inclining to splintery and earthy, even sometimes approaching to slaty. Becomes resinous and shining in the streak. Translucent on the edges. Soft. It occurs in the Hartz, disseminated in amygdaloid and greenstone."

Fifth Subspecies.

Green Earth.

Grünerde, *Werner.*

Green Earth, *Kirwan,* vol. i. p. 196.—Grünerde, *Emm.* b. i. s. 353.—La Terre verte, *Broch.* t. i. p. 445.—Grünerde, *Reuss,* b. ii. s. 157. *Id. Lud.* b. i. s. 126. *Id. Suck.* 1r th. s. 522. *Id. Bert.* s. 214. *Id. Mohs,* b. i. s. 515. *Id. Hab.* s. 39.—Talc Chlorite zographique, *Lucas,* p. 84.—Grünerde, *Leonhard,* Tabel. s. 26.—Chlorite Baldogée, *Brong.* t. i. p. 501.—

Talc Chlorite zographique, *Brard*, p. 198.—Erdiger Chlorit, *Haus.* s. 90.—Grünerde, *Karsten*, Tabel. s. 26.—Talc Chlorite zographique, *Haüy*, Tabl. p. 56.—Grünerde, *Steffens*, b. i. s. 257. *Id. Lenz*, b. ii. s. 621. *Id. Oken*, b. i. s. 277. *Id. Hoff.* b. ii. s. 195.—Green Earth, *Aikin*, p. 201.

External Characters.

Its colour is celandine-green, of various degrees of intensity, which passes into blackish-green, and olive-green.

It occurs massive, seldomer disseminated, more frequently in globular and amygdaloidal-shaped pieces, which are sometimes hollow, in crusts lining the vesicular cavities in amygdaloid, or on the surface of agate balls *.

Internally it is dull.

The fracture is earthy, sometimes small grained uneven.

It is opaque.

It is feebly glistening in the streak, but without any change of colour.

It is very soft, and sectile.

It feels rather greasy.

It adheres slightly to the tongue.

Specific gravity, 2.598, *Karsten.*—2.632, *Kirwan.*—2.606, *Breithaupt.*

Chemical Characters.

Before the blowpipe, it is converted into a black vesicular slag.

Constituent

---

* It is said to occur crystallized in the valley of Fassa in the Tyrol.—Vid. Brocchi's Description of the Valley of Fassa.

Constituent Parts.

	From Cyprus.	From the Veronese.
Silica, - -	51.50	53.0
Oxide of Iron,	20.50	28.0
Magnesia, -	1.50	2.0
Potash, - -	18.00	10.0
Water, - -	8.00	6.0
Loss, -	0.50	
	100.00	99.0
	Klaproth, Beit. b. iv. p. 244.	*Klaproth*, Id. s. 241.

Geognostic Situation.

It occurs principally in the amygdaloidal cavities of amygdaloid, and incrusting the agates found in that rock. It also occasionally colours sandstone, and is disseminated in porphyry.

Geographic Situation.

It is a frequent mineral in the amygdaloid of Scotland; it occurs also in that of England and Ireland. It is found in the amygdaloid of Iceland and the Faroe Islands; and on the Continent of Europe, it occurs in Saxony, Bohemia, near Verona, the Tyrol, and Hungary.

Uses.

It is used as a pigment in water-painting, and is the *mountain-green* of painters. It is very durable in the air, but rarely affords tints equal to those obtained from copper. Before using, it must be ground, and well washed, to free it from impurities. When exposed to a moderate heat, the green changes into a beautiful reddish-brown colour, which is very durable, and it is then used as a water-colour. Of all the known varieties, that of Verona is the most highly esteemed, and is known in trade under the name *Green Earth of Verona*. That of Cyprus is also an article of commerce. It is brought to Holland as ballast, wrapped up in palm-leaves, in hampers. The green earth of Bohemia is also known in trade, but is not so highly esteemed as the Veronese and Cyprian. A colouring matter of this description appears to have been known to the Romans.

Observations.

1. This mineral was first established as a distinct mineral by Werner. It is distinguished by its colour, shape, fracture, streak, hardness, and geognostic situation.

2. The intimate combination of this mineral and calcedony forms Heliotrope, and also the greater number of the pretended specimens of Plasma.

Sixth Subspecies.

Talc *.

Talk, *Werner*.

THIS species is divided into two kinds, viz. Common Talc and Indurated Talc.

First

* The [illegible] well known: some derive it from the Germ[illegible] [illegible]ay feel, or from the Swedish word *talga*, [illegible] [illegible]aintain that it is of Asiatic origin.

First Kind.

Common Talc.

Gemeiner Talk, *Werner.*

Talcum albicans, lamellis subpellucidis flexis, *Waller*. gen. 27. spec. 180.—Gemeiner Talk, *Wid.* s. 441.—Common Talc, or Venetian Talc, *Kirw.* vol. i. p. 150.—Gemeiner Talk, *Estner*, b. ii. s. 824. *Id. Emm.* b. i. s. 391.—Talco compatto, *Nap.* p. 293.—Talc cailleux, *Lam.* t. ii. p. 342.—Talc 'laminaire, *Haüy*, t. iii. p. 252.—Le Talc commun, *Broch.* t. i. p. 487.—Gemeiner Talc, *Reuss*, b. ii. 2. s. 229. *Id. Lud.* b. i. s. 136. *Id. Suck.* 1r th. s. 571. *Id. Bert.* s. 139. *Id. Hab.* s. 64.—Talc laminaire, *Lucas*, p. 83.—Gemeiner Talk, *Leonhard*, Tabel. s. 29.—Talc laminaire, *Brong.* t. i. p. 503. *Id. Brard*, p. 197.—Blättricher Talk, *Haus.* s. 91.—Gemeiner Talk, *Karst.* Tabel. s. 42.—Laminated and Venetian Talk, *Kid*, vol. i. p. 107. & 108.—Talc hexagonal, laminaire, ecailleux, *Haüy*, Tabl. p. 56.—Gemeiner Talk, *Steffens*, b. i. s. 228. *Id. Lenz*, b. ii. s. 665. *Id. Oken*, b. i. s. 389. *Id. Hoff.* b. ii. s. 270.—Blättricher Talk, *Haus.* Handb. b. ii. s. 498.

External Characters.

Its most common colour is greenish-white; it also occurs silver-white, apple-green, asparagus-green, and leek-green, which latter colour passes into duck-blue. The apple-green, sometimes passes into emerald-green.

It occurs massive, disseminated, in plates, reniform, and botryoidal: in distinct concretions, which are large, coarse and small granular; also narrow or broad and stellular or promiscuous radiated, which are again collected into other concretions, having a wedge-shaped prismatic form. It is sometimes

sometimes crystallized in small six-sided tables, which are in druses.

It is generally splendent, or shining, and is pearly, or semi-metallic.

It has a distinct single cleavage, in which the folia are generally curved.

The fragments are wedge-shaped, seldom splintery.

It is translucent; in thin folia transparent.

It is flexible, but not elastic.

It is very soft, or yields easily to the nail.

It is perfectly sectile.

It feels very greasy.

Specific gravity, 2.695, 2.795, *Kirwan.*—2.770, *Karsten.*—2.771, *Breithaupt.*

Chemical Characters.

It becomes white before the blowpipe, and at length, with difficulty, affords a small globule of enamel.

Constituent Parts.

Silica, -	62.00	Silica, -	61.75
Magnesia,	27.00	Magnesia,	30.50
Alumina,	1.50	Potash, -	2.75
Oxide of Iron,	3.50	Oxide of Iron,	2.50
Water, -	6.00	Water, -	0.25
		Loss, -	2.25
Vauquelin, Jour. d. Min. N. 88. p. 243.		*Klaproth*, Karst. Tab. s. 48.	

Geognostic Situation.

It occurs in beds in mica-slate and clay-slate, and in a similar situation in granular limestone and dolomite; also in

cotemporaneous

cotemporaneous veins, in beds of indurated talc, serpentine, and porphyry; and in the reniform external shape, in tin-stone veins.

Geographic Situation.

Europe.—It is found in Aberdeenshire, Banffshire, and Perthshire; and on the Continent of Europe, in Norway, Sweden, Saxony, Bohemia, Switzerland, the Tyrol, and Salzburg. The finest specimens of common talc are found in Salzburg, the Tyrol, and in St Gothard in Switzerland. The beautiful duck-blue variety is brought from the Taberg in Wermeland in Sweden.

Asia.—Persia, China, India.

America.—Maryland, Pennsylvania, Connecticut, Massachusets, and Maine *.

Uses.

It enters into the composition of the cosmetic named *rouge*. This substance is prepared by rubbing together in a warm mortar, generally of serpentine, certain proportions of carmine and finely powdered talc, with a small portion of oil of benzoin. This cosmetic communicates a remarkable degree of softness to the skin, and is not pernicious. The Romans prepared a beautiful blue or purple colour, by combining this substance with the colouring fluid of particular kinds of testaceous animals †; and the flesh polish is given to gypsum figures, by rubbing them with talc. The Persians, according to Tavernier, whiten the walls of their houses and gardens by means of lime-water,

---

* Cleaveland's Mineralogy.

† The Buccinum reticulatum and Buccinum lapillus, that abound on the coasts of the Mediterranean.

water, and then powder them with silver-white coloured talc, which is said to give them a beautiful appearance. The Chinese burn talc, mix it with wine, and use it internally, as a cordial for curing diseases, and procuring long life: even European physicians, at one period, prescribed the powder of talc in dysenteric and hæmorrhoidal affections. It was known in the Materia Medica under the name *Talcum Venetum*.

Observations.

1. The light green colours, distinct concretions, strong lustre, fracture, considerable translucency, softness, perfect sectility, want of elasticity, and very greasy feel, are the principal characters of Common Talc.

2. Common Talc is often confounded with *Mica*, but is distinguished from it by its sectility, greasy feel, inferior hardness, want of elasticity, and colour. It is distinguished from *Chlorite* by its colours.

Second Kind.

Indurated Talc, or Talc-Slate.

Verhärteter Talk, *Werner*.

Verhærteter Talk, *Estner*, b. ii. s. 828. *Id. Emm.* b. iii. s. 280. —Le Talc endurcie, *Broch.* t. i. p. 489.—Verhärteter Talk, *Reuss*, b. ii. 2. s. 233. *Id. Lud.* b. i. s. 136. *Id. Suck.* 1r th. s. 573. *Id. Bert.* s. 140. *Id. Mohs*, b. i. s. 565. *Id. Leonhard*, Tabel. s. 29.—Talc endurcie, *Brong.* t. i. p. 504.—Verhärteter Talc, *Karsten*, Tabel. s. 42.—Indurated Talc, *Kid*, vol. i. p. 109.—Verhärteter Talc, *Steffens*, b. i. s. 230. *Id. Lenz*, b. ii. s. 669. *Id. Oken*, b. i. s. 390. *Id. Hoff.* b. ii. s. 275.—Schiefriger Talk, *Haus.* Handb. b. ii. s. 497.

External

External Characters.

Its colour is greenish-grey, of various degrees of intensity, which sometimes passes into greenish-white and yellowish-white: it very rarely inclines to olive-green.

It occurs massive, and rarely in fibrous distinct concretions.

Its lustre is shining, passing to glistening, and is pearly.

The fracture is curved slaty, passing into imperfect foliated.

The fragments are tabular.

It is strongly translucent on the edges, and sometimes feebly translucent.

It is soft, and the streak is white.

It is rather sectile.

It is rather easily frangible.

It is not flexible.

It feels greasy.

Specific gravity, 2.700, 2.800, *Kirwan.*—2.780, 2.793, *Breithaupt.*

Geognostic Situation.

It occurs in primitive mountains, where it forms beds in clay-slate and serpentine, and is associated with amianthus, chlorite, rhomb-spar, garnet, actynolite, quartz, kyanite and grenatite.

Geographic Situation.

It occurs in Perthshire, Banffshire, the Shetland islands; and on the Continent of Europe, in Sweden, Saxony, Silesia, the Tyrol, Austria, and Switzerland.

Uses.

Uses.

It is employed for drawing lines by carpenters, tailors, hat-makers, and glaziers. The lines are not so easily effaced as those made by chalk, and besides remain unaltered under water. Dr Kid remarks: "If lines be traced on glass by means of a piece of indurated talc, they remain invisible, or are scarcely perceptible by the naked eye, till breathed on. I have not met with an explanation of the effect produced in this instance; but it may perhaps in part depend on the comparative softness of the substance with which the impression is made: the condensation of the breath taking place more readily on the glass, than on the talc covering the glass, and the impression of the talc becoming more apparent by the simple contrast *." It is sometimes made into culinary vessels; and when reduced to powder, may be employed for the purpose of removing from silk stains occasioned by grease.

Observations.

1. It is distinguished from *Common Talc* by its inferior lustre, slaty fracture, inferior translucency, and rather greater hardness and weight; from *Potstone* by its colour-suite, superior lustre, more perfect slaty fracture, and greater translucency; and from *Axe-Stone* by inferior hardness and weight.

2. It passes into potstone, axestone, and steatite.

Seventh

* Kid's Mineralogy, vol. i. p. 109.

Seventh Subspecies.

Nacrite.

Nacrite, *Brongniart.*

Erdiger Talk, *Werner.*

Erdiger Talk, *Wid.* s. 439.—Talcite, *Kirw.* vol. i. p. 149.—Erdiger Talk, *Estner,* b. ii. s. 821. *Id. Emm.* b. i. s. 389.—Talco terroso, *Nap.* p. 293.—Le Talc terreux, *Broch.* t. i. p. 486.—Erdiger Talk, *Reuss,* b. ii. s. 227. *Id. Lud.* b. i. s. 135. *Id. Suck.* 1r th. s. 570. *Id. Bert.* s. 299. *Id. Mohs,* b. i. s. 560. *Id. Leonhard,* Tabel. s. 29.—Nacrite, *Brong.* t. i. p. 505.—Schuppiger Thon, *Karsten,* Tabel. s. 28. *Id. Haus.* s. 85.—Talc granuleux, *Haüy,* Tabl. p. 67.—Schuppiger Thon, *Steffens,* b. i. s. 202.—Erdiger Talc, *Hoff.* b. ii. s. 267.—Schuppiger Talc, *Haus.* Handb. b. ii. s. 498.

External Characters.

Its colours are greenish-white, and greenish-grey.

It consists of scaly parts, which are more or less compacted; the most compact varieties have a thick or curved slaty fracture.

It is strongly glimmering, and is pearly, inclining to resinous.

It is friable.

It feels very greasy.

It soils.

Chemical Characters.

It melts easily before the blowpipe.

Constituent

Constituent Parts.

Alumina,	81.75
Magnesia,	0.75
Lime,	4.00
Potash,	0.50
Water,	13.50
	100.50 *John.*

Geognostic and Geographic Situations.

This is a very rare mineral; it occurs in veins with sparry ironstone, galena, iron-pyrites and quartz, in the mining district of Freyberg in Saxony; Gieren, in Silesia; and Sylva in Piedmont.

Observations.

This rare mineral is named *Nacrite* by Brongniart, which name is here adopted, in preference to that of Earthy Talc, the name given to it by Werner.

Eighth Subspecies.

Potstone, or Lapis ollaris *.

Topfstein, *Werner.*

Lapis comensis, *Plin.* Hist. Nat. xxxvi. 22. p. 44.—Steatites, Lapis ollaris, *Wall.* Syst. Min. i. 387.—Potstone, *Kirw.* vol. i. p. 155.

* So named, on account of the facility with which it can be cut into vessels or pots of different kinds.

p. 155.—Topfstein, *Reuss*, b. ii. 2. s. 236. *Id. Lud.* b. i. s. 115. *Id. Suck.* 1r th. s. 576. *Id. Hab.* s. 30. *Id. Leonhard*, Tabel. s. 26.—Serpentine ollaire, *Brong.* t. i. p. 486.—Topfstein, *Karsten*, Tabel. s. 42.—Talc ollaire, *Haüy*, Tabl. p. 56.—Topfstein, *Steffens*, b. i. s. 231. *Id. Lenz*, b. ii. s. 598. *Id. Oken*, b. i. s. 381. *Id. Hoff.* b. ii. s. 131. *Id. Haus.* Handb. b. ii. s. 496.—Potstone, *Aikin*, p. 234.

External Characters.

Its colour is greenish-grey, of different degrees of intensity; the darker varieties incline to leek-green, and blackish-green.

It occurs massive, and in granular concretions, which are indistinct.

Internally it is glistening, inclining to shining, and is pearly, inclining to resinous.

The fracture is curved, and imperfect foliated, which passes into slaty.

The fragments are indeterminate angular, or slaty.

It is translucent on the edges.

It affords a white-coloured streak.

It is soft, passing into very soft.

It is perfectly sectile.

It feels greasy.

It is rather difficultly frangible.

Specific gravity, 2.800, *Saussure* and *Karsten*.

Chemical Character.

It is infusible before the blowpipe.

Constituent

Constituent Parts.

		Potstone of Chiavenne.	
Silica, - -	39	Silica, -	38.12
Magnesia, -	16	Magnesia,	38.54
Oxide of Iron,	10	Alumina,	6.66
Carbonic Acid,	20	Lime, -	0.41
Water, -	10	Iron, -	16.62
	——	Fluoric Acid,	0.41
	95		——
	Tromsdorf.	*Wiegleb.*	99.76

Geognostic Situation.

It occurs in thick beds, in primitive clay-slate.

Geographic Situation.

Europe.—It occurs abundantly on the shores of the Lake Como in Lombardy, and at Chiavenna in the Valteline: also in different parts in Norway, Sweden, and Finland.

Africa.—It is said to occur in Upper Egypt.

America.—It is found in the country around Hudson's Bay; and in Greenland.

Uses.

When newly extracted from the quarry, it is very soft and tenacious, so that it is frequently fashioned into various kinds of culinary vessels, which harden in drying, and are very refractory in the fire. These vessels do not communicate any taste to the food boiled in them, and have been used for culinary purposes for ages. Pliny mentions them, and describes the mode of making them, and the changes they experi-

ence

ence by using. In those times, potstone was named *Lapis Comensis*, and *Lapis Siphnius*, from the island of Siphnus, (the present Siphanto), where it was found. In Upper Egypt, this mineral is named *Pierre de Baram*, and is used for culinary vessels. Quarries of potstone were worked on the banks of the Lake Como, from the beginning of the Christian era to the 25th of August 1618, when they fell in and destroyed the neighbouring town of Pleurs. It was there used for culinary vessels and oven-soles, both of which were uncommonly durable. In proof of this, it is mentioned, that an oven at Liddus, in the Valais, stood unimpaired for several hundred years. The town of Pleurs drew annually from those quarries, stone to the value of 60,000 ducats. In Greenland and Hudson's Bay, culinary vessels and lamps are made of potstone; and in Norway and Sweden it is used for lining stoves, ovens, and furnaces.

Observations.

It is very nearly allied to Indurated Talc, from which it is distinguished by its deeper grey colour, higher lustre, kind of fracture, distinct concretions, and lower degree of translucency.

Ninth Subspecies.

Steatite, or Soapstone.

Speckstein, *Werner*.

Creta Hispanica, *Wall.* t. i. p. 396.; Creta Briansonia, *Wall.* t. i. p. 390.—Speckstein, *Wid.* s. 451.—Semi-indurated Steatites, *Kirw.*

Kirw. vol. i. p. 151.—Speckstein, *Estner,* b. ii. s. 791. *Id. Emm.* b. i. s. 363.—Steatite compatta, *Nap.* p. 296.—Steatite, *Lam.* t. ii. p. 343.—La Steatite commune, *Broch.* t. i. p. 474. —Talc Steatite, *Haüy,* t. iii. p. 252.—Speckstein, *Reuss,* b. ii. s. 176. *Id. Lud.* b. i. s. 132. *Id. Suck.* 1r th. s. 544. *Id. Bert.* s. 141. *Id. Mohs,* b. i. s. 541. *Id. Hab.* s. 66. *Id. Leonhard,* Tabel. s. 27.—Talc Steatite, *Lucas,* p. 84.—Steatite commune, *Brong.* t. i. p. 496.—Talc Steatite, *Brard,* p. 198.—Dichter Speckstein, *Haus.* s. 100.—Speckstein, *Karst.* Tabel. s. 44.—Steatite, *Kid,* vol. i. p. 96.—Speckstein, *Steffens,* b. i. s. 293. *Id. Lenz,* b. ii. s. 644. *Id. Oken,* b. i. s. 380. *Id. Hoff.* b. ii. s. 236. *Id. Haus.* Handb. b. ii. s. 749.—Soapstone, *Aikin,* p. 235.

External Characters.

Its principal colour is white, of which it presents the following varieties: greyish, greenish, seldom yellowish, and reddish-white; the reddish-white borders on flesh-red; the greenish-white passes into mountain, oil, and, lastly, into siskin-green; and the yellowish-grey into pale isabella-yellow. It is sometimes marked with spotted, and dendritic greyish-black delineations.

It occurs massive, disseminated, in crusts, reniform; and also in the following figures:

1. Equiangular six-sided prism, acutely acuminated on both extremities with six planes.
2. Acute double six-sided pyramid.
3. Rhomboid.

The six-sided prism, and six-sided pyramid, are from rock-crystal, and the rhomboid from calcareous-spar. Both appear to be supposititious.

They

Worcester, in the manu-
fullers earth and indurated
greasy matter, and hence it is
grease from silk and woollen
in polishing gypsum, serpen-
pounded and slightly burnt, it
tics. It writes readily on
it differs from common chalk,
hence it is used by glaziers, in mark-
re they are cut with diamond. Tai-
in preference to common chalk, for
e the trace it leaves is not readily ef-
y ground and mixed with a pigment, it
pastel colour, used for painting on glass.
in their baths instead of soap, to soften
h savage tribes eat it, either alone, or mix
d, to deceive hunger. M. Labillardiere
the inhabitants of New Caledonia eat con-
tities of a soft steatite, in which Vauquelin
gnesia, 0.36 silica, 0.17 oxide of iron, and
s no nourishing ingredient. Humboldt as-
t the Otomacks, a savage race on the banks of
co, live for nearly three months of the year
y on a kind of potters-clay. Mr Goldberry says,
Negroes near the mouth of the Senegal mix their
th a white steatite, and eat it without inconvenience;
t is well known that Negroes in general eat earthy
stances with great avidity.

As steatite becomes hard in the fire, and does not alter its shape, it has been successfully employed in imitating

 engraved

engraved gems by M. Vilcot, an artist of Leuttich, in the county of Liege. The subjects intended to be represented, are engraved on it with great ease; it is then exposed to a strong heat, when it acquires a considerable degree of hardness. It is afterwards polished, and may be coloured by means of metallic solutions.

Observations.

1. The yellowish-white variety approaches to Lithomarge, the flesh-red to Bole, and the siskin-green and greenish-grey to Fullers-Earth.

2. It is distinguished from *Talc* and *Chlorite* by fracture, and from *Serpentine* by softness.

3. In trade it is known under the names, Spanish-Chalk, Chalk of Briançon, and Soapstone.

4. Weiss and Steffens are of opinion, that steatite is not an original substance, but has been formed from other minerals, particularly felspar and mica, by a process somewhat resembling that which takes place with flesh, when it is converted into a fatty substance. This opinion will be fully considered in the geognostic part of the system.

5. The *Pimelite* of Karsten, which occurs along with chrysoprase at Kosemütz in Silesia, and contains 15.12 of Nickel, is arranged by Werner as a variety of Steatite.

Tenth

Tenth Subspecies.

Figurestone, or Agalmatolite *.

Bildstein, *Werner.*

Agalmatolith, *Klaproth.*

Steatites, particulis impalpabilibus, mollis, semi-pellucidus, lardites, colore flavescente, *Wall.* gen. 28. spec. 186. t. i. p. 399. —Indurated Steatites, *Kirw.* vol. i. p. 153.—La Pierre à Sculpture, *Broch.* t. i. p. 451.—Agalmatolite, *Lud.* b. ii. s. 151. *Id. Suck.* 1r th. s. 503. *Id. Bert.* s. 205. *Id. Leonhard,* Tabel. s. 27. *Id. Haus.* s. 86. *Id. Karst.* Tabel. s. 28. *Id. Kid,* vol. i. p. 181.—Talc graphique, *Haüy,* Tabl. p. 68.—Agalmatolith, *Steffens,* b. i. s. 240.—Bildstein, *Lenz,* b. ii. s. 594. *Id. Oken,* b. i. s. 379. *Id. Hoff.* b. ii. s. 244. *Id. Haus.* Handb. b. ii. s. 440.—Agalmatolite, *Aikin,* p. 202.

External Characters.

Its most common colour is greenish-grey, which on the one side passes into mountain-green, asparagus-green, and ore-green, and sometimes greenish-white; on the other into yellowish-grey, pearl-grey, flesh-red, and a colour intermediate between ochre-yellow and yellowish-brown. These colours are generally pale, and sometimes disposed in flamed delineations.

It occurs massive.

Internally it is dull or feebly glimmering.

The

* *Agalmatolite,* from the Greek words *αγαλμα* and λιθος, which signifies *figure-stone,* because it is cut into figures of different kinds in the countries where it is principally found.

The fracture is large and flat conchoidal in the large, and splintery in the small, and sometimes is imperfect slaty.

The fragments are indeterminate angular, and rather sharp-edged, or imperfect tabular.

It is translucent, sometimes only on the edges.

It becomes feebly resinous in the streak.

It is soft.

It is intermediate between sectile and brittle.

It feels rather greasy.

Specific gravity, 2.785, *Kirwan.*—2.815, *Klaproth.*—2.800, 2.827, *Breithaupt.*

Chemical Characters.

It is infusible before the blowpipe.

Constituent Parts.

Chinese Figurestone.			Figurestone of Nagyag.
Silica,	35.00	54.50	55.00
Alumina,	29.00	34.00	33.00
Lime,	2.00		
Potash,	7.00	6.25	7.00
Iron,	1.00	0.75	0.50
Water,	5.00	4.00	3.00
	99.00	99.50	98.50
	Vauquelin.	*Klaproth*, Beit. b. v. s. 21.	*Klaproth*, Id. s. 21.

Geographic Situation.

It occurs in China, and at Nagyag in Transylvania, but the geognostic situations are unknown.

Uses.

This mineral, owing to its softness, can easily be fashioned

shioned into various shapes with the knife: hence, in China, where it frequently occurs, it is cut into figures, generally of men, also into pagodas, cups, snuff-boxes, &c. Baron Veltheim is of opinion, that the celebrated Roman *Vasa murrhina*, brought from the most distant parts of India, were made of figurestone, whilst other antiquaries maintain that they were of porcelain. Data are wanting for enabling us to decide in regard to these *vasa murrhina*.

Observations.

1. This substance was formerly confounded with Steatite, from which it is distinguished by lustre and fracture. It appears to be intermediate between Steatite and Nephrite.

2. Lenz, in the second volume of his Mineralogy, describes what he considers as a distinct subspecies of figurestone, from Ochsenkopf, near Schneeberg in Saxony, where it occurs along with talc, corundum, and magnetic-ironstone. The following analysis of it has been published by Dr John: Silica, 51.50. Alumina, 32.50. Oxide of Iron, 1.75. Oxide of Manganese, 12.00. Potash, 6.00. Lime, 3.00. Water, 5.15.

* Clay-Slate.

Thonschiefer, *Werner*.

Schistus ardesia tegularis, *Wall.* t. i. p. 351.—Thonschiefer, *Wid.* s. 391.—Argillite, *Kirw.* vol. i. p. 234.—Killas, *Id.* p. 237.

---

* Clay-slate, Whet-slate, Black Chalk, and Alum-slate, are placed immediately after the subspecies of Rhomboidal Mica, on account of their affinity with it.

p. 237. *Id. Emm.* b. i. s. 284. *Id. Estner,* b. ii. s. 667.—Ardoise, *Lam.* t. i. p. 110.—Le Schiste argilleux, *Broch.* t. i. p. 395.—Argile schisteuse tegulaire tabulaire, *Haüy,* t. iv, p. 447.—Thonschiefer, *Reuss,* b. ii. s. 151. *Id. Lud.* b. i. s. 113. *Id. Suck.* 1r th. s. 508. *Id. Bert.* s. 215. *Id. Mohs,* b. i. s. 462. *Id. Hab.* s. 42. *Id. Leonhard,* Tabel. s. 23.—Schiste argilleux, *Brong.* t. i. p. 557.—Thonschiefer, *Haus.* s. 87. *Id. Karst.* Tabel. s. 38.—Schistus, or Slate, *Kid,* vol. i. p. 186.—Thonschiefer, *Steffens,* b. i. s. 210. *Id. Lenz,* b. ii. s. 578. *Id. Oken,* b. i. s. 359. *Id. Hoff.* b. ii. s. 98. *Id. Haus.* Handb. b. ii. s. 478.—Clay-Slate, *Aikin,* p. 243.

External Characters.

Its colours are yellowish, ash, smoke, bluish, pearl, and greenish-grey; from greenish-grey it passes into a colour intermediate between leek-green and blackish-green; from dark smoke-grey into greyish-black and bluish-black; and from pearl-grey into brownish-red * and cherry-red.

It is sometimes spotted, striped, or flamed.

It occurs massive.

Its lustre is pearly, and is glistening, or glimmering.

The fracture is more or less perfect slaty; and some varieties approach to foliated, and others to compact. The slaty is either straight, or undulating curved, and the latter has a twofold obliquely intersecting cleavage.

The fragments are generally tabular, seldom long splintery or trapezoidal.

It is opaque.

It affords a greyish-white dull streak.

It is soft.

It is sectile, and easily split.

It

* Houses roofed with the red variety of clay-slate, appear as if covered with copper.

It feels rather greasy.

Specific gravity, 2.661, *Kirwan.*—2.786, *Karsten.*

Chemical Characters.

It is fusible into a slag before the blowpipe.

Constituent Parts.

Silica, -	48.6
Alumina, -	23.5
Magnesia, -	1.6
Peroxide of Iron,	11.3
Oxide of Manganese,	0.5
Potash, -	4.7
Carbon,	0.3
Sulphur, -	0.1
Water, and Volatile Matter, -	7.6
Loss,	1.8
	100 *Daubuisson.*

Silica, -	38.0
Alumina, -	26.0
Magnesia, -	8.0
Lime, -	4.0
Peroxide of Iron,	14.0
	Kirwan.

Geognostic Situation.

It occurs in primitive and transition mountains: in primitive mountains it often rests on mica-slate, and alternates with it; when the mica-slate is awanting, it rests on gneiss, and alternates with it in the same manner as it does with mica-slate; when the gneiss is awanting, it rests on granite, and also alternates with it. These facts show, that clay-slate is sometimes of cotemporaneous formation with mica-slate, sometimes with gneiss, and even with granite: In transition mountains, it rests on and alternates with grey

grey-wacke, grey-wacke-slate, transition trap, transition limestone, and other rocks of the transition class.

Transition clay-slate is sometimes scarcely to be distinguished from the primitive varieties of this rock, otherwise than by its geognostic characters: Transition clay-slate alternates with, and passes into grey-wacke-slate; Primitive clay-slate alternates with, and passes into mica-slate; and these are some of the geognostic characters by which we are enabled to distinguish the one from the other.

Geographic Situation.

It is a very generally distributed rock throughout the mountainous regions in the different quarters of the globe. It abounds in many of the highland districts in Great Britain and Ireland, and in several of the smaller islands that lie near their coasts. On the Continent of Europe, it forms a considerable portion of the Hartz, the Erzgebirge, the Fichtelgebirge, the Thuringerwaldgebirge, and of many other great groups of mountains.

Uses.

It is principally used for roofing of houses. Those varieties of clay-slate used for roofing houses, are named *Roofing-Slate*, and should possess the following properties.

1. They must split easily and regularly into thin and straight plates of the requisite magnitude. This is only the case, however, with such varieties of clay-slate as possess a regular and perfect slaty fracture, without rents, or intermixed foreign parts. A clay-slate which contains grains,

crystals,

crystals, or veins of quartz, garnet, schorl, hornblende, or iron-pyrites, will not split into regular plates or *slates*, because these hard bodies do not yield on splitting the mass, and hence the slate generally breaks at such places. If the clay-slate is very thick-slaty, it cannot be split into slates of sufficient thinness, and hence it is of but little use, because when the slates are beyond a certain thickness they are too heavy for roofs. When the clay-slate is curved-slaty, it does not split into useful slates. It may be noticed, that care must be taken to keep the slate in a damp place, previous to splitting, otherwise, if it becomes dry, it will not split without difficulty. It is therefore advisable to split the masses as soon as possible after separating them from the rock.

2. A good roof-slate must be sufficiently compact, and not porous, so that the rain and snow water may not percolate through and destroy the wooden work of the roof. Some varieties of clay-slate are so porous that they imbibe much water, do not dry easily, and hence afford opportunity for the growth of mosses and lichens, which in time cover the surface of the slate. These plants retain moisture long, and keep the surface, and even the interior of the slate moist, so that during the winter season, by the freezing of the moisture, the slate splits and falls into pieces. In order to ascertain whether or not the slate has the requisite compactness, we have only to dry it completely, then weigh it, afterwards plunge it into water, and allow it to remain for some time. If, after wiping it with a cloth, it has not acquired any considerable increase of weight, it is a proof of its being sufficiently compact; on the contrary, if it absorbs much water, and becomes considerably heavier by immersion, it shews that it is of a porous and loose texture.

It

It is remarked, that the slates in the upper strata in quarries are generally porous and loose in their texture, and hence these are generally thrown away as useless.

3. A good slate must be sufficiently solid, and not brittle and shattery; for such slates break in pieces on the application of but a weak force, and do not form a firm roof. When the slate is too brittle, it flies into pieces during the dressing and boring; if it emits a pretty clear sound when struck with a hammer, it is a proof that it is not over brittle; and if it emits a dull sound, it shows that it is soft and shattery. Lastly, if a slate of inconsiderable thickness breaks easily with the hands, it is a proof of its being too soft.

4. No slate can be used with advantage for roofing houses, which readily decomposes by the action of the weather. The decomposition observed to take place in roof slates is of two kinds: the one is mechanical the other is chemical. The mechanical decomposition is principally caused by the freezing of water in the porous and softer varieties, by which they are split in pieces: the chemical decomposition is caused by the decay of disseminated iron-pyrites, or the increased oxidation of intermixed iron.

5. Lastly, a good slate ought to resist the action of a considerable degree of heat.

The best roof-slates found in Scotland, are those of Easdale, and some neighbouring islands off the coast of Lorn in Argyle, and of Ballihulish in Appin, also in Argyle. The quantity manufactured annually at Easdale and its vicinity, is about five millions, which gives employment to 300 men; and at Ballihulish, it is estimated that about half a million of slates are prepared every year. There are also considerable slate-quarries in the parish of Luss in Dumbartonshire, in Monteath, Strathearn, Strathmore, the Garioch,

Garioch, and other places. The slate principally in use in London, is brought from Wales, from quarries which are worked at Bangor in Caernarvonshire. There are also extensive slate-quarries near Kendal in Westmoreland, and the slates from that quarter, which are of a bluish-green colour, are more highly esteemed by the London builders than those from Wales. They are not of a large size, but they possess great durability, and are well calculated to give a neat appearance to the roof on which they may be placed. French slates were very much in use in London, about seventy years ago; they are of small size, very thin, and consequently light, and therefore much less calculated for the climate of this island than the heavier and more durable slates of England and Scotland.

We shall next mention some other uses of clay-slate.

The dark-coloured, most compact, and solid varieties, named *Table-slate*, are used for writing on, but are previously prepared in the following manner. The plate or slate is first smoothed by means of an iron instrument; it is afterwards ground with sandstone, and slightly polished with tripoli, and, lastly, rubbed with charcoal-powder. It is cut into the required shape, set in a wooden-frame, and is then ready for use. When these table-slates are first taken from the quarry, they are rather soft, hence are easily worked; but they become hard by drying.

The small pieces of slate used for writing with, are obtained from a particular variety of clay-slate, named *Writing-slate*, which, on splitting, falls into prismatic or splintery fragments. In order to form a good writing material, it must be more sectile and softer than table-slate, so that it may leave a coloured streak on its surface, without scratching it. This variety of slate does not occur either frequently or abundantly; and it is remarked, that the

strata

strata in which it is contained are generally traversed by vertical rents, and that the best kinds are found between them. When the slate is separated from the stratum in which it is contained, and laid in heaps, it soon falls into long splintery pieces, which are from a quarter to half an inch thick, and from a few inches to upwards of a foot in length. It is said, that if these pieces are exposed for some time to the action of the sun or frost, they are rendered useless: hence workmen are careful to cover them up, and sprinkle them with water as soon as extracted from the quarry, and preserve them in damp cellars. The pieces are afterwards split, by means of a particular instrument, and then made into the required shape.

In some places in Wales, and also in Germany, clay-slate is used for grave-stones; and it is sometimes turned into vases, and other similar articles. The masses used for grave-stones, are cut smooth with sandstone, polished with tripoli, and, lastly, rubbed with charcoal-powder, or lamp-black, or graphite, in order to deepen the black colour. On account of its softness, it receives but an imperfect polish: hence, in order to give it a higher degree of lustre, it is a practice to dip it into oil, after polishing, by which process its lustre is improved, and it is also rendered more durable. It is remarked, that if a window or door is opened in the apartment where the workmen are turning the clay-slate into any particular form, it very frequently flies in pieces, although after the work is finished, it may be exposed to the usual alternations of temperature without risk of injury.

Pounded or ground clay-slate is used for cleaning the surface of iron, and other kinds of metallic ware. It scarcely acts on the metal, but unites with the adventitious soiling-matter on its surface. Clay-slate, when well ground,

and

and mixed in certain proportions with loam, forms a compound excellently fitted for moulds, as it receives the most delicate impressions, and with the greatest accuracy: hence it is very advantageously employed in cast-iron works. When it is burnt, and afterwards coarsely ground, it may be used in place of sand, in the making of mortar: mortar of this kind is said to become very solid and impermeable under water.

In smelting-houses, it is sometimes employed as a flux, with ores that contain much calcareous earth.

Observations.

It passes into Mica-slate, Chlorite-slate, Talc-slate, Whet-slate, Alum-slate, Drawing-slate, and probably into Compact Felspar.

* Whet-Slate.

Wetzschiefer, *Werner*.

Schistus coticula, *Wall.* t. ii. p. 353.—Wetzschiefer, *Wid.* s. 402.—Novaculite, *Kirw.* vol. i. p. 238.—Wetzschiefer, *Estner*, b. ii. s. 664. *Id. Emm.* b. ii. s. 305.—Pietra cote, *Nap.* p. 270.—Cos, *Lam.* t. ii. p. 105.—Le Schiste à aiguiser, *Broch.* t. i. p. 393.—Argile schisteuse novaculaire, *Haüy*, t. iv. p. 448.—Wetzschiefer, *Reuss*, b. ii. s. 149. *Id. Lud.* b. i. s. 112. *Id. Suck.* 1r th. s. 506. *Id. Bert.* s. 216. *Id. Mohs*, b. i. s. 460. *Id. Hab.* s. 42. *Id. Leonhard*, Tabel. s. 23.—Schiste coticule, *Brong.* t. i. p. 558.—Wetzschiefer, *Haus.* s. 87. *Id. Karsten*, Tabel. s. 38.—Novaculite, or Honestone, *Kid*, vol. i. p. 216.—Wetzschiefer, *Steffens*, b. i. s. 211. *Id. Lenz*, b. ii. s. 576. *Id. Oken*, b. i. s. 359. *Id. Hoff.* b. ii. s. 95. *Id. Haus.* Handb. b. ii. s. 477.—Whet-slate, *Aikin*, p. 245.

External

External Characters.

Its most common colour is greenish-grey; it is found also mountain, asparagus, olive, and oil green.

It occurs massive.

Internally it is feebly glimmering.

The fracture in the large is straight slaty; in the small, splintery.

The fragments are tabular.

It is translucent on the edges.

The streak is greyish-white.

It is soft in a low degree.

It feels rather greasy.

Specific gravity 2.722, *Karsten*.

Geognostic Situation.

It occurs in beds in primitive and transition clay-slate.

Geographic Situation.

It is found at Seifersdorf, near Freyberg; at Launstein and Sonnenberg, in the district of Meinengen; and also in the Hartz, and in Stiria and Siberia. Very fine varieties are brought from Turkey.

Uses.

When cut and polished, it is used for sharping iron and steel instruments. For these purposes, it is necessary that it contain no intermixed hard minerals, such as quartz. The light-green coloured varieties, from the Levant, are the most highly prized: those from Bohemia are also much esteemed in commerce. The Levant whet-slate is brought in masses to Marseilles, and is there cut into pieces of various sizes. It is ground by means of sand or sandstone, and

and polished with pumice and tripoli. These whet-stones, or *hones*, as they are called, ought to be kept in damp and cool places; for when much exposed to the sun, they become too hard and dry for many purposes.

The powder of whet-slate is used for cutting and polishing metals, and is by artists considered as a variety of emery.

Observations.

1. It is distinguished from other minerals by colour, fracture, transparency, and hardness.

2. This subspecies does not include every kind of mineral used as whet-stone; for some varieties of clay-slate, of sandstone, and of slate-clay, are used for that purpose.

* Drawing-Slate, or Black Chalk.

Zeichenschiefer, *Werner*.

Schistus pictorius nigrica, *Wall.* t. i. p. 358.—Zeichenschiefer, *Wern.* Cronst. s. 208.—Black Chalk, *Kirw.* vol. i. p. 195.—Schwarze Kreide, *Estner*, b. ii. s. 661.—Zeichenschiefer, *Emm.* b. i. s. 303.—Schisto pittorio, *Nap.* p. 269.—Melantirite, ou Crayon noire, *Lam.* t. ii. p. 112.—Argile schisteuse graphique, *Haüy*, t. iv. p. 447.—Le Schiste à dessiner, *Broch.* t. i. p. 391.—Zeichenschiefer, *Reuss*, b. ii. s. 146. *Id. Lud.* b. i. s. 112. *Id. Suck.* 1r th. s. 505. *Id. Bert.* s. 217. *Id. Mohs*, b. i. s. 458. *Id. Hab.* s. 43. *Id. Leonhard*, Tabel. s. 23.—Ampelite graphique, *Brong.* t. i. p. 563.—Zeichenschiefer, *Haus.* s. 85. *Id. Karsten*, Tabel. s. 36.—Black Crayon, *Kid*, vol. i. p. 190.—Zeichenschiefer, *Steffens*, b. i. s. 208. *Id. Lenz*, b. ii. s. 575. *Id. Oken*, b. i. s. 361. *Id. Hoff.* b. ii. s. 91. *Id. Haus.* Handb. b. ii. s. 475.—Black Chalk, *Aikin*, p. 242.

 External

External Characters.

Its colour is intermediate between bluish and greyish black, but rather more inclining to the latter colour.

It is massive.

The lustre of the principal fracture is glimmering, of the cross fracture dull.

The principal fracture is slaty, generally straight, sometimes curved; the cross fracture fine earthy.

The fragments are partly tabular, partly long splintery.

It is opaque.

It soils slightly, and writes.

It retains its colour in the streak, and becomes glistening.

It is very soft.

It is sectile.

It is easily frangible.

It adheres slightly to the tongue.

It feels fine, but meagre.

Specific gravity, 2.110, *Kirwan.*—2.111, *Karsten.*

Chemical Character.

It is infusible.

Constituent Parts.

Silica, - -	64.06
Alumina,	11.00
Carbon,	11.00
Water, - -	7.20
Iron, -	2.75

According to *Wiegleb*, Crell's Ann. 1797, s. 485.

Geognostic

Geognostic Situation.

It occurs in beds in primitive and transition clay-slate; also in secondary or flœtz formations.

Geographic Situation.

It is found at Marvilla in Spain, Brittany in France, and in Italy; also in Germany, as in the mountains of Bareuth; and in the coal-formation in Scotland.

Uses.

It is used for drawing, and also as a black colour in painting. When used for drawing, it is cut into square pencils, which are sometimes inclosed in wooden cases, like pencils of graphite or black-lead. We must select for this purpose those varieties having the darkest colour, the finest earthy fracture, and which are free of quartzy particles and veins. It has been found, that these pencils become dry, hard, and unfit for drawing by long keeping. To prevent this evil, the pencils should be kept in a moist place; or, what is better, the slate should be ground, and mixed with gum-water, and run into moulds; and pencils of this kind, if well prepared, will remain long fit for use. We must be careful that too much gum-water is not added, otherwise the particles will be so closely aggregated, that the pencils will not leave a trace on the paper; and on the other hand, we must see that too little gum is not added; for if this be the case, the pencil will soil the paper, and no regular or well formed trace will be left on it.

When black chalk is used for painting, it is first pounded and ground, and then mixed with oil or size, and is used as a black paint. It is, however, not much valued, as it is at best but a coarse colour. Certain varieties burn red, or

 reddish-

reddish-brown, and these are sometimes used for red or brown colours.

Observations.

1. Some varieties of bituminous Shale have been confounded with Black Chalk; but a comparison of their trace on paper, enables us at once to distinguish them: the trace of Bituminous Shale being brownish and irregular, whereas that of Black Chalk is regular and black.

2. The most highly prized varieties of this mineral, are those found in Spain, Italy, and France.

* Alum-Slate.

Alum-Slate is divided into two kinds, viz. Common Alum-Slate, and Glossy Alum-Slate.

First Kind.

Common Alum-Slate.

Gemeiner Alaunschiefer, *Werner*.

Schistus aluminaris? *Wall.* t. ii. p. 32.—Var. of Alaunschiefer, *Wid.* s. 396. *Id.* *Estner*, b. ii. s. 651.—Gemeiner Alaunschiefer, *Emm.* b. i. s. 296.—Schisto aluminoso, *Nap.* p. 264.—Varieté de l'Argile schisteuse, *Haüy*.—Le Schiste alumineux commune, *Broch.* t. i. p. 386.—Gemeiner Alaunschiefer, *Reuss*, b. ii. s. 143.—Alaunschiefer, *Lud.* b. i. s. 110. *Id.* *Suck.* 1r th. s. 529.—Schiefriger Aluminit, *Bert.* s. 219.—Alaunschiefer, *Mohs*, b. i. s. 454.—Gemeiner Alaunschiefer, *Leonhard*, Tabel. s. 22.—Alaunschiefer, *Haus.* s. 86.—Gemeiner Alaunschiefer, *Karsten*, Tabel. s. 36. *Id.* *Steffens*, b. i. s. 205. *Id.* *Lenz*,

Lenz, b. ii. s. 571. *Id. Oken*, b. i. s. 362. *Id. Hoff.* b. ii. s. 84. *Id. Haus.* b. ii. s. 481.

External Characters.

Its colour is intermediate between bluish and iron black.

It occurs massive, and sometimes in roundish balls, which are imbedded in the massive varieties.

Its lustre is more or less glimmering.

The fracture is nearly perfect straight slaty.

The fragments are tabular.

It is opaque.

It does not soil.

It retains its colour in the streak, but becomes glistening.

It is intermediate between soft and semihard.

It is easily frangible, and rather brittle.

Specific gravity 2.384, *Kirwan*.

Second Kind.

Glossy Alum-Slate.

Glänzender Alaunschiefer, *Werner*.

Var. Alaunschiefer, *Wid.* s. 395.—Glänzender Alaunschiefer, *Emm.* b. i. s. 297.—Alaunschiefer, *Estner*, b. ii. s. 651.—Variété de l'Argile schisteuse, *Haüy*.—La Schiste alumineux eclatante, *Broch.* t. i. p. 388.—Glänzender Alaunschiefer, *Reuss*, b. ii. s. 145. *Id. Hab.* s. 49. *Id. Leonhard*, Tabel. s. 22. *Id. Karsten*, Tabel. s. 36. *Id. Steffens*, b. i. s. 206. *Id. Lenz*, b. ii. s. 572. *Id. Oken*, b. i. s. 362. *Id. Hoff.* b. ii. s. 85. *Id. Haus.* Handb. b. ii. s. 481.

External

External Characters.

Its colour is intermediate between bluish and iron black, and it sometimes exhibits on the surface of fissures the pavonine, columbine, or temper-steel tarnish.

It occurs massive.

Its lustre is semi-metallic and splendent, shining, or glistening, on the principal fracture, and glimmering or dull on the cross fracture.

The principal fracture is generally undulating curved and short slaty; seldom inclines to straight slaty. Cross fracture is earthy.

The fragments are tabular, and these run into wedge-shaped fragments.

Specific gravity 2.588, 2.389, *Kirwan.*

In all the other characters it agrees with the preceding subspecies.

Geognostic Situation.

Both subspecies agree in geognostic situation: they occur in primitive, and also in transition clay-slate, and more rarely in veins traversing these rocks. Some varieties of alum-slate have been observed associated with secondary rocks.

Geographic Situation.

It occurs along with greywacke and greywacke-slate in the vicinity of Moffat, in Dumfriesshire; in the transition districts of Lanarkshire, particularly in the neighbourhood of Lead Hills; and near the Ferry-town of Cree in Galloway: there are considerable beds of alum-slate on the Continent of Europe, as in Saxony, Bohemia, France, and Hungary. Esmark observed a vein of alum-slate, about

two

two fathoms wide, at Telkobanya in Hungary; and similar veins are to be seen near Freyberg in Saxony.

Uses.

This mineral, when roasted and lixiviated, affords alum.

Observations.

1. Alum-Slate is distinguished from *Clay-Slate*, by its streak always remaining unaltered in the colour.

2. The two kinds were distinguished by Wallerius and Cronstedt.

The following minerals are placed immediately after the genus Mica, on account of the general affinity to it. Their present situation is not to be considered as fixed, but only temporary.

*

1. Native Magnesia, 2. Magnesite, 3. Meerschuum.

**

4. Nephrite, 5. Serpentine, 6. Fullers Earth.

1. Native Magnesia.

Native Magnesia, *Bruce.*

Bruce on Native Magnesia from New Jersey, American Mineralogical Journal, vol. i. p. 26.–30.

External Characters.

Its colour is snow-white, passing into greenish-white.

It.

It occurs massive, and in granular and prismatic concretions.

Its lustre is pearly.

It is semi-transparent in the mass, transparent in single folia.

It is soft, and somewhat elastic.

It adheres slightly to the tongue.

Specific gravity 2.13.

Chemical Characters.

Before the blowpipe, it becomes opaque and friable, and loses weight. It is soluble in the sulphuric, nitric, and muriatic acids.

Constituent Parts.

Magnesia, - -	70
Water of crystallization,	30
	100

Bruce, American Min. Journal, vol. i. p. 30.

Geognostic and Geographic Situations.

It occurs in small veins in serpentine, at Hoboken in New Jersey.

Observations.

It was discovered by the late Dr Bruce, Professor of Mineralogy in New York, to whom America is deeply indebted for the present flourishing state of mineralogy in that country.

2. Magnesite.

2. Magnesite.

Reine oder Natürliche Talkerde, *Werner*.

Magnesie native, *Broch*. t. ii. p. 499.—Reine Talkerde, *Reuss*, b. ii. s. 223. *Id. Lud.* b. i. s. 154. *Id. Suck.* 1r th. s. 539. —Luftsaure Bittererde, *Bert.* s. 136.—Reine Talkerde, *Mohs*, b. i. s. 528. *Id. Hab.* s. 68. *Id. Leonhard*, Tabel. s. 27.—Magnesite de Mitchell, *Brong*. t. i. p. 490.—Magnesit, *Karsten*, Tabel. s. 48.—Magnesie carbonatée, *Haüy*, Tabl. p. 16.—Magnesit, *Steffens*, b. i. s. 243. *Id. Lenz*, b. ii. s. 631. *Id. Oken*, b. i. s. 386.—Reine Talkerde, *Hoff*. b. ii. s. 216.—Magnesite, *Haus*. Handb. b. iii. s. 824.

External Characters.

Its colour is yellowish-grey or yellowish-white, passing into cream-yellow. It is marked with yellowish and ash-grey spots, and also with bluish-grey dots, and dendritic delineations.

It occurs massive, tuberose, reniform, and in a shape which is intermediate between vesicular and perforated; and the walls of the vesicles are rough and uneven.

It has a rough surface.

Internally it is dull.

The fracture is large and flat conchoidal, which passes into fine earthy.

The fragments are rather sharp-edged.

It is nearly opaque.

It is scratched by fluor-spar, but it scratches calcareous-spar.

It adheres pretty strongly to the tongue.

It feels rather meagre.

It

It is dull in the streak.
It is rather easily frangible.
Specific gravity 2.881, *Haberle.*

Chemical Characters.

It is infusible; but before the blowpipe it becomes so hard as to scratch glass.

Constituent Parts.

Magnesia,	48.00	46.00	45.42
Carbonic Acid, -	52.00	51.00	47.00
Silica, -			4.50
Alumina, -	Trace.	1.00	0.50
Ferruginous Manganese,	Trace.	0.25	0.50
Lime, - -	Trace.	0.16	0.08
Water, - -		1.00	2.00
	Bucholz.	*Bucholz.*	*Bucholz.*

Geognostic and Geographic Situations.

It is found at Hrubschitz in Moravia, in serpentine rocks, along with meerschaum, common and earthy talc, mountain-cork, and rhomb-spar; also at Gulfen, near Kraubat in Upper Stiria, where it occurs in serpentine, along with bronzite; and in serpentine, at Baudissero and Castella-Monte in Italy.

Uses.

The mineral of Baudissero is used in the manufacture of porcelain.

Observations.

Observations.

1. It is characterized by its yellowish colour, denditric delineations, rough surface, dull streak, conchoidal fracture, and hardness.

2. It is distinguished from *Meerschaum*, with which it has been confounded, by its colour, external shape, fracture, meagre feel, and weight.

2. It was first discovered by that excellent mineralogist, the late Dr Mitchell of Belfast.

3. Meerschaum *.

Meerschaum, *Werner*.

Meerschaum, *Wid.* s. 456.—Keffekill, *Kirw.* vol. i. p. 144.—Meerschaum, *Emm.* b. i. s. 378.—Schiuma di Mare, *Nap.* p. 307.—Varieté de Talc, *Lam.* t. i. p. 342.—L'Ecume de Mer, *Broch.* t. i. p. 462.—Meerschaum, *Reuss*, b. ii. s. 219. *Id. Lüd.* b. i. s. 129. *Id. Suck.* 1r th. s. 566. *Id. Bert.* s. 139. *Id. Mohs*, b. i. s. 529. *Id. Hab.* s. 69. *Id. Leonhard*, Tabel. s. 27. *Id. Kid*, vol. i. p. 99. *Id. Karst.* Tabel. s. 42. *Id. Steffens*, b. i. s. 241. *Id. Lenz*, b. ii. s. 626. *Id. Oken*, b. i. s. 386. *Id. Hoff.* b. ii. s. 220. *Id. Haus.* b. ii. s. 744.

External Characters.

Its colours are yellowish and greyish white, seldom snow-white.

It occurs massive.

Internally

* *Meerschaum* in German, signifies *sea-froth*, and is by some philologists alleged to have been applied to this mineral on account of its general aspect and lightness; while others derive it from the Natolian word *myreen*.

Internally it is dull.

The fracture is fine earthy, passing on the one side into flat conchoidal, on the other into even.

The fragments are indeterminate angular, and not particularly sharp-edged.

It is opaque; rarely translucent on the edges.

It becomes slightly shining in the streak.

It does not soil.

It is very soft.

It is sectile.

It is rather difficultly frangible.

It adheres strongly to the tongue.

It feels rather greasy.

Specific gravity, 1.209, *Karsten*.—1.600, *Klaproth*.—0.988, 1.279, *Breithaupt*.

Chemical Characters.

Before the blowpipe, it melts on the edges into a white enamel.

Constituent Parts.

Silica, -	41.50
Magnesia, - -	18.25
Lime, - -	0.50
Water and Carbonic Acid,	39.00
	98.25

Klaproth, Beit. b. ii. s. 172.

Geognostic and Geographic Situations.

Europe.—It occurs in veins in the serpentine of Cornwall; in serpentine, at Hrubschitz in Moravia; at Vallecas, near Madrid in Spain, also in serpentine. It is dug at Sebastopol

bastopol and Kaffa, in the Crimea *; and near Thebes in Greece.

Asia.—It occurs in beds immediately under the soil, at Kittisch and Bursa in Natolia; and in the mountains of Esekischehir, also in Natolia, from 600 to 700 men are employed in digging meerschaum.

Uses.

When first dug from the earth, it is soft and greasy. It lathers with water like soap: hence it is used by some nations, as by the Tartars, for washing. In Turkey, it is made into tobacco-pipes. These pipes are manufactured of the meerschaum of Natolia, and that dug near Thebes. It is prepared for that purpose in the following manner: It is first agitated with water in great reservoirs, and is then allowed to remain at rest for some time. The mixture soon passes into a kind of fermentation, resembling that which porcelain-earth experiences when placed in similar circumstances, and a disagreeable odour, resembling that of rotten eggs, is exhaled. As soon as the smell ceases, the mass is farther diluted with water, which is after a time poured off, and fresh water added repeatedly, until the mass is sufficiently washed and purified: what remains is the mass in a pure state. The pure meerschaum is now dried to a considerable degree, is then pressed into a brass mould, and some days afterwards it is hollowed out. The heads formed in this way are then dried in the shade, and, lastly, baked in a furnace constructed for the purpose. The heads in this state are brought to Constantinople, where they are subjected to farther processes: they are first boiled in milk, and next in linseed-oil and wax; when perfectly

* Gallitzin, Descript. Physique de la Contrée de Tauride, p. 85.

fectly cool, they are polished with rushes and leather. The boiling in oil and wax makes them denser, and more capable of receiving a higher polish; and further, when thus impregnated, they acquire, by use, various shades of red and brown on their surface, which is thought to add very considerably to their beauty. In Turkey, and even in Germany, pipes which have been much used, are more valued than those newly made, on account of the colouring they possess. Indeed, in those countries, there are people whose sole employment is smoking tobacco-pipes, until they acquire the favourite tints of colour. By long use, the heads become black; but by boiling in milk and soap, they become white again.

When meerschaum is exposed to a very high degree of heat, it becomes so hard as to give sparks with steel. It is alleged that the porcelain of Samos was made of the meerschaum found in that island; and it is supposed that the porcelain knives mentioned by Pliny, as being used by surgeons, were made from this mineral.

In Spain it is used in the manufacture of porcelain.

Observations.

1. It is distinguished from *Magnesite* by its colour, difficult frangibility, strong adhesion to the tongue, inferior hardness and specific gravity. Its want of distinct concretions at once distinguishes it from *Native Magnesia.*

2. It is nearly allied to Magnesite, into which it sometimes passes.

3. The *Kel* of the Tartars, and the *Keffikel* of the Turks, is not, as some suppose, a variety of Meerschaum, it appears rather to be a kind of fullers earth.

4. Nephrite.

4. Nephrite.

Nephrit, *Werner*.

Of this mineral there are two kinds, viz. Common Nephrite, and Axestone.

a. Common Nephrite.

Gemeiner Nephrit, *Werner*.

Fetter Nephrit, *Saussure*.

Jaspis lapis Nephriticus, *Wall.* Syst. Min. i. p. 302.—Jade, *Kirw*. vol. i. p. 171.—Le Nephrite commune, *Broch*. t. i. p. 467.—Gemeiner Nephrit, *Reuss*, b. ii. 2. s. 187. *Id. Lud.* b. i. s. 131. *Id. Suck.* 1r th. s. 551. *Id. Bert.* s. 144. *Id. Mohs*, b. i. s. 335.—Jade nephrite, *Lucas*, p. 197. *Id. Brong.* t. i. p. 347. *Id. Brard*, p. 410.—Gemeiner Nephrit, *Leonhard*, Tabel. s. 28.—Nephrit, *Haus.* s. 100. *Id. Karsten*, Tabel. s. 44.—Nephrite, *Kid*, vol. i. p. 113.—Jade nephretique, *Haüy*, Tabl. p. 61.—Gemeiner Nephrit, *Steffens*, b. i. s. 266. *Id. Lenz*, b. ii. s. 507. *Id. Oken*, b. i. s. 331. *Id. Hoff.* b. ii. s. 250. *Id. Haus.* Handb. b. ii. s. 753.

External Characters.

Its colour is leek-green, of various degrees of intensity, and sometimes passes into mountain-green, greenish-grey, and greenish-white.

It occurs massive, in blunt-edged pieces, and rolled pieces.

Internally dull or glimmering, owing to intermixed talc and asbestus.

The

The fracture is coarse-splintery, and the splinters are greenish-white.

The fragments are indeterminate angular, and rather sharp-edged.

It is strongly translucent.

It is nearly as hard as rock-crystal.

It is difficultly frangible.

It feels rather greasy.

It is rather brittle.

Specific gravity, 2.962, Oriental, according to *Karsten.*—3.020, Mexican, *Karsten.*—2.970, 3.071, *Saussure*, the Father.—2.957, *Saussure* the Son.—2.989, 3.024, *Breithaupt* *.

Chemical Characters.

Before the blowpipe, it melts into a white enamel.

Constituent Parts.

Silica, - -	50.50
Magnesia,	31.00
Alumina,	10.00
Iron,	5.50
Chrome, -	0.05
Water, -	2.75

Karsten.

Geognostic and Geographic Situations.

Europe.—In Switzerland, nephrite occurs in granite and gneiss; in the Hartz, in veins that traverse primitive greenstone; and in rolled masses near Leipsic in Saxony.

Asia.

* Clarke Abel gives the following specific gravities of the *Yu* stone of the Chinese, which appears to belong to this mineral, 2.858, 3.19, 3.33, 3.4.

Asia.—The most beautiful varieties of this mineral are brought from Persia and Egypt, from the mines of Seminowski, near Kolyvan in Siberia; and from China.

America.—It is found on the banks of the River of Amazons, and near Tlascala in Mexico.

Uses.

Nephrite, when cut and polished, has always an oily and muddy aspect, yet it is prized as an ornamental stone. The Turks cut it into handles for sabres and daggers. Artists sometimes engrave figures of different kinds on it; and it is said to be highly esteemed as a talisman by the savage tribes of the countries where it is found. It was formerly believed to be useful in alleviating or preventing nephritic complaints: hence it has been called *Nephritic Stone.*

The stone called *Yu* by the Chinese, and which is so highly prized by them, appears to be a variety of nephrite *. It is worked by the Chinese artists into a variety of forms; into beautifully carved rings, worn on the thumbs of archers, to defend them from the friction of the bow-string, and into fine chains, cups, and vases. Mr Clarke Abel saw in China a beautiful vase of a greenish-white *Yu.* The handle represented a lizard, with all its parts minutely displayed. Figures of the same animal were sculptured in high relief on its sides, some crawling up, and others overlooking the rim of the vessel. Whatever part of the exterior surface they left unoccupied, was filled with Chinese characters deeply engraved. Its price was one hundred and twenty Spanish dollars. A sceptre of the whitish

* Clarke Abel's Journey into China, p. 132.

variety was sent from the Emperor of China to the Prince Regent *.

Observations.

1. This mineral is characterized by colour, coarse-splintery fracture, white-coloured splinters, resinous aspect, and considerable hardness and weight.

2. It is the *Pietra d'Egitto* of antiquaries.

3. The *Omphax* of Theophrastus appears to be our nephrite.

4. The South American variety is sometimes named *Amazon-stone.*

b. Axestone.

Beilstein, *Werner.*

Panamustein, *Blumenbach.*

Beilstein, *Estner,* b. ii. s. 851. *Id. Emm.* b. iii. s. 351.—La Pierre de Hache, *Broch.* t. i. p. 470.—Beilstein, *Reuss,* b. ii. 2. s. 120. *Id. Leonhard,* Tabel. s. 28.—Jade axinien, *Brong.* t. i. p. 349.—Neuseelandischer Nephrit, *Oken,* b. i. s. 331.—Beilstein, *Hoff.* b. ii. s. 248.—Schaaliger Serpentin, *Haus.* Handb. b. ii. s. 755.

External Characters.

Its colour is intermediate between grass-green and leek-green, and passes into mountain-green, oil-green, and greenish-grey.

It

* In the article *Prehnite* in this work, the Chinese nephrite is stated, on the authority of Bournon, to be a variety of that mineral, but from the lately published observations of Abel, it appears to belong to nephrite.

It occurs massive.

Internally its lustre is strongly glimmering, inclining to glistening.

The fracture is slaty in the great, and more or less distinctly splintery in the small.

The fragments are tabular.

It is translucent, or only strongly translucent on the edges.

It is semi-hard, approaching to hard.

It is softer than common nephrite.

It is rather difficultly frangible.

Specific gravity, 3.008, 3.000, *Karsten*.—3.007, *Lichtenberg*.—2.932, *Breithaupt*.

Geographic Situation.

It occurs in New Zealand, and several of the islands in the South Sea. Also in Saxony; and at Gothaab in Greenland in primitive rocks.

Uses.

It is used by the natives of New Zealand, and other islanders in the South Sea, for hatchets and ear-drops.

Observations.

1. It is nearly allied to common nephrite, indurated talc, and serpentine.

2. It was first brought to Europe by Captain Cook, and into Germany by Dr Forster, who accompanied that illustrious commander in his second voyage round the world.

5. Serpentine.

Serpentin, *Werner.*

There are two kinds of this mineral, viz. Common Serpentine, and Precious Serpentine.

a. Common Serpentine.

Gemeiner Serpentin, *Werner.*

Steatites serpentinus, *Wall.* t. i. p. 156.—Serpentin, *Wid.* s. 462. *Id. Kirwan,* vol. i. p. 156. *Id. Emm.* b. i. s. 384. *Id. Estner,* b. ii. s. 855.—La Serpentine, *Broch.* t. i. p. 481.—Roche serpentineuse, *Haüy,* t. iv. p. 436.—Gemeiner Serpentin, *Reuss,* b. ii. 2. s. 210. *Id. Lud.* b. i. s. 133. *Id. Suck.* 1r th. s. 561. *Id. Bert.* s. 146. *Id. Mohs,* b. i. s. 551. *Id. Hab.* s. 58. *Id. Leonhard,* Tabel. s. 28.—Serpentine commune, *Brong.* t. i. p. 486.—Gemeiner Serpentin, *Haus.* s. 100. *Id. Karsten,* Tabel. s. 42.—Serpentine, *Kid,* vol. i. p. 93.—Gemeiner Serpentin, *Steffens,* b. i. s. 268. *Id. Lenz,* b. ii. s. 651. *Id. Oken,* b. i. s. 378. *Id. Hoff.* b. ii. s. 255. *Id. Haus.* Handb. b. ii. s. 758.—Common Serpentine, *Aikin,* p. 235.

External Characters.

Its principal colour is green, of which it presents the following varieties: leek, oil, and olive green; from oil-green it passes into mountain-green and greenish-grey; from leek-green it passes into greenish-black; from greenish-black into blackish-green; sometimes it occurs straw-yellow, and rarely yellowish-brown, and liver-brown: further, red, of which the following varieties occur; blood-red, brownish-red, peach-blossom-red, and scarlet-red. The peach-

peach-blossom and scarlet red are the rarest. The colour is either uniform, or veined, spotted, dotted, and clouded; and frequently several of these delineations occur together.

It occurs massive.

Internally it is dull, or glimmering, owing to intermixed foreign parts.

The fracture is small and fine splintery, sometimes small and fine grained uneven, which sometimes passes into even; and it is occasionally large, and flat conchoidal.

The fragments are rather sharp-edged.

It is translucent on the edges, or opaque.

It is soft. It does not yield to the nail, but is scratched by calcareous-spar.

It is rather sectile.

It is rather difficultly frangible.

It feels somewhat greasy.

Specific gravity, 2.348, *Karsten*.—2.587, *Brisson*.—2.561, 2.574, *Kirwan*.—2.560, 2.604, *Breithaupt*.

Physical Characters.

Some varieties of serpentine not only move the magnetic needle, but even possess magnetic poles.

Chemical Characters.

It is infusible before the blowpipe, but on exposure to a higher temperature, it melts with difficulty into an enamel.

Constituent Parts.

Silica, - -	31.50	28.00	Silica, - -	32.00
Magnesia, -	47.25	34.50	Magnesia, -	37.24
Alumina, -	3.00	23.00	Alumina, -	0.50
Lime, - -	0.50	0.50	Lime, - -	10.60
Iron, - -	5.50	4.50	Iron, - -	0.66
Oxide of Manganese,	1.50		Volatile matter, and	
Water, -	10.50	10.50	Carbonic Acid,	14.16
	John, Chem. Untersuch. ii. s. 94.	*Rose*.		*Hisinger*, Afhandlingar, i Fysik, iii p. 303

Richter and Rose discovered a small portion of Chrome in the serpentine of Saxony.

Geognostic Situation.

Serpentine occurs in primitive, transition, and secondary rocks. In primitive mountains, it occurs in beds, often of great thickness, in gneiss, mica-slate, and clay-slate; in transition rocks, it is associated with clay-slate; and in secondary rocks, it is imbedded in greenstone, into which it seems to pass. These beds, particularly those that occur in primitive mountains, contain many of the minerals of the talc and steatite kinds, and not unfrequently ores, particularly of magnetic ironstone, and veins of native copper.

Geographic Situation.

Europe.—In Scotland, it occurs in the islands of Unst and Fetlar, in Shetland; Isle of Glass in the Hebrides *; at Portsoy in Banffshire; near Drimnadrochit, and the [illegible] of Inverness in Inverness-shire; at the Bridge of Cor[illegible] in Forfarshire; between Ballantrae and Girvan, in [illegible]; and near Burntisland in Fifeshire. It abounds in

* [illegible]

in some districts in Cornwall in England; and it occurs at Cloghan Lee, on the west coast of Ireland, in the county of Donnegal *. On the Continent of Europe, it occurs in Saxony, Bohemia, Silesia, Bavaria, Salzburg, Tyrol, Austria, Switzerland, Savoy, Italy, and the island of Corsica.

Asia.—It is found in different districts in Siberia; and in New Holland.

America.—In the Bare Hills near Baltimore in Maryland; in several counties in Pennsylvania; along with limestone in Massachusets; and near Newport in Rhode Island; also in the Island of Cuba.

Uses.

As it is soft and sectile, and takes a good polish, it is cut and turned into vessels and ornaments of various kinds. But it must be used soon after it is quarried, otherwise it becomes harder, and is not so easily turned. At Zöblitz in Upper Saxony, many people are employed in quarrying, cutting, turning, and polishing the serpentine which occurs in that neighbourhood; and the various articles into which it is there manufactured are carried all over Germany. At Portsoy in Banffshire, the serpentine is also turned into a variety of elegant ornamental articles, which, on account of the beauty of the stone, are sold at a high price. The serpentine of Portsoy much exceeds that of Zöblitz in beauty and variety of colour, and hence is deservedly more esteemed. Those varieties which have an intermixture of blood-red, peach-blossom-red, and scarlet-red, and yellowish-green, are the most highly prized: indeed, in Saxony, they are in such estimation, as to be arranged with the precious stones, and claimed as the property of the State. In ancient

* Greenough.

ancient times, serpentine was an article of the materia medica: it was prescribed with wine as a remedy for the stone, recommended as a certain cure for the bite of serpents, and was considered as possessing talismanic powers in lethargy, small-pox, poisoning, and madness. Boetius de Boot gravely remarks, that serpentine has a repulsion for poison of every kind, so that the moment the poisoning liquid is poured into a vessel of this mineral, it begins to foam, and is expelled from it.

Observations.

1. It is distinguished from *Precious Serpentine* by its numerous colours, want of lustre, uneven or splintery fracture, inferior translucency, and inferior hardness.

2. It passes into Steatite, and from thence into Talc, Asbestus, and Amianthus.

3. The greenish-black, with white or red veins, is named *Verde di Prato;* the green with white veins *Verde di Susa.*

4. It is worthy of remark, that Common Serpentine passes on the one hand into Greenstone, and on the other into Asbestus, which passes into Actynolite or Hornblende.

b. Precious Serpentine.

Edler Serpentin, *Werner.*

La Serpentine noble, *Broch.* t. i. p. 484.—Edler Serpentin, *Reuss,* b. ii. 2. s. 210. *Id. Lud.* b. i. s. 134. *Id. Suck.* 1r th. s. 563. *Id. Bert.* s. 147. *Id. Leonhard,* Tabel. s. 28.—Serpentine Noble, *Brong.* t. i. p. 485.—Edler Serpentin, *Haus.* *Id. Karst.* Tabel. s. 42.—Noble Serpentine, *Kid,* vol. i.

vol. i. p. 94.—Edler Serpentin, *Steffens*, b. i. s. 271. *Id.* *Lenz*, b. ii. s. 656. *Id. Oken*, b. i. s. 331. *Id. Hoff.* b. ii. s. 261. *Id. Haus.* Handb. b. ii. s. 756.

This mineral is divided into two sub-kinds, viz. Splintery Precious Serpentine, and Conchoidal Precious Serpentine.

α. Splintery Precious Serpentine.

Edler Splittriger Serpentin, *Werner.*

External Characters.

Its colour is dark leek-green.

It occurs massive.

Internally it is feebly glimmering.

The fracture is coarse and long splintery, and sometimes inclines to slaty in the large.

The fragments are rather sharp-edged.

It is feebly translucent.

It is soft, passing into semi-hard.

Specific gravity, 2.704, *Breithaupt.*

In other characters it agrees with Common Serpentine.

Geognostic and Geographic Situations.

It occurs in the Island of Corsica, and in Bareuth.

Use.

In Corsica it is cut into snuff-boxes, and other similar articles.

Observations.

1. Its inferior lustre, and flat splintery fracture, distinguish it from conchoidal serpentine.

2. It is a rare mineral.

β. Conchoidal

β. Conchoidal Precious Serpentine.

Edler muschlicher Serpentin, *Werner*.

External Characters.

Its colour is leek-green, which sometimes passes into blackish-green; seldom into pistachio-green, siskin-green, and oil-green.

It occurs massive, and disseminated.

Its lustre is glistening, passing into glimmering, and is resinous.

The fracture is flat conchoidal.

The fragments are sharp-edged.

It is translucent, but only translucent on the edges in the dark varieties.

It is intermediate between soft and semi-hard.

Specific gravity 2.561, 2.643, *Breithaupt*.

In other characters it agrees with the foregoing.

Constituent Parts.

Silica, -	42.50
Magnesia,	38.63
Lime,	0.25
Alumina, - -	1.00
Oxide of Iron, -	1.50
Oxide of Manganese,	0.62
Oxide of Chrome, -	0.25
Water, -	15.20

John, Chem. Untersuchungen, b. ii. s. 218.

Geognostic

Geognostic Situation.

It generally occurs intermixed with foliated granular limestone in beds subordinate to gneiss, mica-slate and other primitive rocks. It sometimes occurs in cotemporaneous masses in common serpentine, and then it occasionally contains scales of mica.

Geographic Situation.

It occurs at Portsoy in Banffshire, and in the Shetland islands. In the island of Holyhead. At Sala in Sweden; at Waldheim and Zöblitz in Saxony; at Chambane near Aosta in Italy; Kerchenstein in Silesia; and at Dobschau in Upper Hungary.

Uses.

It receives a finer polish than common serpentine, and was much used by the ancients for pillars and other similar ornamental purposes. At present it is also in great esteem as an ornamental stone.

Observations.

1. The distinctive characters of this kind of serpentine are its simple colours, fracture, lustre, considerable translucency and hardness. Its higher lustre, and conchoidal fracture, distinguish it from *Splintery Serpentine.*

2. Many mineralogists are of opinion, that the Ophites (Οφιτης) of the ancients is precious serpentine. Dr John of Berlin controverts this opinion, and maintains that it is common, not precious serpentine. But the passages in Dioscorides, v. 162. and in Pliny, Hist. Nat. xxxvi. 7. do not countenance either opinion, but show that the ancient name

name Ophites was applied to a mixture of precious serpentine and foliated granular limestone, which is known to artists under the name *Verde Antico* and *Polzevera*, and is found not only in Italy, but also in Sweden, isle of Anglesey, and in Scotland.

6. Fullers Earth.

Walkerde, *Werner*.

Walkerde, *Wid.* s. 429.—Fullers Earth, *Kirw.* vol. i. p. 184.—Walkerde, *Estner*, b. ii. s. 777. *Id. Emm.* b. ii. s. 375.—Terra da Follone, *Nap.* p. 258.—La Terre à foulon, *Broch.* t. i. p. 464.—Argile smectique, *Haüy*, t. iv. p. 443.—Walkerde, *Reuss*, b. ii. s. 111. *Id. Lud.* b. i. s. 130. *Id. Mohs*, b. i. s. 532. *Id. Hab.* s. 39. *Id. Leonhard*, Tabel. s. 27.—Argile smectique, *Brong.* t. i. p. 522.—Walkthon, *Haus.* s. 86.—Walkerde, *Karsten*, Tabel. s. 28.—Fullers Earth, *Kid*, vol. i. p. 175.—Walkerde, *Steffens*, b. i. s. 250. *Id. Lenz*, b. ii. s. 640. *Id. Oken*, b. i. s. 385. *Id. Hoff.* b. ii. s. 230.—Walkthon, *Haus.* Handb. b. ii. s. 461.—Fullers Earth, *Aikin*, p. 239.

External Characters.

Its colours are greenish-white, greenish-grey, olive-green, and oil-green. Some varieties exhibit clouded and striped colour-delineations.

It occurs massive.

It is dull.

The fracture is coarse and fine grained uneven; some varieties are large conchoidal; and others incline to slaty.

The fragments are blunt-edged, and occasionally incline to slaty.

It

It is opaque; but when it inclines to steatite it is translucent on the edges.

It becomes shining and resinous in the streak.

It is very soft, sometimes nearly friable.

It is sectile.

It scarcely adheres to the tongue.

It feels greasy.

Specific gravity, 1.72, *Karsten.*—1.198, *Hoffmann.*—2.198, *Breithaupt.*

Chemical Characters.

It falls into powder in water, without the crackling noise which accompanies the disintegration of bole.

It melts into a brown spongy scoria before the blowpipe.

Constituent Parts.

	Fullers Earth of Rygate.	
Silica, - -	53.00	51.8
Alumina,	10.00	25.0
Magnesia,	1.25	0.7
Lime, -	0.50	3.3
Muriat of Soda, -	0.10	
Trace of Potash.		
Oxide of Iron, -	9.75	0.7
Water, -	24.00	15.5
	98.60	*
	Klaproth, Beit. b. iv. s. 338.	*Bergmann*, Opusc. t. iv. p. 156.

Geognostic

* Gehlen found Chrome in fullers earth.

Geognostic and Geographic Situations.

In England, it occurs in beds, sometimes below, sometimes above the chalk formation; at Rosswein, in Upper Saxony, under strata of greenstone-slate; and in different places in Austria, Bavaria, and Moravia, it is found immediately under the soil.

Uses.

This mineral was employed by the ancients for cleaning woollen, and also linen cloth, and they named it *Terra Fullonum*, and *Creta Fullonum;* hence the name *Fullers Earth.* The *Morochtus* of Dioscorides, which he celebrates on account of its remarkable saponaceous properties, is conjectured to have been a variety of fullers earth. Some ancient writers describe it under the name *Galactites*, because it communicates to water a milk-white colour; also *Mellilites*, from the fancied sweet taste it communicates to water. The fullers earth of different countries varies in goodness: the most celebrated, and the best, is that found in Buckinghamshire and Surry. Good fullers earth has a greenish-white or greenish-grey colour, falls into powder in water, appears to melt on the tongue like butter, communicates a milky colour to water, and deposites very little sand when mixed with boiling water. The remarkable detersive property of this substance depends on the alumina it contains; and it appears that the proportion of this should not be less than a fourth or fifth of the whole mass. It should not, however, be much more, for in that case the fullers earth would be so tenacious that it would not diffuse itself through water *. Before the general use of

* Kid's Mineralogy, vol. i. p. 476.

of soap, this substance was very universally employed for cleansing woollen cloth; but in consequence of the general substitution of soap, it is now much less used than formerly *. It is also used for extracting greasy stains or spots from woollen cloth, and from silk. When we wish to remove a greasy stain or spot, the fullers earth is scraped down, and then diffused in hot water; in this state it is applied to the cloth or silk, allowed to dry, and afterwards brushed off.

Observations.

1. Fullers earth, although nearly allied to Steatite, is distinguished from it by colour, fracture, opacity, and inferior specific gravity. Some varieties of steatite, particularly the greenish-grey, pass into fullers earth.

2. Werner is of opinion that the fullers earth of Rosswein in Saxony, is formed by the decomposition of greenstone-

* Although the demand for fullers earth is not now nearly so great as it was formerly, in consequence of many of the clothiers using soap instead of it, yet there is still a considerable demand for it, especially for that which is procured in Surry. Mr Malcolm, in his Agricultural Report of that county, says, that he endeavoured to ascertain the annual consumption of the kingdom, and that as nearly as might be, he found it to be about 6300 tons; of which quantity, about 4000 tons were sent from Surry. The price at the pit in 1805, was about *5s.* or 6s. *per* ton, whereas in 1744, the price was 4d. *per* bushel, which is after the rate of 8s. *per* ton,—a proof either that the supply had increased, or that the demand had diminished. Fullers earth was deemed by the Legislature of so much consequence to our woollen manufactures, that a special act was passed in the 28th year of the reign of his present Majesty, prohibiting the exportation of fullers earth and fulling clay, under a heavy penalty, and obliging the dealers and buyers of it to enter into bonds, to prevent its exportation; and certainly, whatever be the opinion and practice now, the great and acknowledged superiority of English cloth was formerly ascribed, both at home and abroad, to the use of fullers earth. —Edinburgh Encyclopædia, art. *England*, p. 742, 743.

stone-slate, as it is covered by it, and we can trace the gradation from the fully formed fullers earth to the fresh greenstone-slate. Steffens conjectures it to have been formed from previously existing strata, by a proceess analogous to that by which muscular fibre is converted into a kind of spermaceti: hence he says it is of newer formation than the bounding rocks.—May it not be an original deposition of greenstone, in a loose state of aggregation, resembling the disintegrated felspar in certain granites?

ORDER *IV.*

Chemical Characters.

Before the blowpipe, it becomes first black, then brown, but is infusible: on the addition of borax, it melts rapidly, and effervesces, tinging the flame green, and is reduced to the metallic state. In diluted muriatic acid, it effervesces slightly; the oxide of copper dissolves, and there remains behind a nearly colourless and often semi-gelatinous mass of silica, of the same size as the original specimen.—*Aikin.*

Constituent Parts.

Copper,	40.00	42.00
Oxygen, -	10.00	7.63
Carbonic Acid, -	7.00	3.00
Water,	17.00	17.50
Silica, - -	26.00	28.37
Sulphat of Lime -		1.50
	100.00	100.00
	Klaproth, Beit. b. i. s. 36.	*John*, Chem. Unters. b. ii. s. 260.

Geognostic Situation.

It is met with in the same geognostic situations as malachite, and is usually associated with copper-pyrites, tile-ore, grey copper-ore, malachite, brown ironstone, and other ores.

Geographic Situation.

Europe.—It occurs in Cornwall, along with olivenite, and also in the vale of Newlands, near Keswick. It is found

β. Conchoidal Precious Serpentine.

Edler muschlicher Serpentin, *Werner*.

External Characters.

Its colour is leek-green, which sometimes passes into blackish-green; seldom into pistachio-green, siskin-green, and oil-green.

It occurs massive, and disseminated.

Its lustre is glistening, passing into glimmering, and is resinous.

The fracture is flat conchoidal.

The fragments are sharp-edged.

It is translucent, but only translucent on the edges in the dark varieties.

It is intermediate between soft and semi-hard.

Specific gravity 2.561, 2.643, *Breithaupt*.

In other characters it agrees with the foregoing.

Constituent Parts.

Silica, -	42.50
Magnesia,	38.63
Lime,	0.25
Alumina, - -	1.00
Oxide of Iron, -	1.50
Oxide of Manganese,	0.62
Oxide of Chrome, -	0.25
Water, -	15.20

John, Chem. Untersuchungen, b. ii. s. 218.

Geognostic

Geognostic Situation.

It generally occurs intermixed with foliated granular limestone in beds subordinate to gneiss, mica-slate and other primitive rocks. It sometimes occurs in cotemporaneous masses in common serpentine, and then it occasionally contains scales of mica.

Geographic Situation.

It occurs at Portsoy in Banffshire, and in the Shetland islands. In the island of Holyhead. At Sala in Sweden; at Waldheim and Zöblitz in Saxony; at Chambane near Aosta in Italy; Kerchenstein in Silesia; and at Dobschau in Upper Hungary.

Uses.

It receives a finer polish than common serpentine, and was much used by the ancients for pillars and other similar ornamental purposes. At present it is also in great esteem as an ornamental stone.

Observations.

1. The distinctive characters of this kind of serpentine are its simple colours, fracture, lustre, considerable translucency and hardness. Its higher lustre, and conchoidal fracture, distinguish it from *Splintery Serpentine.*

2. Many mineralogists are of opinion, that the Ophites (Οφιτης) of the ancients is precious serpentine. Dr John of Berlin controverts this opinion, and maintains that it is common, not precious serpentine. But the passages in Dioscorides, v. 162. and in Pliny, Hist. Nat. xxxvi. 7. do not countenance either opinion, but show that the ancient

name

e Ophites was applied to a mixture of precious serpen- and foliated granular limestone, which is known to ar- under the name *Verde Antico* and *Polzevera*, and is nd not only in Italy, but also in Sweden, isle of Angle- , and in Scotland.

6. Fullers Earth.

Walkerde, *Werner*.

Walkerde, *Wid.* s. 429.—Fullers Earth, *Kirw.* vol. i. p. 184.— Walkerde, *Estner*, b. ii. s, 777. *Id. Emm.* b. ii. s. 375.— Terra da Follone, *Nap.* p. 258.—La Terre à foulon, *Brock.* t. i. p. 464.—Argile smectique, *Haüy*, t. iv. p. 443.—Walk- erde, *Reuss*, b. ii. s. 111. *Id. Lud.* b. i. s. 130. *Id. Mohs*, b. i. s. 532. *Id. Hab.* s. 39. *Id. Leonhard*, Tabel. s. 27.— Argile smectique, *Brong.* t. i. p. 522.—Walkthon, *Haus.* s. 86. —Walkerde, *Karsten*, Tabel. s. 28.—Fullers Earth, *Kid*, vol. i. p. 175.—Walkerde, *Steffens*, b. i. s. 250. *Id. Lenz*, b. ii. s. 640. *Id. Oken*, b. i. s. 385. *Id. Hoff.* b. ii. s. 230.—Walk- thon, *Haus.* Handb. b. ii. s. 461.—Fullers Earth, *Aikin*, p. 239.

External Characters.

Its colours are greenish-white, greenish-grey, olive-green, and oil-green. Some varieties exhibit clouded and striped colour-delineations.

It occurs massive.

It is dull.

The fracture is coarse and fine grained uneven; so varieties are large conchoidal; and others incline to slaty.

The fragments are blunt-edged, and occasionally incline to slaty.

It

p. 224.—Schlackiges eisenschüssiges Kupfergrün, *Karsten*, Tabel. s. 62.—Muschlicher Pharmakochalzit, *Haus.* s. 136.—Cuivre hydraté silicifere resinite, *Vauquelin*, Journ. des Mines, t. xxxiii. p. 341.—Schlakkiges eisenschüssiges Kupfergrün, *Hoff.* b. iv. s. 157.

External Characters.

It is blackish-green, and dark pistachio-green.

It occurs massive and disseminated.

Internally it is shining and glistening, and the lustre is resinous.

The fracture is small conchoidal.

The fragments are indeterminate angular, and more or less sharp-edged.

It is opaque.

Its colour becomes paler in the streak.

It is soft, verging into very soft.

It is easily frangible.

Constituent Parts.

It is probably a compound of Conchoidal Copper-Green and Oxide of Iron.

Geognostic Situation.

Both subspecies usually occur together, and they frequently pass into each other. They are usually accompanied with copper-green, blue copper, and malachite; frequently also with grey copper, foliated copper-glance, tile-ore, ochry and compact brown iron-ore, compact red copper-ore, quartz, and straight lamellar heavy-spar.

Geographic

Geognostic and Geographic Situations.

In England, it occurs in beds, sometimes below, sometimes above the chalk formation; at Rosswein, in Upper Saxony, under strata of greenstone-slate; and in different places in Austria, Bavaria, and Moravia, it is found immediately under the soil.

Uses.

This mineral was employed by the ancients for cleaning woollen, and also linen cloth, and they named it *Terra Fullonum*, and *Creta Fullonum;* hence the name *Fullers Earth.* The *Morochtus* of Dioscorides, which he celebrates on account of its remarkable saponaceous properties, is conjectured to have been a variety of fullers earth. Some ancient writers describe it under the name *Galactites*, because it communicates to water a milk-white colour; also *Mellilites*, from the fancied sweet taste it communicates to water. The fullers earth of different countries varies in goodness: the most celebrated, and the best, is that found in Buckinghamshire and Surry. Good fullers earth has a greenish-white or greenish-grey colour, falls into powder in water, appears to melt on the tongue like butter, communicates a milky colour to water, and deposites very little sand when mixed with boiling water. The remarkable detersive property of this substance depends on the alumina it contains; and it appears that the proportion of this should not be less than a fourth or fifth of the whole mass. It should not, however, be much more, for in that case the fullers earth would be so tenacious that it would not diffuse itself through water *. Before the general use of

* Kid's Mineralogy, vol. i. p. 476.

of soap, this substance was very universally employed for cleansing woollen cloth; but in consequence of the general substitution of soap, it is now much less used than formerly *. It is also used for extracting greasy stains or spots from woollen cloth, and from silk. When we wish to remove a greasy stain or spot, the fullers earth is scraped down, and then diffused in hot water; in this state it is applied to the cloth or silk, allowed to dry, and afterwards brushed off.

Observations.

1. Fullers earth, although nearly allied to Steatite, is distinguished from it by colour, fracture, opacity, and inferior specific gravity. Some varieties of steatite, particularly the greenish-grey, pass into fullers earth.

2. Werner is of opinion that the fullers earth of Rosswein in Saxony, is formed by the decomposition of greenstone-

* Although the demand for fullers earth is not now nearly so great as it was formerly, in consequence of many of the clothiers using soap instead of it, yet there is still a considerable demand for it, especially for that which is procured in Surry. Mr Malcolm, in his Agricultural Report of that county, says, that he endeavoured to ascertain the annual consumption of the kingdom, and that as nearly as might be, he found it to be about 6300 tons; of which quantity, about 4000 tons were sent from Surry. The price at the pit in 1805, was about 5s. or 6s. *per* ton, whereas in 1744, the price was 4d. *per* bushel, which is after the rate of 8s. *per* ton,—a proof either that the supply had increased, or that the demand had diminished. Fullers earth was deemed by the Legislature of so much consequence to our woollen manufactures, that a special act was passed in the 28th year of the reign of his present Majesty, prohibiting the exportation of fullers earth and fulling clay, under a heavy penalty, and obliging the dealers and buyers of it to enter into bonds, to prevent its exportation; and certainly, whatever be the opinion and practice now, the great and acknowledged superiority of English cloth was formerly ascribed, both at home and abroad, to the use of fullers earth. —Edinburgh Encyclopædia, art. *England*, p. 742, 743.

stone-slate, as it is covered by it, and we can trace the gradation from the fully formed fullers earth to the fresh greenstone-slate. Steffens conjectures it to have been formed from previously existing strata, by a proceess analogous to that by which muscular fibre is converted into a kind of spermaceti: hence he says it is of newer formation than the bounding rocks.—May it not be an original deposition of greenstone, in a loose state of aggregation, resembling the disintegrated felspar in certain granites?

ORDER IV.

d. Two opposite planes, so much larger than the others, that the crystal has a tabular form.

e. Rectangular four-sided prism, or eight-sided prism, acuminated with four planes, which are set on the lateral planes.

The crystals are small and very small, seldom middle-sized. They are sometimes aggregated in globular and botryoidal forms; other crystals occur in druses, or singly superimposed.

The external surface of the particular external shapes is drusy and glimmering; that of the crystals sometimes smooth and splendent; sometimes the lateral planes of the rectangular four-sided prism is obliquely streaked.

Externally the crystallized varieties are shining, but the massive and particular external shapes are dull.

Internally it is shining and glistening, and the lustre is intermediate between vitreous and resinous.

Its cleavage is threefold; two of the cleavages are parallel with the sides of an oblique four-sided prism, the third one with the shorter diagonal of the prism.

The fracture is small and imperfect conchoidal.

The fragments of the prismatic or radiated varieties are wedge-shaped; those of the foliated and conchoidal splintery.

The crystals are translucent, passing into semi-transparent, and are sometimes only translucent on the edges.

The colour becomes lighter in the streak.

It is harder than calcareous-spar, but not so hard as fluor-spar.

It is brittle, and rather easily frangible.

Specific gravity, 3.6082, *Brisson*.—3.652, *Breithaupt*.—3.5,—3.7, *Mohs*.

Chemical

β. Conchoidal Precious Serpentine.

Edler muschlicher Serpentin, *Werner.*

External Characters.

Its colour is leek-green, which sometimes passes into blackish-green; seldom into pistachio-green, siskin-green, and oil-green.

It occurs massive, and disseminated.

Its lustre is glistening, passing into glimmering, and is resinous.

The fracture is flat conchoidal.

The fragments are sharp-edged.

It is translucent, but only translucent on the edges in the dark varieties.

It is intermediate between soft and semi-hard.

Specific gravity 2.561, 2.643, *Breithaupt.*

In other characters it agrees with the foregoing.

Constituent Parts.

Silica, -	42.50
Magnesia,	38.63
Lime,	0.25
Alumina, - -	1.00
Oxide of Iron, -	1.50
Oxide of Manganese,	0.62
Oxide of Chrome, -	0.25
Water, -	15.20

John, Chem. Untersuchungen, b. ii. s. 218.

Geognostic

Geognostic Situation.

It generally occurs intermixed with foliated granular limestone in beds subordinate to gneiss, mica-slate and other primitive rocks. It sometimes occurs in cotemporaneous masses in common serpentine, and then it occasionally contains scales of mica.

Geographic Situation.

It occurs at Portsoy in Banffshire, and in the Shetland islands. In the island of Holyhead. At Sala in Sweden; at Waldheim and Zöblitz in Saxony; at Chambane near Aosta in Italy; Kerchenstein in Silesia; and at Dobschau in Upper Hungary.

Uses.

It receives a finer polish than common serpentine, and was much used by the ancients for pillars and other similar ornamental purposes. At present it is also in great esteem as an ornamental stone.

Observations.

1. The distinctive characters of this kind of serpentine are its simple colours, fracture, lustre, considerable translucency and hardness. Its higher lustre, and conchoidal fracture, distinguish it from *Splintery Serpentine*.

2. Many mineralogists are of opinion, that the Ophites (οφιτης) of the ancients is precious serpentine. Dr John of Berlin controverts this opinion, and maintains that it is common, not precious serpentine. But the passages in Dioscorides, v. 162. and in Pliny, Hist. Nat. xxxvi. 7. do not countenance either opinion, but show that the ancient name

name Ophites was applied to a mixture of precious serpentine and foliated granular limestone, which is known to artists under the name *Verde Antico* and *Polzevera*, and is found not only in Italy, but also in Sweden, isle of Anglesey, and in Scotland.

6. Fullers Earth.

Walkerde, *Werner*.

Walkerde, *Wid.* s. 429.—Fullers Earth, *Kirw.* vol. i. p. 184.—Walkerde, *Estner*, b. ii. s. 777. *Id. Emm.* b. ii. s. 375.—Terra da Follone, *Nap.* p. 258.—La Terre à foulon, *Broch.* t. i. p. 464.—Argile smectique, *Haüy*, t. iv. p. 443.—Walkerde, *Reuss*, b. ii. s. 111. *Id. Lud.* b. i. s. 130. *Id. Mohs*, b. i. s. 532. *Id. Hab.* s. 39. *Id. Leonhard*, Tabel. s. 27.—Argile smectique, *Brong.* t. i. p. 522.—Walkthon, *Haus.* s. 86.—Walkerde, *Karsten*, Tabel. s. 28.—Fullers Earth, *Kid*, vol. i. p. 175.—Walkerde, *Steffens*, b. i. s. 250. *Id. Lenz*, b. ii. s. 640. *Id. Oken*, b. i. s. 385. *Id. Hoff.* b. ii. s. 230.—Walkthon, *Haus.* Handb. b. ii. s. 461.—Fullers Earth, *Aikin*, p. 239.

External Characters.

Its colours are greenish-white, greenish-grey, olive-green, and oil-green. Some varieties exhibit clouded and striped colour-delineations.

It occurs massive.

It is dull.

The fracture is coarse and fine grained uneven; some varieties are large conchoidal; and others incline to slaty.

The fragments are blunt-edged, and occasionally incline to slaty.

It

red copper-ore, tile-ore, and probably grey copper and copper-pyrites; in another with white and green lead-spar; and in a third, with earthy cobalt-ochre, and straight lamellar heavy-spar.

At Leadhills, in Lanarkshire, it is accompanied with galena or lead-glance, ochre of manganese, earthy lead-spar, sparry ironstone, calamine, ochry brown iron-ore, brown hematite, iron-pyrites, green lead-spar, white lead-spar, and lead-vitriol or sulphate of lead.

The most beautiful specimens of this mineral are those found in the mines of Chessy, near Lyons in France.

In Hungary, it is associated with copper-pyrites, malachite, copper-green, grey copper, and iron-ochre; in the Bannat, which produces very beautiful specimens, it occurs along with fibrous malachite, earthy tile-ore, compact red copper-ore, iron-ochre, copper-green, and asbestous actynolite; near Laak in Upper Carniola, with quartz and malachite; in the district of Kamsdorf in Saxony, it is accompanied with yellow and brown iron-ochre, ironshot copper-green, and several ores of copper; at Saalfeld in Thuringia, with straight lamellar heavy-spar, grey copper, malachite, ironshot copper-green, tile-ore, and iron-ochre; at Kupferberg in Silesia, with brown-spar and malachite: the Siberian, which rivals that of the Bannat in beauty, is accompanied with copper-green, malachite, tile-ore, green and white lead-spars, brown iron-ore, heavy-spar, and quartz.

Geographic Situation.

Europe.—It occurs at Leadhills in Dumfriesshire, and Wanlockhead in Lanarkshire; Huel-Virgin and Carharrack in Cornwall. On the Continent, it is met with in the iron-mines at Arendal in Norway; in the government of Olnetz in Russia; in the Hartz; Thalitter in Hessia; Moschelandsberg in Deux-

Geognostic and Geographic Situations.

In England, it occurs in beds, sometimes below, sometimes above the chalk formation; at Rosswein, in Upper Saxony, under strata of greenstone-slate; and in different places in Austria, Bavaria, and Moravia, it is found immediately under the soil.

Uses.

This mineral was employed by the ancients for cleaning woollen, and also linen cloth, and they named it *Terra Fullonum*, and *Creta Fullonum;* hence the name *Fullers Earth.* The *Morochtus* of Dioscorides, which he celebrates on account of its remarkable saponaceous properties, is conjectured to have been a variety of fullers earth. Some ancient writers describe it under the name *Galactites*, because it communicates to water a milk-white colour; also *Mellilites*, from the fancied sweet taste it communicates to water. The fullers earth of different countries varies in goodness: the most celebrated, and the best, is that found in Buckinghamshire and Surry. Good fullers earth has a greenish-white or greenish-grey colour, falls into powder in water, appears to melt on the tongue like butter, communicates a milky colour to water, and deposites very little sand when mixed with boiling water. The remarkable detersive property of this substance depends on the alumina it contains; and it appears that the proportion of this should not be less than a fourth or fifth of the whole mass. It should not, however, be much more, for in that case the fullers earth would be so tenacious that it would not diffuse itself through water *. Before the general use of

* Kid's Mineralogy, vol. i. p. 476.

of soap, this substance was very universally employed for cleansing woollen cloth; but in consequence of the general substitution of soap, it is now much less used than formerly *. It is also used for extracting greasy stains or spots from woollen cloth, and from silk. When we wish to remove a greasy stain or spot, the fullers earth is scraped down, and then diffused in hot water; in this state it is applied to the cloth or silk, allowed to dry, and afterwards brushed off.

Observations.

1. Fullers earth, although nearly allied to Steatite, is distinguished from it by colour, fracture, opacity, and inferior specific gravity. Some varieties of steatite, particularly the greenish-grey, pass into fullers earth.

2. Werner is of opinion that the fullers earth of Rosswein in Saxony, is formed by the decomposition of greenstone-

* Although the demand for fullers earth is not now nearly so great as it was formerly, in consequence of many of the clothiers using soap instead of it, yet there is still a considerable demand for it, especially for that which is procured in Surry. Mr Malcolm, in his Agricultural Report of that county, says, that he endeavoured to ascertain the annual consumption of the kingdom, and that as nearly as might be, he found it to be about 6300 tons; of which quantity, about 4000 tons were sent from Surry. The price at the pit in 1805, was about 5s. or 6s. *per* ton, whereas in 1744, the price was 4d. *per* bushel, which is after the rate of 8s. *per* ton,—a proof either that the supply had increased, or that the demand had diminished. Fullers earth was deemed by the Legislature of so much consequence to our woollen manufactures, that a special act was passed in the 28th year of the reign of his present Majesty, prohibiting the exportation of fullers earth and fulling clay, under a heavy penalty, and obliging the dealers and buyers of it to enter into bonds, to prevent its exportation; and certainly, whatever be the opinion and practice now, the great and acknowledged superiority of English cloth was formerly ascribed, both at home and abroad, to the use of fullers earth. —Edinburgh Encyclopædia, art. *England*, p. 742, 743.

stone-slate, as it is covered by it, and we can trace the gradation from the fully formed fullers earth to the fresh greenstone-slate. Steffens conjectures it to have been formed from previously existing strata, by a proceess analogous to that by which muscular fibre is converted into a kind of spermaceti: hence he says it is of newer formation than the bounding rocks.—May it not be an original deposition of greenstone, in a loose state of aggregation, resembling the disintegrated felspar in certain granites?

ORDER IV.

1. Rather oblique four-sided prism, bevelled on the extremities, the bevelling planes set on the obtuse lateral edges *.
2. The preceding figure truncated on the obtuse lateral edges, which thus forms a six-sided prism, in which the bevelling planes are set on two opposite lateral planes.
3. Acute angular three-sided prism, in which the terminal planes are set on, either straight or oblique.

The crystals are generally short, capillary, and acicular. When very short, they form velvety drusy pellicles; and when longer, they are scopiformly aggregated.

Internally it is intermediate between glistening and glimmering, and the lustre is pearly or silky.

The fragments are wedge-shaped and splintery.

The crystals are translucent, but the massive varieties only translucent on the edges, or opaque,

It is softer than blue copper.

The colour of the streak is pale-green.

It is brittle, inclining to sectile, and easily frangible.

Specific gravity, 3.5718, *Brisson*.—3.661, 3.712, *Breithaupt*.

Chemical Characters.

Before the blowpipe, it decrepitates and becomes black, and is partly infusible, partly reduced to a black slag. It melts with borax, to which it communicates a dark yellowish-green colour, and readily affords with it a bead of copper.

* According to Bournon, the lateral planes of the prism meet under angles of 103° and 77°; whereas those in the oblique four-sided prism of blue copper, are said to meet under angles of 116° and 56°.

β. Conchoidal Precious Serpentine.

Edler muschlicher Serpentin, *Werner*.

External Characters.

Its colour is leek-green, which sometimes passes into blackish-green; seldom into pistachio-green, siskin-green, and oil-green.

It occurs massive, and disseminated.

Its lustre is glistening, passing into glimmering, and is resinous.

The fracture is flat conchoidal.

The fragments are sharp-edged.

It is translucent, but only translucent on the edges in the dark varieties.

It is intermediate between soft and semi-hard.

Specific gravity 2.561, 2.643, *Breithaupt*.

In other characters it agrees with the foregoing.

Constituent Parts.

Silica,	42.50
Magnesia,	38.63
Lime,	0.25
Alumina, - -	1.00
Oxide of Iron, -	1.50
Oxide of Manganese,	0.62
Oxide of Chrome, -	0.25
Water, -	15.20

John, Chem. Untersuchungen, b. ii. s. 218.

Geognostic

Geognostic Situation.

It generally occurs intermixed with foliated granular limestone in beds subordinate to gneiss, mica-slate and other primitive rocks. It sometimes occurs in cotemporaneous masses in common serpentine, and then it occasionally contains scales of mica.

Geographic Situation.

It occurs at Portsoy in Banffshire, and in the Shetland islands. In the island of Holyhead. At Sala in Sweden; at Waldheim and Zöblitz in Saxony; at Chambane near Aosta in Italy; Kerchenstein in Silesia; and at Dobschau in Upper Hungary.

Uses.

It receives a finer polish than common serpentine, and was much used by the ancients for pillars and other similar ornamental purposes. At present it is also in great esteem as an ornamental stone.

Observations.

1. The distinctive characters of this kind of serpentine are its simple colours, fracture, lustre, considerable translucency and hardness. Its higher lustre, and conchoidal fracture, distinguish it from *Splintery Serpentine.*

2. Many mineralogists are of opinion, that the Ophites (οφιτης) of the ancients is precious serpentine. Dr John of Berlin controverts this opinion, and maintains that it is common, not precious serpentine. But the passages in Dioscorides, v. 162. and in Pliny, Hist. Nat. xxxvi. 7. do not countenance either opinion, but show that the ancient name

Geographic Situation.

Europe.—It occurs in Cornwall, along with olivenite; at Saalfeldt in Thuringia, it is associated with malachite, blue copper, copper-green, copper-pyrites, grey copper, yellow and brown cobalt-ochre, red cobalt, and straight lamellar heavy-spar; at Lauterberg in the Hartz, along with blue copper, malachite, and grey copper; at Schwatz in the Tyrol, along with foliated copper-glance, copper-green, fibrous malachite, and blue copper; at Saska in the Bannat, along with copper-ore, green and red copper-ore.

Asia.—In the Gumashevsk mines in Siberia, associated with compact and ochry brown iron-ore, tile-ore, malachite, blue copper, grey copper, red copper-ore, white and yellow lead spars, and native silver, with quartz; also in other mines in Siberia.

America.—Chili.

Observations.

Mr Kirwan suspects, from its olive-green colour, that it may contain arsenic acid. Heergen says, that he convinced himself of the presence of this acid in many varieties, and proposes to consider it as a subspecies of olivenite *; but the experiments of Vauquelin already mentioned, show that it cannot be considered as an arseniate of copper.

GENUS II.

* Heergen, Descripcion y annuncio de varios Minerales del Regno de Chile; and in the Anales de Ciencias Naturales, mes. d. Julio 1801, n. 11. t. 4. p. 198; also in Von Moll's Annalen der Berg und Huttenküinde, 1r B. 2te Lieferung, s. 150.

It is opaque; but when it inclines to steatite it is translucent on the edges.

It becomes shining and resinous in the streak.

It is very soft, sometimes nearly friable.

It is sectile.

It scarcely adheres to the tongue.

It feels greasy.

Specific gravity, 1.72, *Karsten.*—1.198, *Hoffmann.*—2.198, *Breithaupt.*

Chemical Characters.

It falls into powder in water, without the crackling noise which accompanies the disintegration of bole.

It melts into a brown spongy scoria before the blowpipe.

Constituent Parts.

	Fullers Earth of Rygate.	
Silica, - -	53.00	51.8
Alumina,	10.00	25.0
Magnesia,	1.25	0.7
Lime, -	0.50	3.3
Muriat of Soda, -	0.10	
Trace of Potash.		
Oxide of Iron, -	9.75	0.7
Water, -	24.00	15.5
	98.60	*
	Klaproth, Beit. b. iv. s. 338.	*Bergmann*, Opusc. t. iv. p. 156.

Geognostic

* Gehlen found Chrome in fullers earth.

Cuivre carbonaté bleu, *Brard*, p. 293.—Strahlige Kupferlazur, *Karsten*, Tabel. s. 62.—Edler Kupferlazur, *Haus.* s. 135.—Cuivre carbonaté bleu, *Haüy*, Tabl. p. 89.—Feste Kupferlazur, *Hoff.* b. iv. s. 137.—Edler Kupferlazur, *Haus.* Handb. b. iii. s. 1021.—Blue Copper, *Aikin*, p. 30.

External Characters.

Its principal colour is azure-blue, which often passes into blackish-blue, seldomer into Berlin-blue and smalt-blue.

It occurs massive, disseminated, in plates, in crusts; also globular, botryoidal, reniform, stalactitic, and cellular; in prismatic distinct concretions, which are straight, narrow, scopiform, and stellular, and these are again traversed by others which are curved lamellar. Sometimes there is a tendency to granular concretions. It is very frequently crystallized. Its primitive form is an oblique prism, the dimensions of which have not yet been determined. The following are some of its secondary forms:

1. Oblique four-sided prism, rather acutely bevelled on the terminal planes, the bevelling planes set on the acuter lateral edges. It exhibits the following varieties:
   a. The angles which the bevelling planes make with two lateral planes, or with the obtuse lateral edges, truncated.
   b. The acute lateral edges bevelled, the edges of the bevelment, and the obtuse lateral edges, truncated: when these planes meet, an eight-sided prism is formed.
   c. The proper edge of the bevelment on the terminal planes more or less deeply truncated; sometimes so deeply, that the figure appears as a simple oblique four-sided prism.

d. Two

d. Two opposite planes, so much larger than the others, that the crystal has a tabular form.

e. Rectangular four-sided prism, or eight-sided prism, acuminated with four planes, which are set on the lateral planes.

The crystals are small and very small, seldom middle-sized. They are sometimes aggregated in globular and botryoidal forms; other crystals occur in druses, or singly superimposed.

The external surface of the particular external shapes is drusy and glimmering; that of the crystals sometimes smooth and splendent; sometimes the lateral planes of the rectangular four-sided prism is obliquely streaked.

Externally the crystallized varieties are shining, but the massive and particular external shapes are dull.

Internally it is shining and glistening, and the lustre is intermediate between vitreous and resinous.

Its cleavage is threefold; two of the cleavages are parallel with the sides of an oblique four-sided prism, the third one with the shorter diagonal of the prism.

The fracture is small and imperfect conchoidal.

The fragments of the prismatic or radiated varieties are wedge-shaped; those of the foliated and conchoidal splintery.

The crystals are translucent, passing into semi-transparent, and are sometimes only translucent on the edges.

The colour becomes lighter in the streak.

It is harder than calcareous-spar, but not so hard as fluor-spar.

It is brittle, and rather easily frangible.

Specific gravity, 3.6082, *Brisson.*—3.652, *Breithaupt.*—3.5,—3.7, *Mohs.*

Chemical

Cuivre carbonaté bleu, *Brard*, p. 293.—Strahlige Kupferlazur, *Karsten*, Tabel. s. 62.—Edler Kupferlazur, *Haus.* s. 135.—Cuivre carbonaté bleu, *Haüy*, Tabl. p. 89.—Feste Kupferlazur, *Hoff.* b. iv. s. 137.—Edler Kupferlazur, *Haus.* Handb. b. iii. s. 1021.—Blue Copper, *Aikin*, p. 30.

External Characters.

Its principal colour is azure-blue, which often passes into blackish-blue, seldomer into Berlin-blue and smalt-blue.

It occurs massive, disseminated, in plates, in crusts; also globular, botryoidal, reniform, stalactitic, and cellular; in prismatic distinct concretions, which are straight, narrow, scopiform, and stellular, and these are again traversed by others which are curved lamellar. Sometimes there is a tendency to granular concretions. It is very frequently crystallized. Its primitive form is an oblique prism, the dimensions of which have not yet been determined. The following are some of its secondary forms:

1. Oblique four-sided prism, rather acutely bevelled on the terminal planes, the bevelling planes set on the acuter lateral edges. It exhibits the following varieties:
   - *a.* The angles which the bevelling planes make with two lateral planes, or with the obtuse lateral edges, truncated.
   - *b.* The acute lateral edges bevelled, the edges of the bevelment, and the obtuse lateral edges, truncated: when these planes meet, an eight-sided prism is formed.
   - *c.* The proper edge of the bevelment on the terminal planes more or less deeply truncated; sometimes so deeply, that the figure appears as a simple oblique four-sided prism.

d. Two

d. Two opposite planes, so much larger than the others, that the crystal has a tabular form.

e. Rectangular four-sided prism, or eight-sided prism, acuminated with four planes, which are set on the lateral planes.

The crystals are small and very small, seldom middle-sized. They are sometimes aggregated in globular and botryoidal forms; other crystals occur in druses, or singly superimposed.

The external surface of the particular external shapes is drusy and glimmering; that of the crystals sometimes smooth and splendent; sometimes the lateral planes of the rectangular four-sided prism is obliquely streaked.

Externally the crystallized varieties are shining, but the massive and particular external shapes are dull.

Internally it is shining and glistening, and the lustre is intermediate between vitreous and resinous.

Its cleavage is threefold; two of the cleavages are parallel with the sides of an oblique four-sided prism, the third one with the shorter diagonal of the prism.

The fracture is small and imperfect conchoidal.

The fragments of the prismatic or radiated varieties are wedge-shaped; those of the foliated and conchoidal splintery.

The crystals are translucent, passing into semi-transparent, and are sometimes only translucent on the edges.

The colour becomes lighter in the streak.

It is harder than calcareous-spar, but not so hard as fluor-spar.

It is brittle, and rather easily frangible.

Specific gravity, 3.6082, *Brisson.*—3.652, *Breithaupt.*—3.5,—3.7, *Mohs.*

Chemical

Cuivre carbonaté bleu, *Brard*, p. 293.—Strahlige Kupferlasur, *Karsten*, Tabel. s. 62.—Edler Kupferlasur, *Haus*. s. 135. —Cuivre carbonaté bleu, *Haüy*, Tabl. p. 89.—Feste Kupferlasur, *Hoff*. b. iv. s. 137.—Edler Kupferlasur, *Haus*. Handb. b. iii. s. 1021.—Blue Copper, *Aikin*, p. 30.

External Characters.

Its principal colour is azure-blue, which often passes into blackish-blue, seldomer into Berlin-blue and smalt-blue.

It occurs massive, disseminated, in plates, in crusts; also globular, botryoidal, reniform, stalactitic, and cellular; in prismatic distinct concretions, which are straight, narrow, scopiform, and stellular, and these are again traversed by others which are curved lamellar. Sometimes there is a tendency to granular concretions. It is very frequently crystallised. Its primitive form is an oblique prism, the dimensions of which have not yet been determined. The following are some of its secondary forms:

1. Oblique four-sided prism, rather acutely bevelled on the terminal planes, the bevelling planes set on the acuter lateral edges. It exhibits the following varieties:

a. The angles which the bevelling planes make with two lateral planes, or with the obtuse lateral edges, truncated.

b. The acute lateral edges bevelled, the edges of the bevelment, and the obtuse lateral edges, truncated: when these planes meet, an eight-sided prism is formed.

c. The proper edge of the bevelment on the terminal planes more or less deeply truncated; sometimes so deeply, that the figure appears as a simple oblique four-sided prism.

d. Two

d. Two opposite planes, so much larger than the others, that the crystal has a tabular form.

e. Rectangular four-sided prism, or eight-sided prism, acuminated with four planes, which are set on the lateral planes.

The crystals are small and very small, seldom middle-sized. They are sometimes aggregated in globular and botryoidal forms; other crystals occur in druses, or singly superimposed.

The external surface of the particular external shapes is drusy and glimmering; that of the crystals sometimes smooth and splendent; sometimes the lateral planes of the rectangular four-sided prism is obliquely streaked.

Externally the crystallized varieties are shining, but the massive and particular external shapes are dull.

Internally it is shining and glistening, and the lustre is intermediate between vitreous and resinous.

Its cleavage is threefold; two of the cleavages are parallel with the sides of an oblique four-sided prism, the third one with the shorter diagonal of the prism.

The fracture is small and imperfect conchoidal.

The fragments of the prismatic or radiated varieties are wedge-shaped; those of the foliated and conchoidal splintery.

The crystals are translucent, passing into semi-transparent, and are sometimes only translucent on the edges.

The colour becomes lighter in the streak.

It is harder than calcareous-spar, but not so hard as fluor-spar.

It is brittle, and rather easily frangible.

Specific gravity, 3.6082, *Brisson*.—3.652, *Breithaupt*.—3.5,—3.7, *Mohs*.

Chemical

ramid, in which the lateral planes of the one are set on the lateral planes of the other *.

The crystals are middle-sized and small, and sometimes crystallized in druses.

Externally it is smooth and shining; internally glistening and shining, and pearly, inclining to vitreous.

The cleavage is in the directions of the lateral and bevelling planes of the oblique four-sided prism.

The fracture is small-grained uneven, which sometimes passes into imperfect conchoidal.

The fragments are indeterminate angular, and rather sharp-edged.

It is translucent.

It yields a pale verdigris-green, or sky-blue coloured streak.

It is harder than gypsum, but not so hard as calcareous spar.

It is nearly brittle, and uncommonly easily frangible.

Specific gravity, 2.8, 2.9, *Mohs.*—2.881, *Bournon.*

Chemical Characters.

Before the blowpipe it is converted into a black friable scoria.

Constituent Parts.

Oxide of Copper,	-	49
Arsenic Acid,		14
Water,	-	35
		98

Chenevix in Phil. Trans. for 1801.

Geognostic

* The double four-sided pyramid is so flat, that it has a lenticular aspect; hence the name *Lenticular Copper* given to this species.

Geognostic and Geographic Situations.

It has been hitherto found only in Cornwall, where it is associated with copper-mica, and other cupreous minerals.

Observations.

It is characterized by its colour, form, and hardness. Its colour, form, and inferior hardness, distinguish it from *Blue Copper*; and the latter character distinguishes it from *Azurite*, and other similar blue-coloured minerals.

3. Acicular Olivenite.

Nadelförmiger Oliven-malachit, *Mohs.*

This species is divided into four subspecies, viz. Radiated Acicular Olivenite, Foliated Acicular Olivenite, Fibrous Acicular Olivenite, and Earthy Acicular Olivenite.

First Subspecies.

Radiated Acicular Olivenite.

Strahlerz, *Werner.*

Cupreous Arseniate of Iron, *Bournon*, Phil. Trans. for 1801, part i. p. 191.—Cuivre arseniaté ferrifere, *Brong.* t. ii. p. 232. —Strahlenkupfer, *Karsten*, Tabel. s. 64.—Cuivre arseniaté ferrifere, *Haüy*, Tabl. p. 91.—Strahlenkupfer, *Haus.* Handbuch, b. iii. s. 1050.—Strahlerz, *Hoff.* b. iv. s. 168.—Martial Arseniate of Copper, *Aikin*, p. 93.

External

External Characters.

Externally its colour is dark verdigris-green, sometimes bordering on blackish-green; internally it is pale verdigris-green, either pure, or intermixed with sky-blue.

It occurs massive and flat reniform; also in radiated prismatic concretions, which are straight and scopiform; and crystallized in flat oblique four-sided prisms, acuminated with four planes; sometimes the acute edges are truncated, when the prism appears six-sided, or all the lateral edges are truncated, when it appears eight-sided.

The crystals are generally small, and superimposed.

The external surface of the reniform shape is very drusy.

Internally the lustre is intermediate between shining and glistening, and is pearly.

The fragments are wedge-shaped.

It is translucent on the edges.

It is as hard as calcareous-spar.

It is brittle, and easily frangible.

Specific gravity, 3.400, *Bournon*.

Second Subspecies.

Foliated Acicular Olivenite.

Blättriches Olivenerz, *Werner*.

Arseniate of Copper, in the form of an acute octahedron, *Bournon*, Phil. Trans. part i. for 1801.—Cuivre arseniaté aigue, *Brong.* t. ii. p. 231.—Dichtes Olivenerz, *Karsten*, Tabel. s. 64.—Cuivre arseniaté octaedre aigue, *Haüy*, Tabl. p. 91.—Cuivre arseniaté en prisme tetraedre rhomboidal, *Bournon*, Catalogue Mineralogique, p. 254.—Gemeines Oliven Kupfer, *Haus.*

Haus. Handb. b. iii. s. 1045.—Blättriges Olivenerz, *Hoff.* b. iv. s. 171.—Prismatic Arseniate, *Aikin*, p. 94.

External Characters.

Its colour is dark olive-green, which passes on the one side into pistachio-green, on the other into blackish and leek-green.

It seldom occurs massive, and in angulo-granular concretions, generally in drusy crusts, and in small crystals, which present the following varieties of form:

1. Oblique four-sided prism, acutely bevelled on the extremities, the bevelling planes set on the acute lateral edges.
2. Preceding figure, in which the obtuse lateral edges are more or less deeply truncated.
3. Acute double four-sided pyramid; sometimes the angles on the common base are flatly bevelled; and the bevelling planes are set on the lateral edges.

The crystals are small and very small, and always superimposed.

The planes of the crystals are smooth, shining, and splendent.

Internally it is glistening, and the lustre is resinous, inclining to pearly.

The cleavage or foliated structure is imperfect.

The fracture is small and imperfect conchoidal, which passes into uneven.

The fragments are indeterminate angular, and rather sharp-edged.

It ranges from translucent to translucent on the edges.

It yields an olive-green coloured streak.

It is as hard as calcareous-spar.

It is rather brittle, and easily frangible.

Specific gravity, 4.280, *Bournon*.—4.2, 4.6, *Mohs*.

Chemical Characters.

Before the blowpipe, it first boils, and then gives a hard reddish-brown scoria.

Constituent Parts.

Oxide of Copper, -	60.0
Arsenic Acid, - -	39.7
	99.7

Chenevix, Phil. Trans. 1801.

Geognostic and Geographic Situations.

It has been hitherto found only in the copper-mines of Cornwall.

Third Subspecies.

Fibrous Acicular Olivenite.

Fasriges Olivenerz, *Werner*.

Hæmatitiform and Amianthiform Arseniate, *Bournon*, Phil. Trans. for 1801.—Cuivre arseniaté capillaire et mamelonné, *Brong*. t. ii. p. 231, 232.—Fasriges Olivenerz, *Karsten*, Tabel. s. 64.—Cuivre arseniaté mamelonné fibreux, *Haüy*, Tabl. p. 91.—Cuivre arseniaté en petites masses habituellement fibreuses et mamelonnées, *Bournon*, Catalogue Mineralogique, p. 259.—Fasriges Oliven Kupfer, *Haus*. Handbuch, b. iii. s. 1047.—Fasriges Olivenerz, *Hoff*. b. iv. s. 173.—Hæmatitic and Amianthiform Arseniate, *Aikin*, p. 94.

External

External Characters.

Its colour is olive-green of different degrees of intensity. The darker varieties border on blackish-green, the lighter pass into pistachio-green, straw-yellow, liver-brown, wood-brown, and greenish-white.

The colours are sometimes arranged in curved and striped delineations.

It occurs massive, and reniform; in fibrous concretions which are delicate, straight, and scopiform, and these are collected into coarse or small granular concretions, and sometimes traversed by others, which are curved lamellar; also crystallized in capillary and acicular oblique four-sided prisms, in which the obtuse lateral edges are truncated, and bevelled on the extremities, the bevelling planes being set on the acute edges.

The crystals are small and very small, and sometimes scopiformly aggregated.

Internally the massive varieties are glistening or glimmering, with a pearly or silky lustre.

The fragments are intermediate angular, and wedge-shaped.

It is opaque, seldom translucent on the edges, and only translucent in the crystals.

It is as hard as calcareous-spar.

It is rather brittle.

The fibres are sometimes flexible *.

* The fibres are sometimes so delicate, so short, and so confusedly grouped together, that the whole appears like a dusty cottony mass, the true nature of which is discoverable only by the lens. At other times, this variety appears in thin laminæ, rather flexible, sometimes scarcely perceptible to the naked eye, sometimes tolerably large, and perfectly like Amianthus papyraceus.—*Bournon*, Phil. Trans. for 1801, part i. p. 180.

The streak is brown or yellow.

Specific gravity, between 4.100 and 4.200, *Bournon.*

Constituent Parts.

	Amianthiform.	Hæmatitiform.
Oxide of Copper, -	50	50
Arsenic Acid,	29	29
Water, -	21	21
	100	100

Chenevix, in Phil. Trans. for 1801.

Geognostic and Geographic Situations.

It is associated generally with the other arseniates of copper, and various ores of copper.

It occurs principally in Cornwall; it has been lately discovered in small quantity at Zinwald in Saxony, and, I believe, also in the Kaisersteimel, on the Rhine.

Fourth Subspecies.

Earthy Acicular Olivenite.

Cuivre arseniaté terreux, *Haüy*, Tabl. p. 91.—Erdiches Olivenkupfer, *Haus.* Handb. b. iii. s. 1049.

External Characters.

Its colours are olive-green, verdigris-green, and siskin-green.

It occurs massive, disseminated, and in crusts.

It is dull.

The

The fracture is fine earthy.

It sometimes occurs in concentric lamellar distinct concretions.

It is opaque.

It is soft and very soft.

Geognostic and Geographic Situations.

It occurs along with the other subspecies of olivenite in the copper-mines of Cornwall.

4. Hexahedral Olivenite, or Cube-Ore.

Wurfelerz, *Werner.*

Wurfelerz, *Reuss*, b. iv. s. 163. *Id. Lud.* b. i. s. 183.—Arsenik-saures Eisen, *Suck.* 2ter th. s. 297.—Wurfelerz, *Bert.* s. 420. *Id. Mohs*, b. iii. s. 437.—Fer arseniaté, *Lucas*, p. 148.—Wurfelerz, *Leonhard*, Tabel. s. 68.—Fer arseniaté, *Brong.* t. ii. p. 182. *Id. Brard*, p. 332.—Wurfelerz, *Karsten*, Tabel. s. 66. —Pharmakosiderit, *Haus.* s. 138.—Arseniate of Iron, *Kid*, vol. ii. p. 101.—Fer arseniaté, *Haüy*, Tabl. p. 100.—Pharmakosiderit, *Haus.* Handb. b. iii. s. 1066.—Wurfelerz, *Hoff* b. iv. s. 177.—Arseniate of Iron, *Aikin*, p. 107.

External Characters.

Its colour is pistachio-green, of different degrees of intensity, which passes on the one side into olive-green, on the other into blackish green; it rarely approaches to leek-green.

It occurs massive; and crystallized in the following figures:

1. Perfect cube.

2. Cube

2. Cube, in which four diagonally opposite angles are truncated.
3. Cube truncated on all the edges.
4. Cube truncated on all the edges and angles.

The crystals are small and very small, and always superimposed and in druses.

The planes of the crystals are smooth and splendent.

Internally it is glistening, and the lustre is intermediate between vitreous and resinous.

It has a cleavage which is parallel with the truncations on the angles.

The fragments are indeterminate angular, and rather sharp-edged.

It is translucent, or translucent on the edges.

The streak is straw-yellow.

It is harder than gypsum, but softer than calcareous-spar.

It is rather brittle, and easily frangible.

Specific gravity, 3.000, *Bournon*.—2.9, 3.0, *Mohs*.

Chemical Characters.

Before the blowpipe it melts, and gives out arsenical vapours.

Constituent Parts.

Iron, - -	48	Arsenic Acid,	31.0
Arsenic Acid,	18	Oxide of Iron,	45.5
Water of crystallization, -	32	Oxide of Copper,	9.0
		Silica, - -	4.0
Carbonate of Lime,	2 to 3	Water, -	10.5
	100		100
Vauquelin, in Brong. Min. t. ii. p. 183.		*Chenevix*, in Phil. Trans. for 1801.	

Geognostic

Geognostic Situation.

It is found in veins, accompanied with ironshot quartz, copper-glance or vitreous copper, copper-pyrites, and brown iron-ore.

Geographic Situation.

It occurs in Tincroft, Carrarach, Muttrel, Huel-Gorland, and Gwenap mines in Cornwall; and at St Leonard, in the department of Haut-Vienne in France.

* Atacamite, or Muriate of Copper.

Salzkupfererz, *Werner*.

This species is divided into two subspecies, viz. Compact and Arenaceous.

First Subspecies.

Compact Atacamite, or Muriate of Copper (a).

Festes Salzkupfererz, *Werner*.

Cuivre muriaté massif, *Brong*. t. ii. p. 228.—Gemeines Salzkupfererz, *Karsten*, Tabel. s. 64.—Cuivre muriaté, *Haüy*, Tabl. p. 89.—Blättricher & Strahliger Smaragdochalzit, *Haus*. Handb. b. iii. s. 1039.—Salzkupfererz, *Hoff*. b. iv. s. 180.—Muriate of Copper, *Aikin*, p. 92.

External

(a) I place this mineral immediately after the Genus Olivenite, on account of its resemblance to it; but want of accurate information in regard to it, prevents me including it as a species of that genus.

External Characters.

Its colour is leek-green, which passes on the one side into blackish-green, on the other into pistachio-green.

It occurs massive, disseminated, imperfect reniform, in prismatic distinct concretions, which are short, small and scopiform, also in granular concretions; in crusts or investing; and in short needle-shaped crystals, of the following forms:

1. Oblique four-sided prism, bevelled on the extremities; the bevelling planes set on the acute lateral edges.
2. The preceding figure, in which the acuter lateral edges are deeply truncated, thus forming a six-sided prism.

Internally it is shining and glistening, and pearly.

It has an imperfect cleavage.

The fragments are indeterminate angular.

It is translucent on the edges.

It is soft.

It is brittle, and easily frangible.

Specific gravity, 4.4 ?

Chemical Character.

It tinges the flame of the blowpipe of a bright green and blue, muriatic acid rises in vapours, and a bead of copper remains on the charcoal. It is soluble in nitric acid without effervescence.

Constituent

Constituent Parts.

Oxide of Copper,	73.0	76.595
Water, - -	16.9	12.767
Muriatic Acid, -	10.1	10.638
	100.0	100.000
	Klaproth, Beit. b. iii. s. 200.	*Proust.* in Journ. de Phy. t. 50. p. 63.

Geognostic and Geographic Situations.

It occurs in veins at Los Remolinos, La Soledad, Guasco, Caymas, and Ojanos, in Chili; also at Virneberg near Rheinbreitenbach on the Rhine, and at Schwarzenberg in Saxony. In the fissures of the lavas of Vesuvius, particularly those of the years 1804 and 1805.

Observations.

This mineral was first brought from Chili to Europe, by Mr Christian Heuland, brother of the present Mr Heuland, the first and principal collector of the minerals of that remote and interesting country.

Second Subspecies.

Arenaceous Atacamite, or Copper-Sand.

Kupfersand, *Werner.*

Cuivre muriaté pulverulent, *Haüy*, t. iii. p. 561. *Id. Brong.* t. ii. p. 229.—Sandiges Salzkupfer, *Karsten*, Tabel. s. 64.—Cuivre muriaté pulverulent, *Haüy*, Tabl. p. 89.—Sandiger Smaragdochalzit, *Haus.* Handb. b. iii. s. 1040.

External

External Characters.

Its colour is grass-green, inclining to emerald-green.

It occurs in scaly particles, which are shining, glistening, and pearly.

It does not soil.

It is translucent.

Constituent Parts.

Oxide of Copper, -	63	70.5
Water, -	12	18.1
Muriatic Acid,	10	11.4
Carbonate of Iron, -	1	
Mixed Siliceous Sand,	11	
	97	100.0
	La Rochefoucault, *Berthollet*, and *Fourcroy*, Mem. de l'Acad. 1786, p. 158.	*Proust*, Journ. de Phys. t. 50. p. 63.

Geognostic and Geographic Situations.

It is found in the sand of the river Lipes, 200 leagues beyond Copiapu, in the desart of Atacama, which separates Chili from Peru.

Observations.

It was brought from South America by the traveller Dombey.

GENUS IV.

Genus IV. EMERALD COPPER.

Smaragd Malachit, *Mohs*.

This genus contains but one species, viz. Rhomboidal Emerald Copper, or Dioptase.

1. Rhomboidal Emerald Copper, or Dioptase.

Rhomboedrischer Smaragd-Malachite, *Mohs*.

Kupfer-Schmaragd, *Werner*.

Emeraudine, *Lam.* t. ii. p. 230.—Dioptase, *Haüy*, t. iii. p. 136. *Id. Broch.* t. ii. p. 511.—Achirite, *Hermann*, in Nov. Act. Petrop. xiii. 339.—Kupfersmaragd, *Reuss*, b. iii. s. 472. *Id. Lud.* b. i. s. 233. *Id. Mohs*, b. iii. s. 297. *Id. Leonhard*, Tabel. s. 69.—Cuivre dioptase, *Brong.* t. ii. p. 225.—Dioptase, *Brard*, p. 161. *Id. Karsten*, Tabel. s. 62. *Id. Haus.* s. 136.—Cuivre dioptase, *Haüy*, Tabl. p. 91.—Kupfersmaragd, *Hoff.* b. iv. s. 158.—Dioptas, *Haus.* Handb. b. iii. s. 1032.—Emerald Copper, *Aikin*, p. 91.

External Characters.

Its colour is emerald-green, which sometimes inclines to pistachio and blackish green.

It occurs only crystallized. The primitive form is a rhomboid of 123° 58′. The only secondary form at present known, is the equiangular six-sided prism, which is rather

acutely

acutely acuminated on both extremities by three planes, which are set on the alternate lateral edges.

The lateral planes are smooth.

Internally it is shining, and the lustre is pearly.

It has a threefold cleavage, and the folia are parallel to the faces of the rhomboid, of which, however, only one is very distinct.

The fracture is small conchoidal.

It is translucent, passing to semi-transparent.

It is as hard as apatite.

It is brittle, and easily frangible.

Specific gravity, 3.300, *Haüy*.—3.3, 3.4, *Mohs*.

Chemical Characters.

It becomes of a chesnut-brown colour before the blowpipe, and tinges the flame green, but is infusible; with borax it gives a bead or globule of copper.

Constituent Parts.

Oxide of Copper,	28.57	Oxide of Copper,	55
Carbonate of Lime,	42.85	Silica, -	33
Silica, -	28.57	Water,	12
	99.99		100
	Vauquelin, in Haüy, t. iii. p. 137.		*Lowitz*, in Nova Acta Petrop. xiii.

Geognostic and Geographic Situations.

It is found, according to Hermann, in the land of the Kirguise, 125 leagues from the Russian frontier, where it is associated with fibrous and compact malachite, calcareous-spar, and limestone.

Observations.

Observations.

1. It was brought to Petersburgh about twenty-seven years ago by General Bogdanof, who obtained it from the discoverer, Achir Mahmed, a Bucharian merchant. La Metherie considered it as a kind of emerald, and named it *Emeraudine*. Hermann names it *Achirite*, from Achir Mahmed the merchant, who first brought it into Europe.

2. It is distinguished from *Emerald* by its higher specific gravity, and inferior hardness; and from green *Tourmaline*, by colour, and inferior hardness.

ORDER V.

ORDER V.—*KERATE* *.

GENUS I.—CORNEOUS SILVER.

THIS genus contains one species, viz. Hexahedral Corneous Silver. * Earthy Corneous Silver.

1. Hexahedral Corneous Silver.

Hornerz, *Werner.*

Argent muriaté, *Haüy.*

Minera Argenti cornea, *Wall.* t. ii. p. 331.—Argent corné, *Romé de Lisle,* t. ii. p. 463. *Id. De Born,* t. ii. p. 420.—Hornerz, *Wern.* Pabst. b. i. s. 29. *Id. Wid.* s. 691.—Corneous Silver-ore, *Kirw.* vol. ii. p. 113.—Hornerz, *Estner,* b. iii. s. 348.—*Id. Emm.* b. ii. s. 168.—Argent corné, *Lam.* t. i. p. 130.—La Mine cornée, ou L'Argent corné ou muriaté, *Broch.* t. ii. p. 127.—Argent muriaté, *Haüy,* t. iii. p. 418. 422.—Hornerz, *Reuss,* b. iii. s. 330. *Id. Lud* b. i. s. 212. *Id. Suck.* 2ter th. s. 137. *Id. Bert.* s. 364. *Id. Mohs,* b. iii. s. 134.—Argent muriaté, *Lucas,* p. 108.—Silber Hornerz, *Leonhard,* Tabel. s. 54.—Argent muriaté, *Brong.* t. ii. p. 256. *Id. Brard,* p. 250.—Hornerz, *Karsten,* Tabel. s. 60.—Muriate of Silver, *Kid,* vol. ii. p. 91.—Argent muriaté, *Haüy,* Tabl. p. 76.—Hornsilber, *Haus.* Handb. b. iii. s. 1010.—Hornerz, *Hoff.* b. iii. s. 51.—Horn-silver, *Aikin,* p. 80.

External

---

* *Kerate,* from the Greek word κεϱας, *horn,* given to it on account of the species resembling horn in general aepcct and tenacity.

External Characters.

Its most frequent colour is pearl-grey, from which it passes on the one side into violet-blue and lavender-blue, on the other into greyish, yellowish, and greenish-white, further, into siskin-green, asparagus-green, pistachio-green, and pale leek-green. On exposure to the light it becomes brownish.

It occurs massive, in prismatic and granular concretions, in thick flakes, disseminated, in egg-shaped pieces, which are hollow in the centre, and the hollows lined with crystals. The crystals are the following:

1. Cube.
   - *a.* Perfect.
   - *b.* Truncated on the edges.
   - *c.* Truncated on the angles.
2. Octahedron.
3. Rhomboidal dodecahedron.

The crystals are small and very small, and are occasionally aggregated in rows, or in a scalar-like form.

The external surface is smooth, and sometimes marked with little hollows.

Externally it is shining, but becomes gradually duller on exposure: internally it is intermediate between shining and glistening, and the lustre is resinous.

The fracture is conchoidal, and sometimes inclines to earthy.

The fragments are indeterminate angular and blunt-edged.

It is translucent, or only feebly translucent on the edges.

It retains its colour, and becomes more shining in the streak.

It is very soft.

It

It is malleable.

It is flexible, but not elastic.

Specific gravity 4.6.

Chemical Characters.

It is fusible in the flame of a candle: before the blow-pipe, on charcoal, is reducible to a metallic globule, giving out at the same time vapours of muriatic acid; and when rubbed with a piece of moistened zinc, the surface becomes covered with a thin film of metallic silver.

Constituent Parts.

Silver, -	67.75	Silver, -	76.0
Muriatic Acid,	14.75	Muriatic Acid,	16.4
Oxygen,	6.75	Oxygen,	7.6
Oxide of Iron,	6.00		
Alumina,	1.75		100.0
Sulphuric Acid,	0.25		Id. s. 12.
	97.25		
	Klaproth, Beit. b. iv. s. 13.		

Geognostic Situation.

It occurs in silver veins, and generally in their upper part. These veins traverse gneiss, mica-slate, clay-slate, grey-wacke, porphyry, and limestone, and contain, besides the corneous silver, the following metalliferous and earthy minerals, viz. silver-glance or sulphuretted silver, and iron-ochre; more rarely native silver, earthy cobalt-ochre, red silver, tile-ore, malachite, blue copper, white lead-spar, iron-pyrites, galena, atacamite, copper-green, grey copper, and

and hornstone; also calcareous-spar, heavy-spar, and quartz.

Geographic Situation.

Europe.—At Huel-Mexico in Cornwall: in France and Saxony *, in veins that traverse gneiss, mica-slate, and clay-slate, where it is associated with silver-glance or sulphureted silver, and iron-ochre, and more rarely with native silver; at Schemnitz in Hungary, in massive quartz, along with fibrous malachite, silver-glance or sulphureted silver-ore, and white lead-spar; in the mountain of Chalanches, in calcareous-spar, and accompanied with silver-glance, native silver, and more rarely with earthy cobalt-ochre, and red silver.

Asia.—In Siberia, it occurs along with native gold, common and auriferous native silver, silver-glance, malachite, blue copper, tile-ore, white lead-spar, hornstone, calcareous-spar, heavy-spar, quartz, and sometimes lithomarge. These veins appear to traverse limestone.

America.—This mineral, which is so seldom found in Europe, is very abundant in the mines of Catorce, Fresnillo, and the Cerro San Pedro, near the town of San Luis Potosi. That of Fresnillo is frequently of an olive-green colour, which passes into leek-green. In the veins of Catorce, the corneous silver is accompanied with yellow lead-spar and green lead-spar.

* Large masses of corneous silver were dug out of the Saxon mines in the sixteenth century.

* Earthy Corneous Silver-Ore (a).

Erdiges Hornerz, *Karsten*.

Id. Karsten, in Magazin der Naturforschender Freünde zu Berlin, b. i. s. 159. *Id. Karsten*, Tabel. s. 60.

External Characters.

Internally the colour is pale mountain-green, inclining to greyish-white; externally it has a bluish-grey tarnish.

It occurs in thick crusts.

Internally it is dull.

The fracture is coarse and fine earthy.

The fragments are blunt-angular.

It is very soft, almost friable.

The streak is shining and resinous.

It is sectile.

It is heavy.

Constituent Parts.

Silver, - - -	24.64
Muriatic Acid, - -	8.28
Alumina, with a trace of Copper, - -	67.08
	100.00

Klaproth, Beit. b. i. s. 137.

Geognostic

(a) This mineral appears to be a mechanical mixture of corneous silver and clay, and hence it is placed beside it, but not as a variety of the species.

Geognostic and Geographic Situations.

It is found in veins that traverse transition rocks at Andreasberg in the Hartz.

Observations.

1. It sometimes occurs in a fluid form, in veins and drusy cavities, when it is said to resemble butter-milk: hence the German name *(Buttermilcherze)* given to it.

2. It appears to be an intimate mixture of corneous silver-ore and clay.

3. It was first discovered in the Hartz in the year 1576, and continued to be found until the year 1617; since that period, it has almost disappeared, and therefore is at present a very rare mineral.

GENUS II.—CORNEOUS MERCURY.

Perl Kerate, *Mohs.*

THIS Genus contains one species, viz. Pyramidal Corneous Mercury.

1. Pyramidal

1. Pyramidal Corneous Mercury.

Pyramidales Perl Kerate, *Mohs.*

Quecksilber Hornerz, *Werner.*

Mercure muriaté, *Haüy.*

Woulfe, in Phil. Trans. lxvi. ii. 618.—Mercure corné, ou Mercure doux volatile, *Romé de Lisle*, t. iii. p. 161.—Quecksilber hornerz, *Wern.* Pabst. b. i. s. 7. *Id. Wid.* s. 724.—Mercury mineralized by the Vitriolic and Marine Acids, *Kirw.* vol. ii. p. 266.—Quecksilber hornerz, *Estner*, b. iii. s. 275. *Id. Emm.* b. ii. s. 136.—Mercure corné, *De Born*, t. ii. p. 399. *Id. Lam.* t. i. p. 168.—La Mine de Mercure cornée, ou le Mercure muriaté, *Broch.* t. ii. p. 101.; Mercure muriaté, t. iii. p. 447.—Quecksilber Hornerz, *Reuss*, b. i. s. 277. *Id. Lud.* b. i. s. 206.—Salziges Quecksilber, *Suck.* 2ter th. s. 112.—Quecksilber-hornerz, *Bert.* s. 434. *Id. Mohs*, b. iii. s. 91. *Id. Leonhard*, Tabel. s. 52.—Mercure muriaté, *Brong.* t. ii. p. 244. *Id. Brard*, p. 260.—Quecksilber-hornerz, *Karsten*, Tabel. s. 60.—Horn-quecksilber, *Haus.* s. 134.—Muriate of Quicksilver, *Kid*, vol. ii. p. 97.—Mercure muriaté, *Haüy*, Tabl. p. 78.—Horn-quecksilber, *Haus.* Handb. b. iii. s. 1017. —Quecksilber-hornerz, *Hoff.* b. iii. s. 25.—Horn Quicksilver, *Aikin*, p. 83.

External Characters.

Its colour is ash-grey, of various degrees of intensity, which passes into yellowish-grey, and from this into greyish-white, and even sometimes inclines to greenish-grey.

It occurs very rarely massive, almost always in small vesicles, which are crystallized in the interior.

The crystals are the following:

1. Rectangular four-sided prism, acuminated on the extremities with four planes, which are set on the lateral planes.

2. Rectangular

2. Rectangular four-sided prism, acuminated with four planes, which are set on the lateral edges.
3. Double four-sided pyramid.

The crystals are always so minute, that it is with difficulty their forms can be determined. Their external surface is sometimes smooth, sometimes drusy, and is in general shining and adamantine.

Internally it is shining, with an adamantine lustre.

It has a single cleavage, which appears to be parallel with the terminal plane of the prism.

It is faintly translucent, or only translucent on the edges.

It is soft, approaching to very soft.

It is sectile, and easily frangible.

Chemical Characters.

It is totally volatilized before the blowpipe, and emits a garlic smell. It is soluble in water, and the solution mixed with lime-water gives an orange-coloured precipitate.

Constituent Parts.

Oxidé of Mercury,	76.00
Muriatic Acid,	16.40
Sulphuric Acid,	7.60
	100.00 *Klaproth.*

Geognostic Situation.

In the quicksilver mines of the Palatinate, and the Dutchy of Deux Ponts, it is accompanied with native mercury, cinnabar, ochry brown iron-ore, seldomer with fibrous malachite, massive and crystallized blue copper, massive grey copper, in cavities of an iron-shot clayey sandstone, sometimes in clay ironstone, and red iron-ore. That

at

at Idria, occurs in cavities of an indurated clay, accompanied with crystals of cinnabar; sometimes in slate-clay, which is traversed with small veins of cinnabar.

Geographic Situation.

It occurs at Horzowitz in Bohemia; Moschellandberg in Deux Ponts; Morsfeld in the Palatinate; Rutha in Upper Hessia; and at Almaden in Spain.

Observations.

It was discovered about thirty-five years ago in the mines of the Palatinate by Mr Woulfe.

ORDER VI.

Order VI. BARYTE.

Genus I. LEAD-SPAR.

Blei Baryt, *Mohs*.

This genus contains five species, viz. 1. Tri-prismatic Lead-spar, 2. Pyramidal Lead-spar, 3. Prismatic Lead-spar, 4. Rhomboidal Lead-spar, 5. Di-prismatic Lead-spar. * Corneous Lead. Arseniat of Lead. Native Minium.

1. Tri-prismatic Lead-Spar, or Sulphate of Lead.

Tri-prismatischer Blei Baryt, *Mohs*.

Vitriol Bleierz, *Werner*.

Plomb Sulphaté, *Haüy*.

Native Vitriol of Lead, *Kirw*. vol. ii. p. 211.—Natürlicher Bleivitriol, *Emm*. b. ii. s. 413. & b. iii. s. 366.—Sulphate de Plomb, *Lam*. t. i. p. 211.—Le vitriol de Plomb natif, *Broch*. t. ii. p. 325.—Plomb sulphaté, *Haüy*, t. iii. p. 503.—Bleivitriol, *Reuss*, b. iv. 2. s. 264. *Id. Lud*. b. i. s. 264. *Id. Suck*. 2r th. s. 32. *Id. Bert*. s. 452. *Id. Mohs*, b. iii. s. 547.—Plomb sulphaté, *Lucas*, p. 121.—Bleivitriol, *Leonhard*, Tabel. s. 73. —Plomb sulphaté, *Brong*. t. ii. p. 200. *Id. Brard*, p. 275. —Bleivitriol, *Karsten*, Tabel. s. 68. *Id. Haus*. s. 140.—Sulphat of Lead, *Kid*, vol. i. p. 140.—Plomb sulphaté, *Haüy*, Tabl. p. 83.—Bleivitriol, *Haus*. Handb. b. iii. s. 1113.—Vitriol-bleierz, triolbleierz, *Hoff.* b. iv. s. 41.—Sulphat of Lead, *Aikin*, p. 113.

External Characters.

Its colours are yellowish and greyish white, and these are occasionally stained pale-yellowish, from brown iron-ochre.

It occurs massive, disseminated, and in angulo-granular distinct concretions, but most frequently crystallized.

In the primitive form, the vertical prism is 120°; the horizontal prism in the direction of the longer diagonal 70° 31′, and in the direction of the smaller diagonal 101° 32′. The following are the principal crystallizations, described according to the Wernerian method:

1. Oblique four-sided prism, acutely bevelled on the extremities, and the bevelling planes set on the acuter lateral edges. Fig. 124. Pl. 6.
   - *a.* The obtuse lateral edges more or less deeply truncated. Fig. 125. Pl. 6.
   - *b.* The angles of the bevelment bevelled, and the bevelling planes set on the lateral planes.
   - *c.* The angles of the bevelment truncated.

     When the prism No. 1. becomes lower, so that the acuter lateral edges terminate in angles, there is formed
2. A broad rectangular four-sided pyramid.

The crystals are small and very small, seldom middle sized; and occur in druses, or superimposed.

Externally it is splendent and shining; internally shining, and the lustre adamantine.

Its cleavage is in the direction of the planes of the prisms.

The fracture is small conchoidal.

The

The fragments are indeterminate angular, and rather blunt edged.

It alternates from transparent to translucent.

It is as hard as calcareous-spar.

Its streak is white.

It is rather brittle, and easily frangible.

Specific gravity, 6.300, *Klaproth*.

Chemical Characters.

It decrepitates before the blowpipe, then melts, and is soon reduced to the metallic state.

Constituent Parts.

	From Anglesey.	Wanlockhead.
Oxide of Lead, -	71.0	70.50
Sulphuric Acid,	24.8	25.75
Water of crystallization,	2.0	2.25
Oxide of Iron, -	1.0	
	98.8	98.05

Klaproth, Beit. b. iii. s. 164. & 166.

Zellerfeldt.	
Oxide of Lead,	72.914
Sulphuric Acid,	26.019
Water, -	0.124
Oxide of Iron,	0.115
Oxide of Manganese, -	0.165
Silica and Alumina,	trace.
	99.799

Stromeyer, Gott. Gel. anz. 812. 204.

Geognostic and Geographic Situations.

It occurs in veins along with galena or lead-glance, and different spars of lead, at Wanlockhead in Dumfriesshire, and Lead Hills in Lanarkshire; at Pary's Mine in Anglesey, and Penzance in Cornwall. On the Continent, it is met with at Zellerfeld in the Hartz, in veins that traverse clay-slate and grey-wacke, associated with quartz, calcareous-spar, brown-spar, heavy-spar, brown iron-ore, copper-green, blue copper, green lead-spar, and white lead-spar; in

in the Westerwald mountains; and in lead-mines in Andalusia in Spain.

Asia.—In Siberia.

America.—In the neighbourhood of Southampton in the United States *.

Observations.

Form, lustre, and weight, are the principal characters of this mineral. It is distinguished from *White Lead-spar* by its crystallizations and inferior specific gravity; from *Columnar Heavy Spar* and *Celestine* by its greater weight.

2. The *Bleiglas* of Dr John, in most of its varieties, appears to belong to this species.

3. A reniform variety of this mineral, named *Bleinierè*, occurs along with reniform arseniate of lead, at Nertschinsky in Siberia.

2. Pyramidal Lead-Spar, or Yellow Lead-Spar.

Pyramidal Blei-Baryt, *Mohs.*

Gelb Bleierz, *Werner.*

Plomb molybdaté, *Haüy.*

Oxide de Plomb spathique jaune, *De Born*, t. ii. p. 379.—Yellow Lead-spar, *Kirw.* vol. ii. p. 212.—Gelbes Bleierz, *Emm.* b. ii. s. 403.—Plomb molybdaté, *Haüy*, t. iii. p. 498.—La Mine de Plomb jaune, ou Le Plomb jaune, *Broch.* t. ii. p. 322.—Gelb Bleierz, *Reuss*, b. iv. s. 236. *Id. Mohs*, b. iii. s. 535. Plomb molybdaté, *Brong.* t. ii. p. 205.—Molybdate of Lead, *Kid.* vol. ii. p. 139.—Bleigelb, *Haus.* Handb. b. iii. s. 1100.—Gelb-bleierz, *Hoff.* b. iv. s. 36.—Molybdate of Lead, *Aikin*, p. 115.

External

* Bruce's American Mineralogical Journal, p. 150.

External Characters.

Its most frequent colour is wax-yellow; from which it passes, on the one side, into lemon-yellow and orange-yellow, on the other side, into yellowish-brown and yellowish-grey; sometimes of a colour which is intermediate between yellowish-white and greyish-white.

It occurs massive, in crusts, cellular; and crystallized in the following figures:

Its primitive form is a pyramid, in which the angles are 99° 40′, and 131° 45′ *. Fig. 126. Pl. 6. The following are its secondary figures:

1. The pyramid truncated on the angles and summits †. Fig 127. Pl. 6.
2. The pyramid so deeply truncated on the summits, and on the common base, that the original faces disappear, when there is formed a rectangular parallelopiped, which is either tabular, or in the form of a cubical prism ‡, as in fig. 128. Pl. 6.
3. The pyramid so deeply truncated in all the angles, and on the common base, that the original faces disappear, when there is formed a regular eight-sided table ||. Fig. 129. Pl. 7., which is sometimes so thick as to appear as an eight-sided prism. Sometimes four of the terminal edges are truncated, when a twelve-sided table is formed. Fig. 130. Pl. 7.

4. Pyramid

---

* Plomb molybdaté primitif, Haüy.

† Plomb molybdaté epointé, Haüy.

‡ Plomb molybdaté bis-unitaire, Haüy.

|| Plomb molybdaté tri-unitaire, Haüy.

4. Pyramid deeply truncated on the summits, and on the common base, and the angles of the common base bevelled, which gives rise to the rectangular four-sided table, bevelled on the terminal edges.
5. Pyramid truncated on the lateral edges, which gives rise to the double eight-sided pyramid. When this figure is deeply truncated on the summits, there is formed
6. A regular eight-sided table, bevelled on the terminal planes.

The tables are usually broad and thin, and alternate from small to very small, but are seldom middle-sized. They frequently intersect each other, and are often closely aggregated.

Externally it is generally splendent or shining; internally it is shining or glistening, and the lustre resinous, inclining to adamantine.

The cleavage is fourfold, and the folia are in the direction of the sides of the primitive form. There is what is termed a fifth cleavage, in the direction of the common base.

The fracture is small and fine-grained uneven, or small conchoidal.

The fragments are indeterminate angular, and rather sharp edged.

It is generally translucent, or only translucent on the edges; some rare crystals are semitransparent.

It is as hard as calcareous-spar.

It is rather brittle, and easily frangible.

Specific gravity, 5.706, *Hatchet.*—6.5, 6.8, *Mohs.*

Chemical

Chemical Characters.

It decrepitates before the blowpipe, and then melts into a dark greyish-coloured mass, in which the globules of reduced lead are dispersed. With borax, it forms a brownish-yellow bead; but when in small proportions, and heated by the interior flame, it occasionally produces a glass, which is greenish-blue, and sometimes deep blue.

Constituent Parts.

Klaproth was the first who made us acquainted with the chemical composition of this ore; but we are indebted to our celebrated countryman Hatchett, for the most complete and accurate analysis of it.

Oxide of Lead,	64.42
Molybdic Acid,	34.25
	98.67

Klaproth, Beit. b. ii. s. 275.

Oxide of Lead,	58.40
Molybdic Acid,	38.00
Oxide of Iron,	2.08
Silica, - -	0.28
	96.66

Hatchett, Phil. Trans. for 1796.

Geognostic and Geographic Situations.

Europe.—It occurs at Bleiberg in Carinthia, in a compact limestone, which is much traversed by veins of calcareous spar, and is associated with galena, white, black, and green lead-spar, calamine, malachite, calcareous-spar, and fluor-spar; also in the Maukeriz, near Brixlegg in the Tyrol, along with brown iron-ore, and red copper-ore; at Annaberg in Austria, and Rezbanya in Transylvania.

America.

America.—In Pennsylvania, at the Perkiomen lead-mine, and at the Southampton lead-mine in Massachusets *; in compact limestone at Zimapan in Mexico.

Observations.

This mineral is well characterized by its colour, crystallization, fracture, lustre, hardness, and weight. It is distinguished from the yellow varieties of *Green Lead-spar*, by form of its crystals, and inferior specific gravity, and fracture; from *White Lead-spar*, by the form of its crystals, and colours.

3. Prismatic Lead-Spar, or Red Lead-Spar.

Prismatischer Blei-Baryt, *Mohs.*

Roth-Bleierz, *Werner.*

Plomb chromaté, *Haüy.*

Minera Plumbi rubra, *Wall.* t. ii. p. 309.—Rothes Bleierz, *Werner*, Pabst. b. i. s. 127. *Id. Wid.* s. 861.—Red Lead-spar, *Kirw.* vol. ii. p. 214.—Oxide de Plomb spathique rouge, *De Born*, t. ii. p. 376.—Rothes Bleierz, *Emm.* b. ii. s. 399.—Oxide rouge de Plomb, *Lam.* t. i. p. 287.—Plomb chromaté, *Haüy*, t. iii. p. 476.—La Mine de Plomb rouge, ou Le Plomb rouge, *Broch.* t. ii. p. 318.—Rothbleierz, *Mohs*, b. iii. s. 527. Plomb chromé, *Brong.* t. ii. p. 205.—Kallochrom, *Haus.* s. 139.—Chromate of Lead, *Kid*, vol. ii. p. 143.—Plomb chromaté, *Haüy*, Tabl. p. 81.—Kallochrom, *Haus.* Handb. b. iii. s. 1084.—Rothbleierz, *Hoff.* b. iv. s. 33.—Chromate of Lead, *Aikin*, p. [illegible]

External Characters.

Its only colour is hyacinth-red, which is more or less [illegible].

It

* [illegible] Mineralogy.

It seldom occurs massive, generally in flakes; and crystallized:

1. Long slightly oblique four-sided prism.
2. The preceding figure, in which the terminal planes are set obliquely on the lateral edges.
3. The prism acutely and obliquely bevelled on the extremities, the bevelling planes set on the lateral edges.
4. The prism acuminated with four planes, which are set on the lateral planes.
5. The prism truncated on the lateral edges; sometimes bevelled on two opposite lateral edges.

The crystals are generally small, thin, and always superimposed.

The lateral planes formed by bevelment are longitudinally streaked, the other planes smooth, and are shining or splendent.

Internally it is shining or splendent, and the lustre is adamantine.

The cleavage is double, and the folia appear to be in the direction of the planes of an oblique four-sided prism.

The fracture is small-grained uneven, sometimes passing into imperfect and small conchoidal.

The fragments are indeterminate angular, and rather sharp-edged.

It is more or less translucent.

It gives a streak, which is of a colour intermediate between lemon-yellow and orange-yellow.

It is harder than gypsum, but softer than calcareous spar.

It is almost sectile, and easily frangible.

Specific gravity, 6.056, *Brisson.*—6.0–6.1, *Mohs.*

Chemical

Chemical Characters.

Before the blowpipe, it crackles and melts into a grey slag. With borax, it is partly reduced. It does not effervesce with acids.

Constituent Parts.

Oxide of Lead,	63.96
Chromic Acid,	36.40
	100.36

Vauquelin, Journ. des Mines, n. 34. 737

Geognostic and Geographic Situations.

It occurs in veins in gneiss, in the gold mines of Beresofsk, in the Uralian Mountains. In these mines, it is associated with brown iron-ore, cubes of iron-pyrites, native gold, green lead-spar, galena, and quartz. It is reported to have been found at Tarnowitz in Silesia, in secondary rocks; and I understand it has been brought from Mexico and Brazil *.

Use.

In Russia, a very beautiful and costly orange-yellow colour is prepared from it, and which is used by painters.

Observations.

1. It is distinguished from *Red Orpiment*, by the form of its crystals, its colour, cleavage, and superior weight; from *Red Silver*, by its colour, form, inferior specific gravity,

* [illegible]

vity, and yellow-coloured streak; and from *Cinnabar*, by the colour of its streak.

2. It was first made known to European naturalists in the year 1766 by M. Laxman.

3. The red lead-spar is sometimes accompanied with a mineral having a yellowish or liver-brown colour, with botryoidal and stalactitic forms, and is also a compound of oxide of lead and chromic acid. Green-coloured crystals sometimes accompany the red lead-spar, which are said to be compounds of oxides of lead and of chrome.

4. Rhomboidal Lead-Spar.

Rhomboedrisches Blei Baryt, *Mohs.*

THIS species contains two subspecies, viz. Green Lead-Spar and Brown Lead-Spar.

First Subspecies.

Green Lead-Spar.

Grün Bleierz, *Werner.*

Plomb phosphaté, *Haüy.*

Minera Plumbi viridis, *Wall.* t. ii. p. 308.—Grün Bleyerz, *Wern.* Pabst. b. i. s. 123. *Id. Wid.* s. 857.—Phosphorated Lead-ore, *Kirw.* vol. ii. p. 207.—Oxide de Plomb spathique verte, Phosphaté de Plomb, *De Born,* t. ii. p. 377.—Grün Bleyerz, *Emm.* b. ii. s. 394.—Plomb phosphaté, *Haüy,* t. iii. p. 490.—La Mine de Plomb verte, ou le Plomb verte, *Broch.* t. ii. p. 314.—Grünbleierz, *Reuss,* b. iv. s. 216. *Id. Lud.* b. i. s. 262.

Id. Suck. 2ter th. s. 331. *Id. Bert.* s. 455. *Id. Mohs*, b. iii. s. 517. *Id. Leonhard*, Tabel. s. 72.—Plomb phosphaté, *Brong.* t. ii. p. 200. *Id. Brard*, p. 271.—Gemeines Phosphorblei, (in part), *Karst.* Tabel. s. 68.—Phosphate of Lead, *Kid*, vol. ii. p. 141.—Plomb phosphaté, *Haüy*, Tabl. p. 82.—Grünbleierz, *Hoff.* b. iv. s. 27.—Phosphate of Lead, *Aikin*, p. 112.

External Characters.

Its colour is grass green, which passes on the one side through pistachio-green, blackish, olive, oil, and siskin green, into sulphur-yellow; on the other side, through asparagus-green, yellowish-white, into greenish-white. Some varieties approach to leek-green. The olive and pistachio green colours are the most common. Sometimes several colours occur together in the same specimen, even in the same crystal.

It seldom occurs massive, sometimes stalactitic, reniform and botryoidal, sometimes in distinct concretions, which are granular or prismatic; but most commonly crystallized. The primitive form is a di-rhomboid, or a flat equiangular double six-sided pyramid, in which the lateral planes of the one are set on the lateral planes of the other. The angles are 141° 47′ and 81° 46′. The following are the secondary forms:

1. Equiangular six-sided prism *. Fig. 131. Pl. 7.
2. Six-sided prism, truncated on all the lateral edges, thus forming a twelve-sided prism †. Fig. 132. Pl. 7.
3. Six-sided prism, flatly acuminated on the extremities

* Plomb phosphaté prismatique, Haüy.

† Plomb phosphaté peridodecaedre, Haüy.

ties with six planes, which are set on the lateral planes *. Fig. 133. Pl. 7.

4. Six-sided prism, in which the terminal edges are truncated †. Fig. 134. Pl. 7.
5. When the six-sided prism becomes very low, an equiangular six-sided table is formed. The lateral edges of the table are sometimes truncated, and then the table appears acutely bevelled on the terminal planes, and has sometimes a lenticular appearance.
6. Sometimes the lateral planes of the six-sided prism are bent towards the extremities of the prism: when the prism is short, it appears bulging, but when long, it has an acute pyramidal form. These crystals are hollow at their ends.

The crystals are small and very small, seldom middle-sized; they are superimposed, in druses, or scalarwise or rose-like aggregated. Sometimes they form velvety or moss-like drusy crusts.

Externally it is smooth and shining, or splendent; internally glistening, and the lustre is resinous.

Cleavages are to be observed in the direction of the planes of the di-rhomboid, and also parallel with the planes of the six-sided prism.

The fracture is small-grained uneven, passing on the one hand into splintery, on the other into conchoidal.

The fragments are indeterminate angular, and blunt-edged.

It is more or less translucent, seldom nearly transparent, and is sometimes only translucent on the edges.

 It

* Plomb phosphaté trihexaedre, Haüy.

† Plomb phosphaté annulaire, Haüy.

It is harder, than calcareous-spar, and sometimes as hard as fluor-spar.

It is brittle, and easily frangible.

Specific gravity, 6.9411, from the Breisgau, *Haüy*.—6.9–7.2, *Mohs*.

Chemical Characters.

It dissolves in acids without effervescence. Before the blowpipe, on charcoal, it usually decrepitates, then melts, and on cooling, forms a polyhedral globule, the faces of which present concentric polygons: if this globule be pulverized, and mixed with borax, and again heated, a milk-white opaque enamel is partly formed; on continuance of the heat, the globule effervesces, and at length becomes perfectly transparent, the lower part of it being studded with globules of metallic lead.

Constituent Parts.

	Zschoppau.	Hoffsgrund.	Wanlockhead.
Oxide of Lead,	78.40	77.10	80.00
Phosphoric Acid,	18.37	19.00	18.00
Muriatic Acid,	1.70	1.54	1.62
Oxide of Iron,	0.10	0.10	a trace.
	98.57	97.74	99.96

Klaproth, Beit. b. iii. s. 153.—161.

Johangeorgenstadt.	
Oxide of Lead,	76.8
Phosphoric Acid,	9.0
Muriatic Acid,	7.0
Arsenic Acid,	4.0
Water, -	1.5
	98.3

Laugier, Ann. du Mus. t. vi. p. 171.

Geognostic Situation.

It occurs in veins and beds in primitive, transition, and secondary rocks. It generally occupies the upper part of the veins; and it is associated with brown iron-ochre, galena or lead-glance, white lead-spar, heavy-spar, quartz, and

and other minerals. It appears to be a newer formation than galena, or even white lead-spar, and seems to belong principally to that formation of galena which contains but little silver.

Geographic Situation.

Europe.—It occurs along with galena or lead-glance, and other ores of lead, at Leadhills and Wanlockhead. In England, it is met with at Alston in Cumberland, Allonhead, Grasshill, and Teesdale, in Durham, and Nithisdale in Yorkshire. On the Continent, it is found in several of the mines in the Hartz; also at Zschoppau in Saxony; Prizbram in Bohemia; Hoffsgrund in the Breisgau; and Erlenbach in Alsace.

Asia.—In the lead-mines in Siberia; and in those of Beresof.

America.—In the Perkiomen mine in Pennsylvania.

Observations.

1. Green lead-spar, when it has a very pale greenish-white colour, is apt to be confounded with White Lead-spar; but we can distinguish them by the following characters:—*a.* The prisms of green lead-spar are generally equiangular, but those of white lead-spar are unequiangular. *b.* Its lustre is resinous, but that of white lead-spar is adamantine. *c.* It is harder than white lead-spar. *d.* Its crystals are often scalarwise aggregated, which is never the case with white lead-spar. *e.* Its prisms are generally shorter than those of white lead-spar; and this mineral does not effervesce with acids, which is the case with white lead-spar, and is not reduced to the metallic state before the blowpipe without addition.

2. It

2. It is distinguished from *Apatite* by its cleavage, inferior hardness, and greater specific gravity; from *Malachite*, and *Olivenite*, by its green colours, cleavage, and greater weight.

Second Subspecies.

Brown Lead-Spar.

Braun Bleierz, *Werner*.

Plomb phosphaté, *Haüy*.

Id. Wern. Pabst. b. i. s. 115. *Id. Wid.* s. 848.—Brown Lead-ore, *Kirw.* vol. ii. p. 222.—Braun Bleyerz, *Emm.* b. i. s. 283. La Mine de Plomb brune, *Broch.* t. ii. p. 305.—Braun Bleierz, *Reuss*, b. i. s. 212. *Id. Lud.* b. i. s. 260. *Id. Suck.* 2ter th. s. 323. *Id. Bert.* s. 454. *Id. Mohs*, b. iii. s. 489. *Id. Leonhard*, Tabel. s. 71.—Gemeiner Phosphorblei, *Karsten*, Tabel. (in part), s. 68.—Plomb phosphaté, *Haüy*, Tabl. p. 82.—Gemeiner Pyromorphit, *Haus.* b. iii. s. 1090.—Braun Bleierz, *Hoff.* b. iv. s. 15.—Brown Phosphate of Lead, *Aikin*, p. 113.

External Characters.

Its colour is clove-brown, of different degrees of intensity, rarely approaching to liver-brown, sometimes so pale that it inclines to white.

It occurs massive, also in distinct concretions, which are thin prismatic, and curved lamellar; and crystallized in the following figures:

1. Equiangular six-sided prism, which is sometimes bulging.
2. Preceding figure, in which the lateral planes are alternately

ternately broad and narrow, and sometimes the lateral edges are truncated.

3. Six-sided prism, converging towards both ends, and thus inclining to the pyramidal form.
4. Acute double three-sided pyramid, in which the lateral planes of the one are set on the lateral planes of the other, and in which the common basis is sometimes more or less deeply truncated. It originates from the bulging six-sided prism, in which the alternate lateral planes have disappeared.

The crystals are middle-sized, and small, are sometimes short and acicular, and singly imbedded, or scopiformly or globularly aggregated.

The surface of the crystals is sometimes blackish or yellowish brown, and rough.

Internally it is glistening, and the lustre is resinous.

The fracture is small and fine-grained uneven, and sometimes passes into small splintery.

The fragments are indeterminate angular.

It is feebly translucent, or translucent on the edges.

It is as hard as green lead-spar.

The streak is greyish-white.

It is rather brittle, and easily frangible.

Specific gravity, 6.974, *Wiedeman.*—6.909, from Huelgoët, *Haüy.*

Chemical Characters.

It melts pretty easily before the blowpipe without being reduced, and during cooling shoots into acicular crystals. It does not effervesce with nitric acid, but is soluble in it.

Constituents

[illegible]

[illegible]	[illegible]
[illegible]	[illegible]
[illegible]	[illegible]
[illegible]	[illegible]
	[illegible]

[illegible]

Geognostic Situation.

[illegible]

Geographic Situation.

Europe.—It is found at Mies in Bohemia; near Schemnitz in Hungary; [illegible] in the [illegible]; Zschoppau in Saxony; Huelgoet and Poullaouen in Lower Brittany.

America.—Zimapan in Mexico.

5. Di-prismatic Lead-Spar.

Di-prismatischer Blei-Baryt, *Mohs.*

This species is divided into three subspecies, viz. White Lead-Spar, Black Lead-Spar, and Earthy Lead-Spar.

First

First Subspecies.

White Lead-Spar.

Weiss-Bleierz, *Werner*.

Plomb carbonaté, *Haüy*.

Minera Plumbi alba spathosa, *Wall.* t. ii. p. 307.—Mine de Plomb blanche, *Romé de Lisle,* t. iii. p. 380.—Weiss Bleyerz, *Wern.* Pabst. b. i. s. 118. *Id. Wid.* s. 852.—Plomb spathique blanc, *De Born,* t. ii. p. 368.—White Lead-ore, *Kirw.* vol. ii. p. 203.—Weiss Bleyerz, *Emm.* b. ii. s. 388.—Plomb blanc, *Lam.* t. i. p. 305.—Plomb carbonaté, *Haüy,* t. iii. p. 475.—La Mine de Plomb blanche, ou le Plomb blanc, *Broch.* t. ii. p. 309.—Weissbleierz, *Reuss,* b. iv. s. 245. *Id. Lud.* b. i. s. 261. *Id. Suck.* 2ter th. s. 326. *Id. Bert.* s. 459. *Id. Mohs,* b. iii. s. 493.—Kohlenstoffsaures Bleierz, *Hab.* s. 128.—Plomb carbonaté, *Lucas,* p. 117.—Weissbleierz, *Leonhard,* Tabel. s. 71. —Plomb carbonaté, *Brong.* t. ii. p. 198. *Id. Brard,* p. 268. —Lichter Bleispath, *Karsten,* Tabel. s. 68.—Spathiges Bleiweiss, *Haus.* s. 114.—Crystallized Carbonate of Lead, *Kid,* vol. ii. p. 136.—Plomb carbonaté, *Haüy,* Tabl. p. 81.—Bleiweiss, *Haus.* Handb. b. iii. s. 1107.—Weissbleierz, *Hoff.* b. iv. s. 21.—Carbonate of Lead, *Aikin,* p. 110.

External Characters.

Its colours are snow-white, greyish-white, greenish-white, and yellowish-white; from yellowish-white it passes into wine-yellow, isabella-yellow, and clove brown; and from greyish-white into pale-yellow, and ash-grey. It has sometimes a tempered steel tarnish. It is sometimes coloured externally yellow or brown, by yellow or brown iron-ochre; occasionally green, by earthy malachite, and blue, by earthy blue copper.

It

Constituent Parts.

	From Huelgoët in Brittany.
Oxide of Lead, -	78.58
Phosphoric Acid, -	19.73
Muriatic Acid, - -	1.65
	99.96

Klaproth. Beit. b. iii. s. 157.

Geognostic Situation.

It occurs in veins that traverse gneiss, clay-slate, and porphyry. The veins generally contain lead and silver ores, also native silver, iron and copper pyrites, malachite, blende, ochry ironstone, heavy-spar, and quartz.

Geographic Situation.

Europe.—It is found at Miess in Bohemia; near Schemnitz in Hungary; Saska in the Bannat; Zschoppau in Saxony; Huelgoët and Poullaouen in Lower Brittanny.

America.—Zimapan in Mexico.

5. Di-prismatic Lead-Spar.

Di-prismatischer Blei-Baryt, *Mohs.*

This species is divided into three subspecies, viz. White Lead-Spar, Black Lead-Spar, and Earthy Lead-Spar.

First

planes of the other. It is the preceding figure without the prism.

7. Long acicular and capillary crystals, which are columnarly aggregated.

9. It occurs also in twin and triple crystals.

The crystals are usually small and very small; seldom middle-sized; are often long and acicular, also broad and tabular.

The crystals occur superimposed, and either single or in druses; more frequently columnarly and scopiformly, or promiscuously aggregated.

Externally, it alternates from specular splendent to glistening.

Internally, it alternates from shining to glistening, and the lustre is adamantine, sometimes inclining to semimetallic, sometimes to resinous.

Its cleavage is in the direction of the lateral planes of the horizontal and perpendicular prisms.

The fracture is small conchoidal, which sometimes passes into uneven and splintery.

The fragments are indeterminate angular, and rather sharp-edged.

It alternates from translucent to transparent; and it refracts double in a high degree.

It is harder than calcareous-spar, but not so hard as fluor-spar; it is softer than rhomboidal lead-spar.

It is brittle, and very easily frangible.

Specific gravity, 6.2, 6.6, *Mohs.*—6.480, from Leadhills, according to *Klaproth.*—6.5586, *Haüy.*—6.255, *Karsten.*

Chemical

Chemical Characters.

It is insoluble in water. It dissolves with effervescence in muriatic and nitric acids. Before the blowpipe it decrepitates, becomes yellow, then red, and is soon reduced to a metallic globule.

Constituent Parts.

	Leadhills.	Nertschinsk. Transparent.	Nertschinsk. Translucent.
Oxide of Lead,	82	84.5	73.50
Carbonic Acid,	16	15.5	15.00
Silica, -	—	—	8.00
Alumina, -	—	—	} 2.66
Oxide of Iron, -	—	—	} 2.66
Water, -	2		
	100	100.0	100.0

Klaproth, Beit. b. iii. s. 168. — *John's* Chem. Unters. b. ii. s. 233. 236.

Geognostic Situation.

It occurs in veins, and sometimes also in beds, in gneiss, mica-slate, and clay-slate, foliated granular limestone, grey-wacke, and grey-wacke-slate, and secondary limestone. The veins in which it is found are generally lead veins, which contain besides this spar, galena or lead-glance, green, black and yellow lead spars, sulphat of lead, earthy lead-spar, copper and iron pyrites, malachite, blue copper, grey manganese-ore, copper-green, brown iron-ore, sparry iron, native copper, white silver-ore, blende, and calamine; and the following vein-stones, heavy-spar, fluor, calcareous spar, quartz, and sometimes mountain-cork. Of all the accompanying minerals, galena is the most frequent, and when

green

green lead-spar is associated with these, the white lead rests on the galena, and the green lead on the white lead.

Geographic Situation.

Europe.—It occurs at Leadhills in Lanarkshire, in veins that traverse transition rocks, in which it is associated with galena or lead-glance, earthy white lead, green lead-spar, lead-vitriol or sulphat of lead, sparry iron, iron-pyrites, brown hematite, calamine, and blue copper; and the vein-stones are quartz, lamellar heavy-spar, calcareous-spar, brown-spar, and mountain-cork. It is found also with galena or lead-glance at Allonhead and Teesdale in Durham; with the same ore at Alston in Cumberland, and Snailback in Shropshire.

On the Continent, it it met with in several mines in the Hartz; also at Johanngeorgenstadt in Saxony; Prizbram in Bohemia; Tarnowitz in Silesia; Freiburg in the Breisgau; Schemnitz in Hungary; Bleiberg in Carinthia; Huelgoët and Poullaouen in Brittany; Saska and Dognatska in the Bannat; and in the Crimea.

Asia.—In several mines, particularly those of Gazimour in Siberia, where specimens of great beauty are found.

America.—It is met with in the mines of Chili, and at the Perkiomen mine in Pennsylvania; and on Conestoga Creek, near Lancaste *.

Observations.

It is distinguished from *Calcareous-Spar* by its greater specific gravity; from *Heavy-Spar* and *Celestine*, by its form,

* Cleaveland's Mineralogy, p. 517.

form, cleavage, lustre, and superior specific gravity; a from *Arragonite*, by its higher specific gravity.

2. Several of the varieties of Bleiglas appear to belong this subspecies of lead-spar.

Second Subspecies.

Black Lead-Spar.

Schwarz Bleierz, *Werner*.

Id. Werner, Pabst. b. i. s. 116. *Id. Wid.* s. 850.—Black Lea Ore, *Kirw.* vol. ii. p. 221.—Schwarz Bleyerz, *Emm.* b. s. 385.—La Mine de Plomb noire, *Broch.* t. ii. p. 307.- Schwarz Bleyerz, *Reuss*, b. iv. s. 241. *Id. Lud.* b. i. s. 261. *I Suck.* 2ter th. s. 324. *Id. Bert.* s. 461. *Id. Mohs*, b. iii. s. 49 *Id. Leonhard*, Tabel. s. 71.—Plomb noire, *Brong.* t. ii. p. 19 —Dunkler Bleispath, *Karsten*, Tabel. s. 68.—Plomb carb naté noire, *Haüy*, Tabl. p. 82.—Bleischwärtze, *Haus.* Handl b. iii. s. 1111.—Schwarzbleierz, *Hoff.* b. iv. s. 18.

External Characters.

Its colour is greyish-black of different degrees of intensi ty, which sometimes passes into ash-grey.

It occurs massive, disseminated, corroded, cellular, an seldom crystallized, in small and very small six-side prisms.

The surface of the crystals is sometimes drusy, sometime smooth, and sometimes longitudinally streaked.

Externally it is generally splendent, and sometimes shin ing.

Internally

Internally it is only shining, sometimes passing into glistening, and the lustre is metallo-adamantine.

The fracture is small-grained uneven, which sometimes passes into imperfect conchoidal.

It alternates from translucent to opaque.

Its streak is whitish-grey.

In other characters agrees with the preceding.

Constituent Parts.

Oxide of Lead,	-	79	78.5
Carbonic Acid,	-	18	18.1
Carbon,	-	2	1.5
		99	

Lampadius, Handb. Zu. Chem. Anal.

Geognostic Situation.

It generally occurs in the upper part of veins, associated with white lead-spar, and galena or lead-glance. Frequently this ore incrusts galena, and has resting upon it white lead-spar, and sometimes even green lead-spar. We often observe a nucleus of galena incrusted with black lead-spar, or black lead-spar forms a nucleus, which is incrusted with white lead-spar.

Geographic Situation.

Europe.—It occurs at Leadhills; at Fair Hill and Flow Edge, Durham. On the Continent, it is met with at Miess and Prizbram in Bohemia; Freyberg and Zschopau in Saxony; Schwarzleogang in Salzburg; Poullaouen in Lower Brittany in France.

Asia.—Schlangenberg in Siberia.

Third

Third Subspecies.

Earthy Lead-Spar.

Bleierde, *Werner.*

This subspecies is divided into two kinds, viz. Indurated Earthy Lead-Spar, and Friable Earthy Lead-Spar.

First Kind.

Indurated Earthy Lead-Spar.

Verhärtete Bleierde, *Werner.*

Id. Wid. s. 868.—Le Plomb terreux endurci, *Broch.* t. ii. p. 329. —Verhärtete gelb und grau Bleyerde, *Reuss,* b. iv. s. 270. & 272.—Bleierde, *Lud.* b. i. s. 265. *Id. Suck.* 2ter th. s. 345. *Id. Bert.* s. 462. *Id. Mohs,* b. iii. s. 553. *Id. Leonhard,* Tabel. s. 70.—Plomb oxydé terreux, *Brong.* t. ii. p. 197. *Id. Brard,* p. 270.—Verhärtete Bleierde, *Karsten,* Tabel. s. 68.—Erdiches Bleiweiss, *Haus.* s. 114.—Earthy Carbonate of Lead, *Kid,* vol. ii. p. 138.—Plomb carbonaté terreux, *Haüy,* Tabl. p. 82.—Feste Bleierde, *Haus.* Handb. b. iii. s. 1109.—Verhärtete Bleierde, *Hoff.* b. iv. s. 46.—Earthy Carbonate of Lead, *Aikin,* p. 111.

External Characters.

Its most frequent colour is yellowish-grey, from which it passes, on the one side, into straw-yellow and cream-yellow; on the other, into yellowish-brown. It occurs also smoke-grey, bluish-grey, and light-brownish red.

It occurs massive.

Internally

Internally it is glimmering, inclining to glistening; and the lustre is resinous *.

The fracture is small and fine-grained uneven, which passes on the one side into fine splintery, on the other into earthy.

The fragments are indeterminate angular, and blunt-edged.

Is usually opaque, or extremely faintly translucent on the edges.

It yields a brown-coloured streak.

It is soft, passing into very soft, even into friable, particularly the yellowish-grey, and yellow varieties.

Specific gravity, 5.579, *John*.

Chemical Characters.

It is very easily reduced before the blowpipe; effervesces with acids, and becomes black with sulphuret of ammonia.

Constituent Parts.

	Tarnowitz.
Oxide of Lead,	66.00
Carbonic Acid,	12.00
Water, -	2.25
Silica, -	10.50
Alumina, -	4.75
Iron and Oxide of Manganese,	2.25
	97.75

John, Chem. Unt. b. ii. s. 229.

* This lustre is [illegible] to be owing to intermixed white lead-ore or lead-vitriol.

Geognostic Situation.

The yellow-coloured varieties occur in a bed in prim tive limestone, accompanied with galena or lead-glance ar other ores of lead, in the Bannat; the grey-coloured vari ties occur sometimes in veins, sometimes in beds, and eith in transition or secondary rocks, and are usually accomp nied with galena or lead-glance, white lead-spar, iron-p rites, malachite, and quartz.

Geographic Situation.

Europe.—It is found in the lead veins of Wanlockhe and Leadhills; also at Grassfield Mine near Nenthead Durham, and in Derbyshire. On the Continent, it is m with at Andreasberg and Zellerfeld in the Hartz, Johan georgenstadt in Saxony; Tarnowitz in Silesia; Chentz in Poland; in the country of Salzburg; and at Saska in tl Bannat.

Asia.—Nertschinsk in Siberia.

Second Kind.

Friable Earthy Lead-Spar.

Zerreibliche Bleierde, *Werner.*

Le Plomb terreux friable, *Broch.* t. ii. p. 328.—Zerreiblicl gelbe Bleierde, und zerreibliche grüne Bleierde, *Reu* b. iv. s. 268, 269. 271, 272. *Id. Mohs,* b. iii. s. 356. *Leonhard,* Tabel. s. 73. *Id. Karsten,* Tabl. s. 68.—Zerreiblicl Bleiweiss, *Haus.* s. 114.—Zerreibliche Bleierde, *Haus.* Hai b. iii. s. 1110. *Id. Hoff.* b. iv. s. 45.

Extern

External Characters.

Its colour is yellowish-grey and straw-yellow, which sometimes approaches to sulphur-yellow and lemon-yellow.

It occurs massive, disseminated, and in crusts.

It is composed of dull dusty particles, which are feebly cohering.

It soils feebly.

It is meagre, and rough to the feel.

It is heavy.

Geognostic Situation.

It occurs on the surface or in the hollows of other minerals, and is usually accompanied with galena or lead-glance and lead-spars

Geographic Situation.

Europe.—It is found at Wanlockhead and Leadhills; Zellerfeld in the Hartz; Zschopau, and also near Freyberg in the kingdom of Saxony; in the mountains of Kracau in Poland: and at La Croix in Lothringen in France.

Asia.—The mines of Nertschinsk and Berseowskoi in Siberia.

Observations.

It is distinguished from other friable minerals by colour, meagre feel, and weight.

* Corneous Lead (a).

Hornblei, *Werner*.

Plomb corné, ou Muriate de Plomb natif. *Broch.* t. ii. p. 539.—547, 548.—Hornblei, *Karsten*, Tab. 1. Ausg. s. 78.—*Chenevix*, in Nicholson's Journal, vol. v. p. 219.—Hornblei, *Reuss*, Min. b. ii. s. 261. *Id. Lud.* b. ii. s. 197. *Id. Suck.* 2ter th. s. 344. *Id. Bert.* s. 453. *Id. Leonhard.* Tabel. s. 73.—Plomb muriaté, *Brong.* t. ii. p. 203.—Hornblei, *Karst.* Tabel. 2te Ausg. s. 68.—Hornblei, *Haus.* Handb. b. iii. s. 1104.—Muriate of Lead, *Kid*, vol. ii. p. 145. *Id. Aikin*, p. 111.

External Characters.

Its colours are greyish-white, and yellowish-grey, passing into pale wine-yellow.

It occurs crystallized, in the following figures:

1. Oblique four-sided prism.
   - *a.* Truncated on the angles.
   - *b.* Truncated on the lateral edges.
   - *c.* Bevelled on the lateral edges.
   - *d.* Truncated on the terminal edges.
   - *e.* Acuminated with four planes, which are set on the lateral planes.

Internally it is splendent, and the lustre is adamantine.

It has a threefold cleavage, the cleavages parallel to the planes of the four-sided prism.

The fracture is conchoidal.

It is more or less transparent.

It is soft; rather softer than white lead-spar.

It is sectile, and easily frangible.

Specific gravity, 6.065, *Chenevix*.

Chemical

(a) The minerals marked *, are not yet included in the Genus Lead-spar, as their characters have not been completely ascertained.

Chemical Characters.

On exposure to the blowpipe or charcoal, it melts into an orange-coloured globule, and appears reticular externally, and of a white colour when solid; when again melted it becomes white; and on increase of the heat the acid flies off, and minute globules of lead remain behind.

Constituent Parts.

Oxide of Lead,	85.5
Muriatic Acid,	8.5
Carbonic Acid,	6.0
	100.0

Klaproth, Beit. b. iii. s. 144.

Geographic Situation.

Europe.—In Cromford Level, near Matlock in Derbyshire; and at Hausbaden, near Badweiler in Germany *.

America.—In the neighbourhood of Southampton in the United States †.

Observations.

1. It is a very rare mineral. A good many years ago, a few specimens of it, the only ones hitherto collected in England, were found in Cromford Level, which was soon afterwards filled with water, and the spot which afforded the specimens hid from view.

2. It will probably prove but a variety of white lead-spar.

* Arseniate

* Leonhard's Taschenbuch for 1815, p. 338.

† Found by William Meade, M. D. as mentioned at p. 152. of Bruce's Mineralogical Journal.

* Arseniate of Lead.

Bleiblüthe, *Hausmann*.

This Species is divided into three Subspecies, viz. Reniform arseniate of Lead, Filamentous Arseniate of Lead, and Earthy Arseniate of Lead.

First Subspecies.

Reniform Arseniate of Lead.

Bleiniere, *Hausmann*.

Blenicire, *Reuss*, b. ii. 4. s. 225. *Id. Leonhard*, Tabel. s. 73.—Plomb arsenié, & Plomb reniforme, *Brong*. t. ii. p. 202.—Bleiniere, *Karsten*, Tabel. s. 68.—Plomb arsenié concretionné-mamelonné et compacte, *Haüy*, Tabl. p. 80.—Bleiniere, *Haus*. Handb. b. iii. s. 1097.—Reniform Arseniate of Lead, *Aikin*, p. 114.

External Characters.

Its colours on the fresh fracture are reddish-brown and brownish-red; externally ochre-yellow, and straw-yellow.

It occurs reniform and tuberose; also in curved lamellar concretions.

Internally it is shining and resinous.

The fracture is conchoidal, sometimes inclining to even and uneven.

It is opaque.

It is soft and brittle.

Specific gravity 3.933, *Karsten*.

Chemical

Chemical Characters.

It is insoluble in water. Before the blowpipe on charcoal it gives out arsenical vapours, and is more or less perfectly reduced. It colours glass of borax lemon-yellow.

Constituent Parts.

Oxide of Lead,	35.00
Arsenic acid,	25.00
Water, - -	10.00
Oxide of Iron,	14.00
Silver, -	1.15
Silica,	7.00
Alumina, -	2.00
	95.15

Bindheim, in Beob. u. Endeck. de Berl. Ges. Natf. Fr. iv. s. 374.

Geographic Situation.

It has been hitherto found only in one mine near Nertschinsky in Siberia.

Second Subspecies.

Filamentous Arseniate of Lead.

Flockenerz, *Karsten.*

Plomb arsenié filamenteux, *Haüy*, t. iii. p. 465.—Flockenerz, *Karsten*, Tabel. s. 68.—Flockige Bleiblüthe, *Haus.* Handb. b. iii. s. 1098.—Arseniate of Lead, *Aikin*, p. 114.

External

External Characters.

Its colours are grass-green, wine-yellow, wax-yellow, and lemon-yellow.

It occurs massive, in granular concretions, and either in small acicular six-sided prisms, which are collected into flakes, or in very delicate capillary silky fibres, which are transparent, slightly flexible, and easily frangible.

Specific gravity 5.0, 6.4.

Constituent Parts,

Oxide of Lead,	69.76
Arsenic Acid,	26.4
Muriatic Acid,	1.58

Gregor.

Geographic Situation.

It occurs in the mine of Huel Unity in Gwennap in Cornwall; at St Prix, in the Department of the Soane and Loire in France.

Third Subspecies.

Earthy Arseniate of Lead.

Erdige Bleiblüthe, *Hausmann.*

Plomb arsenié terreux, *Lucas,* t. ii. p. 315.

External Characters.

Its colour is yellow.

It occurs in crusts.

Its

Its fracture is earthy.

It is friable.

Geognostic and Geographic Situations.

It occurs along with filamentous arseniate of lead at St Prix; and also near St Oisans.

* Native Minium, or Native Red Oxide of Lead.

Natürliche Menninge, } *Hausmann.*
Roth Bleioxyd, }

Smithson, in Nicholson's Journal, xvi. p. 127.—*Hänle*, in Magaz. d. Gesel. Natf. Fr. zu Berlin, iii. s. 235.—Plomb oxydé rouge, *Haüy*, Tabl. p. 80. Note 120.—Das rothe Bleioxyd, *Haus.* Handb. b. i. s. 351.—Native Minium, *Aikin*, p. 110.

External Characters.

Its colour is scarlet-red.

It occurs massive, amorphous, and pulverulent; but when examined by the lens, exhibits a crystalline structure, like that of galena, on which it generally rests.

Chemical Characters.

Before the blowpipe, on charcoal, it is first converted into litharge, and then into metallic lead.

Geognostic and Geographic Situations.

It is found in Grassington Moor, Craven; Grasshill Chapel, Wierdale, Yorkshire. On the Continent, it is found in the mine of Hausbaden, near Badenweiler, on galena, and associated with quartz.

Observations.

Observations.

This mineral, in the opinion of Mr Smithson, is produced by the decay of galena or lead-glance; and he adduces in confirmation of this idea, the description of a specimen, which is galena in the centre, but native minium towards the surface.

GENUS II. BARYTE.

Hal-Baryt, *Mohs.*

THIS Genus contains four species, viz. 1. Rhomboidal Baryte, 2. Prismatic Baryte, 3. Di-Prismatic Baryte, 4. Axifrangible Baryte.

1. Rhomboidal Baryte, or Witherite.

Rhomboedrischer Hal-Baryt, *Mohs.*

Witherit, *Werner.*

Baryte aërée, *De Born,* t. i. p. 267.—Witherit, *Wid.* s. 554.—Barolite, *Kirw.* vol. i. p. 134.—Luft oder Kohlensaurer Baryt, *Estner,* b. ii. s. 1124.—Witerite, *Nap.* p. 387. *Id. Lam.* t. ii. p. 20.—Baryte carbonatée, *Haüy,* t. ii. p. 309.—La Witherite, *Broch.* t. i. p. 613.—Witherit, *Reuss,* b. ii. 2. s. 430. *Id. Lud.* b. i. s. 167. *Id. Suck.* 1r th. s. 693. *Id. Bert.* s. 120. *Id. Mohs,* b. ii. s. 200. *Id. Hab.* s. 93.—Baryte carbonatée, *Lucas,* p. 16.—Witherite, *Leonhard,* Tabel. s. 39.—Baryt carbonaté, *Brong.* t. i. p. 255. *Id. Brard,* p. 60.—Witherit, *Karsten,* Tabel. s. 54.

s. 54. *Id. Haus.* s. 132. *Id. Kid,* vol. i. p. 86.—Baryte carbonatée, *Haüy,* Tabl. p. 13.—Witherit, *Lenz,* b. ii. s. 881. *Id. Oken,* b. i. s. 412. *Id. Haus.* Handb. b. iii. s. 1003. *Id. Hoff.* b. iii. s. 150. *Id. Aikin,* p. 166.

External Characters.

Its colours are greyish and yellowish white, also pale bluish-grey, yellowish-grey, and pale wine-yellow.

It occurs massive, disseminated, in crusts, cellular, corroded, large globular, reniform, botryoidal, stalactitic; also in distinct concretions, which are wedge-shaped, sometimes scopiform radiated, and occasionally pass into coarse granular. More rarely crystallized.

The primitive form is a rhomboid of 88° 6′ and 91° 54′. The following are the secondary forms:

1. Equiangular six-sided prism.
   a. Truncated on the terminal edges.
   b. Acutely acuminated on the extremities with six planes, which are set on the lateral planes. Sometimes the alternate acuminating planes are larger and smaller. The apices of the acuminations are sometimes more or less deeply truncated, and in some crystals, the edges between the acuminating and lateral planes are truncated.
2. Acute double six-sided pyramid, in which the lateral planes of the one are set on the lateral planes of the other.

The crystals are small, and very small, seldom middle-sized.

The prisms are sometimes scopiformly grouped, or they are in druses.

Externally

Observations.

This mineral, in the opinion of Mr Smithson, is produced by the decay of galena or lead-glance; and he adduces in confirmation of this idea, the description of a specimen, which is galena in the centre, but native minium towards the surface.

Genus II. BARYTE.

Hal-Baryt, *Mohs.*

This Genus contains four species, viz. 1. Rhomboidal Baryte, 2. Prismatic Baryte, 3. Di-Prismatic Baryte, 4. Axifrangible Baryte.

1. Rhomboidal Baryte, or Witherite.

Rhomboedrischer Hal-Baryt, *Mohs.*

Witherit, *Werner.*

Baryte aërée, *De Born,* t. i. p. 267.—Witherit, *Wid.* s. 554.—Barolite, *Kirw.* vol. i. p. 134.—Luft oder Kohlensaurer Baryt, *Estner,* b. ii. s. 1124.—Witerite, *Nap.* p. 387. *Id. Lam.* t. ii. p. 20.—Baryte carbonatée, *Haüy,* t. ii. p. 309.—La Witherite, *Broch.* t. i. p. 613.—Witherit, *Reuss,* b. ii. 2. s. 430. *Id. Lud.* b. i. s. 167. *Id. Suck.* 1r th. s. 693. *Id. Bert.* s. 120. *Id. Mohs,* b. ii. s. 200. *Id. Hab.* s. 93.—Baryte carbonatée, *Lucas,* p. 16. —Witherite, *Leonhard,* Tabel. s. 39.—Baryt carbonaté, *Brong.* t. i. p. 255. *Id. Brard,* p. 60.—Witherit, *Karsten,* Tabel. s. 54.

Geognostic and Geographic Situations.

It occurs in Cumberland and Durham, in lead-veins that traverse a secondary limestone, which rests on red sandstone, and in these it is associated with coralloidal arragonite, brown-spar, earthy fluor-spar, heavy-spar, and galena or lead-glance, white lead-spar, green lead-spar, copper-pyrites, blue copper, malachite, iron-pyrites, sparry iron, calamine, and blende. In these counties, it is met with at Aldstone in Cumberland; Arkendale, Welhope, and Dufton in Durham. It also occurs at Merton Fell in Westmoreland; Snailback mine in Shropshire *, and at Anglesark in Lancashire, in a vein of galena, along with heavy-spar. It is associated with sparry iron near Steinbauer, not far from Neuberg and Mariazel in Stiria: in granite in Hungary: in the Leogang in Salzburg, along with sparry iron, and copper-pyrites: in the sulphur mines of Azaro and Radussa, and in the river of Nisi, also in Sicily, along with lead-ore.

Asia.—At Schlangenberg and Zincof, in the Altain Mountains.

Uses.

It is a very active poison, and in some districts, as in Cumberland, it is employed for the purpose of destroying rats. When dissolved by muriatic acid, the solution thus obtained, is said to prove serviceable in scrofula.

Observations.

---

* Mr Aikin informs us, that the Witherite of Snailback occurs in masses, varying from 4 lb. to 2 or 3 cwt. imbedded in heavy-spar, contained in a thick vein of galena or lead-glance.—Vid. Geol. Trans. vol. iv. part ii. p. 438. &c.

Observations.

Cleavage, fracture, and weight, distinguish this mineral from calcareous spar, rhomb-spar, apatite, and gypsum.

2. Prismatic Baryte, or Heavy-Spar.

Prismatischer Hal-Baryt, *Mohs.*

Schwerspath, *Werner.*

This species is divided into nine subspecies, viz. 1. Earthy Heavy-Spar, 2. Compact Heavy-Spar, 3. Granular Heavy-Spar, 4. Curved Lamellar Heavy-Spar, 5. Straight Lamellar Heavy-Spar, 6. Fibrous Heavy-Spar, 7. Radiated Heavy-Spar, 8. Columnar Heavy-Spar, and, 9. Prismatic Heavy-Spar.

First Subspecies.

Earthy Heavy-Spar.

Schwerspath Erde, *Werner.*

Baryte vitriolée terreuse, *De Born*, t. i. p. 268.—Schwerspath-erde, *Wid.* s. 558.—Earthy Baroselenite, *Kirw.* vol. i. p. 138.—Schwerspath-erde, *Estner*, b. ii. s. 1143. *Id. Emm.* b. i. s. 550.—Baryta vitriolata terrea, *Nap.* p. 402.—Le Spath pesant terreux, *Broch.* t. i. p. 617.—Erdiger Baryt, *Reuss*, b. ii. 2. s. 437. *Id. Lud.* b. i. s. 168. *Id. Suk.* 1r th. s. 697. *Id. Bert.* s. 122.—Baryterde, *Mohs*, b. ii. s. 106.—Erdiger Baryt, *Leonhard*, Tabel. s. 39.—Baryte sulphatée terreux, *Brong.* t. i. p. 252.—Erdiger Baryt, *Haus.* s. 134. *Id. Karsten*, Tabel. s. 54.—Earthy Sulphate of Baryt, *Kid*, vol. i. p. 87.—Baryt-erde, *Lenz*, b. ii. s. 389.—Schwerspatherde, *Hoff.* b. iii. s. 156.

External

External Characters.

Its colours are yellowish and reddish white.

It is of friable consistence, and consists of feebly glimmering, nearly dull, particles, which are intermediate between scaly and dusty, that soil feebly, and are generally loose, or but feebly cohering.

It feels meagre, and rather rough.

Specific gravity, 4.0.

Constituent Parts.

It is Sulphate of Barytes.

Geognostic and Geographic Situations.

It occurs in drusy cavities in veins of heavy-spar, in Staffordshire and Derbyshire; at Freyberg in Saxony; Riegelsdorf in Hessia; and Mies in Bohemia.

Observations.

1. It is distinguished from all other earthy minerals, by its great specific gravity.

2. Some mineralogists are of opinion, that it is disintegrated compact Heavy-Spar, while others maintain that it is an original formation. The latter opinion is countenanced in those instances where the earthy heavy-spar occurs in close cavities.

Second

Second Subspecies.

Compact Heavy-Spar.

Dichter Schwerspath, *Werner.*

Baryte vitriolata compacte, *De Born,* t. i. p. 268.—Dichter Schwerspath, *Wid.* s. 559.—Compact Baroselenite, *Kirw.* vol. i. p. 138.—Dichter Schwerspath, *Estner,* b. ii. s. 1146. *Id. Emm.* b. i. s. 552.—Barite vitriolata compatta, *Nap.* p. 400. —Le Spath pesant compacte, *Broch.* t. i. p. 618.—Dichter Schwerspath, *Reuss,* b. ii. 2. s. 438. *Id. Lud.* b. i. s. 169. *Id. Suck.* 1r th. s. 698. *Id. Bert.* s. 123. *Id. Mohs,* b. ii. s. 206. *Id. Leonhard,* Tabel. s. 39.—Baryte sulphatée compacte, *Brong.* t. i. p. 252.—Dichter Baryte, *Karsten,* Tabel. s. 54. *Id. Haus.* s. 133.—Baryte sulphatée compacte, *Haüy,* Tabl. p. 13.—Dichter Baryte, *Lenz,* b. ii. s. 390.—Dichter Schwerspath, *Hoff.* b. iii. s. 158.

External Characters.

Its colours are yellowish-white, and greyish-white, which pass into yellowish-grey, and ash-grey.

It occurs massive, disseminated, reniform, semi-globular, tuberose, with cubic impressions; and in curved lamellar concretions.

Internally it is glimmering.

The fracture is intermediate, between coarse earthy, and fine-grained uneven, which sometimes passes into imperfect foliated, and more rarely into splintery.

The fragments are indeterminate angular, and blunt-edged.

It is opaque, or translucent on the edges.

It is soft.

It

It is rather sectile, and easily frangible.

Specific gravity, 4.84.

Constituent Parts.

Sulphate of Barytes, -	83.0
Silica, -	6.0
Alumina,	1.0
Water, -	2.0
Oxide of Iron,	4.0
	96.0

Westrumb, in Bergbaukunde, ii. s. 47.

Geognostic and Geographic Situations.

It is found in the mines of Staffordshire and Derbyshire, where it is named *Cawk*. It also occurs at Meis in Bohemia, Freyberg in Saxony, in the Hartz, and the Breisgau: also in clay-slate, near Servos in Savoy; and in Austria and Stiria.

Third Subspecies.

Granular Heavy-Spar.

Körniger Schwerspath, *Werner*.

Blättriger Schwerspath, *Wid.* s. 561.—Körniger Schwerspath, *Emm.* b. i. s. 556.—Le Spath pesant grenue, *Broch.* t. i. p. 620.—Körniger Baryt, *Reuss*, b. ii. 2. s. 441. *Id. Lud.* b. i. s. 169. *Id. Suck.* 1r th. s. 701. *Id. Bert.* s. 124. *Id. Mohs*, b. ii. s. 206. *Id. Leonhard*, Tabel. s. 39.—Baryte sulphatée grenue, *Brong.* t. i. p. 253.—Schuppiger Baryt, *Haus.* s, 133. —Körniger Baryt, *Karst.* Tabel. s. 54.—Baryte sulphatée granulaire,

granulaire, *Haüy*, Tabl. p. 13.—Körniger Baryt, *Lenz*, b. ii. s. 891.—Körniger Schwerspath, *Hoff.* b. iii. s. 160.—Granular Heavy-Spar, *Aikin*, p. 170.

External Characters.

The colours are snow, yellowish, greyish, and reddish white, and sometimes dark ash-grey. It is occasionally spotted brown and yellow on the surface.

It occurs massive, and in fine granular concretions, which are sometimes so minute as scarcely to be discernible.

Internally it is glistening, approaching to shining, and is pearly.

The fragments are indeterminate angular, and blunt-edged.

It is feebly translucent.

It is soft.

It is rather brittle, and easily frangible.

Specific gravity, 4.380, *Klaproth*.

Constituent Parts.

Sulphat of Barytes, -	90
Silica, -	10
	100

Klaproth, Beit. b. ii. s. 72.

Geognostic Situation.

It occurs principally in beds, along with galena, blende, copper-pyrites, and iron-pyrites.

Geographic Situation.

It occurs in beds, along with galena, blende, copper-pyrites, and iron-pyrites, at Peggau in Stiria; also in the Hartz,

Hartz, in beds, along with copper and iron pyrites, galena, and blende; and at Schlangenberg in Siberia, where it is associated with copper-green, and native copper.

Observations.

1. It bears a striking resemblance to Foliated Granular Limestone, from which, however, it is distinguished by the following characters:

1st, It has a lower degree of lustre.

2d, When the distinct concretions are of the same size as in the foliated granular limestone, they are not so well defined.

3d, It is more easily broken in pieces than the limestone, owing to the concretions being less intimately connected together.

4th, It is much heavier.

2. Foliated Granular Limestone, Granular Heavy-spar, and Granular Gypsum, may be distinguished from each other by the relative distinctness of the concretions: in the Foliated Granular Limestone they are well defined; in Granular Heavy-spar less so; and in Granular Gypsum still more indistinct.

Fourth Subspecies.

Curved Lamellar Heavy-Spar.

Krummschaaliger Schwerspath, *Werner*.

Le Spath pesant testacé courbe, ou le Spath lamelleux, *Broch.* t. i. p. 621.—Krummschaaliger Baryt, *Reuss*, b. ii. 2. s. 443. *Id. Lud.* b. i. s. 170. *Id. Suck.* 1r th. s. 700. *Id. Bert.* s. 124. *Id.*

Id. Mohs, b. ii. s. 207. *Id. Leonhard,* Tabel. s. 40.—Blättriger Baryt, *Karsten,* Tabel. s. 54.—Baryte sulphatée cretée, *Haüy,* Tabl. p. 13.—Krummschaaliger Schwerspath, *Hoff.* b. iii. s. 162.

External Characters.

Its principal colours are white, grey, and red: the white varieties are yellowish, greyish, and reddish white; the grey varieties are smoke and pearl grey, and there is a transition from pearl-grey into flesh-red and blood-red, and from yellowish-grey into yellowish-brown, and liver-brown. Sometimes several colours occur together, and are arranged in broad stripes.

It generally occurs massive, more frequently reniform, and long globular, with a drusy surface; the drusy surface is formed of very small, thin, and longish four-sided tables; also in reniform curved lamellar concretions, which are frequently floriform, and these are again composed of prismatic concretions. It is rarely marked with cubical impressions.

Internally it is intermediate between shining and glistening, and the lustre is pearly, inclining to resinous.

The fracture is curved foliated, which sometimes inclines to splintery, and thus approaches to the compact subspecies.

The fragments are indeterminate angular, and rather blunt-edged.

It is translucent on the edges.

It scratches calcareous-spar, but does not affect fluor.

It is brittle, and easily frangible.

Specific gravity, 4.307, *Breithaupt.*

Geognostic

Geognostic and Geographic Situations.

It is one of the most common subspecies of heavy-spar. In Scotland, it occurs in trap and sandstone rocks: in Derbyshire, it occurs in secondary limestone: it characterises a particular venigenous formation at Freyberg in Saxony, where it is associated with radiated pyrites, argentiferous galena, brown blende, calcareous-spar, and fluor-spar. It occurs in Sweden, Carinthia, and other countries.

Fifth Subspecies.

Straight Lamellar Heavy-Spar.

It is divided into three kinds, viz. Fresh Straight Lamellar Heavy-Spar, Disintegrated Straight Lamellar Heavy-Spar, and Fetid Straight Lamellar Heavy-Spar.

First Kind.

Fresh Straight Lamellar Heavy-Spar.

Geradschaaliger Schwerspath, *Werner.*

Gypsum spathosum, *Wall.* t. i. p. 168.—Spath pesant ou seleniteux, *Romé de Lisle*, t. i. p. 577.—Baryte vitriolée spathique, *De Born*, t. i. p. 270.—Var. of Blättriger Schwerspath, *Wid.* s. 561.—Gemeiner Schwerspath, *Emm.* b. i. s. 557.—Baryta vitriolata lamellare, *Nap.* p. 395.—Foliated Baroselenite, *Kirw.* vol. i. p. 140.—Le Spath pesant testacé à lames droites, ou Le Spath pesant commun, *Broch.* t. i. p. 624.—Geradschaaliger Baryt, *Reuss*, b. ii. 2. s. 445.—Frischer Geradschaaliger Baryt, *Lud.* b. i. s. 170. *Id. Suck.* 1r th. s. 702. *Id. Bert.* s. 125. *Id. Mohs*, b. ii. s. 209. *Id. Leonhard*, Tabel. s. 40. —Baryte

—Baryte sulphatée pure crystallisée, *Brong.* t. i. p. 250.—Gemeiner Baryt, *Karst.* Tabel. s. 54. *Id. Haus.* s. 133.—Baryte sulphatée, en formes determinables, *Haüy*, Tabl. p. 12.—Geradschaaliger Baryt, *Lenz*, b. ii. s. 894.—Frischer Geradschaaliger Schwerspath, *Hoff.* b. iii. s. 165.

External Characters.

Its colours are snow, milk, reddish, yellowish, and greenish white; greyish, ash, smoke, and bluish-grey; greyish-black; smalt-blue, pale sky-blue, and muddy indigo-blue; verdigris-green and olive-green; cream, honey, wax, and wine yellow; brick-red, blood-red, and brownish-red; and yellowish-brown.

It occurs generally massive; also in distinct concretions, which are straight and thin lamellar; and again collected into others which are coarse granular; and also crystallized. The primitive form is an oblique four-sided prism of 101° 53'. The following are the secondary figures:

1. Rectangular four-sided table.
   a. Perfect.
   b. In which the terminal planes are bevelled *, fig. 140. Pl. 7.
   c. In which the angles of the bevelment are truncated †, fig. 141. Pl. 7.
2. Oblique four-sided table.
   a. Perfect.
   b. Truncated on the lateral edges.
   c. Truncated on the acute terminal edges, and sometimes also on the acute angles.

d. Truncated

* Baryte sulphatée trapezienne, Haüy.

† Baryte sulphatée epointée, Haüy.

d. Truncated on the obtuse angles *.

e. Truncated on the obtuse angles and terminal edges.

f. Bevelled on the obtuse terminal edges.

3. Longish six-sided table.

a. Perfect †, fig. 142. Pl. 7.

b. Bevelled on the terminal planes.

c. Bevelled on the terminal and lateral edges.

d. Bevelled on the lateral planes, and truncated on the bevelling edges.

e. The lateral edges acutely bevelled, and the acute angles truncated.

4. Eight-sided table.

a. Perfect.

b. Bevelled on the terminal planes.

c. Slightly truncated on the lateral and terminal edges.

d. Bevelled on the lateral and terminal planes.

The crystals vary in size, from large to small; and rest on one another, or intersect one another.

Externally they are smooth and splendent; internally shining and splendent, and the lustre intermediate between resinous and pearly.

It has a distinct cleavage, in which the folia are parallel with the planes of the primitive prism, and of these, that parallel to the terminal plane is the most distinct.

The fragments are tabular and rhomboidal.

It is translucent, or transparent, and refracts double.

It scratches calcareous-spar, but is scratched by fluor-spar.

It

* Baryte sulphatée apophane, Haüy.

† Baryte sulphatée retrecée, Haüy.

It is brittle, and easily frangible.

Specific gravity, 4.1, 4.6.

Chemical Characters.

It decrepitates briskly before the blowpipe, and, by continuance of the heat, melts into a hard white enamel. A piece exposed for a short time to the blowpipe, and then laid on the tongue, gives the flavour of sulphuretted hydrogen. When pounded, and thrown on glowing coals, it phosphoresces with a yellow light.

Constituent Parts.

Sulphat of Barytes,	97.60
Sulphat of Strontian,	0.85
Water, -	0.10
Oxide of Iron,	0.80
Alumina,	0.05

Klaproth, Beit. b. ii. s. 78.

Geognostic Situation.

It is found almost always in veins, which occur in granite, gneiss, mica-slate, clay-slate, grey-wacke, limestone, and sandstone. It is often accompanied with ores, particularly the flesh-red variety, and these are, native silver, silver-glance or sulphuretted silver, copper-pyrites, lead-glance, white cobalt-ore, light red silver, native arsenic, earthy cobalt, cobalt-bloom or red cobalt, antimony, and manganese. It occurs sometimes in beds, and encrusting the walls of drusy cavities.

Geographic

Geographic Situation.

In this island, it occurs in veins in different primitive and transition rocks, and also in secondary limestone, sandstone, and trap. Beautiful crystallized varieties are found in the lead-mines of Cumberland, Durham, and Westmoreland. It is very frequent on the Continent of Europe, and also in America, particularly in mining districts.

Uses.

It is said to form a good manure for clover fields. When burnt, and finely ground, it is used in place of bone-ashes for cupels. The white varieties are employed as white colours in painting, and for pastil-pencils; and it is sometimes used as a flux for ores of particular kinds.

Second Kind.

Disintegrated Straight Lamellar Heavy-Spar.

Mulmicher oder mürber geradschaaliger Schwerspath, *Werner*.

Mulmiger Baryt, *Reuss*, b. ii. 2. s. 455. *Id. Leonhard*, Tabel. s. 40. *Id. Karsten*, Tabel. s. 54.—Aufgelöster Baryt, *Lenz*, b. ii. s. 899.—Mulmiger geradschaaliges Schwerspath, *Hoff*. b. iii. s. 173.

External Characters.

Its colours are greyish, greenish, yellowish, and reddish white.

It

It occurs massive.

It is glistening and pearly.

The cleavage and concretions the same as in the preceding species.

It is opaque, or faintly translucent on the edges.

It is soft, passing into very soft.

It is very easily frangible.

In other characters same as the preceding.

Geognostic and Geographic Situations.

It was formerly met with in considerable quantity at Freyberg in Saxony, in a mixture of galena, blende, and iron-pyrites.

Third Kind.

Fetid Straight Lamellar Heavy-Spar, or Hepatite.

Gypsum, Lapis hepaticus, *Wall.* Syst. i. s. 165.—Baryte sulphatée fetide, *Haüy*, t. ii. p. 304.—Hepatit, *Reuss*, b. ii. 2. s. 463. *Id. Lud.* b. ii. s. 157. *Id. Suck.* 1r th. s. 714. *Id. Bert.* s. 131. *Id. Mohs*, b. ii. s. 228.—Baryte sulphatée fetide, *Lucas*, p. 15.—Hepatit, *Leonhard*, Tabel. s. 40.—Baryte sulphatée fetide, *Brong.* t. i. p. 253. *Id. Brard*, p. 59. —Hepatit, *Karsten*, Tabel. s. 54. *Id. Haus.* s. 134.—Baryte sulphatée fetide, *Haüy*, Tabl. p. 13.—Hepatit, *Lenz*, b. ii. s. 908. *Id. Oken*, b. i. s. 404. *Id. Aikin*, p. 171.

External Characters.

Its colours are greyish-white, yellowish and smoke grey, greyish and brownish black.

It

It occurs massive, disseminated, and in globular or elliptical pieces, from an inch to a foot and upwards in diameter; also in lamellar concretions, which are generally straight, sometimes curved and floriform; sometimes there is a tendency to wedge-shaped and radiated concretions.

Externally it is feebly glimmering; internally shining, and intermediate between pearly and resinous.

The fragments are indeterminate angular, and blunt-edged.

It is opaque, or translucent on the edges.

It is nearly as hard as straight lamellar heavy-spar.

It affords a greyish-white coloured streak.

It is heavy.

Chemical Characters.

It burns white before the blowpipe; and when rubbed or heated, gives out a fetid sulphureous odour.

Constituent Parts.

Sulphate of Barytes, with a trace of Sulphate of Strontian,	93.58	Sulphate of Barytes,	92.75	Sulphate of Barytes,	85.25
Sulphate of Lime,	3.58	Carbon and Bitumen,	2.00	Carbon,	0.50
Oxide of Iron,	0.87	Sulphate of Lime,	2.00	Sulphate of Lime,	6.00
Water, Carbonaceous Matter, Sulphur, and Alumina,	2.00	Oxide of Iron,	1 50	Oxide of Iron,	5.00
		Water,	1.25	Alumina,	1.00
		Sulphur, Oxide of Manganese, Chromic Acid, and Alumina, very small quantity.		Loss, including Moisture and Sulphur,	2.25
	100.00		99 05		100.00
John, Chem. Unt. b. ii. s. 73.		*John*, Chem. Unt. b. ii. s. 69.		*Klaproth*, Beit. b. v. s. 121.	

Geognostic

Geognostic and Geographic Situations.

It occurs at Buxton in Derbyshire; at Kongsberg in Norway, in veins that traverse mica-slate and hornblende-slate, along with native silver, heavy-spar, coal-blende, and iron-pyrites; and at Andrarum in Schonen, in transition alum-slate, in the form of balls. These balls are sometimes impregnated with iron-pyrites, or the iron-pyrites forms the central part.

Observations.

1. It is named *Hepatite*, from the disagreeable sulphureous odour it exhales when rubbed or exposed to heat.

2. Marggraf, Linnæus, and Cronstedt, arrange this mineral with Limestone.

Sixth Subspecies.

Fibrous Heavy-Spar.

Fasriger Schwerspath, *Werner.*

Fasriger Baryt, *Leonhard,* Tabel. s. 40. *Id. Karsten,* Tabel. s. 54. *Id. Haus.* s. 133.—Baryte sulphatée concretionnée-fibreuse, *Haüy,* Tabl. p. 13.—Fasriger Baryt, *Lenz,* b. ii. s. 900.—Fasriger Schwerspath, *Hoff.* b. iii. s. 183.

External Characters.

Its colour is pale-yellowish, and wood-brown, which sometimes passes into yellowish-grey.

It

It occurs massive and reniform; also in distinct concretions, which are scopiform prismatic or fibrous, sometimes collected into others, which are curved lamellar, and sometimes into coarse angulo-granular concretions.

Internally it is shining, and the lustre is resinous.

The fragments are splintery, and wedge-shaped.

It is translucent on the edges.

Specific gravity, 4.080, *Klaproth.*—4.289, *Noeggerath.*

Constituent Parts.

Sulphate of Barytes,	-	99.0
Trace of Iron.		
		99.0

Klaproth, Beit. b. iii. s. 288.

Geognostic and Geographic Situations.

It is found at Neu-Leiningen in the Palatinate; also in an ironstone mine in clay-slate, at Chaud-Fontaine, near Lüttich, in the Ourthe department; and at Miess in Bohemia.

Observations.

It was first described by Karsten, and analysed by Klaproth. It was sent to Klaproth as a rare variety of calamine.

Seventh

Seventh Subspecies.

Radiated Heavy-Spar, or Bolognese Spar.

Bologneser Spath, *Werner.*

Gypsum spathosum, opacum, semi-pellucidum, *Wall.* t. i. p. 169. —Var. of Blættriger Schwerspath, *Wid.* s. 561.—Bologneserstein, *Emm.* b. iv. s. 572.—Litheosphore, *Lam.* t. i. p. 24.—Baryte sulphatée rayonnée, *Haüy*, t. ii. p. 302.—La Spath de Bologne, ou La Pierre de Bologne, *Broch.* t. i. p. 633.—Strahliger Baryt, *Reuss*, b. ii. 2. s. 460.—Bologneser Spath, *Lud.* b. i. s. 172. *Id. Suck.* 1r th. s. 712.—Kieselerdiger schwefelsaurer Baryt, *Bert.* s. 130.—Bologneser Spath, *Mohs*, b. ii. s. 227.—Strahliger Baryt, *Leonhard*, Tabel. s. 40.—Baryte sulphatée pure radiée, *Brong.* t. i. p. 251.—Strahliger Baryt, *Haus.* s. 133. *Id. Karsten*, Tabel. s. 54.—Bologna Stone, *Kid*, vol. i. p. 89.—Baryte sulphatée radiée, *Haüy*, Tabl. p. 13. —Strahliger Baryt, *Lenz*, b. ii. s. 901.—Bologneser Spath, *Hoff.* b. ii. s. 180.

External Characters.

Its principal colour is smoke-grey, which passes into ash-grey and yellowish-grey.

It occurs in roundish pieces, which have a lenticular aspect and uneven surface; also in distinct concretions, which are parallel and scopiform prismatic, and also granular.

Internally it is shining or glistening, and the lustre is pearly, inclining to resinous.

The fragments are splintery, or wedge-shaped.

It is translucent.

In other characters it agrees with the preceding.

Chemical

Chemical Characters.

It is remarkably phosphorescent when heated. This property was first observed in the year 1630, by a shoemaker named Vincenzo Casciarolo, during his search after the philosopher's stone. When the mineral is calcined, pulverised, and made into cakes, it acquires a strong phosphorescent property by exposure to light; the phosphorescence is visible, upon taking it into a dark place.

Constituent Parts.

Sulphate of Barytes,	62.00
Lime, - -	2.00
Silica, -	16.00
Alumina, -	14.75
Oxide of Iron,	0.25
Water, -	2.00
	97.00 *Afzelius.*

Geognostic and Geographic Situations.

It occurs imbedded in marl in Monte Paterno, near Bologna: also at Rimini; and in Jutland.

Observations.

1. Its uneven surface, shows that the rounded pieces are of cotemporaneous formation with the marl in which they are contained, and not rolled pieces.

2. When rendered phosphorescent, it is known under the name of *Bolognian Phosphorus.*

Eighth

Eighth Subspecies.

Columnar Heavy-Spar.

Stängenspath, *Werner.*

Var. Blættriger Schwerspath, *Wid.* s. 561.—Stangenspath, *Emm.* b. i. s. 569.—Le Spath pesant en barres, *Broch.* t. i. p. 631. —Baryte sulphatée bacillaire, *Haüy,* t. ii. p. 302.—Stänglicher Baryt, *Reuss,* b. ii. 2. s. 458.—Stangenspath, *Lud.* b. i. s. 172. *Id. Suck.* 1r th. s. 711. *Id. Bert.* s. 130. *Id. Mohs,* b. ii. s. 225.—Stänglicher Baryt, *Leonhard,* Tabel. s. 40.—Baryte sulphatée pure bacillaire, *Brong.* t. i. p. 251. —Stänglicher Baryt, *Karsten,* Tabel. s. 54.—Stangenspath, s. 133.—Baryte sulphatée bacillaire, *Haüy,* Tabl. p. 13.— Stänglicher Baryt, *Lenz,* b. ii. s. 903.—Stangenspath, *Hoff.* b. iii. s. 178.—Columnar Heavy-spar, *Aikin,* p. 170.

External Characters.

Its colours are yellowish, greyish, and greenish white.

It occurs crystallised, in acicular oblique four-sided prisms, which are always columnarly aggregated, and intersect each other.

Externally it is frequently invested with iron-ochre, but when unsoiled, it is shining and pearly.

Its cleavage is the same as that of lamellar heavy-spar.

The fragments are indeterminate angular, and rather sharp-edged.

It is translucent.

Specific gravity 4.500.

Constituent

Constituent Parts.

Barytes, -	63.00
Sulphuric Acid,	33.00
Strontian Earth,	3.10
Oxide of Iron,	1.50
Water,	1.20

Lampadius.

Geognostic and Geographic Situations.

It was formerly found in the vein of Lorenzgegentrum, near Freyberg in Saxony, along with ores of different kinds, and also fluor-spar, quartz, and straight and curved lamellar heavy-spar. It is also mentioned as occurring in Derbyshire.

Observations.

It has been sometimes confounded with White Lead-spar, but is distinguished from that mineral by the following characters: White Lead-Spar has an adamantine lustre, its fracture is small conchoidal, and its specific gravity is 6.558; whereas Columnar Heavy-spar has a pearly lustre, distinct cleavage, and a specific gravity of 4.500.

Ninth Subspecies.

Prismatic Heavy-Spar.

Saulenspath, *Werner.*

Sauliger Baryt, *Reuss,* b. ii. 2. s. 455.—Saulenspath, *Lud.* b. i. s. 172. *Id. Mohs,* b. ii. s. 226.—Sauliger Baryt, *Leonhard,* Tabel. s. 40.—Saulenspath, *Lenz,* b. ii. s. 905.—Saulenschwerspath, *Hoff.* b. iii. s. 174.

External Characters.

Its principal colours are smoke and yellowish grey, seldomer greenish and pearl grey; from yellowish-grey it sometimes passes into dark greyish and yellowish white; and in some crystals there are transitions from greenish-grey into olive-green, and from pearl-grey into flesh-red, and still seldomer from smoke-grey into a kind of indigo-blue.

It seldom occurs massive, or in angulo-granular, and promiscuous prismatic concretions, generally crystallized, and in the following figures:

1. Slightly oblique four-sided prism, rather acutely bevelled on the extremities, the bevelling planes set on the acuter lateral edges. Fig. 143. Pl. 7.
   a. The obtuse edges truncated.
   b. The angles of the bevelment truncated; when these truncating planes increase in magnitude, there is formed
2. An oblique four-sided prism, rather acutely acuminated on the extremities with four planes, which are set on the lateral edges. The obtuse lateral edges of the prism are sometimes truncated. Sometimes

Sometimes this acumination is surmounted by another flatter four-planed acumination. When the truncating planes on the obtuse lateral edges of the varieties (1) and (2) increase, there is formed

3. An unequiangular six-sided prism, with two opposite acuter lateral edges, and with the same terminal bevelment and acuminations as in figures 1. and 2.

When this prism becomes broad, it passes into the tabular form. When the prism in the variety (*b*) disappears, there is formed

4. A rather flat double four-sided pyramid, in which the lateral planes of the one are set on the lateral planes of the other.

The crystals are middle-sized and small, and are generally promiscuously aggregated.

The surface of the crystals is splendent, and the lateral planes are transversely streaked.

Internally it is shining, or splendent, and the lustre is pearly, inclining to resinous.

The cleavage is more or less perfect.

It alternates from translucent to semi-transparent.

Specific gravity 4.471, *Breithaupt*.

Geognostic Situation.

It occurs in veins, along with fluor-spar, and ores of silver and cobalt; in gneiss, mica-slate, and other primitive rocks. It is rare in clay-slate, very rare in secondary rocks.

Geognostic Situation.

It occurs at Kongsberg in Norway; Mies in Bohemia;

 and

and Freyberg, Marienberg, and Ehrenfriedersdorf in Saxony; Roya in Auvergne.

Observations.

The prismatic crystallizations, granular and prismatic concretions, and resinous lustre, are the principal characters of this subspecies.

3. Di-Prismatic Baryte, or Strontianite.

Di-prismatischer, Hal-Baryt, *Mohs.*

Strontian, *Werner.*

Strontian, *Wid.* s. 571. *Id. Kirw.* vol. i. p. 332. *Id. Estner,* b. ii. s. 48.—Kohlensaurer Strontianit, *Emm.* b. i. s. 310.—Strontianite, *Nap.* p. 391.—Strontites, *Hope,* Edin. Trans. for 1790. *Id. Lam.* t. ii. p. 130.—Strontiane carbonatée, *Haüy,* t. ii. p. 327—La Strontianite, *Broch.* t. i. p. 637.—Strontianit, *Reuss,* b. ii. 2. s. 416. *Id. Lud.* b. i. s. 174. *Id. Suck.* 1r th. s. 684. *Id. Bert.* s. 133. *Id. Mohs,* b. ii. s. 198. —Strontiane carbonatée, *Lucas,* p. 18.—Kohlensaurer Strontianit, *Leonhard,* Tabel. s. 41.—Strontiane carbonatée, *Brong.* t. i. p. 259. *Id. Brard,* p. 64.—Strontian, *Karsten,* Tabel. s. 54.—Strontianite, *Haus.* s. 131. *Id. Kid,* vol. i. p. 82.—Strontiane carbonatée, *Haüy,* Tabl. p. 15.—Strontianite, *Lenz,* b. ii. s. 915. *Id. Oken,* b. i. s. 411. *Id. Haus.* Handb. b. iii. s. 979.—Strontian, *Hoff.* b. iii. s. 186. *Id. Aikin,* p. 166.

External Characters.

Its colour is pale asparagus-green, which sometimes inclines to apple-green, sometimes to yellowish-white and greenish-grey. The greenish-grey variety sometimes passes into milk and yellowish white, and pale straw-yellow.

It

It occurs massive, in distinct concretions, which are scopiform, radiated and fibrous, and crystallized. The primitive form is an oblique four-sided prism, bevelled on the extremities. The vertical prism is 117° 19′; the horizontal is not determined. The secondary figures are the following:

1. Acicular six-sided prism, acutely acuminated with six planes, which are set on the lateral planes.
2. Acicular acute double six-sided pyramid.

The crystals are sometimes scopiformly and manipularly aggregated.

The lustre of the distinct concretions is shining or glistening; of the fracture glistening, and is pearly.

The cleavage is in the direction of the lateral planes of the primitive form.

The fracture is fine-grained uneven.

The fragments are wedge-shaped, or splintery.

It is more or less translucent, and sometimes semi-transparent.

It is harder than calcareous-spar, but not so hard as fluor-spar.

It is brittle, and easily frangible.

Specific gravity, 3.675, *Klaproth.*—3.644, *Kirwan.*—3.6583, 3.675, *Haüy.*—3.6, 3.8, *Mohs.*

Chemical Characters.

It is infusible before the blowpipe, but becomes white and opaque, and tinges the flame of a dark purple colour. It is soluble, with effervescence, in muriatic or nitric acid; and paper dipped in the solutions thus produced, burns with a purple flame.

Constituent

Constituent Parts.

Strontian,	61.21	69.5	62.0	74.0
Carbonic Acid,	30.20	30.0	30.0	25.0
Water,	8.50	0 5	8.0	0.5
	100.00	100.0	100.0	99.5
	Hope, Edin. Trans. for 1790.	*Klaproth*, Beit. b. i. s. 270.	*Pelletier*, Jour. des Mines, N. 21. p. 46.	*Buchols*, in Lenz Min. b. ii. s. 916.

Geognostic and Geographic Situations.

It occurs at Strontian in Argyleshire, in veins that traverse gneiss, along with galena or lead-glance, heavy-spar, and calcareous-spar; very rarely at Leogang in Salzburg; also at Braunsdorf in Saxony, along with calcareous-spar, iron and copper pyrites; and at Pisope, near Popyan in Peru.

Observations.

1. The peculiar earth which characterises this mineral was discovered by Dr Hope, and its various properties were made known to the public in his excellent memoir on Strontites, inserted in the Transactions of the Royal Society of Edinburgh for the year 1790.

2. It is characterised by its green colours, acicular crystallizations, prismatic concretions, and weight; it is distinguished from *Arragonite* by colour and greater specific gravity; from *Witherite* by colour, and inferior specific gravity.

3. According to Blumenbach, it is not poisonous like Witherite.

4. Axifrangible

4. Axifrangible Baryte, or Celestine.

Axentheilender Hal-Baryt, *Mohs.*

Zölestin, *Werner.*

This Species is divided into five Subspecies, viz. Foliated Celestine, Prismatic Celestine, Fibrous Celestine, Radiated Celestine, and Fine Granular Celestine.

First Subspecies.

Foliated Celestine.

Blättricher Celestin, *Karsten.*

Schaaliger Zölestin, *Werner.*

Strontiane sulphatée, *Haüy,* t. ii. 318.—Blättricher Schützit, *Reuss,* b. ii. 2. s. 423.—Blättricher Celestin, *Suck.* 1r th. s. 688. *Id. Bert.* s. 134. *Id. Mohs,* b. ii. s. 230.—Blättriger schwefelsaurer Strontianit, *Leonhard,* Tabel. s. 41.—Blättriger Celestin, *Karsten,* Tabel. s. 54.—Strontian sulphatée laminaire, *Haüy,* Tabl. p. 14.—Blättriger Celestin, *Lenz,* b. ii. s. 923. —Schaliger Zölestin, *Hoff.* b. iii. s. 195.

External Characters.

Its colours are milk-white, bluish-grey, smalt-blue, sky-blue; also yellowish-white, and rarely reddish-white, and pale flesh-red.

It occurs massive; also in lamellar distinct concretions, which are generally straight, or slightly curved, and in which the surfaces are smooth and shining; and crystallized in the following figures:

1. Rectangular

1. Rectangular four-sided table, in which the terminal planes are bevelled; of these, two opposite are acute, and other two opposite flatly bevelled, and the terminal edges truncated. Sometimes the edges between the bevelling planes and the lateral planes are truncated.
2. Rectangular four-sided table, bevelled on the terminal edges.

The crystals are middle-sized and small, and frequently rest on each other, or intersect each other.

Externally it is shining and splendent; internally it is shining and pearly, inclining to vitreous.

It has a threefold cleavage in which the folia are parallel with the planes of the primitive figure, and of these, that parallel with the terminal planes is the most distinct.

The fracture is uneven.

The fragments are rhomboidal, or indeterminate angular, and rather sharp-edged.

It is translucent, semi-transparent, or transparent.

It scratches calcareous-spar, but is scratched by fluor-spar.

It is rather sectile, and is very easily frangible.

Specific gravity, 3.960, *Clayfield*.—3.967, *Karsten*.—3.6–4.0, *Mohs*.

Chemical Characters.

It melts before the blowpipe into a white friable enamel, without very sensibly tinging the flame: after a short exposure to heat, it becomes opaque, and has then acquired a somewhat caustic acrid flavour, very different from that of sulphuretted hydrogen, which heavy-spar acquires in similar circumstances.—*Aikin*.

These

These characters apply also to the other subspecies.

Constituent Parts.

Strontian, 57.64 Sulphuric Acid, 43.00 ——— 100.64 *Rose*, in Karsten's Tabellen, s. 55.	Strontian and Sulphuric Acid, 97.208 Sulphate of Barytes, - 2.222 Silica, - 0.254 Oxide of Iron, 0.116 Water, - 0.190 Petroleum, a minute portion. ——— 99.099 *Stromeyer*, in Gött. Gel. Anz. 1811, 188.	Strontian, and Sulphuric Acid, 97.601 Sulphate of Barytes, - 00.975 Silica, - 00.107 Oxide of Iron, and intermixed Hydrate of Iron, 00.646 Water, - 00.248 ——— 99.577 *Stromeyer*, in Gött. Gel. Anz. 1812, 12. 114.

Geognostic and Geographic Situations.

It occurs in trap-tuff in the Calton Hill at Edinburgh *, and in red sandstone at Inverness. It is frequent along with some of the other subspecies at Aust Passage, and elsewhere in the neighbourhood of Bristol, and in the islands in the Bristol Channel, particularly in Barry Island, on the coast of Glamorganshire; also in amygdaloid at Bechely in Gloucestershire †; and it has been found on the banks of the Nidd, near Knaresborough, Yorkshire. It forms a bed, about one-fourth of a fathom thick, in a coal-mine, which appears to be connected with shell limestone, at Süntel in Hanover; and also near Karlshütte, on the road from Göttingen to Hanover; in the Canton of Aargau in Switzerland;

* It was discovered in the Calton Hill by my pupil Mr Sievright of Meggetland.

† It was discovered in the Becheley amygdaloid by my pupil Dr Daubeny.

1. Rectangular four-sided table, in which the terminal planes are bevelled; of these, two opposite are acute, and other two opposite flatly bevelled, and the terminal edges truncated. Sometimes the edges between the bevelling planes and the lateral planes are truncated.
2. Rectangular four-sided table, bevelled on the terminal edges.

The crystals are middle-sized and small, and frequently rest on each other, or intersect each other.

Externally it is shining and splendent; internally it is shining and pearly, inclining to vitreous.

It has a threefold cleavage in which the folia are parallel with the planes of the primitive figure, and of these, that parallel with the terminal planes is the most distinct.

The fracture is uneven.

The fragments are rhomboidal, or indeterminate angular, and rather sharp-edged.

It is translucent, semi-transparent, or transparent.

It scratches calcareous-spar, but is scratched by fluor-spar.

It is rather sectile, and is very easily frangible.

Specific gravity, 3.960, *Clayfield*.—3.967, *Karsten*.—3.6–4.0, *Mohs*.

Chemical Characters.

It melts before the blowpipe into a white friable enamel, without very sensibly tinging the flame: after a short exposure to heat, it becomes opaque, and has then acquired a somewhat caustic acrid flavour, very different from that of sulphuretted hydrogen, which heavy-spar acquires in similar circumstances.—*Aikin.*

These

lateral edges *, Fig. 144. Pl. 7. Sometimes the acute edges are truncated †, Fig. 145. Pl. 7.

2. Sometimes the angles between the bevelling and lateral planes are more or less deeply truncated, and thus form a four-planed acumination, in which the acuminating planes are set on the lateral edges ‡, Fig. 146. Pl. 7.

3. Sometimes the acute edges of the preceding figure are truncated, and thus a six-sided prism is formed ‖, Fig. 147. Pl. 7.

The crystals are middle-sized, and scopiformly aggregated, under an acute angle, and forming druses.

Externally it is smooth, splendent, and resinous.

Internally it is glistening and pearly, inclining to resinous.

The cleavage is the same as in the foliated subspecies.

The fracture is uneven.

The fragments are wedge-shaped and indeterminate angular.

It is translucent, or transparent.

In other characters, it agrees with the preceding subspecies.

Constituent Parts.

	Sicily.
Strontian, -	54
Sulphuric Acid, - -	46
	100 *Vauquelin.*

Geognostic

* Strontiane sulphatée unitaire, Haüy.

† Strontiane sulphatée emousée, Haüy.

‡ Strontiane sulphatée dodecaedre, Haüy.

‖ Strontiane sulphatée epointée, Haüy.

Geognostic and Geographic Situations.

It occurs in drusy cavities in a bed of sulphur, which is associated with gypsum and marl, in the valleys of Noto and Mazzara, in Sicily; in amygdaloid, along with calcareous-spar, in the neighbourhood of Greden, in the circle of the Inn; and in gypsum, near Cadiz.

Observations.

It has much the appearance of Prismatic Heavy-spar, but is distinguished from that mineral by its more distinct cleavage, inferior lustre, less regular concretions, greater sectility, easier frangibility, and inferior weight.

Third Subspecies.

Fibrous Celestine.

Fasriger Zölestin, *Werner.*

Fasriger Cœlestin, *Karsten,* Tabel. s. 54.—Strontiane sulphatée fibreuse-conjointe, *Haüy,* Tabl. p. 14.—Fasriger Cœlestin, *Lenz,* b. ii. s. 931.—Fasriger Zölestin, *Hoff.* b. iii. s. 191.

External Characters.

Its colour is intermediate between smalt-blue and pale indigo-blue, which passes on the one-side into bluish-grey, on the other into milk-white.

It occurs massive, also in distinct concretions, which are straight, parallel, and sometimes curved, fibrous.

Internally it is glistening and pearly.

It has an indistinct cleavage.

The

The fragments are splintery.

It is translucent.

Specific gravity, 3.721, *Karsten*.—3.830, *Klaproth*.

In other characters it agrees with the preceding subspecies.

Constituent Parts.

Strontian,	56.0
Sulphuric Acid,	42.0
Trace of Oxide of Iron.	
	98.0

Klaproth, Beit. b. ii. s. 97.

Geognostic and Geographic Situations.

It occurs in the red sandstone formation near Bristol; imbedded in marl, which is probably connected with gypsum, at Frankstown in Pennsylvania; and at Bouveron, near Toul, in the department of Meurthe in France.

Observations.

It resembles in external aspect *Fibrous Limestone*, *Fibrous Gypsum*, *Fibrous Anhydrite*, but is distinguished from all of them by its blue colours, and greater specific gravity. Its parallel fibrous concretions, and inferior specific gravity, distinguish it from *Fibrous Heavy-spar*.

Fourth

Observations.

Its specific gravity distinguishes it from *Granular Limestone* and *Dolomite*.

Genus III. TUNGSTEN *, or SCHEELIUM †.

Scheel Baryt, *Mohs.*

This genus contains one species, viz. Pyramidal Tungsten.

1. Pyramidal Tungsten.

Pyramidaler Scheel-Baryt, *Mohs.*

Schwerstein, *Werner.*

Minera Ferri lapidea gravissima, *Wall.* t. ii. p. 254.—Wolfram de couleur blanche, *Romé de Lisle,* t. iii. p. 264.—Schwerstein, *Werner,* Pabst. b. i. s. 222.—Weisser Tungsten, *Wid.* s. 980.—Tungsten, *Kirw.* vol. ii. p. 314.—Tungstate calcaire, Mine d'Etaine blanche, *De Born,* t. ii. p. 230.—Schwerstein, *Emm.* b. ii. s. 570.—Tungstene, *Lam.* t. i. p. 402.—Scheelen calcaire,

---

* The name *Tungsten* was given to this mineral by the Swedes, on account of its great weight.

† Werner gave the name *Scheele* to this genus, in honour of the illustrious chemist Scheele, who discovered the peculiar metal which characterises it.

calcaire, *Haüy*, t. iv. p. 320.—La Pierre pesant, ou Le Tungstene, *Broch.* t. ii. p. 453.—Scheelerz, *Reuss*, b. iv. s. 534. *Id. Lud.* b. i. s. 303.—Kalk-Scheel, *Suck.* 2ter th. s. 459.—Schwerstein, *Bert.* s. 509. *Id. Mohs*, b. iii. s. 623.—Scheelin calcaire, *Lucas*, p. 183.—Scheelerz, *Leonhard*, Tabel. s. 81.—Scheelin calcaire, *Brong.* t. ii. p. 93. *Id. Brard*, p. 389.—Scheelerz, *Karsten*, Tabel. s. 74.—Scheelin calcaire, *Haüy*, Tabl. p. 118.—Tungsten, *Kid*, vol. ii. p. 225.—Schwerstein, *Haus.* Handb. b. iii. s. 967. *Id. Hoff.* b. iv. s. 236. *Id. Aikin*, p. 134.

External Characters.

White is the principal colour of this mineral; the following varieties of colour also sometimes occur, viz. plumb-blue, pearl-grey, greenish and ash grey, greyish and yellowish white, and this latter passes into yellowish-grey, and further into clove, broccoli, reddish, and yellowish brown, which sometimes inclines to orange-yellow and hyacinth-red.

It occurs massive, disseminated, also in distinct concretions, which are granular, seldomer wedge-shaped prismatic, and these latter traversed by others which are curved lamellar. It is sometimes crystallized.

The primitive figure is a rather acute double four-sided pyramid, in which the lateral planes of the one are set on the lateral planes of the other. The corresponding planes of the opposite pyramids meet under an angle of 113° 36′, while the adjacent planes of the same pyramid meet under an angle of 107° 26′.

The following are the secondary forms:

1. The primitive figure, in which the angles of the common base are flatly bevelled, and the bevelling planes set on the lateral edges.

Fourth Subspecies.

Radiated Celestine.

Strahliger Zölestin, *Werner.*

Strahliger Zölestin, *Hoff.* b. iii. s. 193.—Strontiane sulphatée fibro-laminaire, *Haüy.*

External Characters.

Its colour is milk-white, which rarely approaches to yellowish and snow white.

It occurs massive; also in prismatic concretions, which are scopiform radiated, collected into others which are wedge-shaped, and these, again, into very large and angulo-granular concretions.

Internally it is shining and splendent, and the lustre pearly, slightly inclining to vitreous.

The fragments are wedge-shaped, and splintery.

It is translucent, or semitransparent.

Specific gravity, 3.785, *Breithaupt.*

In other characters agrees with the other subspecies.

Fifth Subspecies.

Fine Granular Celestine.

Fein Körniger Zölestin, *Werner.*

Dichter Cœlestin, *Karsten,* Tabel. s. 54.—Strontiane sulphatée, calcarifère, *Haüy,* Tabl. p. 14.—Dichter Cœlestin, *Lenz,* b. ii. s. 921.—Feinkorniger Zölestin, *Hoff.* b. iv. s. 132.

External

External Characters.

Its colours are greenish and yellowish grey, and the first inclines sometimes to olive-green.

It occurs massive, in fine granular concretions, in spheroidal or reniform masses, which are often traversed by fissures that divide its surface into quadrangular pieces, which are sometimes lined with minute crystals of celestine. Towards the surface it has a marly aspect.

Internally it is dull or glimmering, and pearly.

The fracture is fine splintery, passing into uneven.

The fragments are blunt-edged.

It is opaque, or translucent on the edges.

It is semihard.

Specific gravity, 3.592, *Haüy*.

In other characters it agrees with the preceding subspecies.

Chemical Characters.

Sulphate of Strontian,	91.42
Carbonate of Lime, -	8.33
Oxide of Iron, - -	0.25
	100.00

Vauquelin, in Brongniart's Mineralogie, t. i. p. 258.

Geognostic and Geographic Situations.

It occurs imbedded in marly clay, with gypsum, at Montmartre, near Paris; and it is said to form a whole bed in Champagna.

Observations.

Geognostic Situation.

It occurs along with tinstone, magnetic iron-ore, and brown iron-ore, in primitive rocks. In the tinstone repositories, it is associated with wolfram, quartz, mica, fluor-spar, and steatite.

Geographic Situation.

It occurs along with wolfram and tin-ore at Pengilly in Breage in Cornwall *: at Bispberg in Sweden, in a bed of magnetic iron-ore: in fine crystals at Schlackenwald in Bohemia: at Zinwald and other places in Saxony: and in the gold-works of Schillgaden at Salzburg.

Observations.

It is distinguished from the white varieties of *Tin-ore*, by its shape, intensity, and kind of lustre, inferior hardness, inferior weight, and its becoming yellow when thrown into nitric acid: from *Yellow Lead-spar*, by colour and higher specific gravity: from *Heavy-spar*, by form, lustre, cleavage, and greater weight, and by the yellow colour it assumes when thrown into nitric acid.

GEN. IV.

* Heuland.

calcaire, *Haüy*, t. iv. p. 320.—La Pierre pesant, ou Le Tungstene, *Broch.* t. ii. p. 453.—Scheelerz, *Reuss*, b. iv. s. 534. *Id. Lud.* b. i. s. 303.—Kalk-Scheel, *Suck.* 2ter th. s. 459.—Schwerstein, *Bert.* s. 509. *Id. Mohs*, b. iii. s. 626.—Scheelin calcaire, *Lucas*, p. 183.—Scheelerz, *Leonhard*, Tabel. s. 81.—Scheelin calcaire, *Brong.* t. ii. p. 93. *Id. Brard*, p. 389.—Scheelerz, *Karsten*, Tabel. s. 74.—Scheelin calcaire, *Haüy*, Tabl. p. 118.—Tungsten, *Kid*, vol. ii. p. 225.—Schwerstein, *Haus.* Handb. b. iii. s. 967. *Id. Hoff.* b. iv. s. 236. *Id. Aikin*, p. 134.

External Characters.

White is the principal colour of this mineral; the following varieties of colour also sometimes occur, viz. plumb-blue, pearl-grey, greenish and ash grey, greyish and yellowish white, and this latter passes into yellowish-grey, and further into clove, broccoli, reddish, and yellowish brown, which sometimes inclines to orange-yellow and hyacinth-red.

It occurs massive, disseminated, also in distinct concretions, which are granular, seldomer wedge-shaped prismatic, and these latter traversed by others which are curved lamellar. It is sometimes crystallized.

The primitive figure is a rather acute double four-sided pyramid, in which the lateral planes of the one are set on the lateral planes of the other. The corresponding planes of the opposite pyramids meet under an angle of 113° 36′, while the adjacent planes of the same pyramid meet under an angle of 107° 26′.

The following are the secondary forms:

1. The primitive figure, in which the angles of the common base are flatly bevelled, and the bevelling planes set on the lateral edges.

 a. Sometimes

ochre yellow, and into yellowish and clove brown. It has sometimes a curved striped colour delineation.

It occurs massive, disseminated, in crusts, stalactitic, reniform, botryoidal, cellular, corroded; also in distinct concretions, which are scopiform, radiated, or fibrous, granular, and curved lamellar.

It is sometimes crystallized.

Its primitive figure is an oblique four-sided prism, bevelled on the extremities, in which the lateral planes (vertical prism) have an angle of 99° 56′, and the bevelling planes (horizontal prism) an angle of 120°. The following are some of the secondary figures.

1. Six-sided prism.
2. Flat six-sided prism, bevelled on the terminal planes; the bevelling planes set on the broader lateral planes. This prism is sometimes so flat, that it appears like a longish rectangular four-sided table bevelled on the terminal planes.
3. Acute double four-sided pyramid, sometimes perfect, sometimes truncated on the summits.
4. Acute double four-sided pyramid, acuminated on both extremities, with four planes, which are set on the lateral planes, and sometimes the summits are truncated.

The crystals are small; and either solitary, or scopiformly aggregated.

Internally, the lustre alternates from glistening to dull, and the lustre is pearly, inclining to adamantine.

The fracture is small and fine-grained uneven.

The cleavage is imperfect, and in the direction of the planes of the primitive prism.

It alternates from transparent to translucent on the edges, and opaque.

The

The fracture is coarse, or small-grained uneven, passing into imperfect conchoidal.

It has a nine-fold cleavage; four of the folia are parallel with the sides of the primitive figure, four parallel with the sides of the acute pyramid, and the ninth parallel with the common base of the double pyramid. Of these the last is generally the most perfect.

The fragments are indeterminate angular, and rather blunt-edged.

It is more or less translucent, seldom semitransparent.

It is harder than fluor-spar, but not so hard as apatite.

It is rather brittle, and easily frangible.

Specific gravity, 6, 6.1, *Mohs.*—6.028, *Kirwan.*—6.000, *Gellert.*—6.015, *Klaproth.*

Chemical Characters.

It crackles before the blowpipe and becomes opaque, but does not melt; with borax it forms a transparent or opaque white glass, according to the proportions of each.

Constituent Parts.

		Schlackenwald.	Cornwall.
Oxide of Tungsten,	65	77.75	75.25
Lime, -	31	17.60	18.70
Silica, -	4	3.00	1.56
Oxide of Iron, -	—	—	1.25
Oxide of Manganese,	—	—	0,75
	100	98.35	97.45
	Scheele, in n. Abhand. d. Schwd. Akad. 1781, 289.	*Klaproth*, Beit. b. iii. s. 47. & 51.	

 Geognostic

Geographic Situation.

It occurs in the lead-mines at Wanlockhead; also in Leicestershire and Flintshire. On the Continent, it is met with at Tarnowitz in Silesia, Tscheren in Bohemia, Rezbanya in Hungary, Bleiberg in Carinthia, Freyburg in the Bresgau, and Stolberg in the Tyrol. It is a rare mineral in northern latitudes, and there it occurs only at Kolywan and Nertschinsky in Siberia.

Observations.

It is distinguished from *White Lead-spar* by its superior hardness, and inferior specific gravity; from *Tabular-spar, Arragonite* and *Zeolite*, by its greater specific gravity. The massive and uncrystallized varieties may be distinguished from *Indurated White Lead-spar* by inferior specific gravity, and from *Brown Iron-ore* by its greater specific gravity.

2. Rhomboidal Calamine.

Galmei, *Werner.*

Rhomboedrischer Zink-Baryt, *Mohs.*

This species is divided into three subspecies, viz. Sparry Rhomboidal Calamine, Compact Rhomboidal Calamine, and Earthy Rhomboidal Calamine.

First

First Subspecies.

Sparry Rhomboidal Calamine.

Späthiger Galmei, *Karsten.*

Minera Zinci vitrea, *Waller.* Syst. Min. t. ii. p. 215. (in part). —Blättricher Galmei, *Reuss,* b. iv. s. 349. (in part).—Spathiger Galmei, *Karsten,* Tabel. s. 70.—Zinc carbonaté, var. 1, 2. *Haüy,* Tabl. p. 103.—Edler Galmei, or Zinkspath, *Haus.* Handb. b. i. s. 345.—Calamine, *Aikin,* p. 119.

External Characters.

Its colours are greyish-white, yellowish-white, bluish-grey, greenish-grey, siskin-green, apple-green, reddish-brown, and clove-brown.

It occurs massive, botryoidal, reniform, stalactitic, tabular, cellular; also in distinct concretions, which are prismatic, granular, and curved lamellar; and crystallized.

The primitive figure is a rhomboid of about 110°. The following are some of its crystallizations:

1. Obtuse rhomboid.
2. Acute rhomboid.
3. Long four-sided table, which is either perfect, or bevelled on the terminal planes; and the angles of the bevelment sometimes more or less deeply truncated.

The crystals are small.

Internally it is shining and pearly.

The cleavages threefold, and in the direction of the planes of the primitive rhomboid.

The

The fragments are rhomboidal.

It alternates from semitransparent to opaque.

It is as hard as apatite.

Specific gravity, 4.2, 4.4, *Mohs.*

Chemical Characters.

It dissolves with effervescence in muriatic acid; it is infusible; loses about 34 *per cent.* by ignition.

Constituent Parts.

		Derbyshire.	Somersetshire.
Oxide of Zinc,	-	65.2	64.8
Carbonic Acid,	-	34.8	35.2
		100	100

Smithson, in Phil. Trans. P. I. for 1803.

Second Subspecies.

Compact Rhomboidal Calamine.

Gemeiner Galmei, *Karsten.*

Lapis calaminaris, *Wall.* Syst. Min. t. ii. p. 216.—Gemeiner Galmei, *Reuss,* b. iv. s. 345. (in part).—Gemeiner Galmei, *Karsten,* Tabel. s. 70.—Zinc carbonaté, var. 3, 4. *Haüy,* Tabl. p. 103.—Gemeiner Galmei, *Haus.* Handb. b. i. s. 347.—Compact Calamine, *Aikin,* p. 119.

External Characters.

Its colours are yellowish, ash, greenish, and smoke-grey; also cream-yellow, straw-yellow, and yellowish-brown.

It

It occurs massive, disseminated, corroded, reniform, stalactitic, and cellular; also in concentric curved lamellar concretions. Rarely in supposititious crystals, or incrusting other crystals.

Internally it is dull, or very feebly glimmering and resinous.

The fracture is coarse-grained uneven, fine splintery, even, and flat conchoidal.

It is opaque, or feebly translucent on the edges.

Chemical Character.

The same as in the preceding subspecies.

Third Subspecies.

Earthy Rhomboidal Calamine.

Zinkblüthe, *Karsten.*

Zinkblüthe, *Karsten*, Tabel. s. 70.—Zinc carbonaté, *Haüy*, Tabl. p. 103. (in part).—Zinkblüthe, *Haus.* Handb. b. i. s. 348.—Earthy Calamine, *Aikin*, p. 120.

External Characters.

Its colours are snow-white, greyish-white, and yellowish-white; sometimes with a yellowish-brown exterior.

It occurs massive, disseminated, botryoidal, flat reniform, and with impressions.

Internally it is dull.

The fracture is fine earthy.

It is opaque.

It yields to the nail.

It

It adheres to the tongue.
Specific gravity 3.358.

Chemical Characters.

The same as in the first subspecies.

Constituent Parts.

	Bleiberg in Carinthia.
Oxide of Zinc,	71.4
Carbonic Acid,	13.5
Water, -	15.1
	100.0

Smithson in Phil. Trans.
P. I. for 1803.

Geognostic Situation of the Species.

It occurs in beds, nests, filling up or lining hollows, in transition limestone, and in secondary or flœtz limestone, and conglomerate rock; also in veins. In these repositories it is generally associated with galena or lead-glance, and occasionally with copper-pyrites, copper-green, malachite, yellow and brown blende, sparry iron, ochry-brown ironstone, brown-spar, calcareous-spar, and quartz.

Geographic Situation of the Species.

Europe.—It occurs in the Mendip Hills, at Shipham, near Cross, Somersetshire; at Allonhead in Durham; at Holywell, and elsewhere in Flintshire; and in Derbyshire. On the Continent, it is met with at Raibel and Bleiberg in Carinthia; Aachen; Namur; Chemnitz in Hungary; Medziana Gora in Poland; Beuthen and Tarnowitz in Silesia; and Iserlohn in the Dutchy of Berg.

Asia.

Asia.—Altai in Siberia.

Uses.

1. Both prismatic and rhomboïdal calamine, when purified and roasted, are used for the fabrication of brass, which is a compound of zinc and copper; and the pure metal is also employed for a variety of other purposes. The use of calamine in the composition of brass was known at a very early period; for it is mentioned by Aristotle, who also makes a distinction between the compound resulting from the mixture of copper and calamine, and that resulting from the mixture of copper and tin *.

2. The compact varieties of prismatic and rhomboidal calamine are sometimes confounded together.

GENUS V.—RED MANGANESE.

THIS Genus contains one species, viz. Rhomboidal Red Manganese.

1. Rhomboidal Red Manganese.

Langaxiger Flinz-Baryt, *Mohs.*

This species is divided into three subspecies, viz. 1. Foliated Rhomboidal Red Manganese, 2. Fibrous Rhomboidal

* Aristot. ed. Paris, 1654, vol. ii. p. 721.

dal Red Manganese, and, 3. Compact Rhomboidal Red Manganese.

First Subspecies.

Foliated Rhomboidal Red Manganese.

Blättriger Braunspath, (in part), *Hoff.* b. iii. s. 111.—Manganspath, *Hoff.* b. iii. s. 155.—Chaux carbonatée manganesifère rose, *Haüy*, Tabl. p. 5.—Manganese oxydé carbonatée, *Haüy*, Tabl. p. 111.—Rosenrod syraatt Manganes, *Hisinger*, in Afhandling. i Fys. Kem. och Min. i. 105.

External Characters.

Its colour is bright rose-red, slightly inclining to flesh-red.

It occurs massive, disseminated, small reniform, globular, with tabular and rhomboidal impressions, and in granular distinct concretions; also crystallized in rhomboids, the dimensions of which have not been accurately determined.

Internally it is shining, inclining sometimes to glistening, sometimes to splendent, and the lustre pearly.

Its cleavage is rhomboidal, or the folia of the cleavage are in the direction of the planes of the rhomboid.

The fragments are indeterminate angular and rather sharp-edged, or rhomboidal.

It is generally translucent on the edges; in some rare varieties translucent.

Its hardness is intermediate between that of calcareous spar and fluor-spar.

It is brittle, and rather easily frangible.

Specific

Specific gravity, 3.3–3.6, *Mohs.*—3.588, *Berzelius.*—3.661–3.685, *Breithaupt.*

Chemical Character.

Before the blowpipe, without addition, it first becomes dark brown, and then melts into a dark reddish-brown bead.

Constituent Parts.

Oxide of Manganese, -	52.60
Silica, - - -	39.60
Oxide of Iron, - -	4.60
Lime, - - - -	1.50
Volatile ingredients, -	2.75
	101.5

Berzelius, in Afh. i Fys. och. Min. i. 110.

Geognostic and Geographic Situations.

It occurs in beds of specular iron-ore and magnetic iron-ore, along with compact garnet and calcareous-spar, in the gneiss hills at Langbanshytta, in Wermeland in Sweden; also at Catharinenburg in Siberia. It is found also in Saxony.

Uses.

The Siberian varieties are cut and polished, and worn as ornamental stones.

Observations.

Its bright rose-red colour, distinct concretions, and fracture, distinguish it from the other subspecies.

Second

It adheres to the tongue.
Specific gravity 3.358.

Chemical Characters.

The same as in the first subspecies.

Constituent Parts.

	Bleiberg in Carinthia.
Oxide of Zinc,	71.4
Carbonic Acid,	13.5
Water, -	15.1
	100.0

Smithson in Phil. Trans.
P. I. for 1803.

Geognostic Situation of the Species.

It occurs in beds, nests, filling up or lining hollows, in transition limestone, and in secondary or flœtz limestone, and conglomerate rock; also in veins. In these repositories it is generally associated with galena or lead-glance, and occasionally with copper-pyrites, copper-green, malachite, yellow and brown blende, sparry iron, ochry-brown ironstone, brown-spar, calcareous-spar, and quartz.

Geographic Situation of the Species.

Europe.—It occurs in the Mendip Hills, at Shipham, near Cross, Somersetshire; at Allonhead in Durham; at Holywell, and elsewhere in Flintshire; and in Derbyshire. On the Continent, it is met with at Raibel and Bleiberg in Carinthia; Aachen; Namur; Chemnitz in Hungary; Medziana Gora in Poland; Beuthen and Tarnowitz in Silesia; and Iserlohn in the Dutchy of Berg.

Asia.

Third Subspecies.

Compact Rhomboidal Red Manganese.

Dichtes Rothbraunstein, *Reuss*, b. iv. s. 470.—Rothstein, *Mohs*, b. ii. s. 122.—Rothbraunstein, *Leonhard*, Tabel. s. 70.—Manganese lithoide, *Brong.* t. ii. p. 110.—Roth Manganerz, *Karsten*, Tabel. s. 72.—Manganese oxydé carbonatée, *Haüy*, Tabl. p. 111. (in part).—Dichtes Rothstein, *Haus.* Handb. b. i. s. 302.—Rother Braunstein, *Hoff.* b. iv. s. 158.

External Characters.

Its principal colour is pale rose-red, which sometimes passes into dark reddish-white. Externally it has sometimes a wood-brown and yellowish-brown colour, owing to the action of the weather.

It occurs massive, disseminated, and sometimes imperfectly reniform.

Internally it is dull or glimmering.

The fracture is even, sometimes inclining to splintery.

The fragments are indeterminate angular, and rather sharp-edged.

Its hardness is intermediate between that of calcareous spar and fluor-spar.

It is brittle, and rather easily frangible.

Specific gravity 3.3–3.9, *Mohs*.

Chemical Characters.

It is infusible before the blowpipe, but becomes black by ignition.

Constituent Parts.

	Siberia.
Oxide of Manganese, -	61
Silica, - - - - -	30
Oxide of Iron, - - -	5
Alumina, - - -	2
	98

Lampadius, in Pract. Chem. Abh. b. ii. s. 209.

Geognostic and Geographic Situations.

It occurs at Kapnik in Transylvania, in veins, along with quartz, black copper-ore, sulphuret of manganese, blende, galena or lead-glance, calcareous-spar, and brown-spar; also at Langbanshytta, in Wermeland in Sweden; and Catharinenburg in Siberia.

Observations.

The pale rose-red colour, want of lustre, compact fracture, and weight, are the distinguishing characters of this mineral. It is distinguished from *Brown-Spar*, by its greater hardness, and weight.

GEN. VI

Genus VI.—SPARRY IRON.

Kurzaxiger Flinz Baryt, *Mohs.*

Spatheisenstein, *Werner.*

Fer oxydé carbonatée, *Haüy.*

This Genus contains one species, viz. Sparry Iron.

1. Sparry Iron.

Spatheisenstein, *Werner.*

Fer oxydé carbonatée, *Haüy.*

Minera Ferri alba, *Wall.* t. ii. p. 251.—*P. J. Hjelm*, Chemisk och Mineralogisk Afhandling om huita Järnmalmer, Upsala, 1774.—Mine de Fer spathique, *Romé de Lisle*, t. iii. p. 281. —Spathiger Eisenstein, *Wern.* Pabst. b. i. s. 164. *Id. Wid.* s. 820.—Calcareous or Sparry Iron-ore, *Kirw.* vol. ii. p. 190. —Fer spathique, ou Mine de Fer blanche, *De Born*, t. ii. p. 290.—Fer spathique, *Lam.* t. i. p. 263.—Chaux carbonatée ferrifère avec Manganese, *Haüy*, t. iv. p. 117, 118.—La Mine de Fer spathique, ou le Fer spathique, *Broch.* t. ii. p. 264.—Spatheisenstein, *Reuss*, b. iv. s. 107. *Id. Lud.* b. i. s. 249. *Id. Suck.* 2ter th. s. 278. *Id. Bert.* s. 428. *Id. Mohs*, b. iii. s. 407. *Id. Hab.* s. 124. *Id. Leonhard*, Tabel. s. 66.—Fer spathique, *Brong.* t. ii. p. 175.—Spath Eisenstein, *Karsten*, Tabel. s. 66.—Eisenspath, *Haus.* s. 129.—Sparry Iron-ore, *Kid*, vol. ii. p. 188.—Fer oxydé carbonatée, *Haüy*, Tabl. p. 99. —Eisenspath, *Haus.* Handb. b. iii. s. 952.—Spatheisenstein, *Hoff.* b. iv. s. 262.—Sparry Iron-ore, *Aikin*, p. 106.

 External

External Characters.

Its colour is pale yellowish-grey, which passes on the one side into pea-yellow and isabella-yellow, and further into yellowish and greyish white; on the other side into yellowish, clove, blackish-brown, and brownish-black. It is rarely ash and greenish grey, or reddish-brown. The lighter fresh colours change on exposure to the air, and become brown, and even black. Probably most of the colours of this mineral have been produced by alterations which have taken place since the period of its formation.

It occurs massive, disseminated, with pyramidal impressions; also in granular distinct concretions; and crystallized.

The primitive form is a rhomboid of 107°. The following are some of the principal forms:

1. Primitive rhomboid.
   a. Perfect, with straight or spherical convex lateral faces.
   b. Truncated on the apices.
   c. Truncated on the terminal edges.
   d. Rounded off on the apices and edges.

   When the truncating planes in the variety 1. *c.* become so large that the original planes disappear, there is formed
2. A still flatter rhomboid.

   From the variety 1. *d.* there arises
3. The spherical lenticular form.

   From the rhomboid with curved faces, there is formed
4. The saddle-shaped lens.

   We sometimes observe the primitive form arranged in rows, so as to form an

5. Equiangular

5. Equiangular six-sided prism, flatly acuminated with three planes, which are set on the alternate lateral planes.

The crystals are middle-sized, small, and very small.

They are seldom singly superimposed, as is the case with the lens, most generally aggregated in druses.

The planes of the lens are delicate drusy, but of all the other forms generally smooth; and the lustre varies from splendent, through shining to glistening.

Internally it is generally glistening, sometimes inclining to shining, and even to splendent; but the black variety is only glimmering, and the lustre is pearly.

It has a threefold cleavage in the direction of the planes of the primitive form, and also of those of the flat rhomboid. The folia are seldom straight, generally spherical curved.

The imperfect foliated fracture is sometimes conjoined with the splintery, and this occurs principally in the greenish-grey varieties.

The fragments are rhomboidal in the foliated varieties, but rather sharp-edged in the compact.

It is generally translucent on the edges, also translucent; but the black varieties are opaque.

The pale varieties afford a white, the darker varieties a yellowish-brown streak.

It is harder than calcareous-spar, but not so hard as fluor.

It is not particularly brittle, and is easily frangible.

Specific gravity, 3.6–3.9, *Mohs*.—3.784, *Gellert*.—3.640–3.810, *Kirwan*.—3.672, *Brisson*.—3.693, *Guyton*.—3.600–3.900, *Collet-Descotils*.

Chemical

Chemical Characters.

It blackens, and becomes magnetic before the blowpipe, but does not melt: it effervesces with muriatic acid. It dissolves with ebullition in glass of borax, and communicates to it an olive-green colour.

Constituent Parts.

	Dankerode.	Baireuth.	Baireuth.		Steinheim.
Oxide of Iron, -	57.50	58.00	59.50		63.75
Carbonic Acid, -	36.00	35.00	36.00		34.00
Oxide of Manganese,	3.50	4.25	a trace		0.75
Lime, - -	1.25	0.50	2.50		...
Magnesia, - - -	-	0.75	-		0.25
Water, - - - -	-	-	2.00	Loss,	1.25
	98.25	98.50	99.00		100.00
	Klaproth, Beit. b. iv. s. 115.	Ibid. s. 115.	*Buchols.*		*Klaproth*, in Magaz. Natf. Fr. b. v. s. [illegible]

Geognostic Situation.

It occurs in veins in granite, gneiss, mica-slate, clay-slate, and grey-wacke, and in these it is associated with ores of lead, cobalt, silver, copper, and seldomer with nickel and bismuth; more frequently with galena, grey copper-ore, iron-pyrites, and grey antimony-ore. In other veins, it is accompanied with brown, red, and black iron-ore, calcareous-spar and quartz. But the most extensive formations of this mineral are in limestone, by some referred to primitive, by others to secondary rocks, in which it is arranged in thick beds. It also occurs filling up amygdaloidal cavities in trap rocks.

Geographic

Geographic Situation.

Europe.—It occurs in small quantities in different places in England, Scotland, and Ireland; also in Saxony, Bohemia, Bayreuth, Upper Palatinate, Silesia, Coburg, Savoy, Switzerland, Sweden, and Norway: but it is only in the following countries where it is found in such quantity as to be employed as an ore of iron:—In the Fichtelgebirge; the black variety occurs in great quantity at Schmalkalden in Hessia, where it has been mined and smelted for many centuries; in the Hartz, as at Clausthal, Iberg, Blankenburg, and Stollberg, it occurs less abundantly; in Westphalia, the light-coloured is mined in great quantity; Eisenerz and Schladinrig in Stiria, affords it in considerable quantity; Hüttenberg in Carinthia, Schwatz in the Tyrol, and Jauberling in Carniola, are well known for mines of sparry iron; in many places in Salzburg, in Hungary, as Schemnitz, Schmolnitz, Dopshau, and Siowinka, it occurs in small quantity; mines of it also exist in Piedmont *, and France; and at Somororstro, in the province of Biscay in Spain, there is a whole hill composed of this species of ironstone, which has been worked for several hundred years. It is there accompanied with red iron-ore, which renders the smelting very advantageous.

Asia.—In the mines of Catharinenburg.

America.—West Greenland; and Mexico.

Uses.

It affords an iron which is excellently suited for steel making. The black variety is said to afford the best kind of iron.

Observations.

---

* Magnificent lenticular crystals are brought from Traversella in Piedmont.

Observations.

1. Colour, cleavage, inferior translucency, hardness, and weight, characterise this mineral as long as it remains unaltered. It is distinguished from *Calcareous-spar*, *Rhomb-spar*, and *Brown-spar*, by its greater specific gravity, and hardness.

2. Cast-iron obtained from this species, or from brown iron-ore, presents a whitish colour and radiated fracture; whereas that obtained from red iron-ore, and several other ores of iron, has a dark grey colour, and a granular fracture. Further, the cast-iron obtained from this species can be converted into steel; but a great portion of that obtained from red iron-ore, &c. passes to the state of malleable iron, long before the mass in the furnace has become steel. The steel obtained from this ore is said to contain a small portion of manganese, which is supposed to be the cause of its durability in the fire, and what renders it less liable to become soft and irony.

3. It generally occurs more or less weathered. By exposure to the air, it experiences a gradual decomposition, which has a great effect on its external aspect. This decomposition at first affects only the external colour, external lustre, and the transparency; but as it advances, it also changes even the structure, hardness, solidity, and weight of the mineral. The oxidation of the iron and manganese destroys the weak combination of these metals with the carbonic acid, and there is formed a hydrate of iron, sometimes also an oxide of iron, and hydrate of manganese. The whole mass is disintegrated by the escape of the carbonic acid. This acid combining with percolating water dissolves the small portion of lime in the ore, and also portions of the still undecomposed carbonate of iron and oxide of manganese.

manganese. A knowledge of these changes enables us to understand the very different results obtained in the analysis of specimens more or less weathered or decomposed, and also throws some light on the different results obtained in the smelting of sparry iron more or less decomposed.

4. The analysis of Hielm, published under the sanction of Bergman, is the earliest we possess of this mineral: it gives as the constituent parts, 22.38 Oxide of Iron: 24.28 Oxide of Manganese: 29.43 Carbonate of Lime; and 6.9 Water. The errors of this analysis have been pointed out and corrected by the labours of Drappier, Descotils, Berthier, Klaproth, and Bucholz.

5. It is described under the names Eisenspath, (Iron-spar), White Ironstone, and Sparry Ironstone.

ORDER *VII.*

Order VII.—HALOIDE.

This Order contains Six Genera, viz. 1. Limestone, 2. Apatite, 3. Fluor, 4. Aluminite, 5. Cryolite, 6. Gypsum.

Genus I.—LIMESTONE.

Kalk-Haloide, *Mohs*.

This Genus contains four Species, viz. 1. Rhomb-Spar, 2. Dolomite, 3. Limestone, 4. Arragonite.

1. Rhomb-Spar.

Langaxiges Kalk-Haloid, *Mohs*.

Bitterspath, *Wid.* s. 518.—Crystallized Muricalcite, *Kirw.* vol. i. p. 92.—Bitterspath, *Emm.* b. iii. s. 353.—Spato Magnesiano, *Nap.* p. 358.—Bitterspath, *Lam.* t. ii. p. 347.—Chaux carbonatée magnesiée, *Haüy*, t. ii. p. 187.—Le Spath magnesien, ou Le Bitterspath, *Broch.* t. i. p. 560.—Bitterspath, *Reuss*, b. ii. 2. s. 330.—Rautenspath, *Lud.* b. i. s. 154.—Gemeiner Bitterspath, *Suck.* 1r th. s. 634.—Rautenspath, *Bert.* s. 113. *Id.* *Mohs*, b. ii. s. 96. 98.—Bitterspath, *Hab.* s. 83.—Gemeiner Bitterspath, *Leonhard*, Tabel. s. 35.—Chaux carbonatée lente, Picrite, *Brong.* t. i. p. 230.—Rhomboedrischer Dolomit, *Karsten*, Tabel. s. 50.—Gemeiner Bitterspath, *Haus.* s. 128.—Chaux carbonatée magnesifere, *Brard*, p. 38.—Rhomb-spar, *Kid*, vol. i. p. 57.—Chaux carbonatée magnesifere primitive, *Haüy*,

Haüy, Tabl. p. 5.—Rautenspath, *Lenz*, b. ii. s. 710. *Id. Oken*, b. i. s. 393.—Rautenspath, *Hoff.* b. iii. s. 60. (in part).

External Characters.

Its colours are greyish-white, yellowish-white, ash-grey and yellowish-grey, which latter passes into pea-yellow, and isabella yellow.

The ash-grey sometimes passes into greyish-black.

It occurs massive, and disseminated; and crystallized in rhomboids, in which the obtuse angle is 106° 15′. These rhomboids are sometimes rounded or truncated on the edges.

The crystals are middle-sized and small; the surface is sometimes smooth, sometimes rough, and either shining or glimmering.

Internally the lustre is splendent, between vitreous and pearly *.

It has a threefold oblique angular cleavage: the alternate angles of which measure 106° 15′ and 73° 45′ †.

The fracture is imperfect conchoidal.

The fragments are rhomboidal.

It is harder than calcareous-spar, and sometimes as hard as fluor-spar.

It is easily frangible, and brittle.

Specific gravity, 2.8, 3.2, *Mohs.*—2.880, 3.000. 2.8901, *Murray* ‡.

Chemical

---

* The lustre in general is stronger than that of calcareous-spar.—*Bournon.*

† Dr Wollaston.

‡ Newton-Stewart, Galloway.

Chemical Characters.

Before the blowpipe it is infusible, without addition: even when pounded, it effervesces but feebly; and dissolves slowly in muriatic acid.

Constituent Parts.

Hall, in the Tyrol.

Carbonate of Lime,	68.00
Carbonate of Magnesia,	25.50
Carbonate of Iron,	1.00
Water,	2.00
Clay intermixed,	2.00
	98.50

Klaproth, Beit. b. iv. s. 238.

Taberg, in Wermeland.

Carbonate of Lime,	73.00
Carbonate of Magnesia,	25.00
Oxide of Iron, mixed with Manganese,	2.25
	100.25

Klaproth, b. i. s. 306.

Near Newton-Stewart in Galloway.

Carbonate of Lime,	56.60
Carbonate of Magnesia,	42.00
	98.60

Or by another result;

Carbonate of Lime,	56.2
Carbonate of Magnesia,	43.5
	98.9

With a trace of Manganese and Iron.—*Murray* *.

Lime,	28.00
Magnesia,	25.05
Carbonic Acid,	48.00
Oxide of Manganese,	1.05
With a trace of Iron.	

Bucholz.

Geognostic

* The above analysis was communicated to me by my friend Dr Murray.

Geognostic Situation.

It occurs imbedded in chlorite-slate, talc-slate, limestone, and serpentine, occasionally associated with asbestus and tremolite; in the salt formation, where it is imbedded in anhydrite and gypsum; in drusy cavities in compact dolomite, and in metalliferous veins.

Geographic Situation.

Europe.—It occurs imbedded in chlorite-slate on the banks of Loch Lomond; in a vein in transition rocks, along with galena, blende, copper-pyrites, and calcareous-spar, near Newton-Stewart in Galloway; in compact dolomite in the Isle of Man and the north of England; in chlorite-slate and talc in the Upper Palatanite; in the mountain of Chalance in Dauphiny, along with asbestus, talc, and chlorite; also at Brienz in Switzerland; in the mountains of Salzburg; in granular limestone, in the silver mines of Sala, and in the Taberg in Wermeland in Sweden.

America.—At Kannioak in North Greenland, imbedded in common and indurated talc; and at Guanuaxuato in Mexico, along with amethyst, common quartz, and felspar.

Observations.

1. This mineral was formerly named *Bitter-Spar*, from the magnesia contained in it, which is denominated Bitter Salt by the Germans, because obtained easily from sulphate of magnesia or Epsom salt. It was named *Muricalcite* by Kirwan, from the magnesia and lime contained in it; magnesia having been called Muriatic Earth, as being the

the base of one of the salts contained in sea-water. Werner named it *Rhomb-Spar*, from its form.

2. It is distinguished from *Calcareous-spar* by the shape of its rhomboid, superior hardness, and specific gravity, and dissolving slowly in the mineral acids.

2. Dolomite.

Kurzaxiges Kalk-Haloide, *Mohs*.

This Species contains three Subspecies, viz. 1. Dolomite, 2. Miemite, 3. Brown Spar.

First Subspecies.

Dolomite.

This Subspecies is divided into three Kinds, viz. Granular Dolomite, Columnar Dolomite, Compact Dolomite.

First Kind.

Granular Dolomite.

This is again divided into White and Brown Granular Dolomite.

* White Granular Dolomite.

Dolomite, *Saussure*, Voyages dans les Alpes, § 1929.—Biegsamer Körniger Kalkstein, *Reuss*, b. ii. 2. s. 281.—Dolomit, *Blumenb*. Nat. s. 617. *Id. Haus*. Handb. b. iii. s. 963. *Id. Hoff*. b. ii. s. 57.—Chaux carbonatée magnesifere granulaire, *Haüy*, Tabl. p. 6.

External

External Characters.

Its colours are snow-white and greyish-white, and rarely pale ash-grey.

It occurs massive; also in small and fine granular distinct concretions, which are frequently so loosely aggregated, that they can be separated by the mere pressure of the finger.

Internally it is glimmering, approaching to glistening, and the lustre is pearly.

The fracture in the large is imperfect and slaty, in the compact varieties small splintery, which passes into uneven.

The fragments are indeterminate angular, and blunt-edged.

It is faintly translucent, or only translucent on the edges.

It is as hard as fluor-spar.

It is brittle, and easily frangible.

Specific gravity, Dolomite of Alps of Carinthia, 2.835, *Klaproth.*—4.913, *Breithaupt.*

Chemical and Physical Characters.

It effervesces very feebly with acids,—a character which distinguishes it from granular limestone.

It in general phosphoresces when placed on heated iron, or when rubbed in the dark; and this property is much stronger in some varieties than in others.

Constituent

Constituent Parts.

	St Gothard.	Apennines.	Carinthia.	Antique.
[illegible] Magnesia,	46.50	35.00	48.00	48.00
[illegible] Lime,	52.08	65.00	52.00	51.50
[illegible] Manganese,	0.25			
[illegible] Iron,	0.50		0.20	
[illegible]	0.75			
	100	100	100.20	99.50
	Klaproth, Beit. b. iv. s. 209.	*Klaproth*, Id. s. 215.	*Klaproth*, Id. s. 219.	*Klaproth*, Id. s. 222.

Iona.	
Carbonic Acid,	48.00
Lime,	31.12
Magnesia,	17.06
Insoluble Matter,	4.00

Tennant, Phil. Trans. for 1799.

Geognostic Situation.

It occurs principally in primitive mountains.

Geographic Situation.

Europe.—Beds of dolomite, containing tremolite, occur in the island of Iona. In the mountain-group of St Gothard, it occurs in beds, often of great thickness, containing imbedded crystals of tremolite, grains of quartz, and scales of mica and talc. In the Apennines, it occurs in imbedded portions, in a dark ash-grey splintery limestone: in Carinthia, it forms whole ranges of mountains: in Bareuth, it occurs in beds along with granular foliated limestone: at Sala in Sweden, it is mixed with mica, talc, and quartz: a beautiful white variety, used by ancient sculptors is found in the isle of Tenedos: veins of it are said

to

to occur traversing granite, in the valley of Sesia in Italy: and it is found loose on Monte Somma.

America.—Province of New-York, with tremolite *.

Asia.—Bengal †, with imbedded tremolite; also in Siberia.

Uses.

It appears to have been used by ancient sculptors in their finest works.

Observations.

1. It is named Dolomite, in honour of the celebrated French geologist Dolomieu.

2. The only mineral with which it is likely to be confounded, is granular foliated limestone; but a simple chemical test at once distinguishes them:—a drop of mineral acid causes a violent effervescence, when poured on granular foliated limestone, but a very feeble one with dolomite.

3. The flexible variety of dolomite was first noticed in the Borghese Palace in Rome, by Ferber: it was afterwards found on the mountain of Campo Longo, in the St Gothard group, by Fleuriau de Bellvue. It was sold at a very high price, until the publication of Fleuriau de Bellvue's experiments, by which it appeared, that the other varieties of dolomite, and also common granular limestone, could be rendered flexible, by exposing them in thin and long slabs, for six hours, to a heat of 200° of Reaumur.

* Brown

* Dr Bruce. † Sir John Murray.

* Brown Dolomite, or Magnesian Limestone of *Tennant.*

Tennant, Transactions of Royal Society of London for 1799.—*Thomson*, Annals of Philosophy for December 1814.

External Characters.

Its colours are yellowish-grey, yellowish-brown, and a colour intermediate between chesnut-brown and yellowish-brown; seldom bluish-grey.

It occurs massive, and in minute granular concretions.

Internally it is glistening or glimmering, and the lustre is between pearly and vitreous.

The fracture is splintery, and sometimes flat conchoidal.

The fragments are indeterminate angular, and rather blunt-edged.

It is translucent, or translucent on the edges.

It is semi-hard; it is harder than calcareous-spar.

It is brittle.

Specific gravity of the crystals, 2.823, *Tennent.*—2.771, 2.820. *Berger.*—2.791, *Thomson.*

Chemical Characters.

It dissolves slowly, and with but feeble effervescence, in nitrous acid. When deprived by heat of its carbonic acid, it is much longer of re-absorbing it from the atmosphere than common limestone.

Constituent

Constituent Parts.

Yorkshire.	Building Hill, near Sunderland.	Humbleton Hill, near Sunderland.
Lime, 29.5 to 31.07	Carbonate of Lime, 56.80	Carbonate of Lime, 51.50
Magnesia, 20.3 to 22.05	Carbonate of Magnesia, - 40.84	Carbonate of Magnesia, - 44.84
Carbonic Acid, 47.2	Carbonate of Iron, 0.36	Insoluble matter, 1.60
Alumina & Iron, 0.8 to 1.24	Insoluble matter, 2.00	Loss, - 2.06
	100.00	100.00
Tennant, Phil. Tr. for 1799.	*Thomson*, Annals of Philosophy, vol. iv. p. 416.	*Thomson*, ib. p. 417.

Geognostic Situation.

In the north of England it occurs in beds of considerable thickness, and great extent, and appears to rest on the Newcastle coal-formation; but in the Isle of Man, it occurs in a limestone which rests on grey-wacke, and contains imbedded portions of quartz, rhomb-spar, and sparry iron *. It occurs in trap-rocks in Fifeshire.

Geographic Situation.

It occurs in Nottinghamshire, Derbyshire, Northamptonshire, Leicestershire, Northumberland, and Durham †: also in Ireland, at Portumna in Galway, Ballyshannon in Donncgal, Castle Island near Killarney ‡. It has been observed among the limestone rocks near Erbefeld and Gemarek, in Westphalia ‖. It also occurs in veins,

 as

* Berger, Geological Society Transactions, vol. ii. p. 44.

† Greenough. ‡ Greenough.

‖ Bournon, Traité de Mineralogie, t. i. p. 368.

Constituent Parts.

	St Gothard.	Apennines.	Carinthia.	Antique.
Carbonate of Magnesia,	46.50	35.00	48.00	48.00
Carbonate of Lime, -	52.08	65.00	52.00	51.50
Oxide of Manganese, -	0.25			
Oxide of Iron, - -	0.50		0.20	
Loss, - - - - -	0.75			
	100	100	100.20	99.50
	Klaproth, Beit. b. iv. s. 209.	*Klaproth*, Id. s. 215.	*Klaproth*, Id. s. 219.	*Klaproth*, Id. s. 222.

	Iona.
Carbonic Acid, - -	48.00
Lime, - - -	31.12
Magnesia, - -	17.06
Insoluble Matter, -	4.00

Tennant, Phil. Trans. for 1799.

Geognostic Situation.

It occurs principally in primitive mountains.

Geographic Situation.

Europe.—Beds of dolomite, containing tremolite, occur in the island of Iona. In the mountain-group of St Gothard, it occurs in beds, often of great thickness, containing imbedded crystals of tremolite, grains of quartz, and scales of mica and talc. In the Apennines, it occurs in imbedded portions, in a dark ash-grey splintery limestone: in Carinthia, it forms whole ranges of mountains: in Bareuth, it occurs in beds along with granular foliated limestone: at Sala in Sweden, it is mixed with mica, talc, and quartz: a beautiful white variety, used by ancient sculptors is found in the isle of Tenedos: veins of it are said to

Flexible Dolomite.

External Characters.

Its colour is yellowish-grey, passing into cream-yellow. It occurs massive. It is dull. The fracture is earthy in the small, and slaty in the large. It is opaque. It yields readily to the knife, but with difficulty to the nail. In thin plates it is uncommonly flexible. Specific gravity, 2.544, *Thomson*. This is probably below the truth, as the stone is porous.

Chemical Characters.

It dissolves in acids as readily as common carbonate of lime.

Constituent Parts.

Carbonate of Lime, -	62.00
Carbonate of Magnesia,	35.96
Insoluble matter, -	1.60
Loss, - - - - -	0.44
	100.00

Thomson, Annals of Phil. vol. iv. p. 418.

Geographic Situation.

It occurs about three miles from Tinmouth Castle.

Observations.

This curious mineral was discovered by my intelligent friend Mr Nicol, Lecturer on Natural Philosophy. To that gentleman I am indebted for the following particulars in

in regard to it. He finds, that its flexibility is considerably influenced by the quantity of water contained in it. When saturated with water, it is remarkably flexible; as the evaporation goes on, it becomes more and more rigid, until the water be reduced to a certain limit, when the flexibility becomes scarcely distinguishable. From this point, however, the flexibility gradually increases, as the moisture diminishes; and as soon as the water is completely exhaled, it becomes nearly as flexible as it was when saturated with that fluid.

Second Kind.

Columnar Dolomite.

Stänglicher Dolomit, *Klaproth.*

Stänglicher Dolomit, *Klaproth*, Mag. der Gesellsch. Naturf. Freunde, b. v. s. 402.

External Characters.

Its colour is pale greyish-white.

It occurs massive, and in thin, long, and straight prismatic concretions.

It has an imperfect cleavage.

The fracture is uneven.

The lustre is vitreous, inclining to pearly.

It breaks into acicular-shaped fragments.

It is feebly translucent.

It is brittle.

Specific gravity 2.765.

Constituent

Constituent Parts.

From the Mine Tschistagowskoy.

Carbonate of Lime, -	51
Carbonate of Magnesia, -	47
Carbonated Hydrate of Iron,	1
	99

Klaproth, Chem. Abhandl. s. 328.

Geognostic and Geographic Situations.

It occurs in serpentine in the mine Tschistagowskoy, on the river Mjafs, in the Government of Orenburg in Russia.

Observations.

It was at one time considered to be a variety of Strontianite; but in external characters, it is much more nearly allied to Tremolite.

Third Kind.

Compact Dolomite, or Gurhofite.

Gurhofian, *Karsten.*

Gurhofian, *Klaproth,* in Magazin der Gesellch. der Naturf. Freünde, b. i. s. 257.—Gurofian, *Karsten,* Tabel. s. 50. *Id.* *Klap.* Beit. b. v. s. 103. *Id. Lenz,* b. ii. s. 724.

External Characters.

Its colour is snow-white.

It occurs massive.

It

It is dull.

The fracture is flat conchoidal, passing to even.

The fragments are indeterminate angular, and sharp-edged.

It is slightly translucent on the edges.

It is hard, bordering on semihard.

It is brittle, and rather difficultly frangible.

Specific gravity, 2.7600, *Karsten*.

Chemical Characters.

When pounded, and thrown into diluted and heated nitrous acid, it is completely dissolved with effervescence.

Constituent Parts.

Carbonate of Lime, -	70.50
Carbonate of Magnesia,	29.50
	100.00

Klaproth, Gesellsch. N. Fr. b. i. s. 258.

Geognostic and Geographic Situations.

It occurs in veins in serpentine rocks, between Gurhof and Aggsbach, in Lower Austria.

Observations.

1. The name Gurhofit, sometimes given to this mineral, is from the place near which it was found.

2. It was at one time considered as a variety of semi-opal; but its greater weight distinguishes it from that mineral.

Second

Flexible Dolomite.

External Characters.

Its colour is yellowish-grey, passing into cream-yellow. It occurs massive. It is dull. The fracture is earthy in the small, and slaty in the large. It is opaque. It yields readily to the knife, but with difficulty to the nail. In thin plates it is uncommonly flexible. Specific gravity, 2.544, *Thomson*. This is probably below the truth, as the stone is porous.

Chemical Characters.

It dissolves in acids as readily as common carbonate of lime.

Constituent Parts.

Carbonate of Lime, -	62.00
Carbonate of Magnesia,	35.96
Insoluble matter, -	1.60
Loss, - - - - -	0.44
	100.00

Thomson, Annals of Phil. vol. iv. p. 418.

Geographic Situation.

It occurs about three miles from Tinmouth Castle.

Observations.

This curious mineral was discovered by my intelligent friend Mr Nicol, Lecturer on Natural Philosophy. To that gentleman I am indebted for the following particulars in

The fragments are rather blunt-edged.

It is translucent.

It is semihard.

It is brittle.

Specific gravity 2.885.

Chemical Characters.

It dissolves slowly, and with little effervescence, in nitrous acid; but more rapidly, and with increased effervescence, when the acid is heated.

Constituent Parts.

Carbonate of Lime,	53.00
Carbonate of Magnesia, -	42.50
Carbonate of Iron, with a little Manganese, -	3.00
	98.50

Klaproth, Beit. b. iii. s. 296.

Geognostic and Geographic Situations.

It is found at Miemo in Tuscany, imbedded in gypsum; at Hall in the Tyrol, imbedded in muriate of soda; and Gieseké met with it in kidneys, along with wavellite, arragonite, and calcedony, in decomposed wacke, at Kannioak, in Omenaksfiord in Greenland.

Observations.

This mineral was first observed by the late Dr Thompson of Naples, who sent specimens of it to Klaproth for analysis. It is named *Miemite*, after the place where it was discovered.

Second

Second Kind.

Prismatic Miemite.

Stänglicher Bitterspath, *Klaproth.*

Strahliger Kalkstein, *Von Schlottheim*, Hoff's Magaz. für die Gesammte Mineralogie, b. i. s. 156.—Stänglicher Bitterspath, *Klaproth*, b. iii. s. 297. *Id. Leonhard*, Tabel. s. 36. *Id. Haus*, s. 128. *Id. Lenz*, b. ii. s. 712. *Id. Oken*, b. i. s. 393.

External Characters.

Its colour is asparagus-green, olive-green, and oil-green.

It occurs in prismatic distinct concretions, and crystallized in flat rhomboids, which are deeply truncated on all the edges.

The crystals are small, and very small, and sometimes they form only drusy crusts.

Internally it is shining and vitreous.

The fracture passes from concealed foliated to splintery.

The fragments are rather blunt-edged.

It is strongly translucent.

It is as hard as the granular miemite.

Specific gravity 2.885, *Karsten.*

Chemical Characters.

It dissolves slowly, and with but feeble effervescence, in nitrous acid.

Constituent

Constituent Parts.

Lime,	-	33.00
Magnesia,		14.50
Oxide of Iron,		2.50
Carbonic Acid,	-	47.25
Water and Loss,	-	2.75
		100

Klaproth, Beit. b. iii. s. 303.

Geognostic and Geographic Situations.

It occurs in cobalt veins that traverse sandstone at Glücksbrunn in Gotha, and at Beska in Servia, on the frontier of Turkey.

Third Subspecies.

Brown-Spar, or Pearl-Spar.

Braunspath, *Werner.*

This species is divided into two kinds, viz. Foliated Brown-Spar, and Columnar Brown-Spar.

First Kind.

Foliated Brown-Spar.

Blättriger Braunspath, *Werner.*

Spath perlé, *Romé de Lisle*, t. i. p. 605.—Braunspath, *Wid.* s. 515.—Sidero-calcite, *Kirw.* vol. i. p. 105.—Braunspath, *Estner*, b. ii. s. 999. *Id. Emm.* b. i. s. 79.—Brunispato, *Nap.* p. 356.

p. 356.—Le Spath brunissant, ou le Braunspath, *Broch.* t. i. p. 563.—Gemeiner Braunspath, *Reuss,* b. i. s. 50. *Id. Lud.* b. i. s. 153. *Id. Suck.* 1r th. s. 630. *Id. Bert.* s. 118. *Id. Mohs,* b. ii. s. 108.—Spathiger Braunkalk, *Hab.* s. 82.—Chaux carbonatée manganesifere, *Lucas,* p. 8.—Spathiger Braunkalk, *Leonhard,* Tabel. s. 85.—Chaux carbonatée brunissante, *Brong.* t. i. p. 237.—Gemeiner Braunspath, *Karsten,* Tabel. s. 50.—Chaux carbonatée ferro-mangmnesienne, *Bournon,* Traité, t. i. p. 277.—Pearl-Spar, *Kid,* vol. i. p. 56.—Chaux carbonatée ferro-manganesifere, *Haüy,* Tabl. p. 5.—Gemeiner Braunkalk, *Lenz,* b. ii. s. 717.—Gemeiner Braunspath, *Oken,* b. i. s. 394.—Blättriger Braunspath, *Hoff.* b. iii. s. 49.

External Characters.

Its colours are flesh-red and brownish-red.

It often occurs massive, also disseminated, seldom globular, stalactitic, reniform, with tabular and pyramidal impressions; also in distinct concretions, which are granular, and rarely thin and straight lamellar; and frequently crystallized.

Its primitive form is a rhomboid, in which the obtuse angle is 107° 22′. The following are the secondary figures:

1. Rhomboid, in which the faces are sometimes cylindrically convex, sometimes cylindrically concave.
2. Lens, both common and saddle-shaped.

It also occurs in rhomboidal six-sided pyramidal supposititious crystals.

The true crystals are generally small and very small; the supposititious crystals large and middle-sized, and are either hollow, or lined with calcareous-spar.

The

as in those of Derbyshire, where it is associated with galena *.

Use.

Like common limestone, it is burnt and made into mortar, but it remains much longer caustic than quicklime from common limestone; and this is the cause of a very important difference between magnesian and common limestone, with regard to their employment in agriculture: Lime, from magnesian limestone, is termed *hot*, and when spread upon land in the same proportion as is generally practised with common quicklime, greatly impairs the fertility of the soil; and when used in a greater quantity, it is said by Mr Tennant to prevent all vegetation †.

Observations.

A flexible variety of Dolomite occurs in England. The following account contains all the information I possess in regard to it:—

Flexible

* Bournon, Traité de Minéralogie, t. i. p. 805.

† In regard to this limestone, Dr Thomson has the following remark. "This magnesian limestone has long been burnt in prodigious quantity in the neighbourhood of Sunderland, and sent coastwise, both to the north and to the south. It goes in great abundance to Aberdeenshire. As no complaints have ever been made of its being injurious, when employed as a manure, it would be curious to know whether this circumstance be owing to the soil on which it is put, or to the small quantity of it used, in consequence of its price, occasioned by its long carriage: for it appears, from Mr Tennant's statement, that at Ferrybridge, the farmers are aware that it does not answer as a manure so well as pure carbonate of lime."—*Annals of Philosophy*, vol. iv. p. 418.

Flexible Dolomite.

External Characters.

Its colour is yellowish-grey, passing into cream-yellow. It occurs massive. It is dull. The fracture is earthy in the small, and slaty in the large. It is opaque. It yields readily to the knife, but with difficulty to the nail. In thin plates it is uncommonly flexible. Specific gravity, 2.544, *Thomson*. This is probably below the truth, as the stone is porous.

Chemical Characters.

It dissolves in acids as readily as common carbonate of lime.

Constituent Parts.

Carbonate of Lime, -	62.00
Carbonate of Magnesia,	35.96
Insoluble matter, -	1.60
Loss, - - - - -	0.44
	100.00

Thomson, Annals of Phil. vol. iv. p. 418.

Geographic Situation.

It occurs about three miles from Tinmouth Castle.

Observations.

This curious mineral was discovered by my intelligent friend Mr Nicol, Lecturer on Natural Philosophy. To that gentleman I am indebted for the following particulars in

Constituent Parts.

Lime, -	33.00
Magnesia,	14.50
Oxide of Iron, -	2.50
Carbonic Acid, -	47.25
Water and Loss, -	2.75
	100

Klaproth, Beit. b. iii. s. 303.

Geognostic and Geographic Situations.

It occurs in cobalt veins that traverse sandstone at Glücksbrunn in Gotha, and at Beska in Servia, on the frontier of Turkey.

Third Subspecies.

Brown-Spar, or Pearl-Spar.

Braunspath, *Werner.*

This species is divided into two kinds, viz. Foliated Brown-Spar, and Columnar Brown-Spar.

First Kind.

Foliated Brown-Spar.

Blättriger Braunspath, *Werner.*

Spath perlé, *Romé de Lisle,* t. i. p. 605.—Braunspath, *Wid.* s. 515.—Sidero-calcite, *Kirw.* vol. i. p. 105.—Braunspath, *Estner,* b. ii. s. 999. *Id. Emm.* b. i. s. 79.—Brunispato, *Nap.* p. 356.

Constituent Parts.

From the Mine Tschistagowskoy.

Carbonate of Lime, -	51
Carbonate of Magnesia, -	47
Carbonated Hydrate of Iron,	1
	99

Klaproth, Chem. Abhandl. s. 328.

Geognostic and Geographic Situations.

It occurs in serpentine in the mine Tschistagowskoy, on the river Mjafs, in the Government of Orenburg in Russia.

Observations.

It was at one time considered to be a variety of Strontianite; but in external characters, it is much more nearly allied to Tremolite.

Third Kind.

Compact Dolomite, or Gurhofite.

Gurhofian, *Karsten.*

Gurhofian, *Klaproth,* in Magazin der Gesellch. der Naturf. Freünde, b. i. s. 257.—Gurofian, *Karsten,* Tabel. s. 50. *Id. Klap.* Beit. b. v. s. 103. *Id. Lenz,* b. ii. s. 724.

External Characters.

Its colour is snow-white.

[illegible]urs massive.

It

It is dull.

The fracture is flat conchoidal, passing to even.

The fragments are indeterminate angular, and sharp-edged.

It is slightly translucent on the edges.

It is hard, bordering on semihard.

It is brittle, and rather difficultly frangible.

Specific gravity, 2.7600, *Karsten*.

Chemical Characters.

When pounded, and thrown into diluted and heated nitrous acid, it is completely dissolved with effervescence.

Constituent Parts.

Carbonate of Lime, -	70.50
Carbonate of Magnesia,	29.50
	100.00

Klaproth, Gesellsch. N. Fr. b. i. s. 258.

Geognostic and Geographic Situations.

It occurs in veins in serpentine rocks, between Gurhof and Aggsbach, in Lower Austria.

Observations.

1. The name Gurhofit, sometimes given to this mineral, is from the place near which it was found.

2. It was at one time considered as a variety of semi-opal; but its greater weight distinguishes it from that mineral.

Second

The most frequent accompanying vein-stone is calcareous spar; besides which, it is often associated with heavy-spar, fluor-spar, quartz, sparry-iron, galena, iron-pyrites, native silver, and various ores of silver. Very often it rests on all the minerals of which the vein is composed: hence it is said to be the newest mineral in the vein; and we frequently observe thin crusts of it investing the surface of crystals, as of calcareous-spar, fluor-spar, heavy-spar, quartz, galena, &c. These crusts seldom invest the whole crystal, generally covering only a part of it; and it is observed, that it is the same side in all the crystals of the same cavity which are encrusted with the brown-spar; and also, that when the whole side is not covered, the crust has the same height, or is on the same level in all the crystals.

Geographic Situation.

It occurs along with galena, and other ores of lead, in the lead-mines of Lead Hills and Wanlockhead in Lanarkshire; in the mines of Cumberland, Northumberland, and Derbyshire. On the Continent, it is found in Norway, Sweden, Saxony, Suabia, Piedmont, France, Hungary, and Transylvania.

Observations.

1. It is distinguished from *Calcareous-spar*, with which it has been confounded, by its colours, cleavage, inferior transparency, perfect pearly lustre, greater hardness, and higher specific gravity. It also in general effervesces less briskly with acids than calcareous-spar.

2. The straight lamellar variety has been mistaken for Heavy-spar, from which, however, it is distinguished, not only by its inferior weight, but also by its concretions being

ing very closely aggregated, which is not the case with heavy-spar.

3. In many instances the foliated Brown-spar of authors includes also varieties of rhomboidal red manganese, and of sparry-iron.

Second Kind.

Columnar Brown-Spar.

Stänglicher Braunspath, *Klaproth*.

Stänglicher Braunspath, *Klaproth*, Beit. b. iv. s. 199. *Id.* *Karsten*, Tabel. s. 50. *Id. Lenz*, b. ii. s. 723.

External Characters.

Its colours are reddish-white, rose-red, and pearl-grey.

It occurs in distinct concretions, which are wedge-shaped, columnar or prismatic, and have glimmering and longitudinally streaked surfaces.

It is splendent, and appears pearly on the fracture-surface.

It has an imperfect cleavage.

The fragments are wedge-shaped.

It is translucent.

It is brittle.

It is easily frangible.

Constituent

Second Kind.

Prismatic Miemite.

Stänglicher Bitterspath, *Klaproth.*

Strahliger Kalkstein, *Von Schlottheim*, Hoff's Magaz. fur die Gesammte Mineralogie, b. i. s. 156.—Stänglicher Bitterspath, *Klaproth*, b. iii. s. 297. *Id. Leonhard*, Tabel. s. 36. *Id. Haus*, s. 128. *Id. Lenz*, b. ii. s. 712. *Id. Oken*, b. i. s. 393.

External Characters.

Its colour is asparagus-green, olive-green, and oil-green.

It occurs in prismatic distinct concretions, and crystallized in flat rhomboids, which are deeply truncated on all the edges.

The crystals are small, and very small, and sometimes they form only drusy crusts.

Internally it is shining and vitreous.

The fracture passes from concealed foliated to splintery.

The fragments are rather blunt-edged.

It is strongly translucent.

It is as hard as the granular miemite.

Specific gravity 2.885, *Karsten.*

Chemical Characters.

It dissolves slowly, and with but feeble effervescence, in nitrous acid.

Constituent

First Subspecies.

Foliated Limestone.

Blättriger Kalkstein, *Werner.*

This subspecies is divided into two kinds, viz. Calcareous-spar, and Foliated Granular Limestone.

First Kind.

Calcareous-Spar, or Calc-Spar.

Kalkspath, *Werner.*

Spathum, *Wall.* t. i. p. 140.—Körniger Kalkstein, var. *Wid.* s. 427.—Common Spar, *Kirw.* vol. i. p. 86.—Kalkspath, *Estner,* b. ii. s. 941. *Id. Emm.* b. i. s. 456.—Spatho calcareo, *Nap.* p. 341.—Calcaire cristallisée, *Lam.* t. i. p. 29.—Chaux carbonatée cristallisée, *Haüy,* t. ii. p. 127.—Le Spath calcaire, *Broch.* t. i. p. 536.—Spathiger Kalkstein, *Reuss,* b. ii. 2. s. 284.—Kalkspath, *Lud.* b. i. s. 149. *Id. Suck.* 1r th. s. 600. —Grossblättricher Kalkstein, *Bert.* s. 90.—Kalkspath, *Mohs,* b. ii. s. 31. *Id. Hab.* s. 76.—Chaux carbonatée, *Lucas,* p. 3. —Gemeiner spathiger Kalkstein, *Leonhard,* Tabel. s. 33.— —Chaux carbonatée pure spathique, *Brong.* t. i. p. 189.— Chaux carbonatee, *Brard,* p. 26.—Kalkspath, *Haus.* s. 125.— Spathiger Kalkstein, *Karst.* Tabel. s. 50.—Crystallized Carbonate of Lime, *Kid,* vol. i. p. 50.—Chaux carbonatée, *Haüy,* Tabl. p. 2.—Kalkspath, *Lenz,* b. ii. s. 742. *Id. Haus.* Handb. b. iii. s. 900. *Id. Hoff.* b. iii. s. 17.—Calcareous-spar, *Aikin,* p. 158.

External

p. 356.—Le Spath brunissant, ou le Braunspath, *Broch.* t. i. p. 563.—Gemeiner Braunspath, *Reuss,* b. i. s. 50. *Id. Lud.* b. i. s. 153. *Id. Suck.* 1r th. s. 630. *Id. Bert.* s. 118. *Id. Mohs,* b. ii. s. 108.—Spathiger Braunkalk, *Hab.* s. 82.—Chaux carbonatée manganesifere, *Lucas,* p. 8.—Spathiger Braunkalk, *Leonhard,* Tabel. s. 85.—Chaux carbonatée brunissante, *Brong.* t. i. p. 237.—Gemeiner Braunspath, *Karsten,* Tabel. s. 50.—Chaux carbonatée ferro-manganesienne, *Bournon,* Traité, t. i. p. 277.—Pearl-Spar, *Kid,* vol. i. p. 56.—Chaux carbonatée ferro-manganesifere, *Haüy,* Tabl. p. 5.—Gemeiner Braunkalk, *Lenz,* b. ii. s. 717.—Gemeiner Braunspath, *Oken,* b. i. s. 394.—Blättriger Braunspath, *Hoff.* b. iii. s. 49.

External Characters.

Its colours are flesh-red and brownish-red.

It often occurs massive, also disseminated, seldom globular, stalactitic, reniform, with tabular and pyramidal impressions; also in distinct concretions, which are granular, and rarely thin and straight lamellar; and frequently crystallized.

Its primitive form is a rhomboid, in which the obtuse angle is 107° 22′. The following are the secondary figures:

1. Rhomboid, in which the faces are sometimes cylindrically convex, sometimes cylindrically concave.
2. Lens, both common and saddle-shaped.

It also occurs in rhomboidal six-sided pyramidal supposititious crystals.

The true crystals are generally small and very small; the supposititious crystals large and middle-sized, and are either hollow, or lined with calcareous-spar.

The

as in those of Derbyshire, where it is associated with galena *.

Use.

Like common limestone, it is burnt and made into mortar, but it remains much longer caustic than quicklime from common limestone; and this is the cause of a very important difference between magnesian and common limestone, with regard to their employment in agriculture: Lime, from magnesian limestone, is termed *hot*, and when spread upon land in the same proportion as is generally practised with common quicklime, greatly impairs the fertility of the soil; and when used in a greater quantity, is said by Mr Tennant to prevent all vegetation †.

Observations.

A flexible variety of Dolomite occurs in England. The following account contains all the information I possess in regard to it:—

Flexible

* Bournon, Traité de Mineralogie, t. i. p. 208.

† In regard to this limestone, Dr Thomson has the following remarks: "This magnesian limestone has long been burnt in prodigious quantity in the neighbourhood of Sunderland, and sent coastwise, both to the north and to the south. It goes in great abundance to Aberdeenshire. As no complaints have ever been made of its being injurious, when employed as a manure, it would be curious to know whether this circumstance be owing to the soil on which it is put, or to the small quantity of it used, in consequence of its price, occasioned by its long carriage; for it appears, from Mr Tennant's statement, that at Ferrybridge, the farmers are aware that it does not answer as a manure so well as pure carbonate of lime."—*Annals of Philosophy*, vol. iv. p. 418.

Flexible Dolomite.

External Characters.

Its colour is yellowish-grey, passing into cream-yellow. It occurs massive. It is dull. The fracture is earthy in the small, and slaty in the large. It is opaque. It yields readily to the knife, but with difficulty to the nail. In thin plates it is uncommonly flexible. Specific gravity, 2.544, *Thomson*. This is probably below the truth, as the stone is porous.

Chemical Characters.

It dissolves in acids as readily as common carbonate of lime.

Constituent Parts.

Carbonate of Lime, -	62.00
Carbonate of Magnesia,	35.96
Insoluble matter, -	1.60
Loss, - - - - -	0.44
	100.00

Thomson, Annals of Phil. vol. iv. p. 418.

Geographic Situation.

It occurs about three miles from Tinmouth Castle.

Observations.

This curious mineral was discovered by my intelligent friend Mr Nicol, Lecturer on Natural Philosophy. To that gentleman I am indebted for the following particulars in

the edges which the bevelling planes make with the broad lateral planes, truncated.

In other varieties, the apices of the acuminations are more or less deeply truncated, and sometimes so deeply, that the acuminating planes appear as truncating planes on the angles of the prism.

3. The preceding figure, in which the six-planed acumination is *flatly acuminated* with three planes, which are set on the acute edges of the six-planed acumination.

4. Six-sided prism *acutely acuminated* with three planes which are set on the alternate lateral planes. The apex of the acumination is sometimes more or less deeply truncated. Sometimes the truncation is so deep, that the remains of the acuminating planes appear as truncations on the alternate terminal edges. In other varieties, the prism becomes so short, that the acuminating planes meet and form an acute double three-sided pyramid.

5. When the planes of the flat three-planed acumination N° 3. increase so much that those of the six-planed acumination disappear, a six-sided prism is formed, *flatly acuminated* with three planes, which are set on the alternate lateral planes in an unconformable position. When the prism disappears, there is formed an obtuse double three-sided pyramid.

These prisms are often pyramidally aggregated.

6. When the prism becomes very low, it may be viewed as an equiangular six-sided table, which is sometimes aggregated in a rose-like form.

7. Sometimes the six-sided prism is truncated on the lateral edges, and thus forms a twelve-sided prism.

The

Constituent Parts.

From the Mine Tschistagowskoy.

Carbonate of Lime,	-	51
Carbonate of Magnesia,	-	47
Carbonated Hydrate of Iron,		1
		99

Klaproth, Chem. Abhandl. s. 328.

Geognostic and Geographic Situations.

It occurs in serpentine in the mine Tschistagowskoy, on the river Mjafs, in the Government of Orenburg in Russia.

Observations.

It was at one time considered to be a variety of Strontianite; but in external characters, it is much more nearly allied to Tremolite.

Third Kind.

Compact Dolomite, or Gurhofite.

Gurhofian, *Karsten.*

Gurhofian, *Klaproth,* in Magazin der Gesellch. der Naturf. Freünde, b. i. s. 257.—Gurofian, *Karsten,* Tabel. s. 50. *Id. Klap.* Beit. b. v. s. 103. *Id. Lenz,* b. ii. s. 724.

External Characters.

Its colour is snow-white.

It occurs massive.

It

cleavage is threefold, and in the direction of the planes of the primitive rhomboid; there are others less distinct in the direction of the planes of the obtuse rhomboid or double three-sided pyramid; and in the direction of the alternate lateral planes of the regular six-sided prism.

The fracture is perfect conchoidal.

The fragments are indeterminate angular, and rather sharp-edged, or they are rhomboidal.

It occurs transparent, semi-transparent, and occasionally only translucent. It refracts double *.

It is semi-hard; it scratches gypsum, but is scratched by fluor-spar.

It is brittle, and very easily frangible.

Specific gravity, 2.5, 2.8, *Mohs.*

Chemical Characters.

It is infusible before the blowpipe, but it becomes caustic, losing by complete calcination about 43 *per cent.*; effervesces violently with acids.

Constituent Parts.

	Iceland Spar.	Iceland Spar.	Iceland Spar.	From Andreasberg.
Lime,	56.15	55.50	56.50	55.9802
Carbonic Acid,	43.70	44.00	43.00	43.5635
Water,		0.50	0.50	0.1000
Oxide of Manganese, with trace of Iron,	0.15			0.3562
	100.00	100.00	100.00	100.0000
	Stromyer, Gilbert's Annalen for 1813, p. 217.	*Philips*, Phil. Mag. xiv. 290.	*Buchols*, Gehl. Journ. iv. 412.	*Stromyer*, Gilbert's An. for 1813, p. 217.

Geognostic

* The double refracting power of calcareous-spar was first observed by Erasmus Bartholin.

Geognostic Situation.

It never occurs in mountain-masses, but venigenous in almost every rock, from granite to the newest secondary formation. The oldest formation of this mineral is that in veins, where it is accompanied with felspar, rock-crystal, probably also with epidote, sphene, and chlorite. It occurs also in beds, along with augite, hornblende, garnet, and magnetic ironstone; and frequently in veins in different metalliferous formations. Thus, it is associated with nearly all the metallic minerals contained in gneiss, mica-slate, clay-slate, syenite, porphyry; seldomer in granite, more frequently, again, in grey-wacke, and along with cobalt and copper ores in the oldest secondary or flœtz limestone. Veins, almost entirely composed of calcareous-spar, abound in the newest limestone formations; and it is a common mineral, either in veins, or in cotemporaneous masses, in the various rocks of the secondary or flœtz-trap series.

An interesting geognostic character of calcareous-spar, is the uniformity of its crystallizations in particular districts. Thus, in the mines of Derbyshire, the acute six-sided pyramid and its congenerous forms are the most frequent and abundant; at Schneeberg in Saxony, and in the Upper Hartz, the prevailing forms are the regular six-sided prism and table; while in the mines of Freyberg the most frequent forms are the regular six-sided prism, acuminated with three planes, set on the lateral planes, and the flat double three-sided pyramid.

Geographic Situation.

This mineral is so common in every country, as to render any account of its geographic distribution unnecessary.

It

The fragments are rather blunt-edged.

It is translucent.

It is semihard.

It is brittle.

Specific gravity 2.885.

Chemical Characters.

It dissolves slowly, and with little effervescence, in nitrous acid; but more rapidly, and with increased effervescence, when the acid is heated.

Constituent Parts.

Carbonate of Lime,	53.00
Carbonate of Magnesia, -	42.50
Carbonate of Iron, with a little Manganese, -	3.00
	98.50

Klaproth, Beit. b. iii. s. 296.

Geognostic and Geographic Situations.

It is found at Miemo in Tuscany, imbedded in gypsum; at Hall in the Tyrol, imbedded in muriate of soda; and Gieseké met with it in kidneys, along with wavellite, arragonite, and calcedony, in decomposed wacke, at Kannioak, in Omenaksfiord in Greenland.

Observations.

This mineral was first observed by the late Dr Thompson of Naples, who sent specimens of it to Klaproth for analysis. It is named *Miemite*, after the place where it was discovered.

Second

Second Kind.

Prismatic Miemite.

Stänglicher Bitterspath, *Klaproth.*

Strahliger Kalkstein, *Von Schlottheim,* Hoff's Magaz. fur die Gesammte Mineralogie, b. i. s. 156.—Stänglicher Bitterspath, *Klaproth,* b. iii. s. 297. *Id. Leonhard,* Tabel. s. 36. *Id. Haus,* s. 128. *Id. Lenz,* b. ii. s. 712. *Id. Oken,* b. i. s. 393.

External Characters.

Its colour is asparagus-green, olive-green, and oil-green.

It occurs in prismatic distinct concretions, and crystallized in flat rhomboids, which are deeply truncated on all the edges.

The crystals are small, and very small, and sometimes they form only drusy crusts.

Internally it is shining and vitreous.

The fracture passes from concealed foliated to splintery.

The fragments are rather blunt-edged.

It is strongly translucent.

It is as hard as the granular miemite.

Specific gravity 2.885, *Karsten.*

Chemical Characters.

It dissolves slowly, and with but feeble effervescence, in nitrous acid.

Constituent

Constituent Parts.

Lime,	-	33.00
Magnesia,		14.50
Oxide of Iron,		2.50
Carbonic Acid,	-	47.25
Water and Loss,	-	2.75
		100

Klaproth, Beit. b. iii. s. 303.

Geognostic and Geographic Situations.

It occurs in cobalt veins that traverse sandstone at Glücksbrunn in Gotha, and at Beska in Servia, on the frontier of Turkey.

Third Subspecies.

Brown-Spar, or Pearl-Spar.

Braunspath, *Werner*.

This species is divided into two kinds, viz. Foliated Brown-Spar, and Columnar Brown-Spar.

First Kind.

Foliated Brown-Spar.

Blättriger Braunspath, *Werner*.

Spath perlé, *Romé de Lisle*, t. i. p. 605.—Braunspath, *Wid.* s. 515.—Sidero-calcite, *Kirw.* vol. i. p. 105.—Braunspath, *Estner*, b. ii. s. 999. *Id. Emm.* b. i. s. 79.—Brunispato, *Nap.* p. 356.

p. 356.—Le Spath brunissant, ou le Braunspath, *Broch.* t. i. p. 563.—Gemeiner Braunspath, *Reuss,* b. i. s. 50. *Id. Lud.* b. i. s. 153. *Id. Suck.* 1r th. s. 630. *Id. Bert.* s. 118. *Id. Mohs,* b. ii. s. 108.—Spathiger Braunkalk, *Hab.* s. 82.—Chaux carbonatée manganesifere, *Lucas,* p. 8.—Spathiger Braunkalk, *Leonhard,* Tabel. s. 85.—Chaux carbonatée brunissante, *Brong.* t. i. p. 237.—Gemeiner Braunspath, *Karsten,* Tabel. s. 50.—Chaux carbonatée ferro-mangmesienne, *Bournon,* Traité, t. i. p. 277.—Pearl-Spar, *Kid,* vol. i. p. 56.—Chaux carbonatée ferro-manganesifere, *Haüy,* Tabl. p. 5.—Gemeiner Braunkalk, *Lenz,* b. ii. s. 717.—Gemeiner Braunspath, *Oken,* b. i. s. 394.—Blättriger Braunspath, *Hoff.* b. iii. s. 49.

External Characters.

Its colours are flesh-red and brownish-red.

It often occurs massive, also disseminated, seldom globular, stalactitic, reniform, with tabular and pyramidal impressions; also in distinct concretions, which are granular, and rarely thin and straight lamellar; and frequently crystallized.

Its primitive form is a rhomboid, in which the obtuse angle is 107° 22′. The following are the secondary figures:

1. Rhomboid, in which the faces are sometimes cylindrically convex, sometimes cylindrically concave.
2. Lens, both common and saddle-shaped.

It also occurs in rhomboidal six-sided pyramidal supposititious crystals.

The true crystals are generally small and very small; the supposititious crystals large and middle-sized, and are either hollow, or lined with calcareous-spar.

The

as in those of Derbyshire, where it is associated with galena *.

Use.

Like common limestone, it is burnt and made into mortar, but it remains much longer caustic than quicklime from common limestone; and this is the cause of a very important difference between magnesian and common limestone, with regard to their employment in agriculture: Lime, from magnesian limestone, is termed *hot*, and when spread upon land in the same proportion as is generally practised with common quicklime, greatly impairs the fertility of the soil; and when used in a greater quantity, is said by Mr Tennant to prevent all vegetation †.

Observations.

A flexible variety of Dolomite occurs in England. The following account contains all the information I possess in regard to it:—

Flexible

* Bournon, Traité de Mineralogie, t. i. p. 208.

† In regard to this limestone, Dr Thomson has the following remarks: "This magnesian limestone has long been burnt in prodigious quantity in the neighbourhood of Sunderland, and sent coastwise, both to the north and to the south. It goes in great abundance to Aberdeenshire. As no complaints have ever been made of its being injurious, when employed as a manure, it would be curious to know whether this circumstance be owing to the soil on which it is put, or to the small quantity of it used, in consequence of its price, occasioned by its long carriage; for it appears, from Mr Tennant's statement, that at Ferrybridge, the farmers are aware that it does not answer as a manure so well as pure carbonate of lime."—*Annals of Philosophy*, vol. iv. p. 418.

Flexible Dolomite.

External Characters.

Its colour is yellowish-grey, passing into cream-yellow. It occurs massive. It is dull. The fracture is earthy in the small, and slaty in the large. It is opaque. It yields readily to the knife, but with difficulty to the nail. In thin plates it is uncommonly flexible. Specific gravity, 2.544, *Thomson*. This is probably below the truth, as the stone is porous.

Chemical Characters.

It dissolves in acids as readily as common carbonate of lime.

Constituent Parts.

Carbonate of Lime, -	62.00
Carbonate of Magnesia,	35.96
Insoluble matter, -	1.60
Loss, - - - - -	0.44
	100.00

Thomson, Annals of Phil. vol. iv. p. 418.

Geographic Situation.

It occurs about three miles from Tinmouth Castle.

Observations.

This curious mineral was discovered by my intelligent friend Mr Nicol, Lecturer on Natural Philosophy. To that gentleman I am indebted for the following particulars in

in regard to it. He finds, that its flexibility is considerably influenced by the quantity of water contained in it. When saturated with water, it is remarkably flexible; as the evaporation goes on, it becomes more and more rigid, until the water be reduced to a certain limit, when the flexibility becomes scarcely distinguishable. From this point, however, the flexibility gradually increases, as the moisture diminishes; and as soon as the water is completely exhaled, it becomes nearly as flexible as it was when saturated with that fluid.

Second Kind.

Columnar Dolomite.

Stänglicher Dolomit, *Klaproth.*

Stänglicher Dolomit, *Klaproth*, Mag. der Gesellsch. Naturf. Freunde, b. v. s. 402.

External Characters.

Its colour is pale greyish-white.

It occurs massive, and in thin, long, and straight prismatic concretions.

It has an imperfect cleavage.

The fracture is uneven.

The lustre is vitreous, inclining to pearly.

It breaks into acicular-shaped fragments.

It is feebly translucent.

It is brittle.

Specific gravity 2.765.

Constituent

Constituent Parts.

From the Mine Tschistagowskoy.

Carbonate of Lime, -	51
Carbonate of Magnesia, -	47
Carbonated Hydrate of Iron,	1
	99

Klaproth, Chem. Abhandl. s. 328.

Geognostic and Geographic Situations.

It occurs in serpentine in the mine Tschistagowskoy, on the river Mjafs, in the Government of Orenburg in Russia.

Observations.

It was at one time considered to be a variety of Strontianite; but in external characters, it is much more nearly allied to Tremolite.

Third Kind.

Compact Dolomite, or Gurhofite.

Gurhofian, *Karsten.*

Gurhofian, *Klaproth,* in Magazin der Gesellch. der Naturf. Freünde, b. i. s. 257.—Gurofian, *Karsten,* Tabel. s. 50. *Id.* *Klap.* Beit. b. v. s. 103. *Id. Lenz,* b. ii. s. 724.

External Characters.

Its colour is snow-white.

It occurs massive.

It

It is dull.

The fracture is flat conchoidal, passing to even.

The fragments are indeterminate angular, and sharp-edged.

It is slightly translucent on the edges.

It is hard, bordering on semihard.

It is brittle, and rather difficultly frangible.

Specific gravity, 2.7600, *Karsten*.

Chemical Characters.

When pounded, and thrown into diluted and heated nitrous acid, it is completely dissolved with effervescence.

Constituent Parts.

Carbonate of Lime, -	70.50
Carbonate of Magnesia,	29.50
	100.00

Klaproth, Gesellsch. N. Fr. b. i. s. 258.

Geognostic and Geographic Situations.

It occurs in veins in serpentine rocks, between Gurhof and Aggsbach, in Lower Austria.

Observations.

1. The name Gurhofit, sometimes given to this mineral, is from the place near which it was found.

2. It was at one time considered as a variety of semi-opal; but its greater weight distinguishes it from that mineral.

Second

b. *White Tiree Marble.*—Its colours are greyish-white and bluish-white: it contains scales of mica, and crystals or grains of common hornblende; which latter, when minutely diffused, give the marble a green or yellowish-green colour, and when very intimately combined with the mass, form beautiful yellowish-green spots.

2. *Iona Marble.*—Its colours are greyish-white and snow-white. Its lustre is glimmering, and fracture minute foliated, combined with splintery. It is harder than most of the other marbles. It is an intimate mixture of limestone and tremolite; for if we immerse it in an acid, the carbonate of lime will be dissolved, and the fibres of tremolite remain unaltered. It is sometimes intermixed with steatite, which gives it a green or yellow colour, in spots. These yellow or green coloured portions receive a considerable polish, and have been erroneously described as nephritic stone, and are known also under the name of *Iona* or *Icolmkill Pebbles*. The marble itself does not receive a high polish: this, with its great hardness, have brought it into disrepute with artists. Several of the varieties of Iona marble are dolomite.

3. *Skye Marble.*—In the Island of Skye, in the property of Lord Macdonald, there are several varieties of marble, deserving of attention, inclosed in porphyry, sandstone, and trap-rocks. One variety is of a greyish inclining to snow-white colour: another greyish-white, veined with ash-grey; and a third is ash-grey, or pale bluish-grey, veined with lemon-yellow or siskin-green *. Dr MacCulloch has described other varieties; and more minute details are expected from his promised work on the "Geology of the Hebrides."

* Mineralogy of Scottish Isles, vol. ii.

The fragments are rather blunt-edged.
It is translucent.
It is semihard.
It is brittle.
Specific gravity 2.885.

Chemical Characters.

It dissolves slowly, and with little effervescence, in nitrous acid; but more rapidly, and with increased effervescence, when the acid is heated.

Constituent Parts.

Carbonate of Lime,	53.00
Carbonate of Magnesia, -	42.50
Carbonate of Iron, with a little Manganese, -	3.00
	98.50

Klaproth, Beit. b. iii. s. 296.

Geognostic and Geographic Situations.

It is found at Miemo in Tuscany, imbedded in gypsum; at Hall in the Tyrol, imbedded in muriate of soda; and Gieseké met with it in kidneys, along with wavellite, arragonite, and calcedony, in decomposed wacke, at Kannioak, in Omenaksfiord in Greenland.

Observations.

This mineral was first observed by the late Dr Thompson of Naples, who sent specimens of it to Klaproth for analysis. It is named *Miemite*, after the place where it was discovered.

Second

Second Kind.

Prismatic Miemite.

Stänglicher Bitterspath, *Klaproth.*

Strahliger Kalkstein, *Von Schlottheim*, Hoff's Magaz. fur die Gesammte Mineralogie, b. i. s. 156.—Stänglicher Bitterspath, *Klaproth*, b. iii. s. 297. *Id. Leonhard*, Tabel. s. 36. *Id. Haus*, s. 128. *Id. Lenz*, b. ii. s. 712. *Id. Oken*, b. i. s. 393.

External Characters.

Its colour is asparagus-green, olive-green, and oil-green.

It occurs in prismatic distinct concretions, and crystallized in flat rhomboids, which are deeply truncated on all the edges.

The crystals are small, and very small, and sometimes they form only drusy crusts.

Internally it is shining and vitreous.

The fracture passes from concealed foliated to splintery.

The fragments are rather blunt-edged.

It is strongly translucent.

It is as hard as the granular miemite.

Specific gravity 2.885, *Karsten.*

Chemical Characters.

It dissolves slowly, and with but feeble effervescence, in nitrous acid.

Constituent

Constituent Parts.

Lime, -	33.00
Magnesia,	14.50
Oxide of Iron,	2.50
Carbonic Acid, -	47.25
Water and Loss, -	2.75
	100

Klaproth, Beit. b. iii. s. 302.

Geognostic and Geographic Situations.

It occurs in cobalt veins that traverse sandstone at Glücksbrunn in Gotha, and at Beska in Servia, on the frontier of Turkey.

Third Subspecies.

Brown-Spar, or Pearl-Spar.

Braunspath, *Werner.*

This species is divided into two kinds, viz. Foliated Brown-Spar, and Columnar Brown-Spar.

First Kind.

Foliated Brown-Spar.

Blättriger Braunspath, *Werner.*

Spath perlé, *Romé de Lisle,* t. i. p. 605.—Braunspath, *Wid.* s. 515.—Sidero-calcite, *Kirw.* vol. i. p. 105.—Braunspath, *Estner,* b. ii. s. 999. *Id. Emm.* b. i. s. 79.—Brunispato, *Nap.* p. 356.

p. 356.—Le Spath brunissant, ou le Braunspath, *Broch.* t. i. p. 563.—Gemeiner Braunspath, *Reuss*, b. i. s. 50. *Id. Lud.* b. i. s. 153. *Id. Suck.* 1r th. s. 630. *Id. Bert.* s. 118. *Id. Mohs*, b. ii. s. 108.—Spathiger Braunkalk, *Hab.* s. 82.—Chaux carbonatée manganesifere, *Lucas*, p. 8.—Spathiger Braunkalk, *Leonhard*, Tabel. s. 85.—Chaux carbonatée brunissante, *Brong.* t. i. p. 237.—Gemeiner Braunspath, *Karsten*, Tabel. s. 50.—Chaux carbonatée ferro-manganesienne, *Bournon*, Traité, t. i. p. 277.—Pearl-Spar, *Kid*, vol. i. p. 56.—Chaux carbonatée ferro-manganesifere, *Haüy*, Tabl. p. 5.—Gemeiner Braunkalk, *Lenz*, b. ii. s. 717.—Gemeiner Braunspath, *Oken*, b. i. s. 394.—Blättriger Braunspath, *Hoff.* b. iii. s. 49.

External Characters.

Its colours are flesh-red and brownish-red.

It often occurs massive, also disseminated, seldom globular, stalactitic, reniform, with tabular and pyramidal impressions; also in distinct concretions, which are granular, and rarely thin and straight lamellar; and frequently crystallized.

Its primitive form is a rhomboid, in which the obtuse angle is 107° 22′. The following are the secondary figures:

1. Rhomboid, in which the faces are sometimes cylindrically convex, sometimes cylindrically concave.
2. Lens, both common and saddle-shaped.

It also occurs in rhomboidal six-sided pyramidal supposititious crystals.

The true crystals are generally small and very small; the supposititious crystals large and middle-sized, and are either hollow, or lined with calcareous-spar.

The

as one of great beauty and splendour. It should not, however, be exposed to the weather, since, by so doing, the talcose substance exfoliates, and leaves hollow spaces, which renders its surface uneven and rough; but it answers extremely well in the interior of buildings, for chimney-pieces, slabs for tables, &c. There are immense quarries of this valuable marble at Campan, near Bagnere, in the High Pyrenees.

4. *Sarencolin Marble.*—It exhibits on its surface large straight zones, and angular spots, of a yellow or blood-red colour, so that at first view it bears some resemblance to the marble called Sicilian. The finer varieties have become very scarce. It is found at Sarencolin, in the High Pyrenees.

5. *Breccia Marble of the Pyrenees.*—One variety contains, in a brownish-red basis, black, grey, and red, middle-sized spots. It admits a good polish. Another variety has an orange-yellow coloured basis, containing small fragments of a snow-white colour. Both varieties are found in the High Pyrenees.

Italian Marbles.

1. *Sienna Marble, or Brocatella di Siena.*—It has a yellowish colour, and disposed in large irregular spots, surrounded with veins of bluish-red, passing sometimes into purple. It is by no means uncommon in Siena. At Montarenti, two leagues from Siena, another yellow marble is found, which is traversed by black and purplish-black veins. This is frequently employed throughout Italy.

2. *Mandelato Marble.*—It is a light red marble, with yellowish-white spots, found at Luggezana in the Veronese. Another variety, bearing the same name, occurs at Preosa. They

Flexible Dolomite.

External Characters.

Its colour is yellowish-grey, passing into cream-yellow. It occurs massive. It is dull. The fracture is earthy in the small, and slaty in the large. It is opaque. It yields readily to the knife, but with difficulty to the nail. In thin plates it is uncommonly flexible. Specific gravity, 2.544, *Thomson*. This is probably below the truth, as the stone is porous.

Chemical Characters.

It dissolves in acids as readily as common carbonate of lime.

Constituent Parts.

Carbonate of Lime, -	62.00
Carbonate of Magnesia,	35.96
Insoluble matter, -	1.60
Loss, - - - - -	0.44
	100.00

Thomson, Annals of Phil. vol. iv. p. 418.

Geographic Situation.

It occurs about three miles from Tinmouth Castle.

Observations.

This curious mineral was discovered by my intelligent friend Mr Nicol, Lecturer on Natural Philosophy. To that gentleman I am indebted for the following particulars in

Sicilian Marbles.

The island of Sicily abounds in marbles. Baron Borch describes upwards of a hundred marbles. Of these, the best known in this country is that named *Sicile Antique*, or by English artists *Sicilian Jasper*. It is red, with large stripes like ribbons, white, red, and sometimes green. Among the Sicilian breccia marbles, are those of the Gallo, the one of a light grey colour, presenting elegant rose-coloured spots, of different shades; and the other also grey, veined yellow, and exhibiting on its surface white translucent spots. The breccia marble of Monte Alcano is light grey, with round rose-coloured spots. That of Taormina has a deep red ground, and presents on its surface yellow and greyish-white spots.

Spanish Marbles.

Spain abounds in beautiful marbles. The vicinity of Valencia, Cadiz, Burgos, Grenada, Molina, and Carthagena, offer a great number of them; and the Tagus, in its course, winds through hills of marble. Hence it is, that the monuments in Spain, those of the middle ages, and of modern times, are profusely decorated with indigenous marbles. The vault of the beautiful theatre at Toledo, is supported by 350 marble columns. The Mosque of Cordova, erected by Caliph Abdoulrahman III. is ornamented with 1200 columns, most of which are of Spanish marble. Among the ruins of ancient Merida, which was built twenty-eight years before the commencement of the Christian era, fragments of fine marbles are still discovered; the Church of the Escurial, and also the Palace are decorated with very beautiful marbles; and the same may be said of the principal churches in Madrid. The

Constituent Parts.

From the Mine Tschistagowskoy.	
Carbonate of Lime, -	51
Carbonate of Magnesia, -	47
Carbonated Hydrate of Iron,	1
	99

Klaproth, Chem. Abhandl. s. 328.

Geognostic and Geographic Situations.

It occurs in serpentine in the mine Tschistagowskoy, on the river Mjafs, in the Government of Orenburg in Russia.

Observations.

It was at one time considered to be a variety of Strontianite; but in external characters, it is much more nearly allied to Tremolite.

Third Kind.

Compact Dolomite, or Gurhofite.

Gurhofian, *Karsten.*

Gurhofian, *Klaproth,* in Magazin der Gesellch. der Naturf. Freünde, b. i. s. 257.—Gurofian, *Karsten,* Tabel. s. 50. *Id.* *Klap.* Beit. b. v. s. 103. *Id. Lenz,* b. ii. s. 724.

External Characters.

Its colour is snow-white.

It occurs massive.

It

It is dull.

The fracture is flat conchoidal, passing to even.

The fragments are indeterminate angular, and sharp-edged.

It is slightly translucent on the edges.

It is hard, bordering on semihard.

It is brittle, and rather difficultly frangible.

Specific gravity, 2.7600, *Karsten.*

Chemical Characters.

When pounded, and thrown into diluted and heated nitrous acid, it is completely dissolved with effervescence.

Constituent Parts.

Carbonate of Lime, -	70.50
Carbonate of Magnesia,	29.50
	100.00

Klaproth, Gesellsch. N. Fr. b. i. s. 258.

Geognostic and Geographic Situations.

It occurs in veins in serpentine rocks, between Gurhof and Aggsbach, in Lower Austria.

Observations.

1. The name Gurhofit, sometimes given to this mineral, is from the place near which it was found.

2. It was at one time considered as a variety of semi-opal; but its greater weight distinguishes it from that mineral.

Second

appearance. The principal marble is that of Fagernech, which is white, with veins of green talc.

Russian and Siberian Marbles.

The vast Empire of Russia affords a great many different kinds of marble. Georgi, in his Description of the Russian Empire, enumerates white, grey, green, blue, yellow, and red varieties; and Patrin gives the following account of the Siberian marbles. "The Uralian Mountains furnish the finest and most variegated marbles. The greater part is taken from the neighbourhood of Catharinenburg, where they are wrought, and from thence transported into Russia, particularly to Petersburgh. The late Empress caused an immense palace to be built in her capital for Orloff, her favourite, which is entirely coated with these fine marbles, both inside and outside. The Empress built the church of Isaac with the same marbles, on a vast space, near the statue of Peter the Great." Patrin found no white statuary marble in the Uralian Mountains; but in that part of the Altain Mountains which is traversed by the river Irtish, he in two places saw immense blocks of marble, perfectly white and pure, from which blocks might be hewn.

Asiatic Marbles.

At present we are very imperfectly acquainted with the marbles of Asia.

Shaw mentions a red marble from Mount Sinai: Russell, in his Natural History of Aleppo, gives an imperfect account of the marbles of Syria; and some Persian marbles are noticed by Chardin. Mr Morier, in his Journey through Persia, mentions a very beautiful marble, under the name *Marble of Tabris*, and informs us, that the tomb of

The fragments are rather blunt-edged.

It is translucent.

It is semihard.

It is brittle.

Specific gravity 2.885.

Chemical Characters.

It dissolves slowly, and with little effervescence, in nitrous acid; but more rapidly, and with increased effervescence, when the acid is heated.

Constituent Parts.

Carbonate of Lime,	53.00
Carbonate of Magnesia, -	42.50
Carbonate of Iron, with a little Manganese, -	3.00
	98.50

Klaproth, Beit. b. iii. s. 296.

Geognostic and Geographic Situations.

It is found at Miemo in Tuscany, imbedded in gypsum; at Hall in the Tyrol, imbedded in muriate of soda; and Gieseké met with it in kidneys, along with wavellite, arragonite, and calcedony, in decomposed wacke, at Kannioak, in Omenaksfiord in Greenland.

Observations.

This mineral was first observed by the late Dr Thompson of Naples, who sent specimens of it to Klaproth for analysis. It is named *Miemite*, after the place where it was discovered.

Second

Second Kind.

Prismatic Miemite.

Stänglicher Bitterspath, *Klaproth.*

Strahliger Kalkstein, *Von Schlottheim*, Hoff's Magaz. fur die Gesammte Mineralogie, b. i. s. 156.—Stänglicher Bitterspath, *Klaproth*, b. iii. s. 297. *Id. Leonhard*, Tabel. s. 36. *Id. Haus*, s. 128. *Id. Lenz*, b. ii. s. 712. *Id. Oken*, b. i. s. 393.

External Characters.

Its colour is asparagus-green, olive-green, and oil-green.

It occurs in prismatic distinct concretions, and crystallized in flat rhomboids, which are deeply truncated on all the edges.

The crystals are small, and very small, and sometimes they form only drusy crusts.

Internally it is shining and vitreous.

The fracture passes from concealed foliated to splintery.

The fragments are rather blunt-edged.

It is strongly translucent.

It is as hard as the granular miemite.

Specific gravity 2.885, *Karsten.*

Chemical Characters.

It dissolves slowly, and with but feeble effervescence, in nitrous acid.

The fragments are rather blunt-edged.

It is translucent.

It is semihard.

It is brittle.

Specific gravity 2.885.

Chemical Characters.

It dissolves slowly, and with little effervescence, in nitrous acid; but more rapidly, and with increased effervescence, when the acid is heated.

Constituent Parts.

Carbonate of Lime,	53.00
Carbonate of Magnesia, -	42.50
Carbonate of Iron, with a little Manganese, -	3.00
	98.50

Klaproth, Beit. b. iii. s. 296.

Geognostic and Geographic Situations.

It is found at Miemo in Tuscany, imbedded in gypsum; at Hall in the Tyrol, imbedded in muriate of soda; and Gieseké met with it in kidneys, along with wavellite, arragonite, and calcedony, in decomposed wacke, at Kannioak, in Omenaksfiord in Greenland.

Observations.

This mineral was first observed by the late Dr Thompson of Naples, who sent specimens of it to Klaproth for analysis. It is named *Miemite*, after the place where it was discovered.

Second

Second Kind.

Prismatic Miemite.

Stänglicher Bitterspath, *Klaproth.*

Strahliger Kalkstein, *Von Schlottheim,* Hoff's Magaz. fur die Gesammte Mineralogie, b. i. s. 156.—Stänglicher Bitterspath, *Klaproth,* b. iii. s. 297. *Id. Leonhard,* Tabel. s. 36. *Id. Haus,* s. 128. *Id. Lenz,* b. ii. s. 712. *Id. Oken,* b. i. s. 393.

External Characters.

Its colour is asparagus-green, olive-green, and oil-green.

It occurs in prismatic distinct concretions, and crystallized in flat rhomboids, which are deeply truncated on all the edges.

The crystals are small, and very small, and sometimes they form only drusy crusts.

Internally it is shining and vitreous.

The fracture passes from concealed foliated to splintery.

The fragments are rather blunt-edged.

It is strongly translucent.

It is as hard as the granular miemite.

Specific gravity 2.885, *Karsten.*

Chemical Characters.

It dissolves slowly, and with but feeble effervescence, in nitrous acid.

Constituent

* Brown Dolomite, or Magnesian Limestone of *Tennant*.

Tennant, Transactions of Royal Society of London for 17[illegible]—*Thomson*, Annals of Philosophy for December 1814.

External Characters.

Its colours are yellowish-grey, yellowish-brown, and a colour intermediate between chesnut-brown and yellowish-brown; seldom bluish-grey.

It occurs massive, and in minute granular concretions.

Internally it is glistening or glimmering, and the lustre is between pearly and vitreous.

The fracture is splintery, and sometimes flat conchoidal.

The fragments are indeterminate angular, and rather blunt-edged.

It is translucent, or translucent on the edges.

It is semi-hard; it is harder than calcareous-spar.

It is brittle.

Specific gravity of the crystals, 2.888, *Tennant*.—2.771, 2.880. *Berger*.—2.791, *Thomson*.

Chemical Characters.

It dissolves slowly, and with but feeble effervescence, in nitrous acid. When deprived by heat of its carbonic acid, it is much longer of re-absorbing it from the atmosphere than common limestone.

Constituent

Constituent Parts.

Yorkshire.	Building Hill, near Sunderland.	Humbleton Hill, near Sunderland.
Lime, 29.5 to 31.07	Carbonate of Lime, 56.80	Carbonate of Lime, 51.50
Magnesia, 20.3 to 22.05	Carbonate of Magnesia, - 40.84	Carbonate of Magnesia, - 44.84
Carbonic Acid, 47.2	Carbonate of Iron, 0.36	Insoluble matter, 1.60
Alumina & Iron, 0.8 to 1.24	Insoluble matter, 2.00	Loss, - 2.06
	100.00	100.00
Tennant, Phil. Tr. for 1799.	*Thomson*, Annals of Philosophy, vol. iv. p. 416.	*Thomson*, ib. p. 417.

Geognostic Situation.

In the north of England it occurs in beds of considerable thickness, and great extent, and appears to rest on the Newcastle coal-formation; but in the Isle of Man, it occurs in a limestone which rests on grey-wacke, and contains imbedded portions of quartz, rhomb-spar, and sparry iron *. It occurs in trap-rocks in Fifeshire.

Geographic Situation.

It occurs in Nottinghamshire, Derbyshire, Northamptonshire, Leicestershire, Northumberland, and Durham †: also in Ireland, at Portumna in Galway, Ballyshannon in Donncgal, Castle Island near Killarney ‡. It has been observed among the limestone rocks near Erbefeld and Gemarek, in Westphalia ||. It also occurs in veins, as

* Berger, Geological Society Transactions, vol. ii. p. 44.

† Greenough. ‡ Greenough.

|| Bournon, Traité de Mineralogie, t. i. p. 268.

as in those of Derbyshire, where it is associated with galena *.

Use.

Like common limestone, it is burnt and made into mortar, but it remains much longer caustic than quicklime from common limestone; and this is the cause of a very important difference between magnesian and common limestone, with regard to their employment in agriculture: Lime, from magnesian limestone, is termed *hot*, and when spread upon land in the same proportion as is generally practised with common quicklime, greatly impairs the fertility of the soil; and when used in a greater quantity, is said by Mr Tennant to prevent all vegetation †.

Observations.

A flexible variety of Dolomite occurs in England. The following account contains all the information I possess in regard to it:—

Flexible

* Bournon, Traité de Mineralogie, t. i. p. 208.

† In regard to this limestone, Dr Thomson has the following remarks: "This magnesian limestone has long been burnt in prodigious quantity in the neighbourhood of Sunderland, and sent coastwise, both to the north and to the south. It goes in great abundance to Aberdeenshire. As no complaints have ever been made of its being injurious, when employed as a manure, it would be curious to know whether this circumstance be owing to the soil on which it is put, or to the small quantity of it used, in consequence of its price, occasioned by its long carriage; for it appears, from Mr Tennant's statement, that at Ferrybridge, the farmers are aware that it does not answer as a manure so well as pure carbonate of lime."—*Annals of Philosophy*, vol. iv. p. 418.

Flexible Dolomite.

External Characters.

Its colour is yellowish-grey, passing into cream-yellow. It occurs massive. It is dull. The fracture is earthy in the small, and slaty in the large. It is opaque. It yields readily to the knife, but with difficulty to the nail. In thin plates it is uncommonly flexible. Specific gravity, 2.544, *Thomson*. This is probably below the truth, as the stone is porous.

Chemical Characters.

It dissolves in acids as readily as common carbonate of lime.

Constituent Parts.

Carbonate of Lime, -	62.00
Carbonate of Magnesia,	35.96
Insoluble matter, -	1.60
Loss, - - - - -	0.44
	100.00

Thomson, Annals of Phil. vol. iv. p. 418.

Geographic Situation.

It occurs about three miles from Tinmouth Castle.

Observations.

This curious mineral was discovered by my intelligent friend Mr Nicol, Lecturer on Natural Philosophy. To that gentleman I am indebted for the following particulars in

in regard to it. He finds, that its flexibility is considerably influenced by the quantity of water contained in it. When saturated with water, it is remarkably flexible; as the evaporation goes on, it becomes more and more rigid, until the water be reduced to a certain limit, when the flexibility becomes scarcely distinguishable. From this point, however, the flexibility gradually increases, as the moisture diminishes; and as soon as the water is completely exhaled, it becomes nearly as flexible as it was when saturated with that fluid.

Second Kind.

Columnar Dolomite.

Stänglicher Dolomit, *Klaproth.*

Stänglicher Dolomit, *Klaproth*, Mag. der Gesellsch. Naturf. Freunde, b. v. s. 402.

External Characters.

Its colour is pale greyish-white.

It occurs massive, and in thin, long, and straight prismatic concretions.

It has an imperfect cleavage.

The fracture is uneven.

The lustre is vitreous, inclining to pearly.

It breaks into acicular-shaped fragments.

It is feebly translucent.

It is brittle.

Specific gravity 2.765.

Constituent

Constituent Parts.

From the Mine Tschistagowskoy.	
Carbonate of Lime, -	51
Carbonate of Magnesia, -	47
Carbonated Hydrate of Iron,	1
	99

Klaproth, Chem. Abhandl. s. 328.

Geognostic and Geographic Situations.

It occurs in serpentine in the mine Tschistagowskoy, on the river Mjafs, in the Government of Orenburg in Russia.

Observations.

It was at one time considered to be a variety of Strontianite; but in external characters, it is much more nearly allied to Tremolite.

Third Kind.

Compact Dolomite, or Gurhofite.

Gurhofian, *Karsten.*

Gurhofian, *Klaproth,* in Magazin der Gesellch. der Naturf. Freünde, b. i. s. 257.—Gurofian, *Karsten,* Tabel. s. 50. *Id. Klap.* Beit. b. v. s. 103. *Id. Lenz,* b. ii. s. 724.

External Characters.

Its colour is snow-white.

It occurs massive.

It

It is dull.

The fracture is flat conchoidal, passing to even.

The fragments are indeterminate angular, and sharp-edged.

It is slightly translucent on the edges.

It is hard, bordering on semihard.

It is brittle, and rather difficultly frangible.

Specific gravity, 2.7600, *Karsten*.

Chemical Characters.

When pounded, and thrown into diluted and heated nitrous acid, it is completely dissolved with effervescence.

Constituent Parts.

Carbonate of Lime, -	70.50
Carbonate of Magnesia,	29.50
	100.00

Klaproth, Gesellsch. N. Fr. b. i. s. 258.

Geognostic and Geographic Situations.

It occurs in veins in serpentine rocks, between Gurhof and Aggsbach, in Lower Austria.

Observations.

1. The name Gurhofit, sometimes given to this mineral, is from the place near which it was found.

2. It was at one time considered as a variety of semi-opal; but its greater weight distinguishes it from that mineral.

Second

correctly copied: from the nature of the stone, we may predict, that its duration will not be longer than that of the original. Both Portland and Bath stone varies much in quality. In buildings constructed of this stone, we may frequently observe some of the stones black, and others white. The black stones are those which are more compact and durable, and preserve their coating of smoke; the white stones are decomposing, and presenting a fresh surface, as if they had been recently scraped *. Roestone is also used as a manure, but when burnt into quicklime, the marly varieties afford rather an indifferent mortar; but those mixed with sand a better mortar.

Observations.

1. It passes into Sandstone, Compact Limestone, and Marl.

2. Some naturalists, as Daubenton, Saussure, Spallanzani, and Gillet Lamont, conjecture, that Roestone is carbonate of lime, which has been granulated in the manner of gunpowder, by the action of water: the most plausible opinion is that which attributes the formation of this mineral to crystallization from a state of solution.

Third Subspecies.

Chalk.

Kreide, *Werner*.

Creta alba, *Wall.* t. i. p. 27.—Kreide, *Wid.* s. 492.—Chalk, *Kirw.* vol. i. p. 77.—Kreide, *Estner,* b. ii. s. 917. *Id. Emm.* b. i.

* Aikin.

The fragments are rather blunt-edged.

It is translucent.

It is semihard.

It is brittle.

Specific gravity 2.885.

Chemical Characters.

It dissolves slowly, and with little effervescence, in nitrous acid; but more rapidly, and with increased effervescence, when the acid is heated.

Constituent Parts.

Carbonate of Lime,	53.00
Carbonate of Magnesia, -	42.50
Carbonate of Iron, with a little Manganese, -	3.00
	98.50

Klaproth, Beit. b. iii. s. 296.

Geognostic and Geographic Situations.

It is found at Miemo in Tuscany, imbedded in gypsum; at Hall in the Tyrol, imbedded in muriate of soda; and Gieseké met with it in kidneys, along with wavellite, arragonite, and calcedony, in decomposed wacke, at Kannioak, in Omenaksfiord in Greenland.

Observations.

This mineral was first observed by the late Dr Thompson of Naples, who sent specimens of it to Klaproth for analysis. It is named *Miemite*, after the place where it was discovered.

Second

Second Kind.

Prismatic Miemite.

Stänglicher Bitterspath, *Klaproth*.

Strahliger Kalkstein, *Von Schlottheim*, Hoff's Magaz. fur die Gesammte Mineralogie, b. i. s. 156.—Stänglicher Bitterspath, *Klaproth*, b. iii. s. 297. *Id. Leonhard*, Tabel. s. 36. *Id. Haus*, s. 128. *Id. Lenz*, b. ii. s. 712. *Id. Oken*, b. i. s. 393.

External Characters.

Its colour is asparagus-green, olive-green, and oil-green.

It occurs in prismatic distinct concretions, and crystallized in flat rhomboids, which are deeply truncated on all the edges.

The crystals are small, and very small, and sometimes they form only drusy crusts.

Internally it is shining and vitreous.

The fracture passes from concealed foliated to splintery.

The fragments are rather blunt-edged.

It is strongly translucent.

It is as hard as the granular miemite.

Specific gravity 2.885, *Karsten*.

Chemical Characters.

It dissolves slowly, and with but feeble effervescence, in nitrous acid.

Constituent

Constituent Parts.

Lime,	-	33.00
Magnesia,		14.50
Oxide of Iron,		2.50
Carbonic Acid,	-	47.25
Water and Loss,	-	2.75
		100

Klaproth, Beit. b. iii. s. 303.

Geognostic and Geographic Situations.

It occurs in cobalt veins that traverse sandstone at Glücksbrunn in Gotha, and at Beska in Servia, on the frontier of Turkey.

Third Subspecies.

Brown-Spar, or Pearl-Spar.

Braunspath, *Werner*.

This species is divided into two kinds, viz. Foliated Brown-Spar, and Columnar Brown-Spar.

First Kind.

Foliated Brown-Spar.

Blättriger Braunspath, *Werner*.

Spath perlé, *Romé de Lisle*, t. i. p. 605.—Braunspath, *Wid.* s. 515.—Sidero-calcite, *Kirw.* vol. i. p. 105.—Braunspath, *Estner*, b. ii. s. 999. *Id. Emm.* b. i. s. 79.—Brunispato, *Nap.* p. 356.

p. 356.—Le Spath brunissant, ou le Braunspath, *Broch.* t. i. p. 563.—Gemeiner Braunspath, *Reuss*, b. i. s. 50. *Id. Lud.* b. i. s. 153. *Id. Suck.* 1r th. s. 630. *Id. Bert.* s. 118. *Id. Mohs*, b. ii. s. 108.—Spathiger Braunkalk, *Hab.* s. 82.—Chaux carbonatée manganesifere, *Lucas*, p. 8.—Spathiger Braunkalk, *Leonhard*, Tabel. s. 35.—Chaux carbonatée brunissante, *Brong.* t. i. p. 237.—Gemeiner Braunspath, *Karsten*, Tabel. s. 50.—Chaux carbonatée ferro-mangamesienne, *Bournon*, Traité, t. i. p. 277.—Pearl-Spar, *Kid*, vol. i. p. 56.—Chaux carbonatée ferro-manganesifere, *Haüy*, Tabl. p. 5.—Gemeiner Braunkalk, *Lenz*, b. ii. s. 717.—Gemeiner Braunspath, *Oken*, b. i. s. 394.—Blättriger Braunspath, *Hoff.* b. iii. s. 49.

External Characters.

Its colours are flesh-red and brownish-red.

It often occurs massive, also disseminated, seldom globular, stalactitic, reniform, with tabular and pyramidal impressions; also in distinct concretions, which are granular, and rarely thin and straight lamellar; and frequently crystallized.

Its primitive form is a rhomboid, in which the obtuse angle is 107° 22′. The following are the secondary figures:

1. Rhomboid, in which the faces are sometimes cylindrically convex, sometimes cylindrically concave.
2. Lens, both common and saddle-shaped.

It also occurs in rhomboidal six-sided pyramidal supposititious crystals.

The true crystals are generally small and very small; the supposititious crystals large and middle-sized, and are either hollow, or lined with calcareous-spar.

The

in regard to it. He finds, that its flexibility is considerably influenced by the quantity of water contained in it. When saturated with water, it is remarkably flexible; as the evaporation goes on, it becomes more and more rigid, until the water be reduced to a certain limit, when the flexibility becomes scarcely distinguishable. From this point, however, the flexibility gradually increases, as the moisture diminishes; and as soon as the water is completely exhaled, it becomes nearly as flexible as it was when saturated with that fluid.

Second Kind.

Columnar Dolomite.

Stänglicher Dolomit, *Klaproth.*

Stänglicher Dolomit, *Klaproth*, Mag. der Gesellsch. Naturf. Freunde, b. v. s. 402.

External Characters.

Its colour is pale greyish-white.

It occurs massive, and in thin, long, and straight prismatic concretions.

It has an imperfect cleavage.

The fracture is uneven.

The lustre is vitreous, inclining to pearly.

It breaks into acicular-shaped fragments.

It is feebly translucent.

It is brittle.

Specific gravity 2.765.

Constituent

Constituent Parts.

From the Mine Tschistagowskoy.

Carbonate of Lime,	-	51
Carbonate of Magnesia,	-	47
Carbonated Hydrate of Iron,		1
		99

Klaproth, Chem. Abhandl. s. 328.

Geognostic and Geographic Situations.

It occurs in serpentine in the mine Tschistagowskoy, on the river Mjafs, in the Government of Orenburg in Russia.

Observations.

It was at one time considered to be a variety of Strontianite; but in external characters, it is much more nearly allied to Tremolite.

Third Kind.

Compact Dolomite, or Gurhofite.

Gurhofian, *Karsten.*

Gurhofian, *Klaproth*, in Magazin der Gesellch. der Naturf. Freünde, b. i. s. 257.—Gurofian, *Karsten*, Tabel. s. 50. *Id.* *Klap*. Beit. b. v. s. 103. *Id. Lenz*, b. ii. s. 724.

External Characters.

Its colour is snow-white.

It occurs massive.

It

in regard to it. He finds, that its flexibility is considerably influenced by the quantity of water contained in it. When saturated with water, it is remarkably flexible; as the evaporation goes on, it becomes more and more rigid, until the water be reduced to a certain limit, when the flexibility becomes scarcely distinguishable. From this point, however, the flexibility gradually increases, as the moisture diminishes; and as soon as the water is completely exhaled, it becomes nearly as flexible as it was when saturated with that fluid.

Second Kind.

Columnar Dolomite.

Stänglicher Dolomit, *Klaproth*.

Stänglicher Dolomit, *Klaproth*, Mag. der Gesellsch. Naturf. Freunde, b. v. s. 402.

External Characters.

Its colour is pale greyish-white.

It occurs massive, and in thin, long, and straight prismatic concretions.

It has an imperfect cleavage.

The fracture is uneven.

The lustre is vitreous, inclining to pearly.

It breaks into acicular-shaped fragments.

It is feebly translucent.

It is brittle.

Specific gravity 2.765.

Constituent

Constituent Parts.

From the Mine Tschistagowskoy.

Carbonate of Lime, -	51
Carbonate of Magnesia, -	47
Carbonated Hydrate of Iron,	1
	99

Klaproth, Chem. Abhandl. s. 328.

Geognostic and Geographic Situations.

It occurs in serpentine in the mine Tschistagowskoy, on the river Mjafs, in the Government of Orenburg in Russia.

Observations.

It was at one time considered to be a variety of Strontianite; but in external characters, it is much more nearly allied to Tremolite.

Third Kind.

Compact Dolomite, or Gurhofite.

Gurhofian, *Karsten.*

Gurhofian, *Klaproth*, in Magazin der Gesellch. der Naturf. Freünde, b. i. s. 257.—Gurofian, *Karsten*, Tabel. s. 50. *Id. Klap.* Beit. b. v. s. 103. *Id. Lenz*, b. ii. s. 724.

External Characters.

Its colour is snow-white.

It occurs massive.

It

Stromeyer says that fibrous limestone contains some *per cents.* of gypsum.

Geognostic and Geographic Situations.

It occurs in thin layers in clay-slate at Aldstone Moore in Cumberland: in layers and veins in the middle district of Scotland, as in Fifeshire. On the Continent, at Potschappel, near Dresden; and at Schneeberg, also in Saxony.

Uses.

It is sometimes cut into necklaces, crosses, and other ornamental articles.

Second Kind.

Fibrous Calc-Sinter *.

Fasriger Kalksinter, *Werner.*

Sintricher fasriger Kalkstein, *Reuss*, b. ii. 2. s. 306.—Kalksinter, *Lud.* b. i. s. 150. *Id. Suck.* 1r th. s. 618. *Id. Bert.* s. 93. *Id. Mohs*, b. ii. s. 86. *Id. Hab.* s. 78.—Fasriger Kalksinter, *Leonhard*, Tabel. s. 34.—Sintriger Kalkstein, *Karsten*, Tabel. s. 50.—Chaux carbonatée concretionnée, *Haüy*, Tabl. p. 4.—Sintricher Kalkstein, *Lenz*, b. ii. s. 751.—Fasriger Kalksinter, *Hoff.* b. iii. s. 32.

External

* This is the Alabaster of the ancients, and is by the moderns named *Calcareous Alabaster*, to distinguish it from another mineral, gypsum, which they name *Gypseous Alabaster*.

External Characters.

The principal colour is white, of which the following are the varieties, viz. yellowish, greenish, greyish, reddish, and snow white; from yellowish-white it passes into wine, wax, and honey yellow, and into a kind of reddish-brown, passing into clove-brown; from greyish-white it passes into yellowish and pearl grey, and from this latter into reddish, seldomer into flesh and peach-blossom red, and brownish-red; lastly, it passes from greenish-white into asparagus, siskin, mountain, verdigris green, and sky-blue.

The peach-blossom colour is owing to cobalt; the flesh-red to manganese; the verdigris-green to copper; the siskin-green to nickel, and the brown to iron.

It is sometimes concentrically and reniformly striped; or it is spotted or clouded.

It occurs massive, stalactitic, globular, tubular, claviform, fruticose, curtain-shaped, cock's-comb-shaped, coralloidal, reniform, and tuberose; also in distinct concretions, which are fibrous, and these are straight, seldom curved, and sometimes scopiform or stellular; also in reniform curved lamellar concretions, and seldom in large and coarse angulo-granular concretions; very rarely we observe the longish external shapes, as the stalactitic, terminated by a three-sided pyramidal crystallization.

The surface is generally rough, and seldom fine drusy.

Internally it is glimmering, which passes on the one side into dull, on the other into glistening; and the lustre is pearly.

The fracture is fine splintery.

The fragments are splintery, or wedge-shaped.

It is translucent, or only translucent on the edges.

The fragments are rather blunt-edged.

It is translucent.

It is semihard.

It is brittle.

Specific gravity 2.885.

Chemical Characters.

It dissolves slowly, and with little effervescence, in nitrous acid; but more rapidly, and with increased effervescence, when the acid is heated.

Constituent Parts.

Carbonate of Lime,	53.00
Carbonate of Magnesia, -	42.50
Carbonate of Iron, with a little Manganese, -	3.00
	98.50

Klaproth, Beit. b. iii. s. 296.

Geognostic and Geographic Situations.

It is found at Micmo in Tuscany, imbedded in gypsum; at Hall in the Tyrol, imbedded in muriate of soda; and Gieseké met with it in kidneys, along with wavellite, arragonite, and calcedony, in decomposed wacke, at Kannioak, in Omenaksfiord in Greenland.

Observations.

This mineral was first observed by the late Dr Thompson of Naples, who sent specimens of it to Klaproth for analysis. It is named *Miemite*, after the place where it was discovered.

Second

Second Kind.

Prismatic Miemite.

Stänglicher Bitterspath, *Klaproth*.

Strahliger Kalkstein, *Von Schlottheim*, Hoff's Magaz. fur die Gesammte Mineralogie, b. i. s. 156.—Stänglicher Bitterspath, *Klaproth*, b. iii. s. 297. *Id. Leonhard*, Tabel. s. 36. *Id. Haus*, s. 128. *Id. Lenz*, b. ii. s. 712. *Id. Oken*, b. i. s. 393.

External Characters.

Its colour is asparagus-green, olive-green, and oil-green.

It occurs in prismatic distinct concretions, and crystallized in flat rhomboids, which are deeply truncated on all the edges.

The crystals are small, and very small, and sometimes they form only drusy crusts.

Internally it is shining and vitreous.

The fracture passes from concealed foliated to splintery.

The fragments are rather blunt-edged.

It is strongly translucent.

It is as hard as the granular miemite.

Specific gravity 2.885, *Karsten*.

Chemical Characters.

It dissolves slowly, and with but feeble effervescence, in nitrous acid.

Constituent

Constituent Parts.

Lime, -	33.00
Magnesia,	14.50
Oxide of Iron,	2.50
Carbonic Acid, -	47.25
Water and Loss, -	2.75
	100

Klaproth, Beit. b. iii. s. 303.

Geognostic and Geographic Situations.

It occurs in cobalt veins that traverse sandstone at Glücksbrunn in Gotha, and at Beska in Servia, on the frontier of Turkey.

Third Subspecies.

Brown-Spar, or Pearl-Spar.

Braunspath, *Werner.*

This species is divided into two kinds, viz. Foliated Brown-Spar, and Columnar Brown-Spar.

First Kind.

Foliated Brown-Spar.

Blättriger Braunspath, *Werner.*

Spath perlé, *Romé de Lisle*, t. i. p. 605.—Braunspath, *Wid.* s. 515.—Sidero-calcite, *Kirw.* vol. i. p. 105.—Braunspath, *Estner*, b. ii. s. 999. *Id. Emm.* b. i. s. 79.—Brunispato, *Nap.* p. 356.

p. 356.—Le Spath brunissant, ou le Braunspath, *Broch.* t. i. p. 563.—Gemeiner Braunspath, *Reuss,* b. i. s. 50. *Id. Lud.* b. i. s. 153. *Id. Suck.* 1r th. s. 630. *Id. Bert.* s. 118. *Id. Mohs,* b. ii. s. 108.—Spathiger Braunkalk, *Hab.* s. 82.—Chaux carbonatée manganesifere, *Lucas,* p. 8.—Spathiger Braunkalk, *Leonhard,* Tabel. s. 35.—Chaux carbonatée brunissante, *Brong.* t. i. p. 237.—Gemeiner Braunspath, *Karsten,* Tabel. s. 50.—Chaux carbonatée ferro-manganesienne, *Bournon,* Traité, t. i. p. 277.—Pearl-Spar, *Kid,* vol. i. p. 56.—Chaux carbonatée ferro-manganesifere, *Haüy,* Tabl. p. 5.—Gemeiner Braunkalk, *Lenz,* b. ii. s. 717.—Gemeiner Braunspath, *Oken,* b. i. s. 394.—Blättriger Braunspath, *Hoff.* b. iii. s. 49.

External Characters.

Its colours are flesh-red and brownish-red.

It often occurs massive, also disseminated, seldom globular, stalactitic, reniform, with tabular and pyramidal impressions; also in distinct concretions, which are granular, and rarely thin and straight lamellar; and frequently crystallized.

Its primitive form is a rhomboid, in which the obtuse angle is 107° 22′. The following are the secondary figures:

1. Rhomboid, in which the faces are sometimes cylindrically convex, sometimes cylindrically concave.
2. Lens, both common and saddle-shaped.

It also occurs in rhomboidal six-sided pyramidal supposititious crystals.

The true crystals are generally small and very small; the supposititious crystals large and middle-sized, and are either hollow, or lined with calcareous-spar.

The

as in those of Derbyshire, where it is associated with galena *.

Use.

Like common limestone, it is burnt and made into mortar, but it remains much longer caustic than quicklime from common limestone; and this is the cause of a very important difference between magnesian and common limestone, with regard to their employment in agriculture: Lime, from magnesian limestone, is termed *hot*, and when spread upon land in the same proportion as is generally practised with common quicklime, greatly impairs the fertility of the soil; and when used in a greater quantity, is said by Mr Tennant to prevent all vegetation †.

Observations.

A flexible variety of Dolomite occurs in England. The following account contains all the information I possess in regard to it:—

Flexible

* Bournon, Traité de Mineralogie, t. i. p. 208.

† In regard to this limestone, Dr Thomson has the following remarks: "This magnesian limestone has long been burnt in prodigious quantity in the neighbourhood of Sunderland, and sent coastwise, both to the north and to the south. It goes in great abundance to Aberdeenshire. As no complaints have ever been made of its being injurious, when employed as a manure, it would be curious to know whether this circumstance be owing to the soil on which it is put, or to the small quantity of it used, in consequence of its price, occasioned by its long carriage; for it appears, from Mr Tennant's statement, that at Ferrybridge, the farmers are aware that it does not answer as a manure so well as pure carbonate of lime."—*Annals of Philosophy*, vol. iv. p. 418.

Flexible Dolomite.

External Characters.

Its colour is yellowish-grey, passing into cream-yellow. It occurs massive. It is dull. The fracture is earthy in the small, and slaty in the large. It is opaque. It yields readily to the knife, but with difficulty to the nail. In thin plates it is uncommonly flexible. Specific gravity, 2.544, *Thomson*. This is probably below the truth, as the stone is porous.

Chemical Characters.

It dissolves in acids as readily as common carbonate of lime.

Constituent Parts.

Carbonate of Lime, -	62.00
Carbonate of Magnesia,	35.96
Insoluble matter, -	1.60
Loss, - - - - -	0.44
	100.00

Thomson, Annals of Phil. vol. iv. p. 418.

Geographic Situation.

It occurs about three miles from Tinmouth Castle.

Observations.

This curious mineral was discovered by my intelligent friend Mr Nicol, Lecturer on Natural Philosophy. To that gentleman I am indebted for the following particulars in

in regard to it. He finds, that its flexibility is considerably influenced by the quantity of water contained in it. When saturated with water, it is remarkably flexible; as the evaporation goes on, it becomes more and more rigid, until the water be reduced to a certain limit, when the flexibility becomes scarcely distinguishable. From this point, however, the flexibility gradually increases, as the moisture diminishes; and as soon as the water is completely exhaled, it becomes nearly as flexible as it was when saturated with that fluid.

Second Kind.

Columnar Dolomite.

Stänglicher Dolomit, *Klaproth.*

Stänglicher Dolomit, *Klaproth*, Mag. der Gesellsch. Naturf. Freunde, b. v. s. 402.

External Characters.

Its colour is pale greyish-white.

It occurs massive, and in thin, long, and straight prismatic concretions.

It has an imperfect cleavage.

The fracture is uneven.

The lustre is vitreous, inclining to pearly.

It breaks into acicular-shaped fragments.

It is feebly translucent.

It is brittle.

Specific gravity 2.765.

Constituent

Constituent Parts.

From the Mine Tschistagowskoy.	
Carbonate of Lime, -	51
Carbonate of Magnesia, -	47
Carbonated Hydrate of Iron,	1
	99

Klaproth, Chem. Abhandl. s. 328.

Geognostic and Geographic Situations.

It occurs in serpentine in the mine Tschistagowskoy, on the river Mjafs, in the Government of Orenburg in Russia.

Observations.

It was at one time considered to be a variety of Strontianite; but in external characters, it is much more nearly allied to Tremolite.

Third Kind.

Compact Dolomite, or Gurhofite.

Gurhofian, *Karsten.*

Gurhofian, *Klaproth*, in Magazin der Gesellch. der Naturf. Freünde, b. i. s. 257.—Gurofian, *Karsten*, Tabel. s. 50. *Id. Klap.* Beit. b. v. s. 103. *Id. Lenz*, b. ii. s. 724.

External Characters.

Its colour is snow-white.

It occurs massive.

It is dull.

The fracture is flat conchoidal, passing to even.

The fragments are indeterminate angular, and sharp-edged.

It is slightly translucent on the edges.

It is hard, bordering on semihard.

It is brittle, and rather difficultly frangible.

Specific gravity, 2.7600, *Karsten*.

Chemical Characters.

When pounded, and thrown into diluted and heated nitrous acid, it is completely dissolved with effervescence.

Constituent Parts.

Carbonate of Lime, -	70.50
Carbonate of Magnesia,	29.50
	100.00

Klaproth, Gesellsch. N. Fr. b. i. s. 258.

Geognostic and Geographic Situations.

It occurs in veins in serpentine rocks, between Gurhof and Aggsbach, in Lower Austria.

Observations.

1. The name Gurhofit, sometimes given to this mineral, is from the place near which it was found.

2. It was at one time considered as a variety of semi-opal; but its greater weight distinguishes it from that mineral.

Second

probable, is the almost constant occurrence of particles of quartz-sand in the centre of these globular concretions. In some rare instances, the centre of the concretions is empty. A mineral resembling peastone, occurs at the Baths of St Philippi in Tuscany; also at Perscheesberg in Silesia; and in Hungary.

Uses.

It is sometimes cut into plates, for ornamental purposes.

Eighth Subspecies.

Slate-Spar.

Schieferspath, *Werner.*

Shieferspath, *Wid.* s. 510.—Argentine, *Kirw.* vol. i. p. 105.—Schisto-spatho, *Nap.* p. 355.—Shifferspath, *Lam.* t. i. p. 385.—Le Spath schisteux, ou le Schieferspath, *Broch.* t. i. p. 558.—Schieferspath, *Reuss,* b. ii. s. 50. *Id. Lud.* b. i. s. 152. *Id. Suck.* 1r th. s. 626. *Id. Bert.* s. 95. *Id. Mohs,* b. ii. s. 3. *Id. Hab.* s. 81. *Id. Leonhard,* Tabel. s. 34.—Chaux carbonatée nacré argentine, *Brong.* t. i. p. 232.—Verhærteter Aphrit, *Karsten,* Tabel. s. 50.—Chaux carbonatée nacré primitive, *Haüy,* Tabl. p. 6.—Schieferspath, *Lenz,* b. ii. s. 761. *Id. Hoff.* b. iii. s. 46.

External Characters.

Its colours are greenish-white, reddish-white, yellowish-white, greyish-white, and snow-white.

It occurs massive, also in distinct concretions, which are generally curved lamellar, and sometimes coarse and large granular.

The

The lustre is intermediate between shining and glistening, and is pearly.

The fragments are either indeterminate angular and blunt-edged, or are tabular.

It is feebly translucent, or only translucent on the edges.

It is soft.

It is intermediate between sectile and brittle.

It is easily frangible.

It feels rather greasy.

Specific gravity, 2.647, *Kirwan.*—2.474, *Blumenbach.*—2.6300, *La Metherie.*—2.611, *Breithaupt.*

Chemical Characters.

It effervesces very violently with acids; but is infusible before the blowpipe.

Constituent Parts.

From Bremsgrün.		From Kongsberg.	
Lime, -	55.00	Lime, -	56.00
Carbonic Acid,	41.66	Carbonic Acid,	39.33
Oxide of Manganese,	3.00	Silica, -	1.66
		Oxide of Iron,	1.00
	Bucholz.	Water, -	2.00
			Sucrsce.

Geognostic Situation.

It occurs in primitive limestone, along with calcareous spar, brown-spar, fluor-spar, and galena; in metalliferous beds, associated with magnetic ironstone, galena, and blende; and in veins, along with tinstone.

Geographic

Second Kind.

Prismatic Miemite.

Stänglicher Bitterspath, *Klaproth.*

Strahliger Kalkstein, *Von Schlottheim,* Hoff's Magaz. fur die Gesammte Mineralogie, b. i. s. 156.—Stänglicher Bitterspath, *Klaproth,* b. iii. s. 297. *Id. Leonhard,* Tabel. s. 36. *Id. Haus,* s. 128. *Id. Lenz,* b. ii. s. 712. *Id. Oken,* b. i. s. 393.

External Characters.

Its colour is asparagus-green, olive-green, and oil-green.

It occurs in prismatic distinct concretions, and crystallized in flat rhomboids, which are deeply truncated on all the edges.

The crystals are small, and very small, and sometimes they form only drusy crusts.

Internally it is shining and vitreous.

The fracture passes from concealed foliated to splintery.

The fragments are rather blunt-edged.

It is strongly translucent.

It is as hard as the granular miemite.

Specific gravity 2.885, *Karsten.*

Chemical Characters.

It dissolves slowly, and with but feeble effervescence, in nitrous acid.

Constituent

as in those of Derbyshire, where it is associated with galena *.

Use.

Like common limestone, it is burnt and made into mortar, but it remains much longer caustic than quicklime from common limestone; and this is the cause of a very important difference between magnesian and common limestone, with regard to their employment in agriculture: Lime, from magnesian limestone, is termed *hot*, and when spread upon land in the same proportion as is generally practised with common quicklime, greatly impairs the fertility of the soil; and when used in a greater quantity, is said by Mr Tennant to prevent all vegetation †.

Observations.

A flexible variety of Dolomite occurs in England. The following account contains all the information I possess in regard to it:—

Flexible

* Bournon, Traité de Mineralogie, t. i. p. 208.

† In regard to this limestone, Dr Thomson has the following remarks: "This magnesian limestone has long been burnt in prodigious quantity in the neighbourhood of Sunderland, and sent coastwise, both to the north and to the south. It goes in great abundance to Aberdeenshire. As no complaints have ever been made of its being injurious, when employed as a manure, it would be curious to know whether this circumstance be owing to the soil on which it is put, or to the small quantity of it used, in consequence of its price, occasioned by its long carriage; for it appears, from Mr Tennant's statement, that at Ferrybridge, the farmers are aware that it does not answer as a manure so well as pure carbonate of lime."—*Annals of Philosophy*, vol. iv. p. 418.

Flexible Dolomite.

External Characters.

Its colour is yellowish-grey, passing into cream-yellow. It occurs massive. It is dull. The fracture is earthy in the small, and slaty in the large. It is opaque. It yields readily to the knife, but with difficulty to the nail. In thin plates it is uncommonly flexible. Specific gravity, 2.544, *Thomson*. This is probably below the truth, as the stone is porous.

Chemical Characters.

It dissolves in acids as readily as common carbonate of lime.

Constituent Parts.

Carbonate of Lime, -	62.00
Carbonate of Magnesia,	35.96
Insoluble matter, -	1.60
Loss, - - - - -	0.44
	100.00

Thomson, Annals of Phil. vol. iv. p. 418.

Geographic Situation.

It occurs about three miles from Tinmouth Castle.

Observations.

This curious mineral was discovered by my intelligent friend Mr Nicol, Lecturer on Natural Philosophy. To that gentleman I am indebted for the following particulars in

in regard to it. He finds, that its flexibility is considerably influenced by the quantity of water contained in it. When saturated with water, it is remarkably flexible; as the evaporation goes on, it becomes more and more rigid, until the water be reduced to a certain limit, when the flexibility becomes scarcely distinguishable. From this point, however, the flexibility gradually increases, as the moisture diminishes; and as soon as the water is completely exhaled, it becomes nearly as flexible as it was when saturated with that fluid.

Second Kind.

Columnar Dolomite.

Stänglicher Dolomit, *Klaproth*.

Stänglicher Dolomit, *Klaproth*, Mag. der Gesellsch. Naturf. Freunde, b. v. s. 402.

External Characters.

Its colour is pale greyish-white.

It occurs massive, and in thin, long, and straight prismatic concretions.

It has an imperfect cleavage.

The fracture is uneven.

The lustre is vitreous, inclining to pearly.

It breaks into acicular-shaped fragments.

It is feebly translucent.

It is brittle.

Specific gravity 2.765.

Con-

Constituent Parts.

From the Mine Tschistagowskoy.

Carbonate of Lime,	-	51
Carbonate of Magnesia,	-	47
Carbonated Hydrate of Iron,		1
		99

Klaproth, Chem. Abhandl. s. 328.

Geognostic and Geographic Situations.

It occurs in serpentine in the mine Tschistagowskoy, on the river Mjafs, in the Government of Orenburg in Russia.

Observations.

It was at one time considered to be a variety of Strontianite; but in external characters, it is much more nearly allied to Tremolite.

Third Kind.

Compact Dolomite, or Gurhofite.

Gurhofian, *Karsten.*

Gurhofian, *Klaproth,* in Magazin der Gesellch. der Naturf. Freünde, b. i. s. 257.—Gurofian, *Karsten,* Tabel. s. 50. *Id. Klap*. Beit. b. v. s. 103. *Id. Lenz,* b. ii. s. 724.

External Characters.

Its colour is snow-white.

It occurs massive.

It

It is dull.

The fracture is flat conchoidal, passing to even.

The fragments are indeterminate angular, and sharp-edged.

It is slightly translucent on the edges.

It is hard, bordering on semihard.

It is brittle, and rather difficultly frangible.

Specific gravity, 2.7600, *Karsten*.

Chemical Characters.

When pounded, and thrown into diluted and heated nitrous acid, it is completely dissolved with effervescence.

Constituent Parts.

Carbonate of Lime, -	70.50
Carbonate of Magnesia,	29.50
	100.00

Klaproth, Gesellsch. N. Fr. b. i. s. 258.

Geognostic and Geographic Situations.

It occurs in veins in serpentine rocks, between Gurhof and Aggsbach, in Lower Austria.

Observations.

1. The name Gurhofit, sometimes given to this mineral, is from the place near which it was found.

2. It was at one time considered as a variety of semi-opal; but its greater weight distinguishes it from that mineral.

Second

First Kind.

Compact Lucullite.

Dichter Lucullan, *John.*

Lapis suillus, *Wall.* t. i. p. 148.—Swinestone, *Kirw.* vol. i. p. 89. —Stinkstein, *Wid.* s. 521. *Id. Estner*, b. ii. s. 1023. *Id. Emm.* b. i. s. 487.—Pierre calcaire puante, ou Pierre puante, *Lam.* t. ii. p. 58.—Chaux carbonatée fetide, *Haüy*, t. ii. p. 188. —La pierre puante, *Broch.* t. i. p. 567.—Gemeiner Stinkstein, *Reuss*, b. ii. 2. s. 335. *Id. Lud.* b. i. s. 155. *Id. Suck.* 1r th. s. 638. *Id. Bert.* s. 111. *Id. Mohs*, b. ii. s. 126.—Gemeiner Stinkstein, *Leonhard*, Tabel. s. 36.—Chaux carbonatée fetide, *Brong.* t. i. p. 236.—Gemeiner Stinkstein, *Haus.* s. 128. *Id. Karsten*, Tabel. s. 50.—Swinestone, *Kid*, vol. i. p. 29.—Chaux carbonatée fetide, *Haüy*, Tabl. p. 6.; et Chaux carbonatée bituminifere, *Id.* p. 6.—Gemeiner Stinkstein, *Lenz*, b. ii. s. 767.—Dichter Stinkstein, *Oken*, b. i. s. 407. —Swinestone, *Aikin*, p. 162.

This kind is divided into Common Compact Lucullite or Black Marble, and Stinkstone.

a. Common Compact Lucullite, or Black Marble.

Dichter Lucullan; Schwarzer Marmor, *John*, Chemisches Laboratorium, t. ii. s. 227. *Id. Lenz*, b. ii. s. 765.

External Characters.

Its colour is greyish-black.

It occurs massive.

Internally it is strongly glimmering, inclining to glistening.

The

The fragments are rather blunt-edged.

It is translucent.

It is semihard.

It is brittle.

Specific gravity 2.885.

Chemical Characters.

It dissolves slowly, and with little effervescence, in nitrous acid; but more rapidly, and with increased effervescence, when the acid is heated.

Constituent Parts.

Carbonate of Lime,	53.00
Carbonate of Magnesia, -	42.50
Carbonate of Iron, with a little Manganese, -	3.00
	98.50

Klaproth, Beit. b. iii. s. 296.

Geognostic and Geographic Situations.

It is found at Miemo in Tuscany, imbedded in gypsum; at Hall in the Tyrol, imbedded in muriate of soda; and Gieseké met with it in kidneys, along with wavellite, arragonite, and calcedony, in decomposed wacke, at Kannioak, in Omenaksfiord in Greenland.

Observations.

This mineral was first observed by the late Dr Thompson of Naples, who sent specimens of it to Klaproth for analysis. It is named *Miemite*, after the place where it was discovered.

Second

Second Kind.

Prismatic Miemite.

Stänglicher Bitterspath, *Klaproth.*

Strahliger Kalkstein, *Von Schlottheim,* Hoff's Magaz. fur die Gesammte Mineralogie, b. i. s. 156.—Stänglicher Bitterspath, *Klaproth,* b. iii. s. 297. *Id. Leonhard,* Tabel. s. 36. *Id. Haus,* s. 128. *Id. Lenz,* b. ii. s. 712. *Id. Oken,* b. i. s. 393.

External Characters.

Its colour is asparagus-green, olive-green, and oil-green.

It occurs in prismatic distinct concretions, and crystallized in flat rhomboids, which are deeply truncated on all the edges.

The crystals are small, and very small, and sometimes they form only drusy crusts.

Internally it is shining and vitreous.

The fracture passes from concealed foliated to splintery.

The fragments are rather blunt-edged.

It is strongly translucent.

It is as hard as the granular miemite.

Specific gravity 2.885, *Karsten.*

Chemical Characters.

It dissolves slowly, and with but feeble effervescence, in nitrous acid.

Constituent

Constituent Parts.

Lime, -	33.00
Magnesia,	14.50
Oxide of Iron,	2.50
Carbonic Acid, -	47.25
Water and Loss, -	2.75
	100

Klaproth, Beit. b. iii. s. 303.

Geognostic and Geographic Situations.

It occurs in cobalt veins that traverse sandstone at Glücksbrunn in Gotha, and at Beska in Servia, on the frontier of Turkey.

Third Subspecies.

Brown-Spar, or Pearl-Spar.

Braunspath, *Werner*.

This species is divided into two kinds, viz. Foliated Brown-Spar, and Columnar Brown-Spar.

First Kind.

Foliated Brown-Spar.

Blättriger Braunspath, *Werner*.

Spath perlé, *Romé de Lisle*, t. i. p. 605.—Braunspath, *Wid.* s. 515.—Sidero-calcite, *Kirw.* vol. i. p. 105.—Braunspath, *Estner*, b. ii. s. 999. *Id. Emm.* b. i. s. 79.—Brunispato, *Nap.* p. 356.

p. 356.—Le Spath brunissant, ou le Braunspath, *Broch.* t. i. p. 563.—Gemeiner Braunspath, *Reuss*, b. i. s. 50. *Id. Lud.* b. i. s. 153. *Id. Suck.* 1r th. s. 630. *Id. Bert.* s. 118. *Id. Mohs*, b. ii. s. 108.—Spathiger Braunkalk, *Hab.* s. 82.—Chaux carbonatée manganesifere, *Lucas*, p. 8.—Spathiger Braunkalk, *Leonhard*, Tabel. s. 85.—Chaux carbonatée brunissante, *Brong.* t. i. p. 237.—Gemeiner Braunspath, *Karsten*, Tabel. s. 50.—Chaux carbonatée ferro-manganesienne, *Bournon*, Traité, t. i. p. 277.—Pearl-Spar, *Kid*, vol. i. p. 56.—Chaux carbonatée ferro-manganesifere, *Haüy*, Tabl. p. 5.—Gemeiner Braunkalk, *Lenz*, b. ii. s. 717.—Gemeiner Braunspath, *Oken*, b. i. s. 394.—Blättriger Braunspath, *Hoff.* b. iii. s. 49.

External Characters.

Its colours are flesh-red and brownish-red.

It often occurs massive, also disseminated, seldom globular, stalactitic, reniform, with tabular and pyramidal impressions; also in distinct concretions, which are granular, and rarely thin and straight lamellar; and frequently crystallized.

Its primitive form is a rhomboid, in which the obtuse angle is 107° 22′. The following are the secondary figures:

1. Rhomboid, in which the faces are sometimes cylindrically convex, sometimes cylindrically concave.
2. Lens, both common and saddle-shaped.

It also occurs in rhomboidal six-sided pyramidal supposititious crystals.

The true crystals are generally small and very small; the supposititious crystals large and middle-sized, and are either hollow, or lined with calcareous-spar.

The

in regard to it. He finds, that its flexibility is considerably influenced by the quantity of water contained in it. When saturated with water, it is remarkably flexible; as the evaporation goes on, it becomes more and more rigid, until the water be reduced to a certain limit, when the flexibility becomes scarcely distinguishable. From this point, however, the flexibility gradually increases, as the moisture diminishes; and as soon as the water is completely exhaled, it becomes nearly as flexible as it was when saturated with that fluid.

Second Kind.

Columnar Dolomite.

Stänglicher Dolomit, *Klaproth.*

Stänglicher Dolomit, *Klaproth,* Mag. der Gesellsch. Naturf. Freunde, b. v. s. 402.

External Characters.

Its colour is pale greyish-white.

It occurs massive, and in thin, long, and straight prismatic concretions.

It has an imperfect cleavage.

The fracture is uneven.

The lustre is vitreous, inclining to pearly.

It breaks into acicular-shaped fragments.

It is feebly translucent.

It is brittle.

Specific gravity 2.765.

Constituent

Constituent Parts.

From the Mine Tschistagowskoy.

Carbonate of Lime,	-	51
Carbonate of Magnesia,	-	47
Carbonated Hydrate of Iron,		1
		99

Klaproth, Chem. Abhandl. s. 328.

Geognostic and Geographic Situations.

It occurs in serpentine in the mine Tschistagowskoy, on the river Mjafs, in the Government of Orenburg in Russia.

Observations.

It was at one time considered to be a variety of Strontianite; but in external characters, it is much more nearly allied to Tremolite.

Third Kind.

Compact Dolomite, or Gurhofite.

Gurhofian, *Karsten.*

Gurhofian, *Klaproth,* in Magazin der Gesellch. der Naturf. Freünde, b. i. s. 257.—Gurofian, *Karsten,* Tabel. s. 50. *Id. Klap*. Beit. b. v. s. 103. *Id. Lenz,* b. ii. s. 724.

External Characters.

Its colour is snow-white.

It occurs massive.

It is dull.

The fracture is flat conchoidal, passing to even.

The fragments are indeterminate angular, and sharp-edged.

It is slightly translucent on the edges.

It is hard, bordering on semihard.

It is brittle, and rather difficultly frangible.

Specific gravity, 2.7600, *Karsten*.

Chemical Characters.

When pounded, and thrown into diluted and heated nitrous acid, it is completely dissolved with effervescence.

Constituent Parts.

Carbonate of Lime, -	70.50
Carbonate of Magnesia,	29.50
	100.00

Klaproth, Gesellsch. N. Fr. b. i. s. 258.

Geognostic and Geographic Situations.

It occurs in veins in serpentine rocks, between Gurhof and Aggsbach, in Lower Austria.

Observations.

1. The name Gurhofit, sometimes given to this mineral, is from the place near which it was found.

2. It was at one time considered as a variety of semiopal; but its greater weight distinguishes it from that mineral.

Secon

It affords a grey-coloured streak.

It is brittle, and easily frangible.

When rubbed, it emits a strongly fetid urinous smell.

Specific gravity, 2.653, 2.688, 2.703, *John*.

Chemical Characters.

When pounded and boiled in water, it gives out a hepatic odour, which continues but for a short time. The filtrated water possesses weak alkaline properties, and contains a small quantity of a muriatic and sulphuric salt. It does not appear to be affected by pure alkalies. It dissolves with effervescence in nitrous and muriatic acids, and leaves behind a coal-black or brownish-coloured residuum.

Constituent Parts.

From Stavern in Norway.		From Greenland.		From Garphytta, in Nericke in Sweden.	
Carbonic Acid,	41.50	Carbonic Acid,	41.53	Carbonic Acid,	41.75
Lime, -	53.37	Lime, -	53.00	Lime, -	54.00
Oxide of Manganese, -	0.75	Oxide of Manganese, -	1.00	Oxide of Manganese, -	0.50
Oxide of Iron,	1.25	Oxide of Iron,	0.75	Oxide of Iron,	0.75
Oxide of Carbon,	1.25	Oxide of Carbon,	1.00	Brown Oxide of Carbon, -	0.75
Sulphur, -	0.25	Sulphur, -	0.50	Sulphur, Alkali, Alkaline Muriate and Sulphate, Water,	2.25
Alumina, -	1.25	Alumina, -	0.75		
Silica, -	1.25	Silica, Alkali, Alkaline Muriate, Water, -	1.47		
Alkali, Alkaline Muriate, Water, Magnesia, Zirconia,	2.13				
	100.00		100.00		100.00
John, Chem. Laborat. b. ii. s. 246.		*John*, ib. s. 248.		*John*, ib. s. 250.	

Geognostic

The fragments are rather blunt-edged.

It is translucent.

It is semihard.

It is brittle.

Specific gravity 2.885.

Chemical Characters.

It dissolves slowly, and with little effervescence, in nitrous acid; but more rapidly, and with increased effervescence, when the acid is heated.

Constituent Parts.

Carbonate of Lime,	53.00
Carbonate of Magnesia, -	42.50
Carbonate of Iron, with a little Manganese, -	3.00
	98.50

Klaproth, Beit. b. iii. s. 296.

Geognostic and Geographic Situations.

It is found at Miemo in Tuscany, imbedded in gypsum; at Hall in the Tyrol, imbedded in muriate of soda; and Gieseké met with it in kidneys, along with wavellite, arragonite, and calcedony, in decomposed wacke, at Kannioak, in Omenaksfiord in Greenland.

Observations.

This mineral was first observed by the late Dr Thompson of Naples, who sent specimens of it to Klaproth for analysis. It is named *Miemite*, after the place where it was discovered.

Second

Second Kind.

Prismatic Miemite.

Stänglicher Bitterspath, *Klaproth.*

Strahliger Kalkstein, *Von Schlottheim,* Hoff's Magaz. fur die Gesammte Mineralogie, b. i. s. 156.—Stänglicher Bitterspath, *Klaproth,* b. iii. s. 297. *Id. Leonhard,* Tabel. s. 36. *Id. Haus,* s. 128. *Id. Lenz,* b. ii. s. 712. *Id. Oken,* b. i. s. 393.

External Characters.

Its colour is asparagus-green, olive-green, and oil-green.

It occurs in prismatic distinct concretions, and crystallized in flat rhomboids, which are deeply truncated on all the edges.

The crystals are small, and very small, and sometimes they form only drusy crusts.

Internally it is shining and vitreous.

The fracture passes from concealed foliated to splintery.

The fragments are rather blunt-edged.

It is strongly translucent.

It is as hard as the granular miemite.

Specific gravity 2.885, *Karsten.*

Chemical Characters.

It dissolves slowly, and with but feeble effervescence, in nitrous acid.

Constituent

Constituent Parts.

Lime,	-	33.00
Magnesia,		14.50
Oxide of Iron,		2.50
Carbonic Acid,	-	47.25
Water and Loss,	-	2.75
		100

Klaproth, Beit. b. iii. s. 303.

Geognostic and Geographic Situations.

It occurs in cobalt veins that traverse sandstone at Glücksbrunn in Gotha, and at Beska in Servia, on the frontier of Turkey.

Third Subspecies.

Brown-Spar, or Pearl-Spar.

Braunspath, *Werner.*

This species is divided into two kinds, viz. Foliated Brown-Spar, and Columnar Brown-Spar.

First Kind.

Foliated Brown-Spar.

Blättriger Braunspath, *Werner.*

Spath perlé, *Romé de Lisle,* t. i. p. 605.—Braunspath, *Wid.* s. 515.—Sidero-calcite, *Kirw.* vol. i. p. 105.—Braunspath, *Estner,* b. ii. s. 999. *Id. Emm.* b. i. s. 79.—Brunispato, *Nap.* p. 356.

p. 356.—Le Spath brunissant, ou le Braunspath, *Broch.* t. i. p. 563.—Gemeiner Braunspath, *Reuss*, b. i. s. 50. *Id. Lud.* b. i. s. 153. *Id. Suck.* 1r th. s. 630. *Id. Bert.* s. 118. *Id. Mohs*, b. ii. s. 108.—Spathiger Braunkalk, *Hab.* s. 82.—Chaux carbonatée manganesifere, *Lucas*, p. 8.—Spathiger Braunkalk, *Leonhard*, Tabel. s. 85.—Chaux carbonatée brunissante, *Brong.* t. i. p. 237.—Gemeiner Braunspath, *Karsten*, Tabel. s. 50.—Chaux carbonatée ferro-manganesienne, *Bournon*, Traité, t. i. p. 277.—Pearl-Spar, *Kid*, vol. i. p. 56.—Chaux carbonatée ferro-manganesifere, *Haüy*, Tabl. p. 5.—Gemeiner Braunkalk, *Lenz*, b. ii. s. 717.—Gemeiner Braunspath, *Oken*, b. i. s. 394.—Blättriger Braunspath, *Hoff.* b. iii. s. 49.

External Characters.

Its colours are flesh-red and brownish-red.

It often occurs massive, also disseminated, seldom globular, stalactitic, reniform, with tabular and pyramidal impressions; also in distinct concretions, which are granular, and rarely thin and straight lamellar; and frequently crystallized.

Its primitive form is a rhomboid, in which the obtuse angle is 107° 22′. The following are the secondary figures:

1. Rhomboid, in which the faces are sometimes cylindrically convex, sometimes cylindrically concave.
2. Lens, both common and saddle-shaped.

It also occurs in rhomboidal six-sided pyramidal supposititious crystals.

The true crystals are generally small and very small; the supposititious crystals large and middle-sized, and are either hollow, or lined with calcareous-spar.

The

as in those of Derbyshire, where it is associated with galena *.

Use.

Like common limestone, it is burnt and made into mortar, but it remains much longer caustic than quicklime from common limestone; and this is the cause of a very important difference between magnesian and common limestone, with regard to their employment in agriculture: Lime, from magnesian limestone, is termed *hot*, and when spread upon land in the same proportion as is generally practised with common quicklime, greatly impairs the fertility of the soil; and when used in a greater quantity, is said by Mr Tennant to prevent all vegetation †.

Observations.

A flexible variety of Dolomite occurs in England. The following account contains all the information I possess in regard to it:—

Flexible

* Bournon, Traité de Mineralogie, t. i. p. 208.

† In regard to this limestone, Dr Thomson has the following remarks: "This magnesian limestone has long been burnt in prodigious quantity in the neighbourhood of Sunderland, and sent coastwise, both to the north and to the south. It goes in great abundance to Aberdeenshire. As no complaints have ever been made of its being injurious, when employed as a manure, it would be curious to know whether this circumstance be owing to the soil on which it is put, or to the small quantity of it used, in consequence of its price, occasioned by its long carriage; for it appears, from Mr Tennant's statement, that at Ferrybridge, the farmers are aware that it does not answer as a manure so well as pure carbonate of lime."—*Annals of Philosophy*, vol. iv. p. 418.

Flexible Dolomite.

External Characters.

Its colour is yellowish-grey, passing into cream-yellow. It occurs massive. It is dull. The fracture is earthy in the small, and slaty in the large. It is opaque. It yields readily to the knife, but with difficulty to the nail. In thin plates it is uncommonly flexible. Specific gravity, 2.544, *Thomson*. This is probably below the truth, as the stone is porous.

Chemical Characters.

It dissolves in acids as readily as common carbonate of lime.

Constituent Parts.

Carbonate of Lime, -	62.00
Carbonate of Magnesia,	35.96
Insoluble matter, -	1.60
Loss, - - - - -	0.44
	100.00

Thomson, Annals of Phil. vol. iv. p. 418.

Geographic Situation.

It occurs about three miles from Tinmouth Castle.

Observations.

This curious mineral was discovered by my intelligent friend Mr Nicol, Lecturer on Natural Philosophy. To that gentleman I am indebted for the following particulars

in

in regard to it. He finds, that its flexibility is considerably influenced by the quantity of water contained in it. When saturated with water, it is remarkably flexible; as the evaporation goes on, it becomes more and more rigid, until the water be reduced to a certain limit, when the flexibility becomes scarcely distinguishable. From this point, however, the flexibility gradually increases, as the moisture diminishes; and as soon as the water is completely exhaled, it becomes nearly as flexible as it was when saturated with that fluid.

Second Kind.

Columnar Dolomite.

Stänglicher Dolomit, *Klaproth.*

Stänglicher Dolomit, *Klaproth,* Mag. der Gesellsch. Naturf. Freunde, b. v. s. 402.

External Characters.

Its colour is pale greyish-white.

It occurs massive, and in thin, long, and straight prismatic concretions.

It has an imperfect cleavage.

The fracture is uneven.

The lustre is vitreous, inclining to pearly.

It breaks into acicular-shaped fragments.

It is feebly translucent.

It is brittle.

Specific gravity 2.765.

Constituent

Constituent Parts.

From the Mine Tschistagowskoy.	
Carbonate of Lime, -	51
Carbonate of Magnesia, -	47
Carbonated Hydrate of Iron,	1
	99

Klaproth, Chem. Abhandl. s. 328.

Geognostic and Geographic Situations.

It occurs in serpentine in the mine Tschistagowskoy, on the river Mjafs, in the Government of Orenburg in Russia.

Observations.

It was at one time considered to be a variety of Strontianite; but in external characters, it is much more nearly allied to Tremolite.

Third Kind.

Compact Dolomite, or Gurhofite.

Gurhofian, *Karsten.*

Gurhofian, *Klaproth,* in Magazin der Gesellch. der Naturf. Freünde, b. i. s. 257.—Gurofian, *Karsten,* Tabel. s. 50. *Id. Klap.* Beit. b. v. s. 103. *Id. Lenz,* b. ii. s. 724.

External Characters.

Its colour is snow-white.

It occurs massive.

It

distinguish it from *Claystone*, and these characters, along with more considerable specific gravity, distinguish it from *Tripoli*.

2. The *Leutrite* of Lenz and Sartorius appears to be a marly sandstone.

3. The *Tutenmergel* or *Nagelkalk* is a variety of marl inclining to compact limestone, disposed in broken conical lamellar concretions, in which the surfaces are transversely streaked. It is found at Gorarp in Sweden.

4. Pliny, Vitruvius, and Varro, describe this mineral under the name *Marga*, and say it was used for improving the soil.

Twelfth Subspecies.

Bituminous Marl-Slate.

Bituminöser Mergelschiefer, *Werner*.

Bituminöser Mergelschiefer, *Wid.* s. 526.—Bituminous Marlite, *Kirw.* vol. i. p. 103.—Bituminöser Mergelschiefer, *Estner*, b. ii. s. 1035. *Id. Emm.* b. i. s. 498.—Schisto marno bituminoso, *Nap.* p. 363.—Le Schiste marneuse bitumineux, *Broch.* t. i. p. 574.—Bituminöser Mergelschiefer, *Reuss*, b. ii. 2. s. 376. *Id. Lud.* b. i. s. 157. *Id. Suck.* 1r th. s. 646. *Id. Bert.* s. 116. *Id. Mohs*, b. ii. s. 132. *Id. Hab.* s. 74 *Id. Leonhard*, Tabel. s. 36. *Id. Karsten*, Tabel. s. 50. *Id. Haus.* s. 127. *Id. Lenz*, b. ii. s. 786. *Id. Oken*, b. i. s. 405. *Id. Hoff.* b. iii. s. 72.

External Characters.

Its colour is intermediate between greyish-black and brownish-black.

It

It occurs massive, and frequently contains impressions of fishes and plants.

Its lustre is glimmering, glistening, or shining, and resinous.

Its fracture is straight, or curved slaty.

The fragments are slaty in the large, but indeterminate and rather sharp angular in the small.

It is opaque.

It is shining and resinous in the streak.

It is soft, and feels meagre.

It is rather sectile, and easily frangible.

Specific gravity 2.631, 2.690, *Breithaupt*.

Constituent Parts.

It is said to be a Carbonate of Lime united with Alumina, Iron, and Bitumen.

Geognostic Situation.

It occurs in secondary or flœtz limestone. It frequently contains cupreous minerals, particularly copper-pyrites, copper-glance, variegated copper-ore, and more rarely, native copper, copper green, and blue copper. It contains abundance of petrified fishes, and these are said to be most numerous in those situations where the strata are basin-shaped. Many attempts have been made to ascertain the genera and species of these animals, but hitherto with but little success. It would appear, that the greater number resemble fresh-water species, and a few the marine species. It also contains fossil remains of lizards, shells, corals, and of cryptogamous fresh-water plants.

Geographic

The fragments are rather blunt-edged.

It is translucent.

It is semihard.

It is brittle.

Specific gravity 2.885.

Chemical Characters.

It dissolves slowly, and with little effervescence, in nitrous acid; but more rapidly, and with increased effervescence, when the acid is heated.

Constituent Parts.

Carbonate of Lime,	53.00
Carbonate of Magnesia, -	42.50
Carbonate of Iron, with a little Manganese, -	3.00
	98.50

Klaproth, Beit. b. iii. s. 296.

Geognostic and Geographic Situations.

It is found at Miemo in Tuscany, imbedded in gypsum; at Hall in the Tyrol, imbedded in muriate of soda; and Gieseké met with it in kidneys, along with wavellite, arragonite, and calcedony, in decomposed wacke, at Kannioak, in Omenaksfiord in Greenland.

Observations.

This mineral was first observed by the late Dr Thompson of Naples, who sent specimens of it to Klaproth for analysis. It is named *Miemite*, after the place where it was discovered.

Second

Second Kind.

Prismatic Miemite.

Stänglicher Bitterspath, *Klaproth.*

Strahliger Kalkstein, *Von Schlottheim,* Hoff's Magaz. fur die Gesammte Mineralogie, b. i. s. 156.—Stänglicher Bitterspath, *Klaproth,* b. iii. s. 297. *Id. Leonhard,* Tabel. s. 36. *Id. Haus,* s. 128. *Id. Lenz,* b. ii. s. 712. *Id. Oken,* b. i. s. 393.

External Characters.

Its colour is asparagus-green, olive-green, and oil-green.

It occurs in prismatic distinct concretions, and crystallized in flat rhomboids, which are deeply truncated on all the edges.

The crystals are small, and very small, and sometimes they form only drusy crusts.

Internally it is shining and vitreous.

The fracture passes from concealed foliated to splintery.

The fragments are rather blunt-edged.

It is strongly translucent.

It is as hard as the granular miemite.

Specific gravity 2.885, *Karsten.*

Chemical Characters.

It dissolves slowly, and with but feeble effervescence, in nitrous acid.

Constituent

rally attached by their terminal planes, seldomer by their lateral planes; sometimes imbedded, and are to be observed intersecting each other.

The lateral planes of the crystals are sometimes smooth, more frequently more or less deeply streaked or grooved. The terminal planes are seldom smooth, generally uneven and rough, and sometimes also deeply notched.

The external lustre varies from dull to shining, and is vitreous: internally it is shining and glistening, and vitreous, inclining to resinous.

It has a fourfold cleavage, in which three of the cleavages are parallel with the lateral planes, and the fourth with the terminal planes.

The fracture is small and imperfect conchoidal, passing into uneven.

The fragments are indeterminate angular, and rather sharp-edged; in the prismatic varieties splintery.

It is translucent, passing into semi-transparent, and refracts double.

It is harder than calcareous-spar, but scarcely as hard as fluor-spar.

It is brittle, and easily frangible.

Specific gravity, 2.6,–3.0, *Mohs.*—2.946, *Haüy.*—2.883, 2.928, *Karsten.*—2.926, *Biot.*—2.891, *Wiedeman.*—2.912, *Bournon.*—3.00, *Kopp.*

Chemical Characters.

If we expose a small fragment to the flame of a candle, it almost immediately splits into white particles, which are dispersed around the flame. This change takes place principally with fragments of transparent crystals, fragments of the other varieties becoming merely white and friable. Fragments

p. 356.—Le Spath brunissant, ou le Braunspath, *Broch.* t. i. p. 563.—Gemeiner Braunspath, *Reuss,* b. i. s. 50. *Id. Lud.* b. i. s. 153. *Id. Suck.* 1r th. s. 630. *Id. Bert.* s. 118. *Id. Mohs,* b. ii. s. 108.—Spathiger Braunkalk, *Hab.* s. 82.—Chaux carbonatée manganesifere, *Lucas,* p. 8.—Spathiger Braunkalk, *Leonhard,* Tabel. s. 35.—Chaux carbonatée brunissante, *Brong.* t. i. p. 237.—Gemeiner Braunspath, *Karsten,* Tabel. s. 50.—Chaux carbonatée ferro-manganesienne, *Bournon,* Traité, t. i. p. 277.—Pearl-Spar, *Kid,* vol. i. p. 56.—Chaux carbonatée ferro-manganesifere, *Haüy,* Tabl. p. 5.—Gemeiner Braunkalk, *Lenz,* b. ii. s. 717.—Gemeiner Braunspath, *Oken,* b. i. s. 394.—Blättriger Braunspath, *Hoff.* b. iii. s. 49.

External Characters.

Its colours are flesh-red and brownish-red.

It often occurs massive, also disseminated, seldom globular, stalactitic, reniform, with tabular and pyramidal impressions; also in distinct concretions, which are granular, and rarely thin and straight lamellar; and frequently crystallized.

Its primitive form is a rhomboid, in which the obtuse angle is 107° 22′. The following are the secondary figures:

1. Rhomboid, in which the faces are sometimes cylindrically convex, sometimes cylindrically concave.
2. Lens, both common and saddle-shaped.

It also occurs in rhomboidal six-sided pyramidal supposititious crystals.

The true crystals are generally small and very small; the supposititious crystals large and middle-sized, and are either hollow, or lined with calcareous-spar.

The

as in those of Derbyshire, where it is associated with galena *.

Use.

Like common limestone, it is burnt and made into mortar, but it remains much longer caustic than quicklime from common limestone; and this is the cause of a very important difference between magnesian and common limestone, with regard to their employment in agriculture: Lime, from magnesian limestone, is termed *hot*, and when spread upon land in the same proportion as is generally practised with common quicklime, greatly impairs the fertility of the soil; and when used in a greater quantity, is said by Mr Tennant to prevent all vegetation †.

Observations.

A flexible variety of Dolomite occurs in England. The following account contains all the information I possess in regard to it:—

Flexible

* Bournon, Traité de Mineralogie, t. i. p. 208.

† In regard to this limestone, Dr Thomson has the following remarks: "This magnesian limestone has long been burnt in prodigious quantity in the neighbourhood of Sunderland, and sent coastwise, both to the north and to the south. It goes in great abundance to Aberdeenshire. As no complaints have ever been made of its being injurious, when employed as a manure, it would be curious to know whether this circumstance be owing to the soil on which it is put, or to the small quantity of it used, in consequence of its price, occasioned by its long carriage; for it appears, from Mr Tennant's statement, that at Ferrybridge, the farmers are aware that it does not answer as a manure so well as pure carbonate of lime."—*Annals of Philosophy*, vol. iv. p. 418.

Flexible Dolomite.

External Characters.

Its colour is yellowish-grey, passing into cream-yellow. It occurs massive. It is dull. The fracture is earthy in the small, and slaty in the large. It is opaque. It yields readily to the knife, but with difficulty to the nail. In thin plates it is uncommonly flexible. Specific gravity, 2.544, *Thomson*. This is probably below the truth, as the stone is porous.

Chemical Characters.

It dissolves in acids as readily as common carbonate of lime.

Constituent Parts.

Carbonate of Lime, -	62.00
Carbonate of Magnesia,	35.96
Insoluble matter, -	1.60
Loss, - - - - -	0.44
	100.00

Thomson, Annals of Phil. vol. iv. p. 418.

Geographic Situation.

It occurs about three miles from Tinmouth Castle.

Observations.

This curious mineral was discovered by my intelligent friend Mr Nicol, Lecturer on Natural Philosophy. To that gentleman I am indebted for the following particulars in

b.

2. *Colour.* White, and sometimes tile-red.

Form. Massive.

Distinct concretions. These are prismatic, sometimes straight, sometimes curved, and always parallel.

Fracture—Not visible.

Fragments. Splintery.

Observations.—This kind of arragonite is described by Haüy under the name Arragonite fibreux, and part of the common fibrous limestone of Hoffmann also belongs here.

2. All the varieties of arragonite are distinguished from *Calcareous-Spar*, by superior hardness, and specific gravity.

3. Arragonite has received different names at different periods: the common prismatic varieties have been named *Arragonian Apatite*, *Arragonian Calc-Spar*, and *Hard Calcareous-Spar;* and the pyramidal varieties have been described under the names *Iglit*, or *Igloit*.

4. The Coralloidal Arragonite has been described under the name *Flos Ferri*.

GENUS II.—APATITE.

THIS Genus contains one Species, viz. Rhomboidal Apatite.

1. Rhomboidal

Constituent Parts.

From the Mine Tschistagowskoy.	
Carbonate of Lime, -	51
Carbonate of Magnesia, -	47
Carbonated Hydrate of Iron,	1
	99

Klaproth, Chem. Abhandl. s. 328.

Geognostic and Geographic Situations.

It occurs in serpentine in the mine Tschistagowskoy, on the river Mjafs, in the Government of Orenburg in Russia.

Observations.

It was at one time considered to be a variety of Strontianite; but in external characters, it is much more nearly allied to Tremolite.

Third Kind.

Compact Dolomite, or Gurhofite.

Gurhofian, *Karsten.*

Gurhofian, *Klaproth,* in Magazin der Gesellch. der Naturf. Freünde, b. i. s. 257.—Gurofian, *Karsten,* Tabel. s. 50. *Id. Klap.* Beit. b. v. s. 103. *Id. Lenz,* b. ii. s. 724.

External Characters.

Its colour is snow-white.

It [illegible]

It

It is dull.

The fracture is flat conchoidal, passing to even.

The fragments are indeterminate angular, and sharp-edged.

It is slightly translucent on the edges.

It is hard, bordering on semihard.

It is brittle, and rather difficultly frangible.

Specific gravity, 2.7600, *Karsten*.

Chemical Characters.

When pounded, and thrown into diluted and heated nitrous acid, it is completely dissolved with effervescence.

Constituent Parts.

Carbonate of Lime, -	70.50
Carbonate of Magnesia,	29.50
	100.00

Klaproth, Gesellsch. N. Fr. b. i. s. 258.

Geognostic and Geographic Situations.

It occurs in veins in serpentine rocks, between Gurhof and Aggsbach, in Lower Austria.

Observations.

1. The name Gurhofit, sometimes given to this mineral, is from the place near which it was found.

2. It was at one time considered as a variety of semi-opal; but its greater weight distinguishes it from that mineral.

Second

h. Flatly acuminated on one extremity with six planes, which are set on the lateral planes. In this figure, the apex of the acumination, all the angles, and the alternate lateral edges, are slightly truncated.

i. Acuminated on both extremities with six planes, the apices, lateral edges, and angles, occasionally truncated.

k. The preceding acumination again very flatly acuminated with six planes, which are set on the planes of the first acumination. The apices of the acuminations truncated.

II. Table.

a. Equiangular six-sided table, in which the edges and angles are sometimes truncated.

b. Eight-sided table, in which four of the terminal edges are truncated.

The crystals are small, very small, and middle-sized; and occur sometimes single, sometimes many irregularly superimposed on each other.

The lateral planes are seldom smooth, generally longitudinally streaked; the truncating and acuminating planes are smooth.

Externally it is splendent or shining; internally glistening, and the lustre is resinous.

It has a fourfold cleavage, in which three of the cleavages are parallel with the lateral planes of the prism, and one (the most perfect), with the terminal planes of the prism.

The fracture is intermediate between uneven and imperfect conchoidal.

The fragments are rather blunt-edged.

It is translucent.

It is semihard.

It is brittle.

Specific gravity 2.885.

Chemical Characters.

It dissolves slowly, and with little effervescence, in nitrous acid; but more rapidly, and with increased effervescence, when the acid is heated.

Constituent Parts.

Carbonate of Lime,	53.00
Carbonate of Magnesia, -	42.50
Carbonate of Iron, with a little Manganese, -	3.00
	98.50

Klaproth, Beit. b. iii. s. 296.

Geognostic and Geographic Situations.

It is found at Miemo in Tuscany, imbedded in gypsum; at Hall in the Tyrol, imbedded in muriate of soda; and Gieseké met with it in kidneys, along with wavellite, arragonite, and calcedony, in decomposed wacke, at Kannioak, in Omenaksfiord in Greenland.

Observations.

This mineral was first observed by the late Dr Thompson of Naples, who sent specimens of it to Klaproth for analysis. It is named *Miemite*, after the place where it was discovered.

Second

Second Kind.

Prismatic Miemite.

Stänglicher Bitterspath, *Klaproth.*

Strahliger Kalkstein, *Von Schlottheim*, Hoff's Magaz. fur die Gesammte Mineralogie, b. i. s. 156.—Stänglicher Bitterspath, *Klaproth*, b. iii. s. 297. *Id. Leonhard*, Tabel. s. 36. *Id. Haus*, s. 128. *Id. Lenz*, b. ii. s. 712. *Id. Oken*, b. i. s. 393.

External Characters.

Its colour is asparagus-green, olive-green, and oil-green.

It occurs in prismatic distinct concretions, and crystallized in flat rhomboids, which are deeply truncated on all the edges.

The crystals are small, and very small, and sometimes they form only drusy crusts.

Internally it is shining and vitreous.

The fracture passes from concealed foliated to splintery.

The fragments are rather blunt-edged.

It is strongly translucent.

It is as hard as the granular miemite.

Specific gravity 2.885, *Karsten.*

Chemical Characters.

It dissolves slowly, and with but feeble effervescence, in nitrous acid.

Constituent

Constituent Parts.

Lime,	-	33.00
Magnesia,		14.50
Oxide of Iron,		2.50
Carbonic Acid,	-	47.25
Water and Loss,	-	2.75
		100

Klaproth, Beit. b. iii. s. 303.

Geognostic and Geographic Situations.

It occurs in cobalt veins that traverse sandstone at Glücksbrunn in Gotha, and at Beska in Servia, on the frontier of Turkey.

Third Subspecies.

Brown-Spar, or Pearl-Spar.

Braunspath, *Werner.*

This species is divided into two kinds, viz. Foliated Brown-Spar, and Columnar Brown-Spar.

First Kind.

Foliated Brown-Spar.

Blättriger Braunspath, *Werner.*

Spath perlé, *Romé de Lisle,* t. i. p. 605.—Braunspath, *Wid.* s. 515.—Sidero-calcite, *Kirw.* vol. i. p. 105.—Braunspath, *Estner,* b. ii. s. 999. *Id. Emm.* b. i. s. 79.—Brunispato, *Nap.* p. 356.

p. 356.—Le Spath brunissant, ou le Braunspath, *Broch.* t. i. p. 563.—Gemeiner Braunspath, *Reuss*, b. i. s. 50. *Id. Lud.* b. i. s. 153. *Id. Suck.* 1r th. s. 630. *Id. Bert.* s. 118. *Id. Mohs*, b. ii. s. 108.—Spathiger Braunkalk, *Hab.* s. 82.—Chaux carbonatée manganesifere, *Lucas*, p. 8.—Spathiger Braunkalk, *Leonhard*, Tabel. s. 35.—Chaux carbonatée brunissante, *Brong.* t. i. p. 237.—Gemeiner Braunspath, *Karsten*, Tabel. s. 50.—Chaux carbonatée ferro-manganesienne, *Bournon*, Traité, t. i. p. 277.—Pearl-Spar, *Kid*, vol. i. p. 56.—Chaux carbonatée ferro-manganesifere, *Haüy*, Tabl. p. 5.—Gemeiner Braunkalk, *Lenz*, b. ii. s. 717.—Gemeiner Braunspath, *Oken*, b. i. s. 394.—Blättriger Braunspath, *Hoff.* b. iii. s. 49.

External Characters.

Its colours are flesh-red and brownish-red.

It often occurs massive, also disseminated, seldom globular, stalactitic, reniform, with tabular and pyramidal impressions; also in distinct concretions, which are granular, and rarely thin and straight lamellar; and frequently crystallized.

Its primitive form is a rhomboid, in which the obtuse angle is 107° 22′. The following are the secondary figures:

1. Rhomboid, in which the faces are sometimes cylindrically convex, sometimes cylindrically concave.
2. Lens, both common and saddle-shaped.

It also occurs in rhomboidal six-sided pyramidal supposititious crystals.

The true crystals are generally small and very small; the supposititious crystals large and middle-sized, and are either hollow, or lined with calcareous-spar.

The

as in those of Derbyshire, where it is associated with galena *.

Use.

Like common limestone, it is burnt and made into mortar, but it remains much longer caustic than quicklime from common limestone; and this is the cause of a very important difference between magnesian and common limestone, with regard to their employment in agriculture: Lime, from magnesian limestone, is termed *hot*, and when spread upon land in the same proportion as is generally practised with common quicklime, greatly impairs the fertility of the soil; and when used in a greater quantity, is said by Mr Tennant to prevent all vegetation †.

Observations.

A flexible variety of Dolomite occurs in England. The following account contains all the information I possess in regard to it:—

Flexible

* Bournon, Traité de Mineralogie, t. i. p. 208.

† In regard to this limestone, Dr Thomson has the following remarks: "This magnesian limestone has long been burnt in prodigious quantity in the neighbourhood of Sunderland, and sent coastwise, both to the north and to the south. It goes in great abundance to Aberdeenshire. As no complaints have ever been made of its being injurious, when employed as a manure, it would be curious to know whether this circumstance be owing to the soil on which it is put, or to the small quantity of it used, in consequence of its price, occasioned by its long carriage; for it appears, from Mr Tennant's statement, that at Ferrybridge, the farmers are aware that it does not answer as a manure so well as pure carbonate of lime."—*Annals of Philosophy*, vol. iv. p. 418.

der the name *Moroxite*. The name moroxite given to this mineral by Karsten, is borrowed from the Morochites of Pliny, concerning which, that author says, "Est gemma, per se porracea viridisque, trita autem candicans."—*Histor. Natur.* l. xxxvii.

3. Apatite is distinguished from Beryl, Schorl, and Chrysolite, by its inferior hardness: its greater hardness and non-effervescence with acids, distinguish it from Calcareous-spar.

Third Subspecies.

Phosphorite.

Phosphorit, *Werner*.

This Subspecies is divided into two Kinds, viz. Common Phosphorite and Earthy Phosphorite.

First Kind.

Common Phosphorite.

Gemeiner Phosphorit, *Karsten*.

Gemeiner Phosphorit, *Haus.* s. 123. *Id. Karsten*, Tabel. s. 52.—Chaux phosphatée terreuse, *Haüy*, Tabl. p. 8.—Gemeiner Phosphorit, *Lenz*, b. ii. s. 801.—Dichter Phosphorit, *Haus.* Handb. b. iii. s. 872.—Phosphorit, *Hoff.* b. iii. s. 92.—Massive Apatite, *Aikin*, p. 172.

External Characters.

Its colour is yellowish-white, sometimes approaching to greyish-

in regard to it. He finds, that its flexibility is considerably influenced by the quantity of water contained in it. When saturated with water, it is remarkably flexible; as the evaporation goes on, it becomes more and more rigid, until the water be reduced to a certain limit, when the flexibility becomes scarcely distinguishable. From this point, however, the flexibility gradually increases, as the moisture diminishes; and as soon as the water is completely exhaled, it becomes nearly as flexible as it was when saturated with that fluid.

Second Kind.

Columnar Dolomite.

Stünglicher Dolomit, *Klaproth.*

Stünglicher Dolomit, *Klaproth*, Mag. der Gesellsch. Naturf. Freunde, b. v. s. 402.

External Characters.

Its colour is pale greyish-white.

It occurs massive, and in thin, long, and straight prismatic concretions.

It has an imperfect cleavage.

The fracture is uneven.

The lustre is vitreous, inclining to pearly.

It breaks into acicular-shaped fragments.

It is feebly translucent.

It is brittle.

Specific gravity 2.765.

Constituents

Constituent Parts.

From the Mine Tschistagowskoy.	
Carbonate of Lime, -	51
Carbonate of Magnesia, -	47
Carbonated Hydrate of Iron,	1
	99

Klaproth, Chem. Abhandl. s. 328.

Geognostic and Geographic Situations.

It occurs in serpentine in the mine Tschistagowskoy, on the river Mjafs, in the Government of Orenburg in Russia.

Observations.

It was at one time considered to be a variety of Strontianite; but in external characters, it is much more nearly allied to Tremolite.

Third Kind.

Compact Dolomite, or Gurhofite.

Gurhofian, *Karsten.*

Gurhofian, *Klaproth*, in Magazin der Gesellch. der Naturf. Freünde, b. i. s. 257.—Gurofian, *Karsten*, Tabel. s. 50. *Id. Klap*. Beit. b. v. s. 103. *Id. Lenz*, b. ii. s. 724.

External Characters.

Its colour is snow-white.

It occurs massive.

It is dull.

The fracture is flat conchoidal, passing to even.

The fragments are indeterminate angular, and sharp-edged.

It is slightly translucent on the edges.

It is hard, bordering on semihard.

It is brittle, and rather difficultly frangible.

Specific gravity, 2.7600, *Karsten.*

Chemical Characters.

When pounded, and thrown into diluted and heated nitrous acid, it is completely dissolved with effervescence.

Constituent Parts.

Carbonate of Lime, -	70.50
Carbonate of Magnesia,	29.50
	100.00

Klaproth, Gesellsch. N. Fr. b. i. s. 258.

Geognostic and Geographic Situations.

It occurs in veins in serpentine rocks, between Gurhof and Aggsbach, in Lower Austria.

Observations.

1. The name Gurhofit, sometimes given to this mineral, is from the place near which it was found.

2. It was at one time considered as a variety of semiopal; but its greater weight distinguishes it from that mineral.

Second

20.00. Muriatic Acid, 2.00. Phosphate of Iron, 3.75. Water, 10.00."

Genus III. FLUOR.

This Genus contains but one species, viz. Octahedral Fluor.

1. Octahedral Fluor.

Flus, *Werner*.

Octaedrisches Flus Haloide, *Mohs*.

It is divided into three subspecies, Compact Fluor, Foliated Fluor, and Earthy Fluor.

First Subspecies.

Compact Fluor.

Dichter Flus, *Werner*.

Fluor solidus, *Wall.* t. i. p. 542.?—Dichter Fluss, *Wid.* s. 542.—Compact Fluor, *Kirw.* vol. i. p. 127.—Dichter Fluss, *Estner*, b. ii. s. 1067. *Id. Emm.* b. i. s. 516.—Fluorite compatta, *Nap.* p. 374.—Le Fluor compacte, *Broch.* t. i. p. 594.—Dichter Fluss, *Reuss*, b. ii. 2. s. 379. *Id. Lud.* b. i. s. 161. *Id. Suck.* 1 th. s. 663. *Id. Bert.* s. 103. *Id. Mohs*, b. ii. s. 150. *Id. Hab.* s. 89.—Chaux fluatée massive compacte, *Lucas*, p. 247.—Dichter Fluss, *Leonhard*, Tabel. s. 38.—Chaux fluatée,

The fragments are rather blunt-edged.

It is translucent.

It is semihard.

It is brittle.

Specific gravity 2.885.

Chemical Characters.

It dissolves slowly, and with little effervescence, in nitrous acid; but more rapidly, and with increased effervescence, when the acid is heated.

Constituent Parts.

Carbonate of Lime,	53.00
Carbonate of Magnesia, -	42.50
Carbonate of Iron, with a little Manganese, -	3.00
	98.50

Klaproth, Beit. b. iii. s. 296.

Geognostic and Geographic Situations.

It is found at Miemo in Tuscany, imbedded in gypsum; at Hall in the Tyrol, imbedded in muriate of soda; and Gieseké met with it in kidneys, along with wavellite, arragonite, and calcedony, in decomposed wacke, at Kannioak, in Omenaksfiord in Greenland.

Observations.

This mineral was first observed by the late Dr Thompson of Naples, who sent specimens of it to Klaproth for analysis. It is named *Miemite*, after the place where it was discovered.

Second

Second Kind.

Prismatic Miemite.

Stänglicher Bitterspath, *Klaproth.*

Strahliger Kalkstein, *Von Schlottheim,* Hoff's Magaz. fur die Gesammte Mineralogie, b. i. s. 156.—Stänglicher Bitterspath, *Klaproth,* b. iii. s. 297. *Id. Leonhard,* Tabel. s. 36. *Id. Haus,* s. 128. *Id. Lenz,* b. ii. s. 712. *Id. Oken,* b. i. s. 393.

External Characters.

Its colour is asparagus-green, olive-green, and oil-green.

It occurs in prismatic distinct concretions, and crystallized in flat rhomboids, which are deeply truncated on all the edges.

The crystals are small, and very small, and sometimes they form only drusy crusts.

Internally it is shining and vitreous.

The fracture passes from concealed foliated to splintery.

The fragments are rather blunt-edged.

It is strongly translucent.

It is as hard as the granular miemite.

Specific gravity 2.885, *Karsten.*

Chemical Characters.

It dissolves slowly, and with but feeble effervescence, in nitrous acid.

Constituent

Constituent Parts.

Lime, -	33.00
Magnesia,	14.50
Oxide of Iron,	2.50
Carbonic Acid, -	47.25
Water and Loss, -	2.75
	100

Klaproth, Beit. b. iii. s. 303.

Geognostic and Geographic Situations.

It occurs in cobalt veins that traverse sandstone at Glücksbrunn in Gotha, and at Beska in Servia, on the frontier of Turkey.

Third Subspecies.

Brown-Spar, or Pearl-Spar.

Braunspath, *Werner.*

This species is divided into two kinds, viz. Foliated Brown-Spar, and Columnar Brown-Spar.

First Kind.

Foliated Brown-Spar.

Blättriger Braunspath, *Werner.*

Spath perlé, *Romé de Lisle,* t. i. p. 605.—Braunspath, *Wid.* s. 515.—Sidero-calcite, *Kirw.* vol. i. p. 105.—Braunspath, *Estner,* b. ii. s. 999. *Id. Emm.* b. i. s. 79.—Brunispato, *Nap.* p. 356.

p. 356.—Le Spath brunissant, ou le Braunspath, *Broch.* t. i. p. 563.—Gemeiner Braunspath, *Reuss*, b. i. s. 50. *Id. Lud.* b. i. s. 153. *Id. Suck.* 1r th. s. 630. *Id. Bert.* s. 118. *Id. Mohs*, b. ii. s. 108.—Spathiger Braunkalk, *Hab.* s. 82.—Chaux carbonatée manganesifere, *Lucas*, p. 8.—Spathiger Braunkalk, *Leonhard*, Tabel. s. 85.—Chaux carbonatée brunissante, *Brong.* t. i. p. 237.—Gemeiner Braunspath, *Karsten*, Tabel. s. 50.—Chaux carbonatée ferro-manganesienne, *Bournon*, Traité, t. i. p. 277.—Pearl-Spar, *Kid*, vol. i. p. 56.—Chaux carbonatée ferro-manganesifere, *Haüy*, Tabl. p. 5.—Gemeiner Braunkalk, *Lenz*, b. ii. s. 717.—Gemeiner Braunspath, *Oken*, b. i. s. 394.—Blättriger Braunspath, *Hoff.* b. iii. s. 49.

External Characters.

Its colours are flesh-red and brownish-red.

It often occurs massive, also disseminated, seldom globular, stalactitic, reniform, with tabular and pyramidal impressions; also in distinct concretions, which are granular, and rarely thin and straight lamellar; and frequently crystallized.

Its primitive form is a rhomboid, in which the obtuse angle is 107° 22′. The following are the secondary figures:

1. Rhomboid, in which the faces are sometimes cylindrically convex, sometimes cylindrically concave.
2. Lens, both common and saddle-shaped.

It also occurs in rhomboidal six-sided pyramidal supposititious crystals.

The true crystals are generally small and very small; the supposititious crystals large and middle-sized, and are either hollow, or lined with calcareous-spar.

The

rough, as in the rhomboidal dodecahedron, and some octahedrons.

Internally the lustre is specular-splendent, or shining, and is vitreous.

It has a fourfold equiangular cleavage, which is parallel with the planes of an octahedron.

The fragments are octahedral or tetrahedral.

It alternates from translucent to transparent, and refracts single.

It is harder than calcareous-spar, but not so hard as apatite.

It is brittle, and easily frangible.

Specific gravity, 3.0943, 3.1911, *Haüy.*—3.148, *Gellert.* —3.092, *Brisson.*—3.156, 3.184, *Muschenbroeck.*—3.138, 3.228, *Karsten.*

Chemical Characters.

Before the blowpipe it generally decrepitates, gradually loses its colour and transparency, and melts, without addition, into a greyish-white glass. When two fragments are rubbed against each other, they become luminous in the dark. When gently heated, or laid on glowing coal, it phosphoresces, (particularly the sky-blue, violet-blue, and green varieties,) partly with a blue, partly with a green light. When brought to a red-heat, it is deprived of its phosphorescent property. The violet-blue variety from Nertschinsky, named *Chlorophane*, when placed on glowing coals, does not decrepitate, but soon throws out a beautiful verdigris-green and apple-green light, which gradually disappears as the mineral cools, but may be again excited, if it is heated; and this may be repeated a dozen of times, provided

Flexible Dolomite.

External Characters.

Its colour is yellowish-grey, passing into cream-yellow. It occurs massive. It is dull. The fracture is earthy in the small, and slaty in the large. It is opaque. It yields readily to the knife, but with difficulty to the nail. In thin plates it is uncommonly flexible. Specific gravity, 2.544, *Thomson*. This is probably below the truth, as the stone is porous.

Chemical Characters.

It dissolves in acids as readily as common carbonate of lime.

Constituent Parts.

Carbonate of Lime, -	62.00
Carbonate of Magnesia,	35.96
Insoluble matter, -	1.60
Loss, - - - - -	0.44
	100.00

Thomson, Annals of Phil. vol. iv. p. 418.

Geographic Situation.

It occurs about three miles from Tinmouth Castle.

Observations.

This curious mineral was discovered by my intelligent friend Mr Nicol, Lecturer on Natural Philosophy. To that gentleman I am indebted for the following particulars in

in regard to it. He finds, that its flexibility is considerably influenced by the quantity of water contained in it. When saturated with water, it is remarkably flexible; as the evaporation goes on, it becomes more and more rigid, until the water be reduced to a certain limit, when the flexibility becomes scarcely distinguishable. From this point, however, the flexibility gradually increases, as the moisture diminishes; and as soon as the water is completely exhaled, it becomes nearly as flexible as it was when saturated with that fluid.

Second Kind.

Columnar Dolomite.

Stänglicher Dolomit, *Klaproth.*

Stänglicher Dolomit, *Klaproth*, Mag. der Gesellsch. Naturf. Freunde, b. v. s. 402.

External Characters.

Its colour is pale greyish-white.

It occurs massive, and in thin, long, and straight prismatic concretions.

It has an imperfect cleavage.

The fracture is uneven.

The lustre is vitreous, inclining to pearly.

It breaks into acicular-shaped fragments.

It is feebly translucent.

It is brittle.

Specific gravity 2.765.

Constituent

Constituent Parts.

From the Mine Tschistagowskoy.		
Carbonate of Lime,	-	51
Carbonate of Magnesia,	-	47
Carbonated Hydrate of Iron,		1
		99

Klaproth, Chem. Abhandl. s. 328.

Geognostic and Geographic Situations.

It occurs in serpentine in the mine Tschistagowskoy, on the river Mjafs, in the Government of Orenburg in Russia.

Observations.

It was at one time considered to be a variety of Strontianite; but in external characters, it is much more nearly allied to Tremolite.

Third Kind.

Compact Dolomite, or Gurhofite.

Gurhofian, *Karsten.*

Gurhofian, *Klaproth,* in Magazin der Gesellch. der Naturf. Freünde, b. i. s. 257.—Gurofian, *Karsten,* Tabel. s. 50. *Id. Klap*. Beit. b. v. s. 103. *Id. Lenz,* b. ii. s. 724.

External Characters.

Its colour is snow-white.

It occurs massive.

It

nandrae; in Auvergne in France; and fluor also occurs in Franconia, Austria, Denmark, Hessia, Silesia; but is very rare in Russia and in Hungary.

Asia.—The chlorophane variety is found at Catharinenburg and Nertschinsky: other varieties are found in granite, in the neighbourhood of the lake of Gussino-Osero, on the Mongol frontier; also at Schlangenberg, in the silver mine Zimeof, in the Altain range. It is also mentioned as a production of the island of Ceylon.

America.—West Greenland; California; Mexico; and in New Jersey, Connecticut, New Hampshire, and Virginia, in the United States *.

Uses.

On account of the variety and beauty of its colours, its transparency, the ease with which it can be worked, and the high polish it receives, it is cut into vases, pyramids, and other ornamental articles. The largest masses, and most beautiful varieties for use, are found in Derbyshire, and it is in that county that all the ornamental articles of fluor-spar are manufactured. It is also used by the metallurgist, as a flux for ores, particularly those of iron and copper; and hence the name *fluor* given to it. The acid it contains has been employed in the way of experiment for engraving upon glass.

Observations.

1. It is distinguished from *Calcareous-spar*, by its greater hardness and weight, and its not effervescing with acids: from *Gypsum*, by its superior hardness, and specific gravity,

* Bruce and Barton, in American Mineralogical Journal, N. I. p. 32, 33.

ty, and its decrepitating in the fire, whilst gypsum exfoliates, and becomes white; and from *Heavy-spar*, by its inferior specific gravity, and greater hardness.

2. The red varieties have been named *False Ruby*, the yellow *False Topaz*, the green *False Emerald*, and the blue *False Sapphire* and *Amethyst*.

3. The name *Chlorophane*, given to the varieties that easily become phosphorescent, is from the green light they exhibit.

Third Subspecies.

Earthy Fluor.

Erdiger Fluss, *Karsten*.

Le Fluor terreuse, *Broch.* t. i. p. 593.—Erdiger Fluss, *Reuss*, b. ii. 2. s. 378. *Id. Lud.* b. i. s. 161. *Id. Suck.* 1r th. s. 662. *Id. Leonhard*, Tabel. s. 38.—Chaux fluatée terreuse, *Brong.* t. i. p. 245. *Id. Brard*, p. 48.—Erdiger Fluor, *Haus.* s. 124. *Id. Karst.* Tabel. s. 52.—Earthy Fluor, *Kid*, vol. i. p. 78.——Chaux fluatée terreuse, *Haüy*, Tabl. p. 9.—Fluserde, *Lenz*, b. ii. s. 829.—Erdiger Fluss, *Haus.* Handb. b. iii. s. 878.

External Characters.

Its colours are greyish-white, and violet-blue, and are sometimes so deep as almost to appear black.

It occurs generally in crusts, investing some other mineral.

It is dull.

It is earthy.

It is friable, passing into very soft.

Constituent

The fragments are rather blunt-edged.

It is translucent.

It is semihard.

It is brittle.

Specific gravity 2.885.

Chemical Characters.

It dissolves slowly, and with little effervescence, in nitrous acid; but more rapidly, and with increased effervescence, when the acid is heated.

Constituent Parts.

Carbonate of Lime,	53.00
Carbonate of Magnesia, -	42.50
Carbonate of Iron, with a little Manganese, -	3.00
	98.50

Klaproth, Beit. b. iii. s. 296.

Geognostic and Geographic Situations.

It is found at Miemo in Tuscany, imbedded in gypsum; at Hall in the Tyrol, imbedded in muriate of soda; and Gieseké met with it in kidneys, along with wavellite, arragonite, and calcedony, in decomposed wacke, at Kannioak, in Omenaksfiord in Greenland.

Observations.

This mineral was first observed by the late Dr Thompson of Naples, who sent specimens of it to Klaproth for analysis. It is named *Miemite*, after the place where it was discovered.

Second

Second Kind.

Prismatic Miemite.

Stänglicher Bitterspath, *Klaproth.*

Strahliger Kalkstein, *Von Schlottheim*, Hoff's Magaz. fur die Gesammte Mineralogie, b. i. s. 156.—Stänglicher Bitterspath, *Klaproth*, b. iii. s. 297. *Id. Leonhard*, Tabel. s. 36. *Id. Haus*, s. 128. *Id. Lenz*, b. ii. s. 712. *Id. Oken*, b. i. s. 393.

External Characters.

Its colour is asparagus-green, olive-green, and oil-green.

It occurs in prismatic distinct concretions, and crystallized in flat rhomboids, which are deeply truncated on all the edges.

The crystals are small, and very small, and sometimes they form only drusy crusts.

Internally it is shining and vitreous.

The fracture passes from concealed foliated to splintery.

The fragments are rather blunt-edged.

It is strongly translucent.

It is as hard as the granular miemite.

Specific gravity 2.885, *Karsten.*

Chemical Characters.

It dissolves slowly, and with but feeble effervescence, in nitrous acid.

Constituent

Constituent Parts.

		Alumstone from Tolfa.	Hungarian Alumstone.
Alumina, -	43.92	19.00	17.50
Silica, - -	24.00	56.50	62.25
Sulphuric Acid,	25.00	16.50	12.50
Potash, - -	3.08	4.00	1.00
Water, - -	4.00	3.00	5.00
		99.00	98.25
	Vauquelin.	*Klaproth*, Beit. b. iv. s. 252.	*Klaproth*, Id. s. 256.

Geognostic and Geographic Situations.

It occurs at Tolfa, near Civita Vecchia, in nests, kidneys, and small veins, in a flœtz or secondary rock. The Hungarian varieties are found in beds at Beregszaz and Nagy-Begany, in the country of Beregher, in Upper Hungary.

Uses.

Alum is obtained from this mineral, by repeatedly roasting it, then lixiviating it, and crystallizing the solution thus obtained. The art of preparing alum is an eastern discovery; and the most ancient known alum-work is that of Rocca, the present Edessa, in Syria. In the middle ages, all the alum of commerce was prepared in the Levant; but in the fifteenth century, some Genoese skilled in the Levant art of alum-making, discovered alumstone in Italy, and immediately began to extract alum from it; and this new source of wealth soon became very considerable

p. 356.—Le Spath brunissant, ou le Braunspath, *Broch.* t. i. p. 563.—Gemeiner Braunspath, *Reuss,* b. i. s. 50. *Id. Lud.* b. i. s. 153. *Id. Suck.* 1r th. s. 630. *Id. Bert.* s. 118. *Id. Mohs,* b. ii. s. 108.—Spathiger Braunkalk, *Hab.* s. 82.—Chaux carbonatée manganesifere, *Lucas,* p. 8.—Spathiger Braunkalk, *Leonhard,* Tabel. s. 35.—Chaux carbonatée brunissante, *Brong.* t. i. p. 237.—Gemeiner Braunspath, *Karsten,* Tabel. s. 50.—Chaux carbonatée ferro-manganesienne, *Bournon,* Traité, t. i. p. 277.—Pearl-Spar, *Kid,* vol. i. p. 56.—Chaux carbonatée ferro-manganesifere, *Haüy,* Tabl. p. 5.—Gemeiner Braunkalk, *Lenz,* b. ii. s. 717.—Gemeiner Braunspath, *Oken,* b. i. s. 394.—Blättriger Braunspath, *Hoff.* b. iii. s. 49.

External Characters.

Its colours are flesh-red and brownish-red.

It often occurs massive, also disseminated, seldom globular, stalactitic, reniform, with tabular and pyramidal impressions; also in distinct concretions, which are granular, and rarely thin and straight lamellar; and frequently crystallized.

Its primitive form is a rhomboid, in which the obtuse angle is 107° 22′. The following are the secondary figures:

1. Rhomboid, in which the faces are sometimes cylindrically convex, sometimes cylindrically concave.
2. Lens, both common and saddle-shaped.

It also occurs in rhomboidal six-sided pyramidal supposititious crystals.

The true crystals are generally small and very small; the supposititious crystals large and middle-sized, and are either hollow, or lined with calcareous-spar.

lith, *Karsten*, Tabel. s. 48. *Id. Haus.* s. 121.—Alumine fluatée alkaline, *Haüy*, Tabl. p. 22.—Kryolith, *Lenz*, b. ii. s. 943. *Id. Haus.* Handb. b. iii. s. 846. *Id. Hoff.* b. iii. s. 204. *Id. Aikin*, p. 174.

External Characters.

Its colours are pale greyish-white, snow-white, yellowish-brown and yellowish-red.

It occurs massive, disseminated, and in straight and thick lamellar distinct concretions.

It is shining, inclining to glistening, and the lustre is vitreous, inclining to pearly.

It has a fourfold cleavage, in which the folia are parallel with an equiangular four-sided pyramid; sometimes the folia are parallel with the diagonals of a rectangular four-sided prism, or with the terminal planes.

The fracture is uneven.

The fragments are cubical or tabular.

It is translucent.

It is harder than gypsum, and sometimes as hard as calcareous-spar.

It is brittle, and easily frangible.

Specific gravity, 2.949, *Haüy*.—2.953, *Karsten*.—2.9, 3.0, *Mohs*.

Chemical Characters.

It becomes more translucent in water, but does not dissolve in it. It melts before it reaches a red heat, and when simply exposed to the flame of a candle. Before the blowpipe, it at first runs into a very liquid fusion, then hardens, and at length assumes the appearance of a slag.

Constituent

Flexible Dolomite.

External Characters.

Its colour is yellowish-grey, passing into cream-yellow. It occurs massive. It is dull. The fracture is earthy in the small, and slaty in the large. It is opaque. It yields readily to the knife, but with difficulty to the nail. In thin plates it is uncommonly flexible. Specific gravity, 2.544, *Thomson*. This is probably below the truth, as the stone is porous.

Chemical Characters.

It dissolves in acids as readily as common carbonate of lime.

Constituent Parts.

Carbonate of Lime, -	62.00
Carbonate of Magnesia,	35.96
Insoluble matter, -	1.60
Loss, - - - - -	0.44
	100.00

Thomson, Annals of Phil. vol. iv. p. 418.

Geographic Situation.

It occurs about three miles from Tinmouth Castle.

Observations.

This curious mineral was discovered by my intelligent friend Mr Nicol, Lecturer on Natural Philosophy. To that gentleman I am indebted for the following particulars

in

2. It has been confounded with Heavy-spar, from which it is distinguished, by inferior specific gravity, and its easy fusibility before the blowpipe: it might also be mistaken for some varieties of Gypsum, but is distinguished from these by superior specific gravity, and its not exfoliating when exposed to the blowpipe.

GENUS VI.—GYPSUM *.

Gyps Haloide, *Mohs.*

THIS Genus contains two species, viz. 1. Prismatic Gypsum, 2. Axifrangible Gypsum.

1. Prismatic

---

* *Gypsum* is from the Greek word Γυψος. The following explanation of the term γυψος, shows that it was applied by the ancients to an earthy substance that had been exposed to the action of fire: Γύψος οἱονεὶ γήεψος τις οὖσα· ἡ ἑψηθεῖσα γῆ (*a*): in which it corresponds with the gypsum of the moderns. The ancient naturalists sometimes seem to apply the term to sulphate of lime, the gypsum of the present day, and sometimes to a calcined carbonate of lime, or quicklime, which they called *calx*. In the following passage, it is applied to a sulphate of lime; " Cognata calci res gypsum est. Qui coquitur lapis non dissimilis alabastritæ esse debet: omnia autem optimum fieri compertum est e *lapide speculari*, squamamve talem habente (*b*):" the term *lapis specularis* applying very closely to our selenite, which is a sulphate of lime. " Gypsoma dicto statim utendum est, quoniam *celerrime* coit (c):" the word *celerrime* being more applicable to the comparatively

(*a*) Vid. Etymolog. Magn.

(*b*) Plin. Hist. Nat. lib. xxxvi.

(*c*) Plin. Hist. Nat. lib. xxxvi.

Constituent Parts.

From the Mine Tschistagowskoy.		
Carbonate of Lime,	-	51
Carbonate of Magnesia,	-	47
Carbonated Hydrate of Iron,		1
		99

Klaproth, Chem. Abhandl. s. 328.

Geognostic and Geographic Situations.

It occurs in serpentine in the mine Tschistagowskoy, on the river Mjafs, in the Government of Orenburg in Russia.

Observations.

It was at one time considered to be a variety of Strontianite; but in external characters, it is much more nearly allied to Tremolite.

Third Kind.

Compact Dolomite, or Gurhofite.

Gurhofian, *Karsten.*

Gurhofian, *Klaproth,* in Magazin der Gesellch. der Naturf. Freünde, b. i. s. 257.—Gurofian, *Karsten,* Tabel. s. 50. *Id.* *Klap*. Beit. b. v. s. 103. *Id. Lenz*, b. ii. s. 724.

External Characters.

Its colour is snow-white.

It occurs massive.

It is dull.

The fracture is flat conchoidal, passing to even.

The fragments are indeterminate angular, and sharp-edged.

It is slightly translucent on the edges.

It is hard, bordering on semihard.

It is brittle, and rather difficultly frangible.

Specific gravity, 2.7600, *Karsten.*

Chemical Characters.

When pounded, and thrown into diluted and heated nitrous acid, it is completely dissolved with effervescence.

Constituent Parts.

Carbonate of Lime, -	70.50
Carbonate of Magnesia,	29.50
	100.00

Klaproth, Gesellsch. N. Fr. b. i. s. 258.

Geognostic and Geographic Situations.

It occurs in veins in serpentine rocks, between Gurhof and Aggsbach, in Lower Austria.

Observations.

1. The name Gurhofit, sometimes given to this mineral, is from the place near which it was found.

2. It was at one time considered as a variety of semi-opal; but its greater weight distinguishes it from that mineral.

Specific gravity, 2.957, *Bournon*.—2.964, *Klaproth*.—2.7, 3.0, *Mohs*.

Chemical Characters.

When exposed to the blowpipe, it does not exfoliate, and melt like gypsum, but becomes glazed over with a white friable enamel.

Constituent Parts.

	From Bern.	From Tyrol.
Lime, - -	40	41.75
Sulphuric Acid, -	60	55.00
Muriate of Soda, -		1.00
	100	97.75
	Haüy, Traité, t. iv. p. 349.	*Klaproth*, Beit. b. iv. s. 235.

Geognostic and Geographic Situations.

It is sometimes met with in the gypsum of Nottinghamshire *. In the salt-mines of Hall in the Tyrol; in those of Bex in Switzerland; in quartz, along with talc, sulphur, and iron-pyrites, in the mine of Pesay, also in Switzerland; Lauterberg in the Hartz; Tiede, near Brunswick; and in the large copper-mine of Fahlun in Sweden.

Second

* Greenough.

It is dull.

The fracture is flat conchoidal, passing to even.

The fragments are indeterminate angular, and sharp-edged.

It is slightly translucent on the edges.

It is hard, bordering on semihard.

It is brittle, and rather difficultly frangible.

Specific gravity, 2.7600, *Karsten*.

Chemical Characters.

When pounded, and thrown into diluted and heated nitrous acid, it is completely dissolved with effervescence.

Constituent Parts.

Carbonate of Lime, -	70.50
Carbonate of Magnesia,	29.50
	100.00

Klaproth, Gesellsch. N. Fr. b. i. s. 258.

Geognostic and Geographic Situations.

It occurs in veins in serpentine rocks, between Gurhof and Aggsbach, in Lower Austria.

Observations.

1. The name Gurhofit, sometimes given to this mineral, is from the place near which it was found.

2. It was at one time considered as a variety of semi-opal; but its greater weight distinguishes it from that mineral.

Second

Third Subspecies.

Fibrous Anhydrite.

Fasriger Muriacit, *Werner.*

Fasriger Muriacit, *Karsten,* Tabel. s. 52.—Fasriger Karstenit, *Haus.* Handb. b. iii. s. 883.—Fasriger Muriacit, *Hoff.* b. iii. s. 136.—Fasriger Anhydrit, *Lenz,* b. ii. s. 724.

External Characters.

Its colours are brick-red, and pale blood-red; also indigo-blue, Berlin-blue, smalt-blue, and smoke-grey.

It occurs massive, and in coarse fibrous concretions, which are straight or curved, and sometimes stellular.

Internally it is glimmering and glistening, and pearly.

The fragments are long splintery.

It is translucent on the edges, or feebly translucent.

It is rather easily frangible.

Specific gravity 3.002, *Breithaupt.*

Geographic Situation.

It is found in the salt-mines of Berchtesgaden, and at Ischel in Upper Austria, at Hall in the Tyrol, Salz on the Neckar, Carinthia, and Tiede near Brunswick.

Uses.

The blue varieties are sometimes cut and polished for ornamental purposes.

Fourth Subspecies.

Convoluted Anhydrite.

Gekröstein, *Werner.*

Chaux sulphatée Anhydre concretionnée, *Haüy.*—Gekröstein, *Hoff.* b. iii. s. 131.

External Characters.

Its colour is dark milk-white.

It occurs massive; also in distinct concretions, which are thick lamellar, and intestinally convoluted or contorted, and these are again composed of others which are thin prismatic.

Internally it is glistening or glimmering, and the lustre is pearly.

The fracture is small and fine splintery.

The fragments are indeterminate angular, and rather sharp-edged.

It is translucent on the edges, or translucent.

Same hardness as other subspecies.

Specific gravity 2.850, *Klaproth.*

Constituent Parts.

Lime, - - -	42.00
Sulphuric Acid, - -	56.50
Muriate of Soda, - -	0.25
	98.75

Klaproth, Beit. b. iv. s. 235

Geognostic and Geographic Situations.

It occurs in the salt-mines of Bochnia, and at Wieliczka in Poland.

Observations.

It was first described as a variety of compact heavy-spar, and is by many named *Pierre de Tripes*, from its convoluted concretions.

Fifth

Second Kind.

Prismatic Miemite.

Stänglicher Bitterspath, *Klaproth.*

Strahliger Kalkstein, *Von Schlottheim,* Hoff's Magaz. fur die Gesammte Mineralogie, b. i. s. 156.—Stänglicher Bitterspath, *Klaproth,* b. iii. s. 297. *Id. Leonhard,* Tabel. s. 36. *Id. Haus,* s. 128. *Id. Lenz,* b. ii. s. 712. *Id. Oken,* b. i. s. 393.

External Characters.

Its colour is asparagus-green, olive-green, and oil-green.

It occurs in prismatic distinct concretions, and crystallized in flat rhomboids, which are deeply truncated on all the edges.

The crystals are small, and very small, and sometimes they form only drusy crusts.

Internally it is shining and vitreous.

The fracture passes from concealed foliated to splintery.

The fragments are rather blunt-edged.

It is strongly translucent.

It is as hard as the granular miemite.

Specific gravity 2.885, *Karsten.*

Chemical Characters.

It dissolves slowly, and with but feeble effervescence, in nitrous acid.

Constituent

Constituent Parts.

Lime, -	33.00
Magnesia,	14.50
Oxide of Iron,	2.50
Carbonic Acid, -	47.25
Water and Loss, -	2.75
	100

Klaproth, Beit. b. iii. s. 303.

Geognostic and Geographic Situations.

It occurs in cobalt veins that traverse sandstone at Glücksbrunn in Gotha, and at Beska in Servia, on the frontier of Turkey.

Third Subspecies.

Brown-Spar, or Pearl-Spar.

Braunspath, *Werner.*

This species is divided into two kinds, viz. Foliated Brown-Spar, and Columnar Brown-Spar.

First Kind.

Foliated Brown-Spar.

Blättriger Braunspath, *Werner.*

Spath perlé, *Romé de Lisle,* t. i. p. 605.—Braunspath, *Wid.* s. 515.—Sidero-calcite, *Kirw.* vol. i. p. 105.—Braunspath, *Estner,* b. ii. s. 999. *Id. Emm.* b. i. s. 79.—Brunispato, *Nap.* p. 356.

p. 356.—Le Spath brunissant, ou le Braunspath, *Broch.* t. i. p. 563.—Gemeiner Braunspath, *Reuss,* b. i. s. 50. *Id. Lud.* b. i. s. 153. *Id. Suck.* 1r th. s. 630. *Id. Bert.* s. 118. *Id. Mohs,* b. ii. s. 108.—Spathiger Braunkalk, *Hab.* s. 82.—Chaux carbonatée manganesifere, *Lucas,* p. 8.—Spathiger Braunkalk, *Leonhard,* Tabel. s. 35.—Chaux carbonatée brunissante, *Brong.* t. i. p. 237.—Gemeiner Braunspath, *Karsten,* Tabel. s. 50.—Chaux carbonatée ferro-manganesienne, *Bournon,* Traité, t. i. p. 277.—Pearl-Spar, *Kid,* vol. i. p. 56.—Chaux carbonatée ferro-manganesifere, *Haüy,* Tabl. p. 5.—Gemeiner Braunkalk, *Lenz,* b. ii. s. 717.—Gemeiner Braunspath, *Oken,* b. i. s. 394.—Blättriger Braunspath, *Hoff.* b. iii. s. 49.

External Characters.

Its colours are flesh-red and brownish-red.

It often occurs massive, also disseminated, seldom globular, stalactitic, reniform, with tabular and pyramidal impressions; also in distinct concretions, which are granular, and rarely thin and straight lamellar; and frequently crystallized.

Its primitive form is a rhomboid, in which the obtuse angle is 107° 22′. The following are the secondary figures:

1. Rhomboid, in which the faces are sometimes cylindrically convex, sometimes cylindrically concave.
2. Lens, both common and saddle-shaped.

It also occurs in rhomboidal six-sided pyramidal suppositious crystals.

The true crystals are generally small and very small; the supposititious crystals large and middle-sized, and are either hollow, or lined with calcareous-spar.

The

It occurs crystallized, in very low oblique four-sided prisms, the lateral edges of which are 104° 28′ and 75° 32′, and in which the terminal planes are set on obliquely.

The crystals occur singly, or in groups.

The lateral planes are transversely streaked; the terminal planes are smooth.

It is shining.

The fracture parallel with the terminal planes and edges is foliated; in other directions it is conchoidal.

It is softer than calcareous-spar.

It is transparent.

It is brittle.

Specific gravity 2.700.

Chemical Characters.

It decrepitates before the blowpipe, and melts into a white enamel. In water it becomes opaque, and is partly soluble.

Constituent Parts.

Dry Sulphate of Lime,	49.0
Dry Sulphate of Soda,	51.0
	100.0

Brongniart, J. des Mines, t. xxiii. p. 17.

Geognostic and Geographic Situations.

It is found imbedded in rock-salt at Villaruba, near Ocana in New Castile, in Spain.

Observations.

It was brought from Spain to Paris by M. Dumeril, and first analysed and described by Brongniart.

2. Axifrangible

The most frequent accompanying vein-stone is calcareous spar; besides which, it is often associated with heavy-spar, fluor-spar, quartz, sparry-iron, galena, iron-pyrites, native silver, and various ores of silver. Very often it rests on all the minerals of which the vein is composed: hence it is said to be the newest mineral in the vein; and we frequently observe thin crusts of it investing the surface of crystals, as of calcareous-spar, fluor-spar, heavy-spar, quartz, galena, &c. These crusts seldom invest the whole crystal, generally covering only a part of it; and it is observed, that it is the same side in all the crystals of the same cavity which are encrusted with the brown-spar; and also, that when the whole side is not covered, the crust has the same height, or is on the same level in all the crystals.

Geographic Situation.

It occurs along with galena, and other ores of lead, in the lead-mines of Lead Hills and Wanlockhead in Lanarkshire; in the mines of Cumberland, Northumberland, and Derbyshire. On the Continent, it is found in Norway, Sweden, Saxony, Suabia, Piedmont, France, Hungary, and Transylvania.

Observations.

1. It is distinguished from *Calcareous-spar*, with which it has been confounded, by its colours, cleavage, inferior transparency, perfect pearly lustre, greater hardness, and higher specific gravity. It also in general effervesces less briskly with acids than calcareous-spar.

2. The straight lamellar variety has been mistaken for Heavy-spar, from which, however, it is distinguished, not only by its inferior weight, but also by its concretions being

greenish-white, and yellowish-white, and also wax-yellow, pale ochre-yellow, and yellowish-brown. Sometimes it is dark-brown, owing to intermixed stinkstone. Some varieties display iridescent colours.

It occurs massive, coarsely disseminated, also in distinct concretions, which are large and coarse granular, and sometimes inclining to thick lamellar; and crystallized.

Its primitive figure is an oblique four-sided prism, in which the angles are 113° 8′, and 66° 52′. The following are some of the secondary figures:

1. Six-sided prism *, generally broad and oblique angular, with two opposite broad, and four smaller lateral planes; or with two opposite very small, and four broader planes; or with alternate broader and narrower lateral planes: the terminal planes or faces are conical or spherical convex, or obtusely bevelled, and the bevelling planes set on obliquely, but parallelly on the broader lateral planes †; or acuminated with four planes, which are set on the smaller lateral planes ‡.
2. Lens.
3. Twin-crystals. These are either formed by two lenses, which are attached by their faces, or by two six-sided prisms pushed into each other in the direction of their breadth, in such a manner, that the united summits at one extremity form a re-entering angle, but at the other a salient angle, or four-planed acumination. When two such twin-crystals

* The primitive figure, according to Haüy, is a four-sided prism, whose bases are oblique parallelograms, with angles of 113° 7′ 48″ and 66° 52′ 12″.

† Chaux sulphatée trapezienne, Haüy.

‡ Chaux sulphatée equivalente, Haüy.

Constituent Parts.

Carbonate of Lime, -	51.50
Carbonate of Magnesia,	32.00
Carbonate of Iron,	7.50
Carbonate of Manganese,	2.00
Water, - - -	5.00
	98.00

Klaproth, Beit. b. iv. s. 208.

Geographic Situation.

It is found at the mine named Segen Gottes at Gersdorf in Saxony; and in that of Valenciana at Guanuaxuato in Mexico.

Observations.

It is distinguished from the other subspecies of *Brown-spar* by distinct concretions, fragments, and transparency.

3. Limestone.

Rhomboedrischer Kalk-Haloide, *Mohs.*

This species is divided into twelve subspecies, viz. 1. Foliated Limestone, 2. Compact Limestone, 3. Chalk, 4. Agaric Mineral, 5. Fibrous Limestone, 6. Calc-Tuff, 7. Peastone, 8. Slate-Spar, 9. Aphrite, 10. Lucullite, 11. Marl, 12. Bituminous Marl Slate.

First Subspecies.

Foliated Limestone.

Blättriger Kalkstein, *Werner.*

This subspecies is divided into two kinds, viz. Calcareous-spar, and Foliated Granular Limestone.

First Kind.

Calcareous-Spar, or Calc-Spar.

Kalkspath, *Werner.*

Spathum, *Wall.* t. i. p. 140.—Körniger Kalkstein, var. *Wid.* s. 427.—Common Spar, *Kirw.* vol. i. p. 86.—Kalkspath, *Estner,* b. ii. s. 941. *Id. Emm.* b. i. s. 456.—Spatho calcareo, *Nap.* p. 341.—Calcaire cristallisée, *Lam.* t. i. p. 29.—Chaux carbonatée cristallisée, *Haüy,* t. ii. p. 127.—Le Spath calcaire, *Broch.* t. i. p. 536.—Spathiger Kalkstein, *Reuss,* b. ii. 2. s. 284.—Kalkspath, *Lud.* b. i. s. 149. *Id. Suck.* 1r th. s. 600.—Grossblättricher Kalkstein, *Bert.* s. 90.—Kalkspath, *Mohs,* b. ii. s. 31. *Id. Hab.* s. 76.—Chaux carbonatée, *Lucas,* p. 3.—Gemeiner spathiger Kalkstein, *Leonhard,* Tabel. s. 33.—Chaux carbonatée pure spathique, *Brong.* t. i. p. 189.—Chaux carbonatée, *Brard,* p. 26.—Kalkspath, *Haus.* s. 125.—Spathiger Kalkstein, *Karst.* Tabel. s. 50.—Crystallized Carbonate of Lime, *Kid,* vol. i. p. 50.—Chaux carbonatée, *Haüy,* Tabl. p. 2.—Kalkspath, *Lenz,* b. ii. s. 742. *Id. Haus.* Handb. b. iii. s. 900. *Id. Hoff.* b. iii. s. 17.—Calcareous-spar, *Aikin,* p. 158.

External

External Characters.

Its most frequent colour is white, of which the following varieties occur, viz. reddish, snow, greyish, greenish, and yellowish white. From reddish-white, it passes on the one side into pearl-grey, brick-red, flesh-red, rose-red, and brownish-red; and on the other side into pale violet-blue: from greyish-white, it passes into smoke-grey, ash-grey, yellowish-grey, and greenish-grey: from greenish-grey, it passes into apple, asparagus, olive, and leek green: from yellowish-grey, it passes into a colour intermediate between wax and ochre yellow, and into honey-yellow; and from honey-yellow, into yellowish-brown, and greyish-black.

The white and grey varieties occur more frequently in the massive, the yellow, green, and red, in those which are crystallized.

The white-coloured transparent varieties are often iridescent.

It occurs massive, disseminated, globular, botryoidal, reniform, tuberose, stalactitic, tubular, cellular, and curtain-shaped; also in distinct concretions, which are large coarse, rarely small, angulo-granular; sometimes very thick, thick and thin, prismatic, generally wedge-shaped prismatic; always straight; sometimes parallel, sometimes scopiform prismatic; and these are intersected by lamellar concretions which are fortification-wise bent, and very frequently crystallized.

Its primitive form is a rhomboid, in which the angles are 105° 5′, and 74° 55′.

The suite of crystallizations of calcareous-spar far exceeds in extent that of any other mineral hitherto discovered,

ed *. The principal varieties are by Werner, according to his method, brought under three classes or subdivisions, which not only form series amongst themselves, but are connected together in such a manner, that the last member of the third class joins with the first member of the first class, and thus the whole forms a very beautiful returning series. Each of these divisions have their characteristic crystalline form, viz.

The first an *acute double six-sided pyramid*; the second an *equiangular six-sided prism*, (including the six-sided table); and the third a *rhomboid* or *three-sided pyramid*.

I. *Acute six-sided Pyramid*.

When perfect, it is always acute, and two and two lateral planes meet under obtuser angles than the others.

It is generally obliquely streaked, but the streaks run from the acute towards the obtuse edges.

It occurs,

A. Single.

B. Double. The lateral planes of the one, set obliquely on the lateral planes of the other, so that the edge of the common base forms a zig-zag line.

These pyramids occur, either perfect, or in the following varieties:

1. The apex acuminated with three planes, which are set on the obtuse lateral edges. These are parallel with the cleavage.
2. The apex flatly acuminated with three convex faces, which are set on the acute lateral edges. The convexity

* Romé de Lisle enumerates 26 varieties of calcareous-spar—Haüy above 150,—and Bournon 642. Many more might be described.

convexity is in the direction of the axis of the double pyramid.

3. The angles on the common base of the double pyramid truncated, thus forming a transition into the six-sided prism.
4. The acute lateral edges of the double pyramid sometimes truncated, and either with straight and smooth planes, or with convex and uneven planes.
5. Twin-crystal.

The double six-sided pyramids apparently pushed into each other, in the direction of their length, in which they are either

(1) *Unchanged* in position, when the acute edges rest on the obtuse edges; or they are

(2) *Turned around* one-sixth of their periphery, so that obtuse edges are set on obtuse edges, and acute edges on acute edges; and the alternate angles on the common base have broken re-entering angles; or the angles on the common basis are truncated, and thus a transition is formed into the next following principal form.

II. *Equiangular Six-sided Prism.*

It is equiangular, but generally with alternate broad and narrow lateral planes. It originates from the pyramid N° 3.; and hence it presents the following varieties:

1. The equiangular six-sided prism, acutely acuminated with six planes, of which two and two meet under obtuse angles, and each is set obliquely on the lateral edges. Sometimes the acute acuminating edges are truncated, or they are bevelled, and the

the edges which the bevelling planes make with the broad lateral planes, truncated.

In other varieties, the apices of the acuminations are more or less deeply truncated, and sometimes so deeply, that the acuminating planes appear as truncating planes on the angles of the prism.

3. The preceding figure, in which the six-planed acumination is *flatly acuminated* with three planes, which are set on the acute edges of the six-planed acumination.

4. Six-sided prism *acutely acuminated* with three planes which are set on the alternate lateral planes. The apex of the acumination is sometimes more or less deeply truncated. Sometimes the truncation is so deep, that the remains of the acuminating planes appear as truncations on the alternate terminal edges. In other varieties, the prism becomes so short, that the acuminating planes meet and form an acute double three-sided pyramid.

5. When the planes of the flat three-planed acumination N° 3. increase so much that those of the six-planed acumination disappear, a six-sided prism is formed, *flatly acuminated* with three planes, which are set on the alternate lateral planes in an unconformable position. When the prism disappears, there is formed an obtuse double three-sided pyramid.

These prisms are often pyramidally aggregated.

6. When the prism becomes very low, it may be viewed as an equiangular six-sided table, which is sometimes aggregated in a rose-like form.

7. Sometimes the six-sided prism is truncated on the lateral edges, and thus forms a twelve-sided prism.

The

The prisms are aggregated in a pyramidal, manipular, scopiform, and tabular manner.

III. *Three-sided Pyramid.*

It is divided, according to the magnitude of the summit-angle, into the following varieties:

1. *Very obtuse three-sided pyramid,* nearly tabular. It is sometimes aggregated in a rose-like form.
2. *Flat three-sided pyramid,* in which the lateral planes of the one are set on the lateral edges of the other. The angles on the common basis are sometimes truncated, and frequently the apices of the pyramid are more or less deeply truncated. When the truncation on the apices is very deep, the crystal appears as a six-sided table, in which the terminal planes are set on alternately oblique.
3. *Acute three-sided pyramid.* This form very nearly resembles the cube.

It would extend this description too much, were we to attempt to give an account of every variety of form exhibited by these crystals; and besides, we have already enumerated the principal ones.

The crystals occur of various magnitudes, as large, small, and very small.

The lateral planes of the prisms and pyramids are generally shining, splendent and smooth; the acuminating planes frequently streaked or drusy, seldom granulated. Sometimes it occurs in extraneous external forms of shells, &c.

Internally it is generally specular splendent, or shining, sometimes glistening, and the lustre is vitreous, which inclines sometimes to resinous, and more rarely to pearly.

It has a distinct cleavage, in which the folia are generally straight, seldom spherically curved. The most distinct cleavage

cleavage is threefold, and in the direction of the planes of the primitive rhomboid; there are others less distinct in the direction of the planes of the obtuse rhomboid or double three-sided pyramid; and in the direction of the alternate lateral planes of the regular six-sided prism.

The fracture is perfect conchoidal.

The fragments are indeterminate angular, and rather sharp-edged, or they are rhomboidal.

It occurs transparent, semi-transparent, and occasionally only translucent. It refracts double *.

It is semi-hard; it scratches gypsum, but is scratched by fluor-spar.

It is brittle, and very easily frangible.

Specific gravity, 2.5, 2.8, *Mohs.*

Chemical Characters.

It is infusible before the blowpipe, but it becomes caustic, losing by complete calcination about 48 *per cent.*; effervesces violently with acids.

Constituent Parts.

	Iceland Spar.	Iceland Spar.	Iceland Spar.	From Andreasberg.
Lime,	56.15	55.50	56.50	55.9802
Carbonic Acid,	43.70	44.00	43.00	43.5835
Water,		0.50	0.50	0.1000
Oxide of Manganese, with trace of Iron,	0.15			0.3362
	100.00	100.00	100.00	100.0000
	Stromyer, Gilbert's Annalen for 1813, p. 217.	*Philips*, Phil. Mag. xiv. 290.	*Buchols*, Gehl. Journ. iv. 412.	*Stromyer*, Gilbert's An. for 1813, p. 217.

Geognostic

* The double refracting power of calcareous-spar was first observed by Erasmus Bartholin.

Geognostic Situation.

It never occurs in mountain-masses, but venigenous in almost every rock, from granite to the newest secondary formation. The oldest formation of this mineral is that in veins, where it is accompanied with felspar, rock-crystal, probably also with epidote, sphene, and chlorite. It occurs also in beds, along with augite, hornblende, garnet, and magnetic ironstone; and frequently in veins in different metalliferous formations. Thus, it is associated with nearly all the metallic minerals contained in gneiss, mica-slate, clay-slate, syenite, porphyry; seldomer in granite, more frequently, again, in grey-wacke, and along with cobalt and copper ores in the oldest secondary or flœtz limestone. Veins, almost entirely composed of calcareous-spar, abound in the newest limestone formations; and it is a common mineral, either in veins, or in cotemporaneous masses, in the various rocks of the secondary or flœtz-trap series.

An interesting geognostic character of calcareous-spar, is the uniformity of its crystallizations in particular districts. Thus, in the mines of Derbyshire, the acute six-sided pyramid and its congenerous forms are the most frequent and abundant; at Schneeberg in Saxony, and in the Upper Hartz, the prevailing forms are the regular six-sided prism and table; while in the mines of Freyberg the most frequent forms are the regular six-sided prism, acuminated with three planes, set on the lateral planes, and the flat double three-sided pyramid.

Geographic Situation.

This mineral is so common in every country, as to render any account of its geographic distribution unnecessary. It

melt into a white enamel: when heated with charcoal, they are converted into sulphuret of lime.

Constituent Parts.

Lime,	-	34	
Sulphuric Acid,	-	48	
Water,	-	18	
		100	*Gerhard.*

Geognostic Situation.

It occurs in beds, along with granular gypsum, selenite, and stinkstone, in the flœtz or secondary class of rocks.

Geographic Situation.

It occurs in the Campsie Hills; Derbyshire; Ferry-bridge in Yorkshire; and Nottinghamshire: on the Continent, in Mansfeldt, Thuringia, Bavaria, Switzerland, and France.

Observations.

Its inferior hardness, and inferior specific gravity, distinguish it from *Compact Limestone.* Its warmer feel also distinguishes it well (when both are polished) from Compact Limestone.

Fourth

Chaux carbonatée sub-granulaire, *Haüy*, Tabl. p. 5.—Körniger Kalkstein, *Lenz*, b. ii. s. 739.—Körnigblättriger Kalkstein, *Hoff.* b. iii. s. 14.—Granular Limestone, *Aikin*, p. 159.

External Characters.

Its most common colour is white, of which it presents the following varieties: snow-white, yellowish-white, greyish-white, and greenish-white, seldom reddish-white: from greyish-white it passes into bluish-grey, greenish-grey, ash-grey, and smoke-grey; from reddish-white it passes into pearl-grey, and flesh-red; from yellowish-white into cream-yellow; and from greenish-white into siskin-green, and olive-green.

It has generally but one colour; sometimes, however, it is spotted, dotted, clouded, striped, and veined.

It occurs massive, and in angulo-granular distinct concretions.

Internally it alternates from shining to glistening and glimmering, and the lustre is intermediate between pearly and vitreous.

The fracture is foliated, but sometimes inclines to splintery.

The fragments are indeterminate angular, and rather blunt-edged.

It is more or less translucent.

It is as hard as calcareous-spar.

It is brittle, and easily frangible.

Specific gravity, Carrara Marble, 2.717. Scottish, 2.716, *Kirwan*.—2.658, 2.711, *Karsten*.

Chemical

Chemical Characters.

It generally phosphoresces when pounded, or when thrown on glowing coals. It is infusible before the blow-pipe. It dissolves with effervescence in acids.

Constituent Parts.

Lime, -	56.50
Carbonic Acid,	43.00
Water, -	0.50
	100

Bucholz, in Neuen Journal der Chem. iv. s. 419.

Geognostic Situation.

This mineral occurs in beds, in granite, gneiss, mica-slate, clay-slate, syenite, greenstone, grey-wacke, and rarely in some of the secondary rocks. It is observed, that the varieties which occur in highly crystallized rocks are in general more crystalline than those which are found in the compact or less crystallised varieties. It frequently contains imbedded minerals of different kinds, such as quartz, mica, hornblende, tremolite, sahlite, asbestus, steatite, serpentine, galena, blende, iron-pyrites, and magnetic ironstone: of these the quartz and mica are the most frequent.

Geographic Situation.

This mineral occurs in all the great ranges of primitive rocks that occur in Europe, and in such as have been examined in Asia, Africa, and America.

Uses.

Uses.

All the varieties of this subspecies may be burnt into quicklime; but it is found, that in many of them, the concretions exfoliate and separate during the volatilization of their carbonic acid, so that by the time when they are rendered perfectly caustic, their cohesion is destroyed, and they fall into a kind of sand,—a circumstance which will always render it improper to use such varieties in a common kiln. But the most important use of this mineral is as marble. The marbles we are now to mention, have in general purer colours, more translucency, and receive a higher polish than those of compact limestone. They have been known from a very early period; and ancient statuaries have immortalised their names, by the master-pieces of art which they have executed in them. To give a full description of all the ancient and modern marbles enumerated by mineralogists, would much exceed the limits of this article; and besides, it would encroach on the more complete economical history of them, intended to be given in another work. We shall here notice only some of the more remarkable ancient marbles, and a few of the modern marbles found in this country, on the Continent of Europe, and in other countries.

Ancient, or Antique Marbles.

Under this head, we include those marbles which were made use of by the ancients, and the quarries of many of which are no longer known.

1. *Parian Marble.*—Its colour is snow-white, inclining to yellowish-white, and it is fine granular, when polished, has somewhat of a waxy appearance. It hardens by exposure

exposure to the air, which enables it to resist decomposition for ages. Varro and Pliny inform us, that it was named *Lychnites* by the ancients, from its being hewn in the quarry by the light of lamps, from λυχνις, *a lamp*; but Hill is of opinion, that the appellation is from the verb λυχνάω, *to be very bright*, or *shining*, from the shining lustre of this marble: the etymological derivation of Varro and Pliny is that which is generally adopted. Dipœnus, Scyllis, Malas, and Micciades, employed Parian marble, and were imitated by their successors. This preference was justified by the excellent qualities of this marble; for it receives with accuracy the most delicate touches of the chisel, and it retains for ages, with all the softness of wax, the mild lustre even of the original polish. The finest Grecian sculpture which has been preserved to the present time, is generally of Parian marble.—The Medicean Venus, the Diana venatrix, the colossal Minerva (called Pallas of Velletri), Ariadne (called Cleopatra), Juno (called Capitolina), &c. It is also a variety of Parian marble on which the celebrated tables at Oxford are inscribed.

2. *Pentelic Marble*, from Mount Pentelicus, near Athens. This marble very closely resembles the preceding, but is more compact, and finer granular, sometimes combined with splintery. At a very early period, when the arts had attained their full splendour, in the age of Pericles, the preference was given by the Greeks, not to the marble of Paros, but to that of Mount Pentelicus, because it was whiter, and also, perhaps, because it was found in the vicinity of Athens. The Parthenon was built entirely of Pentelic marble. Many of the Athenian statues, and the works carried on near to Athens during the administration of Pericles, (as, for example, the temples of Ceres or Eleusis),

were

were executed in the marble of Pentelicus *. Among the statues of this marble in the Royal Museum in Paris, are the Torso; a Bacchus in repose; a Paris; the Discobolus reposing; the bas-relief known by the name of the Sacrifice; the throne of Saturn; and the Tripod of Apollo. It is remarked by Dr Clarke, that while the works executed in Parian marble remain perfect, those which were finished in Pentelican marble have been decomposed, and sometimes exhibit a surface as earthy and as rude as common limestone. This is principally owing to veins of extraneous substances which intersect the Pentelican quarries, and which appear more or less in all the works executed in this kind of stone.

3. *Greek White Marble,—Marmo Greco* of Italian artists. Its colour is snow-white; is fine granular; and is rather harder than the other white marbles; hence it takes a higher polish. This is one of those varieties which being found near the river Coralus in Phrygia, was called *Corallitic* or *Corallic Marble* by the ancients. The Greek marble was obtained from several islands of the Archipelago, such as Scio, Samos, &c.

4. *White Marble of Luni*, on the coast of Tuscany. It is of a snow-white colour, small granular and very compact; it takes a fine polish, and may be employed for the most delicate work: hence it is said to have been preferred by the Grecian sculptors, both to the Parian and Pentelic marbles. It is the general opinion of mineralogists, that the Belvidere Apollo is of Luni marble; but the Roman sculptors look upon it as Greek marble †. The Antinous

* Clarke's Travels, vol. iii.

† Dr Clarke says it is of Parian marble. Vid. Travels, vol. iii.

was also quarried and wrought by t
ries are said to have been opene
Cæsar. In the centre of blocks o
rock-crystals are found which are
mondo[*].

6. *White Marble of Mount Hym*
marble has a greater intermixture o
varieties already mentioned. The
the Royal Museum in Paris, is of t

7. *Translucent White Marble*,-
the Italians.—This marble much r
but differs from it in being more t
at Venice, and in several other t
lumns and altars of this marble. Th
statuario are unknown.

8. *Flexible White Marble.*—Th
greyish-white coloured marble, whic
flexibility. It was dug up in the
In the Borghese Palace in Rome,
tables of it.

These are the chief white mar
used for the purposes of architectur

---

[*] It is said the marble quarries of Carrar
that statuary marble is in future to be procure

9. *Red antique Marble,—Roso antico* of the Italians, —*Ægyptum* of the ancients.—This marble, according to antiquaries, is of a deep blood-red colour, here and there traversed by white veins, and, if closely inspected, appears to be sprinkled over with minute white dots, as if it were strewed with sand. The Egyptian Antinous in the Royal Museum in Paris, is of this marble. But the most highly prized variety of antique red marble, is that of a very deep red, without veins, such as is seen in the Indian Bacchus in the same Museum. The white points, which are never wanting in the true red antique marble, distinguish it from others of the same colour. It is not known from whence the ancients obtained this marble: the conjecture is, that it was brought from Egypt.

10. *Green antique Marble,—Verde antico* of the Italians.—This beautiful marble is an indeterminate mixture of white marble and green serpentine. It was known to the ancients under the name *Marmor Spartam* or *Lacedæmonium*.

11. *Yellow antique Marble,—Giallo antico* of the Italians.—This marble is of a yellowish-brown, sometimes inclining to a cream-yellow colour, and is either of an uniform colour, or marked with black or deep yellow-coloured rings. It is found only in small detached pieces, and in antique inlaid-work. The Sienna marble is a good substitute for it.

12. *Antique Cipolin Marble.*—Cipolin is a name given to all such white marbles as are marked with green-coloured zones, caused by talc or chlorite. It was much used by the ancients. It takes a fine polish, but its green-coloured stripes always remain dull, and are that part of the marble which first decomposes, when exposed to the open air. There are modern Cipolins as fine as those used by the ancients.

b. White Tiree Marble.—Its colours are greyish-white and bluish-white: it contains scales of mica, and crystals or grains of common hornblende; which latter, when minutely diffused, give the marble a green or yellowish-green colour, and when very intimately combined with the mass, form beautiful yellowish-green spots.

2. *Iona Marble.*—Its colours are greyish-white and snow-white. Its lustre is glimmering, and fracture minute foliated, combined with splintery. It is harder than most of the other marbles. It is an intimate mixture of limestone and tremolite; for if we immerse it in an acid, the carbonate of lime will be dissolved, and the fibres of tremolite remain unaltered. It is sometimes intermixed with steatite, which gives it a green or yellow colour, in spots. These yellow or green coloured portions receive a considerable polish, and have been erroneously described as nephritic stone, and are known also under the name of *Iona* or *Icolmkill Pebbles*. The marble itself does not receive a high polish: this, with its great hardness, have brought it into disrepute with artists. Several of the varieties of Iona marble are dolomite.

3. *Skye Marble.*—In the Island of Skye, in the property of Lord Macdonald, there are several varieties of marble, deserving of attention, inclosed in porphyry, sandstone, and trap-rocks. One variety is of a greyish inclining to snow-white colour: another greyish-white, veined with ash-grey; and a third is ash-grey, or pale bluish-grey, veined with lemon-yellow or siskin-green *. Dr MacCulloch has described other varieties; and more minute details are expected from his promised work on the "Geology of the Hebrides."

* Mineralogy of Scottish Isles, vol. ii.

4. *Assynt.*—The following varieties of marble found in Sutherland, have been introduced into commerce by Mr Joplin of Gateshead.

a. White marble, which acquires a smooth surface on the polisher, but remains of a dead hue, like the marble of Iona: hence its uses as an ornamental marble are much circumscribed.

b. White mottled with grey, and capable of receiving a high polish, and is not deficient in beauty.

c. Grey coloured, and highly translucent and crystalline, and capable of being applied to the purposes of ornament in sepulchral sculpture.

d. Dove-coloured, compact, translucent, and receiving a good polish.

e. Pure white, and translucent, and capable of being used in plain ornaments, but too translucent for sculpture.

f. White, with irregular yellow marks, from being intermixed with serpentine. It is very compact.

g. White variety, with layers of slate-spar.

5. *Glen Tilt Marble.*—The limestone of Glen Tilt, first mentioned by Dr Macknight, in his description of that valley *, has of late attracted the notice of the Duke of Athole, through the suggestion of Dr MacCulloch. The marbles are white and grey, and veined or spotted with yellow or green: they vary in the size of the grain or concretion, and also in the degree and kind of polish they receive.

6. *Marble of Ballichulish.*—This marble is of a grey or white colour, and is very compact. It may be raised in blocks of considerable size.

7. *Boyne Marble.*—Its colours are grey or white, and it receives a pretty good polish.

8. *Blairgowrie*

* Wernerian Memoirs, vol. i. p. 362.

8. *Blairgowrie Marble.*—Mr Williams, in his Natural History of the Mineral Kingdom, mentions a beautiful saline marble, of a pure white colour, which occurs near Blairgowrie in Perthshire, not far from the road side. According to him, it may be raised in blocks and slabs, perfectly free of blemishes, and in every respect fit to be employed in statuary and ornamental architecture.

9. *Glenavon Marble,*—is of a white colour, and the concretions are large granular. It is mentioned by Williams as a valuable marble; but he adds, that its situation is remote, and difficult of access.

English Marbles.

Hitherto but few marbles of granular foliated limestone have been quarried in England; the greater number of varieties belonging to the flœtz or secondary limestone. One of the most remarkable of the English marbles of the present class, is that of Anglesey, named *Mona Marble*, which is not unlike the *Verde Antico.* Its colours are greenish-black, leek-green, and sometimes purple, irregularly blended with white; but they are not always seen together in the same piece. The white part is limestone: the green shades are said to be owing to serpentine and asbestus. The Black Marbles found in England, are varieties of Lucullite.

Irish Marbles.

The Black Marbles of Ireland, now so generally used by architects, are Lucullites. In the county of Waterford, different kinds of marble are known; thus at Toreen, there is a fine variegated sort, of various colours, viz. chesnut-brown, white, yellow, and blue, and which takes a good polish: a

grey

grey marble, beautifully clouded with white, susceptible of a good polish, has been found near Kilcrump, in the parish of Whitechurch, in the same county. At Loughlougher, in the county of Tipperary, a fine purple marble is found, which, when polished, is said to be beautiful. Smith describes several variegated marbles in the county of Cork; but whether these, and others now enumerated as Irish marbles, are granular limestone, I cannot discover, as I have neither met with good descriptions of them, nor seen any specimens. Thus, he mentions one with a purplish ground, and white veins and spots, found at Churchtown; a bluish and white marble from the same place; and several fine ash-coloured varieties, as that of Castle Hyde, &c. The county of Kerry affords several variegated marbles, such as that found near Tralee. Marble of various colours is found in the same county, in the islands near Dunkerron, in the river of Kenmare: some are purple and white, intermixed with yellow spots; and some beautiful specimens have been seen, of a purple colour, veined with dark-green.

French Marbles *.

A great many different kinds of marble are quarried in the different Departments of the kingdom of France, and of these we shall mention the following.

1. *Griotte Marble.*—Its colour is deep brown, with blood-red oval spots, produced by shells. This marble has obtained its name from its brownish colour, being similar to that

* As I have not seen all the varieties of foreign marble now to be described, I cannot pretend to say with certainty that the whole of them belong to the Granular Foliated Limestone. The descriptions are from Brard's Treatise on Precious Stones.

that of a variety of cherries, likewise called *griotte;* but it also sometimes contains large white veins, which traverse the other spots, and which, as destroying the harmony of the other tints, are considered as a defect. Some of the ornaments of the Triumphal Arch of the Carousel, are made of griotte; which is now much employed in the decoration of public monuments, and of splendid furniture. It is sold at about 200 francs the cubic foot. It is found in the Department of Herault.

2. *Marble of Languedoc, or of St Beaume.*—It is of a bright red colour, and is marked with white and grey zones, formed by madrepores. The eight columns which adorn the Triumphal Arch in the Carousel at Paris, are of this marble. The quarries are at St Beaume, in the Department of Aude.

3. *Campan Marble.*—This is a mixture of granular foliated limestone and a green talcky mineral, which forms veins on its surface. There are three varieties of Campan, which, however, are often united in the same piece: the first, called *Green Campan,* is of a pale sea-green colour, and exhibits on its surface lines of a much deeper green, and forming a kind of net-work: the second, called *Isabel Campan,* is of a delicate rose-colour, and, like the first, is furnished with undulating veins of green talc: the third variety, the *Red Campan,* is of a deep red colour, with veins of a still deeper red, and in some degree resembles parts of the griotte. In order to form a correct idea of the Campan marble, properly speaking, we must imagine that these three varieties are united, so as to form large stripes, from a few inches, to two, three, or even six feet broad, which produce a very grand and pleasing effect when viewed in large masses. When, therefore, the Campan marble can be employed in the large way, it may be looked upon

as one of great beauty and splendour. It should not, however, be exposed to the weather, since, by so doing, the talcose substance exfoliates, and leaves hollow spaces, which renders its surface uneven and rough; but it answers extremely well in the interior of buildings, for chimney-pieces, slabs for tables, &c. There are immense quarries of this valuable marble at Campan, near Bagnere, in the High Pyrenees.

4. *Sarencolin Marble.*—It exhibits on its surface large straight zones, and angular spots, of a yellow or blood-red colour, so that at first view it bears some resemblance to the marble called Sicilian. The finer varieties have become very scarce. It is found at Sarencolin, in the High Pyrenees.

5. *Breccia Marble of the Pyrenees.*—One variety contains, in a brownish-red basis, black, grey, and red, middle-sized spots. It admits a good polish. Another variety has an orange-yellow coloured basis, containing small fragments of a snow-white colour. Both varieties are found in the High Pyrenees.

Italian Marbles.

1. *Sienna Marble, or Brocatella di Siena.*—It has a yellowish colour, and disposed in large irregular spots, surrounded with veins of bluish-red, passing sometimes into purple. It is by no means uncommon in Siena. At Montarenti, two leagues from Siena, another yellow marble is found, which is traversed by black and purplish-black veins. This is frequently employed throughout Italy.

2. *Mandelato Marble.*—It is a light red marble, with yellowish-white spots, found at Luggezana in the Veronese. Another variety, bearing the same name, occurs at Preosa.

They

They are both employed for columns, and various other works.

3. *Green Marble of Florence.*—It is of a green colour, which it owes to an intermixture of steatite.

4. *Verde di Prado Marble.*—It is a green marble, marked with dark green spots, having greater intensity than the base or ground. It is found near the little town of Prado in Tuscany.

5. *Rovigo Marble.*—It is of a white colour, but is inferior in quality to those of Carrara and Genoa. It is found at Padua.

6. *Luni Marble.*—It is of a white colour, with red-coloured spots and dots. It is found at Luni, on the coast of Tuscany.

7. *Venetian Marble.*—It is white, with red and yellow spots and veins. It is found in the Venetian territory.

8. *Lago Maggiore Marble.*—It is white, with black spots and dots, and is of great beauty. It has been employed for decorating the interior of many churches in the Milanese.

9. *Brèche d'Italie.*—It has a reddish-brown ground, veined with white. It is a beautiful marble, but requires much care in keeping, since it becomes soon spotted, by coming into contact with greasy substances.

10. *Bretonico Marble.*—This beautiful marble is composed of yellow, grey, and rose-coloured portions or fragments. It is found near the village of Bretonico, in the Veronese.

11. *Bergamo Marble.*—It is composed of grey and black fragments, in a green basis.

Sicilian

The following are some of the principal marbles found in Spain.

1. *White Marble.*—Near Cordova, there is a white fine granular marble, which takes a good polish, and is very fit for sculpture. Near Filabres, three leagues from Almeria in Grenada, there is a hill of about a league in circumference, and 2000 feet in height, which is said to be entirely composed of the purest white marble, capable of the finest polish; and the rocks which surround the town of Molina in New Castile, are composed of a white marble, which has been employed in the Palace of the Alhambra at Grenada.

2. *Red Marble.*—There is a beautiful red variety, with shining red and white spots and veins, called *Red Seville Marble.* There is also a flesh-coloured variety, veined with white, from Santiago. A dull red marble, with minute black veins, is found in Meguera in Valencia, and is much used in Spain for tables. The mountains of Guipuscoa afford a red marble, veined with grey, and closely resembling that of Sarencolin.

3. *Tortosa Marble.*—Its basis or ground is violet, and it is spotted with bright yellow.

4. *Grenada Marble.*—It is of a green colour, and very much resembles the Verde Antico. It is found at Grenada.

5. *Spanish Brocatello Marble.*—This is a well known and very beautiful variety of marble.

6. *Breccia Marble.*—Several beautiful varieties of this marble occur in Spain. At Riela in Arragon, there is a beautiful breccia marble, composed of angular portions or fragments of a black marble, imbedded in a reddish-yellow base. The breccia marble of Old Castile is of a bright red, dotted with yellow and black, and incloses middle-sized

appearance. The principal marble is that of Fagernech, which is white, with veins of green talc.

Russian and Siberian Marbles.

The vast Empire of Russia affords a great many different kinds of marble. Georgi, in his Description of the Russian Empire, enumerates white, grey, green, blue, yellow, and red varieties; and Patrin gives the following account of the Siberian marbles. "The Uralian Mountains furnish the finest and most variegated marbles. The greater part is taken from the neighbourhood of Catharinenburg, where they are wrought, and from thence transported into Russia, particularly to Petersburgh. The late Empress caused an immense palace to be built in her capital for Orloff, her favourite, which is entirely coated with these fine marbles, both inside and outside. The Empress built the church of Isaac with the same marbles, on a vast space, near the statue of Peter the Great." Patrin found no white statuary marble in the Uralian Mountains; but in that part of the Altain Mountains which is traversed by the river Irtish, he in two places saw immense blocks of marble, perfectly white and pure, from which blocks might be hewn.

Asiatic Marbles.

At present we are very imperfectly acquainted with the marbles of Asia.

Shaw mentions a red marble from Mount Sinai: Russell, in his Natural History of Aleppo, gives an imperfect account of the marbles of Syria; and some Persian marbles are noticed by Chardin. Mr Morier, in his Journey through Persia, mentions a very beautiful marble, under the name *Marble of Tabriz*, and informs us, that the tomb

of

of the celebrated poet Hafitz is constructed with it, and that the wainscotting of the principal room of the Haft-tan, near Schiraz, is likewise of this marble. Its colours are described as light green, with veins, sometimes of red, sometimes of blue, and it has great translucency. It is cut in large slabs; for Mr Morier saw some that measured nine feet in length, and five feet in breadth. He says, that it is not procured near the city of Tabriz, or taken from a quarry, but is said to be rather a petrifaction, found in large quantities, and in immense blocks, on the borders of the Lake Shahee, near the town of Meraugheh. If it is a mere calcareous deposition, formed in the way of calcareous-alabaster or calc-sinter, it must be considered, not as marble, but a variety of that mineral.

The marbles of Hindostan, Siam, and China, are almost unknown to us. Authors speak of a quarry of white marble in the neighbourhood of Pekin; and of a similar marble in the vicinity of the capital of Siam.

African Marbles.

Beds of marble occur in the Atlas Mountains, and in those ranges that bound the shores of the Red Sea *.

American Marbles.

A good many different marbles have been discovered in the United States. The principal quarries are at Stockbridge and Lanesborough, Massachussets: in Vermont and Pennsylvania: in New-York; and in Virginia. According to Professor Hall, as mentioned by Mr Kœnig, marble has been found in many places on the west side of the Green Mountains in Vermont. A few years sinc[illegible]

* Vid. Murray's interesting and valuable work, "Historic[al Account] of Discoveries and Travels in Africa," for particulars in regard to th[e]

luable quarry was opened in Middleburg, a town situated on Otter Creek, eleven miles above Vergennes. The marble is of different colours in different parts of the bed. The principal colour, however, is bluish-grey. It takes a good polish, and is in general free of admixture of any substance that might affect its polish.

Second Subspecies.

Compact Limestone.

Dichter Kalkstein, *Werner.*

This subspecies is divided into three kinds, viz. Common Compact Limestone, Blue Vesuvian Limestone, and Roestone.

First Kind.

Common Compact Limestone.

Gemeiner Dichter Kalkstein, *Werner.*

Calcareus aequabilis, *Wall.* t. i. p. 122.—Dichter Kalkstein, *Wid.* s. 494.—Compact Limestone, *Kirw.* vol. i. p. 82.—Gemeiner Dichter Kalkstein, *Emm.* b. i. s. 437.—Pietra calcarea compacta, *Nap.* p. 33.—La pierre calcaire compacte commune, *Broch.* t. i. p. 523.—Chaux carbonatée compacte, *Haüy,* t. ii. p. 166.—Gemeiner Dichter Kalkstein, *Reuss,* b. ii. 2. s. 262. *Id. Lud.* b. i. s. 146. *Id. Suck.* 1r th. s. 585. *Id. Bert.* s. 88. *Id. Mohs,* b. ii. s. 14. *Id. Hab.* s. 71. *Id. Leonhard,* Tabel. s. 32.—Chaux carbonatée compacte, *Brong.* t. i. p. 199.—Gemeiner Kalkstein, *Haus.* s. 126.—Dichter Kalkstein, *Karsten,* Tabel. s. 50.—Chaux carbonatée compacte, *Haüy,* Tabl. p. 4.—Dichter gemeiner Kalkstein, *Lenz,* b. ii. s. 732.—Gemeiner Dichter Kalkstein, *Hoff.* b. iii. s. 8.—Common Lime- *Aikin,* p. 160.

External

External Characters.

Its most frequent colour is grey, of which the following varieties have been observed: yellowish, bluish, ash, pearl, greenish, and smoke grey; the ash-grey passes into greyish-black; the yellowish-grey into yellowish-brown, ochre-yellow, and into a colour bordering on cream-yellow. It also occurs blood-red, flesh-red, and peach-blossom-red, which latter colour is very rare.

It frequently exhibits veined, zoned, striped, clouded, and spotted coloured delineations; and sometimes also black and brown coloured arborisations.

It very rarely exhibits a beautiful play of colours, caused by intermixed portions of pearly shells.

It occurs massive, corroded, in large plates, rolled masses, and in various extraneous external shapes, of univalve, bivalve, and multivalve shells, of corals, fishes, and more rarely of vegetables, as of ferns and reeds.

Internally it is dull, seldom glimmering, which is owing to intermixed calcareous-spar.

The fracture is small and fine splintery, which sometimes passes into large and flat conchoidal, sometimes into uneven, inclining to earthy, and it occasionally inclines to straight and thick slaty.

The fragments are indeterminate angular, more or less sharp-edged, but in the slaty variety they are tabular.

It is generally translucent on the edges, sometimes opaque.

It is in general rather softer than granular foliated limestone.

It is brittle, and easily frangible.

Its streak is generally greyish-white.

Specific

Specific gravity, Splintery, 2.600, 2.720, *Brisson.*—Opalescent Shell Marble, 2.673, *Leonhard.*—2.675, *Werner.*

Chemical Characters.

It effervesces with acids, and the greater part is dissolved; and burns to quicklime, without falling to pieces.

Constituent Parts.

Rudersdorf.		Bluish-grey Limestone.		Limestone from Sweden.		Limestone from Ettersberg *.	
Lime,	53.00	Lime,	49.50	Lime,	49.25	Lime,	33.41
Carbon. acid,	42.50	Carbon. acid,	40.00	Carbon. acid,	35.00	Carbon. acid,	42.00
Silica,	1.12	Silica,	5.25	Silica,	8.75	Silica,	10.25
Alumina,	1.00	Alumina,	2.75	Alumina,	2.50	Magnesia,	9.43
Iron,	0.75	Iron,	1.37	Iron,	2.75	Iron,	2.25
Water,	1.03	Water,	1.13	Loss,	1.75	Manganese,	1.25
						Loss,	1.41
	100		100		100		100
Simon, Gehlen's Jour. iv. s. 426.		*Simon*, Ib.		*Simon*, Ib.		*Buchols*, Ib.	

Geognostic Situation.

This mineral occurs in vast abundance in nature, principally in secondary formations, along with sandstone, gypsum, and coal; and in small quantity in primitive mountains. The variegated varieties, which are frequently traversed by veins of calcareous-spar, occur principally in districts composed of grey-wacke and clay-slate. It is distinctly stratified, and the strata vary in thickness, from a few inches to many fathoms, and are from a few fathoms to many miles in extent. The strata generally incline to horizontal

* Some of the limestones in Fifeshire agree in composition with that of Ettersberg.

rizontal; sometimes, however, they are vertical, or variously convoluted, even arranged in concentric layers, thus presenting appearances illustrative of their chemical nature. Petrifactions, both of animals and vegetables, but principally of the former, abound in compact limestone: these are of corals, shells, fishes, and sometimes of amphibious animals. On a general view, it is to be considered as rich in ores of different kinds, particularly ores of lead and zinc: thus, nearly all the rich and valuable lead-mines in England are situated in limestone.

Geographic Situation.

It abounds in the sandstone and coal formations, both in Scotland and England; and in Ireland, it is a very abundant mineral in all the districts where clay-slate and red sandstone rocks occur. On the Continent of Europe, it is a very widely and abundantly distributed mineral; and forms a striking feature in many extensive tracts of country in Asia, Africa and America, as will be particularly described in the Geognostic part of this work.

Uses.

When compact limestone joins to pure and agreeable colours, so considerable a degree of hardness that it takes a good polish, it is by artists considered as a Marble; and if it contains petrifactions mineralized, it is named *shell* or *lumachella*, and *coral* or *zoophytic marble*, according as the organic remains are testaceous or coralline*. In one particular variety of lumachella

* The name *marmor*, is derived from the Greek *μαρμαιρειν*, *to shine*, or *glitter*, and was by the ancients applied, not only to limestone, but also to stones possessing agreeable colours, and receiving a good polish, such as gypsum, jasper, serpentine, and even granite and porphyry.

lumachella or shell marble, found at Bleiberg in Carinthia, the shells and fragments of shells, which belong to the nautilus tribe, are set in a brown-coloured basis, and reflect many beautiful and brilliant pearly inclining to metallic colours, principally the fire-red, green, and blue tints. It is named *opalescent* or *fire marble*. Another lumachella marble from Astracan, contains, in a reddish-brown basis, pearly shells of nautili, that reflect a very brilliant gold-yellow colour. In some compact marbles, the surface presents a beautiful arborescent appearance, and these are named *arborescent* or *dendritic marbles*. Such are those of Papenheim in Bavaria.

The *Florentine Marble*, or *Ruin Marble*, as it is sometimes called, is a compact limestone. It occurs on the Po and the Arno, and is worked into various articles at Florence. It is said to occur in balls. It presents angular figures of a yellowish-brown, on a base of a lighter tint, and which passes to greyish-white. Seen at a distance, slabs of this stone resemble drawings done in bister. "One is amused (says Brard) to observe in it kinds of ruins: there it is a Gothic castle half destroyed, here it presents ruined walls; in another place old bastions; and, what still adds to the illusion, is, that, in these sorts of natural paintings, there exists a kind of aërial perspective, which is very sensibly perceived. The lower part, or what forms the first plane, has a warm and bold tone; the second follows it, and weakens it as it increases the distance; the third becomes still fainter; while the upper part, agreeing with the first, presents in the distance a whitish zone, which terminates the horizon, then blends itself more and more as it rises, and at length reaches the top, where it forms sometimes as it were clouds. But approach close to it, all va-

nishes immediately, and these pretended figures, which, at a distance, seemed so well drawn, are converted into irregular marks, which present nothing to the eye." To the same compact limestone may be referred the variety called *Cottam Marble*, from being found at Cottam, near Bristol. It resembles in many respects the Landscape Marble.

In different parts of Scotland, compact limestone is cut and polished as marble: this was the case in the parish of Cummertrees in Dumfriesshire,—in Cambuslang parish, in Lanarkshire,—in Fifeshire, &c. In England, many compact limestones are cut and polished as marbles; such are the limestones of Derbyshire, Yorkshire, Devonshire, Somersetshire, and Dorsetshire. It is sometimes used as a building stone; and, in want of better materials, for paving streets, and making highways. When, by exposure to a high temperature, it is deprived of its carbonic acid, and converted into quicklime, it is used for mortar; also by the soap-maker, for rendering his alkalies caustic; by the tanner, for cleansing hides, or freeing them from hair, muscular substance, and fat; by the farmer, in the improvement of particular kinds of soil; and by the metallurgist, in the smelting of such ores as are difficultly fusible, owing to an intermixture of silica and alumina.

Second Kind.

Blue Vesuvian Limestone.

Blauer Vesuvischer Kalkstein, *Klaproth.*

Blauer Vesuvischer Kalkstein, *Klaproth*, Beit. b. v. s. 92. *Id. Lenz*, b. ii. s. 737.

External Characters.

Its colour is dark bluish-grey, partly veined with white.

Externally it appears as if it had been rolled; and the surface is uneven.

The fracture fine earthy, passing into splintery.

It is opaque.

It affords a white streak.

It is semi-hard in a low degree.

It is rather heavy.

Constituent Parts.

Lime, - -	58.00
Carbonic Acid, -	28.50
Water, which is somewhat ammoniacal, -	11.00
Magnesia, -	0.50
Oxide of Iron,	0.25
Carbon, - -	0.25
Silica,	1.25
	99.75

Klaproth, Beit. b. v. s. 96.

From

From this analysis, it appears, that the vesuvian limestone differs remarkably in composition from common compact limestone. In common compact limestone, 100 parts of lime are combined with at least 80 parts of carbonic acid; whereas in the vesuvian limestone, 100 parts of limestone are not combined with more than 50 parts of carbonic acid. Secondly, In common limestone, independent of the water which adheres to it accidentally, as far as we know, there is no water of composition; but in the vesuvian limestone, there are 11 parts of water of composition.

Geographic Situation.

This remarkable limestone is found in loose masses amongst unaltered ejected minerals in the neighbourhood of Vesuvius.

Observations.

It is known to some collectors under the name *Compact Blue Lava* of Vesuvius; and is employed by artists in their mosaic work, to represent the sky.

Third Kind.

Roestone, or Oolite *.

Roogenstein, *Werner*.

Hammites, *Plin.* Hist. Nat. xxxvii. 10. s. 60.?—F. E. Bruckmanni, Specimen physicum sistens, Histor. Nat. Oolithe, 1721.

---

* *Roestone*, so named on account of its resemblance in form to the roe of fishes.

1721.—Stalactites oolithus, var. *b. c. d. Wall.* t. ii. p. 384.—Roogenstein, *Wid.* s. 511.—Oviform Limestone, *Kirw.* vol. i. p. 91.—Roogenstein, *Estner*, b. ii. s. 928. *Id. Emm.* b. i. s. 442.—Tufo oolitico, *Nap.* p. 353.—L'Oolite, *Broch.* t. i. p. 529.—Chaux carbonatée compacte globuliforme, *Haüy*, t. ii. p. 171.—Roogenstein, *Reuss*, b. ii. 2. s. 270. *Id. Lud.* b. i. s. 148. *Id. Suck.* 1r th. s. 591. *Id. Bert.* s. 89. *Id. Mohs*, b. ii. s. 26. *Id. Hab.* p. 72. *Id. Leonhard*, Tabel. s. 32.—Chaux carbonatée oolithe, *Brong.* t. i. p. 203.—Chaux carbonatée globuliforme, *Brard*, p. 31.—Erbsförmiger Kalkstein, *Karsten*, Tabel. s. 50.—Roestone, *Kid*, vol. i. p. 26.—Chaux carbonatée globuliforme, *Haüy*, Tabl. p. 4.—Roogenstein, *Lenz*, b. ii. s. 738. *Id. Hoff.* b. iii. s. 12.—Schaaliger Kalkstein, *Haus.* b. iii. s. 912.

External Characters.

Its colours are hair-brown, chesnut-brown, and reddish-brown, and sometimes yellowish-grey, and ash-grey.

It occurs massive, and in distinct concretions, which are round granular, the larger are composed of fine spherical granular, and sometimes of very thin concentric lamellar concretions.

Internally it is dull.

The fracture of the grains is fine splintery; but of the mass is round granular in the small, and slaty in the large.

The fragments in the large are blunt-edged.

It is opaque.

It is semi-hard, approaching to soft.

It is rather brittle, and very easily frangible.

Specific gravity, 2.6829, 2.6190, *Kopp.*—2.585, *Breithaupt.*

Chemical

Chemical Characters.

It dissolves with effervescence in acids.

Geognostic Situation.

It occurs along with red sandstone, and *lias* limestone.

Geographic Situation.

This rock, which, in England, is known under the names Bath-Stone, Ketton-Stone, Portland-Stone, and Oolite, extends, with but little interruption, from Somersetshire to the banks of the Humber in Lincolnshire. On the Continent of Europe, it occurs in Thuringia, the Netherlands, the mountains of Jura, and in other countries.

Uses.

The Oolite, or Roestone, particularly that of Bath and Portland, is very extensively employed in architecture; it can be worked with great ease, and has a light and beautiful appearance; but it is porous, and possesses no great durability, and should not be employed where there is much carved or ornamental work, for the fine chiselling is soon effaced by the action of the atmosphere. On account of the ease and sharpness with which it can be carved, it is much used by the English architects, who appear to have little regard for futurity. St Paul's is built of this stone, also Somerset-House. The Chapel of Henry VIII. affords a striking proof of the inattention of the architects to the choice of the stone. All the beautiful ornamental work of the exterior had mouldered away in the short comparative period of 300 years. It has recently been cased with a new front of Bath-Stone, in which the carving has been correctly

correctly copied: from the nature of the stone, we may predict, that its duration will not be longer than that of the original. Both Portland and Bath stone varies much in quality. In buildings constructed of this stone, we may frequently observe some of the stones black, and others white. The black stones are those which are more compact and durable, and preserve their coating of smoke; the white stones are decomposing, and presenting a fresh surface, as if they had been recently scraped *. Roestone is also used as a manure, but when burnt into quicklime, the marly varieties afford rather an indifferent mortar; but those mixed with sand a better mortar.

Observations.

1. It passes into Sandstone, Compact Limestone, and Marl.

2. Some naturalists, as Daubenton, Saussure, Spallanzani, and Gillet Lamont, conjecture, that Roestone is carbonate of lime, which has been granulated in the manner of gunpowder, by the action of water: the most plausible opinion is that which attributes the formation of this mineral to crystallization from a state of solution.

Third Subspecies.

Chalk.

Kreide, *Werner.*

Creta alba, *Wall.* t. i. p. 27.—Kreide, *Wid.* s. 492.—Chalk, *Kirw.* vol. i. p. 77.—Kreide, *Estner,* b. ii. s. 917. *Id. Emm.* b. i.

* Aikin.

External Characters.

Its most frequent colour is grey, of which the following varieties have been observed: yellowish, bluish, ash, pearl, greenish, and smoke grey; the ash-grey passes into greyish-black; the yellowish-grey into yellowish-brown, ochre-yellow, and into a colour bordering on cream-yellow. It also occurs blood-red, flesh-red, and peach-blossom-red, which latter colour is very rare.

It frequently exhibits veined, zoned, striped, clouded, and spotted coloured delineations; and sometimes also black and brown coloured arborisations.

It very rarely exhibits a beautiful play of colours, caused by intermixed portions of pearly shells.

It occurs massive, corroded, in large plates, rolled masses, and in various extraneous external shapes, of univalve, bivalve, and multivalve shells, of corals, fishes, and more rarely of vegetables, as of ferns and reeds.

Internally it is dull, seldom glimmering, which is owing to intermixed calcareous-spar.

The fracture is small and fine splintery, which sometimes passes into large and flat conchoidal, sometimes into uneven, inclining to earthy, and it occasionally inclines to straight and thick slaty.

The fragments are indeterminate angular, more or less sharp-edged, but in the slaty variety they are tabular.

It is generally translucent on the edges, sometimes opaque.

It is in general rather softer than granular foliated limestone.

It is brittle, and easily frangible.

Its streak is generally greyish-white.

Specific

Specific gravity, Splintery, 2.600, 2.720, *Brisson.*—Opalescent Shell Marble, 2.673, *Leonhard.*—2.675, *Werner.*

Chemical Characters.

It effervesces with acids, and the greater part is dissolved; and burns to quicklime, without falling to pieces.

Constituent Parts.

Rudersdorf.		Bluish-grey Limestone.		Limestone from Sweden.		Limestone from Ettersberg *.	
Lime,	53.00	Lime,	49.50	Lime,	49.25	Lime,	33.41
Carbon. acid,	42.50	Carbon. acid,	40.00	Carbon. acid,	35.00	Carbon. acid,	42.00
Silica,	1.12	Silica,	5.25	Silica,	8.75	Silica,	10.25
Alumina,	1.00	Alumina,	2.75	Alumina,	2.50	Magnesia,	9.43
Iron,	0.75	Iron,	1.37	Iron,	2.75	Iron,	2.25
Water,	1.03	Water,	1.13	Loss,	1.75	Manganese,	1.25
						Loss,	1.41
	100		100		100		100
Simon, Gehlen's Jour. iv. s. 426.		*Simon*, Ib.		*Simon*, Ib.		*Bucholz*, Ib.	

Geognostic Situation.

This mineral occurs in vast abundance in nature, principally in secondary formations, along with sandstone, gypsum, and coal; and in small quantity in primitive mountains. The variegated varieties, which are frequently traversed by veins of calcareous-spar, occur principally in districts composed of grey-wacke and clay-slate. It is distinctly stratified, and the strata vary in thickness, from a few inches to many fathoms, and are from a few fathoms to many miles in extent. The strata generally incline to horizontal

* Some of the limestones in Fifeshire agree in composition with that of Ettersberg.

rizontal; sometimes, however, they are vertical, or variously convoluted, even arranged in concentric layers, thus presenting appearances illustrative of their chemical nature. Petrifactions, both of animals and vegetables, but principally of the former, abound in compact limestone: these are of corals, shells, fishes, and sometimes of amphibious animals. On a general view, it is to be considered as rich in ores of different kinds, particularly ores of lead and zinc: thus, nearly all the rich and valuable lead-mines in England are situated in limestone.

Geographic Situation.

It abounds in the sandstone and coal formations, both in Scotland and England; and in Ireland, it is a very abundant mineral in all the districts where clay-slate and red sandstone rocks occur. On the Continent of Europe, it is a very widely and abundantly distributed mineral; and forms a striking feature in many extensive tracts of country in Asia, Africa and America, as will be particularly described in the Geognostic part of this work.

Uses.

When compact limestone joins to pure and agreeable colours, so considerable a degree of hardness that it takes a good polish, it is by artists considered as a Marble; and if it contains petrifactions mineralized, it is named *shell* or *lumachella*, and *coral* or *zoophytic marble*, according as the organic remains are testaceous or coralline*. In one particular variety of lumachella

* The name *marmor*, is derived from the Greek *μαρμαιρειν*, *to shine*, or *glitter*, and was by the ancients applied, not only to limestone, but also to stones possessing agreeable colours, and receiving a good polish, such as gypsum, jasper, serpentine, and even granite and porphyry.

lumachella or shell marble, found at Bleiberg in Carinthia, the shells and fragments of shells, which belong to the nautilus tribe, are set in a brown-coloured basis, and reflect many beautiful and brilliant pearly inclining to metallic colours, principally the fire-red, green, and blue tints. It is named *opalescent* or *fire marble*. Another lumachella marble from Astracan, contains, in a reddish-brown basis, pearly shells of nautili, that reflect a very brilliant gold-yellow colour. In some compact marbles, the surface presents a beautiful arborescent appearance, and these are named *arborescent* or *dendritic marbles*. Such are those of Papenheim in Bavaria.

The *Florentine Marble*, or *Ruin Marble*, as it is sometimes called, is a compact limestone. It occurs on the Po and the Arno, and is worked into various articles at Florence. It is said to occur in balls. It presents angular figures of a yellowish-brown, on a base of a lighter tint, and which passes to greyish-white. Seen at a distance, slabs of this stone resemble drawings done in bister. "One is amused (says Brard) to observe in it kinds of ruins: there it is a Gothic castle half destroyed, here it presents ruined walls; in another place old bastions; and, what still adds to the illusion, is, that, in these sorts of natural paintings, there exists a kind of aërial perspective, which is very sensibly perceived. The lower part, or what forms the first plane, has a warm and bold tone; the second follows it, and weakens it as it increases the distance; the third becomes still fainter; while the upper part, agreeing with the first, presents in the distance a whitish zone, which terminates the horizon, then blends itself more and more as it rises, and at length reaches the top, where it forms sometimes as it were clouds. But approach close to it, all va-

 nishes

nishes immediately, and these pretended figures, which, at a distance, seemed so well drawn, are converted into irregular marks, which present nothing to the eye." To the same compact limestone may be referred the variety called *Cottam Marble*, from being found at Cottam, near Bristol. It resembles in many respects the Landscape Marble.

In different parts of Scotland, compact limestone is cut and polished as marble: this was the case in the parish of Cummertrees in Dumfriesshire,—in Cambuslang parish, in Lanarkshire,—in Fifeshire, &c. In England, many compact limestones are cut and polished as marbles; such are the limestones of Derbyshire, Yorkshire, Devonshire, Somersetshire, and Dorsetshire. It is sometimes used as a building stone; and, in want of better materials, for paving streets, and making highways. When, by exposure to a high temperature, it is deprived of its carbonic acid, and converted into quicklime, it is used for mortar; also by the soap-maker, for rendering his alkalies caustic; by the tanner, for cleansing hides, or freeing them from hair, muscular substance, and fat; by the farmer, in the improvement of particular kinds of soil; and by the metallurgist, in the smelting of such ores as are difficultly fusible, owing to an intermixture of silica and alumina.

Second

Second Kind.

Blue Vesuvian Limestone.

Blauer Vesuvischer Kalkstein, *Klaproth.*

Blauer Vesuvischer Kalkstein, *Klaproth,* Beit. b. v. s. 92. *Id.* *Lenz,* b. ii. s. 737.

External Characters.

Its colour is dark bluish-grey, partly veined with white.

Externally it appears as if it had been rolled; and the surface is uneven.

The fracture fine earthy, passing into splintery.

It is opaque.

It affords a white streak.

It is semi-hard in a low degree.

It is rather heavy.

Constituent Parts.

Lime, - -	58.00
Carbonic Acid, -	28.50
Water, which is somewhat ammoniacal, -	11.00
Magnesia, -	0.50
Oxide of Iron,	0.25
Carbon, - -	0.25
Silica,	1.25
	99.75

Klaproth, Beit. b. v. s. 96.

From

From this analysis, it appears, that the vesuvian limestone differs remarkably in composition from common compact limestone. In common compact limestone, 100 parts of lime are combined with at least 80 parts of carbonic acid; whereas in the vesuvian limestone, 100 parts of limestone are not combined with more than 50 parts of carbonic acid. Secondly, In common limestone, independent of the water which adheres to it accidentally, as far as we know, there is no water of composition; but in the vesuvian limestone, there are 11 parts of water of composition.

Geographic Situation.

This remarkable limestone is found in loose masses amongst unaltered ejected minerals in the neighbourhood of Vesuvius.

Observations.

It is known to some collectors under the name *Compact Blue Lava* of Vesuvius; and is employed by artists in their mosaic work, to represent the sky.

Third Kind.

Roestone, or Oolite *.

Roogenstein, *Werner*.

Hammites, *Plin.* Hist. Nat. xxxvii. 10. s. 60.?—F. E. Bruckmanni, Specimen physicum sistens, Histor. Nat. Oolithe, 1721.

---

* *Roestone*, so named on account of its resemblance in form to the roe of fishes.

1721.—Stalactites oolithus, var. *b. c. d. Wall.* t. ii. p. 384.—Roogenstein, *Wid.* s. 511.—Oviform Limestone, *Kirw.* vol. i. p. 91.—Roogenstein, *Estner,* b. ii. s. 928. *Id. Emm.* b. i. s. 442.—Tufo oolitico, *Nap.* p. 353.—L'Oolite, *Broch.* t. i. p. 529.—Chaux carbonatée compacte globuliforme, *Haüy,* t. ii. p. 171.—Roogenstein, *Reuss,* b. ii. 2. s. 270. *Id. Lud.* b. i. s. 148. *Id. Suck.* 1r th. s. 591. *Id. Bert.* s. 89. *Id. Mohs,* b. ii. s. 26. *Id. Hab.* p. 72. *Id. Leonhard,* Tabel. s. 32.—Chaux carbonatée oolithe, *Brong.* t. i. p. 203.—Chaux carbonatée globuliforme, *Brard,* p. 31.—Erbsförmiger Kalkstein, *Karsten,* Tabel. s. 50.—Roestone, *Kid,* vol. i. p. 26.—Chaux carbonatée globuliforme, *Haüy,* Tabl. p. 4.—Roogenstein, *Lenz,* b. ii. s. 738. *Id. Hoff.* b. iii. s. 12.—Schaaliger Kalkstein, *Haus.* b. iii. s. 912.

External Characters.

Its colours are hair-brown, chesnut-brown, and reddish-brown, and sometimes yellowish-grey, and ash-grey.

It occurs massive, and in distinct concretions, which are round granular, the larger are composed of fine spherical granular, and sometimes of very thin concentric lamellar concretions.

Internally it is dull.

The fracture of the grains is fine splintery; but of the mass is round granular in the small, and slaty in the large.

The fragments in the large are blunt-edged.

It is opaque.

It is semi-hard, approaching to soft.

It is rather brittle, and very easily frangible.

Specific gravity, 2.6829, 2.6190, *Kopp.*—2.585, *Breithaupt.*

Chemical

Chemical Characters.

It dissolves with effervescence in acids.

Geognostic Situation.

It occurs along with red sandstone, and *lias* limestone.

Geographic Situation.

This rock, which, in England, is known under the names Bath-Stone, Ketton-Stone, Portland-Stone, and Oolite, extends, with but little interruption, from Somersetshire to the banks of the Humber in Lincolnshire. On the Continent of Europe, it occurs in Thuringia, the Netherlands, the mountains of Jura, and in other countries.

Uses.

The Oolite, or Roestone, particularly that of Bath and Portland, is very extensively employed in architecture; it can be worked with great ease, and has a light and beautiful appearance; but it is porous, and possesses no great durability, and should not be employed where there is much carved or ornamental work, for the fine chiselling is soon effaced by the action of the atmosphere. On account of the ease and sharpness with which it can be carved, it is much used by the English architects, who appear to have little regard for futurity. St Paul's is built of this stone, also Somerset-House. The Chapel of Henry VIII. affords a striking proof of the inattention of the architects to the choice of the stone. All the beautiful ornamental work of the exterior had mouldered away in the short comparative period of 300 years. It has recently been cased with a new front of Bath-Stone, in which the carving has been correctly

correctly copied: from the nature of the stone, we may predict, that its duration will not be longer than that of the original. Both Portland and Bath stone varies much in quality. In buildings constructed of this stone, we may frequently observe some of the stones black, and others white. The black stones are those which are more compact and durable, and preserve their coating of smoke; the white stones are decomposing, and presenting a fresh surface, as if they had been recently scraped *. Roestone is also used as a manure, but when burnt into quicklime, the marly varieties afford rather an indifferent mortar; but those mixed with sand a better mortar.

Observations.

1. It passes into Sandstone, Compact Limestone, and Marl.

2. Some naturalists, as Daubenton, Saussure, Spallanzani, and Gillet Lamont, conjecture, that Roestone is carbonate of lime, which has been granulated in the manner of gunpowder, by the action of water: the most plausible opinion is that which attributes the formation of this mineral to crystallization from a state of solution.

Third Subspecies.

Chalk.

Kreide, *Werner.*

Creta alba, *Wall.* t. i. p. 27.—Kreide, *Wid.* s. 492.—Chalk, *Kirw.* vol. i. p. 77.—Kreide, *Estner,* b. ii. s. 917. *Id. Emm.* b. i.

* Aikin.

sized fragments of a pale yellow, brick-red, deep brown, and blackish-grey.

Portuguese Marbles.

Few marbles have hitherto been discovered in Portugal, and none of them equal in beauty the finer varieties found in Spain.

Swiss Marbles.

Granular foliated limestone occurs abundantly in Switzerland, but it has not hitherto been much used as a marble.

German Marbles.

Germany abounds in marbles, and affords many varieties, remarkable either for their beauty or singularity. They are quarried in great quantity, and carried to different parts of that vast country, or are exported into the neighbouring states. The varieties are so numerous, that we cannot, in the very brief view we are now taking, pretend to notice even the more remarkable of them, but must refer, for the particular descriptions, to the economical department of this work.

Norwegian Marbles.

Norway is poor in marbles, almost the only quarry of this stone being that of Gillebeck, in the district of Christiania.

Swedish Marbles.

Sweden does not afford many kinds of marble, and none of them are eminently distinguished for the beauty of their

appearance.

appearance. The principal marble is that of Fagernech, which is white, with veins of green talc.

Russian and Siberian Marbles.

The vast Empire of Russia affords a great many different kinds of marble. Georgi, in his Description of the Russian Empire, enumerates white, grey, green, blue, yellow, and red varieties; and Patrin gives the following account of the Siberian marbles. "The Uralian Mountains furnish the finest and most variegated marbles. The greater part is taken from the neighbourhood of Catharinenburg, where they are wrought, and from thence transported into Russia, particularly to Petersburgh. The late Empress caused an immense palace to be built in her capital for Orloff, her favourite, which is entirely coated with these fine marbles, both inside and outside. The Empress built the church of Isaac with the same marbles, on a vast space, near the statue of Peter the Great." Patrin found no white statuary marble in the Uralian Mountains; but in that part of the Altain Mountains which is traversed by the river Irtish, he in two places saw immense blocks of marble, perfectly white and pure, from which blocks might be hewn.

Asiatic Marbles.

At present we are very imperfectly acquainted with the marbles of Asia.

Shaw mentions a red marble from Mount Sinai: Russell, in his Natural History of Aleppo, gives an imperfect account of the marbles of Syria; and some Persian marbles are noticed by Chardin. Mr Morier, in his Journey through Persia, mentions a very beautiful marble, under the name *Marble of Tabriz,* and informs us, that the tomb

of

of the celebrated poet Hafitz is constructed with it, and that the wainscotting of the principal room of the Haft-tan, near Schiraz, is likewise of this marble. Its colours are described as light green, with veins, sometimes of red, sometimes of blue, and it has great translucency. It is cut in large slabs; for Mr Morier saw some that measured nine feet in length, and five feet in breadth. He says, that it is not procured near the city of Tabriz, or taken from a quarry, but is said to be rather a petrifaction, found in large quantities, and in immense blocks, on the borders of the Lake Shahee, near the town of Meraugheh. If it is a mere calcareous deposition, formed in the way of calcareous-alabaster or calc-sinter, it must be considered, not as marble, but a variety of that mineral.

The marbles of Hindostan, Siam, and China, are almost unknown to us. Authors speak of a quarry of white marble in the neighbourhood of Pekin; and of a similar marble in the vicinity of the capital of Siam.

African Marbles.

Beds of marble occur in the Atlas Mountains, and in those ranges that bound the shores of the Red Sea *.

American Marbles.

A good many different marbles have been discovered in the United States. The principal quarries are at Stockbridge and Lanesborough, Massachussets: in Vermont and Pennsylvania: in New-York; and in Virginia. According to Professor Hall, as mentioned by Mr Kœnig, marble has been found in many places on the west side of the Green Mountains in Vermont. A few years since, a valuable

* Vid. Murray's interesting and valuable work, "Historical of Discoveries and Travels in Africa," for particulars in regard to the

luable quarry was opened in Middleburg, a town situated on Otter Creek, eleven miles above Vergennes. The marble is of different colours in different parts of the bed. The principal colour, however, is bluish-grey. It takes a good polish, and is in general free of admixture of any substance that might affect its polish.

Second Subspecies.

Compact Limestone.

Dichter Kalkstein, *Werner*.

This subspecies is divided into three kinds, viz. Common Compact Limestone, Blue Vesuvian Limestone, and Roestone.

First Kind.

Common Compact Limestone.

Gemeiner Dichter Kalkstein, *Werner*.

Calcareus æquabilis, *Wall.* t. i. p. 122.—Dichter Kalkstein, *Wid.* s. 494.—Compact Limestone, *Kirw.* vol. i. p. 82.—Gemeiner Dichter Kalkstein, *Emm.* b. i. s. 437.—Pietra calcarea compacta, *Nap.* p. 33.—La pierre calcaire compacte commune, *Broch.* t. i. p. 523.—Chaux carbonatée compacte, *Haüy*, t. ii. p. 166.—Gemeiner Dichter Kalkstein, *Reuss*, b. ii. 2. s. 262. *Id. Lud.* b. i. s. 146. *Id. Suck.* 1r th. s. 585. *Id. Bert.* s. 88. *Id. Mohs*, b. ii. s. 14. *Id. Hab.* s. 71. *Id. Leonhard*, Tabel. s. 32.—Chaux carbonatée compacte, *Brong.* t. i. p. 199.—Gemeiner Kalkstein, *Haus.* s. 126.—Dichter Kalkstein, *Karsten*, Tabel. s. 50.—Chaux carbonatée compacte, *Haüy*, Tabl. p. 4.—Dichter gemeiner Kalkstein, *Lenz*, b. ii. s. 732.—Gemeiner Dichter Kalkstein, *Hoff.* b. iii. s. 8.—Common Lime-

External Characters.

Its most frequent colour is grey, of which the following varieties have been observed: yellowish, bluish, ash, pearl, greenish, and smoke grey; the ash-grey passes into greyish-black; the yellowish-grey into yellowish-brown, ochre-yellow, and into a colour bordering on cream-yellow. It also occurs blood-red, flesh-red, and peach-blossom-red, which latter colour is very rare.

It frequently exhibits veined, zoned, striped, clouded, and spotted coloured delineations; and sometimes also black and brown coloured arborisations.

It very rarely exhibits a beautiful play of colours, caused by intermixed portions of pearly shells.

It occurs massive, corroded, in large plates, rolled masses, and in various extraneous external shapes, of univalve, bivalve, and multivalve shells, of corals, fishes, and more rarely of vegetables, as of ferns and reeds.

Internally it is dull, seldom glimmering, which is owing to intermixed calcareous-spar.

The fracture is small and fine splintery, which sometimes passes into large and flat conchoidal, sometimes into uneven, inclining to earthy, and it occasionally inclines to straight and thick slaty.

The fragments are indeterminate angular, more or less sharp-edged, but in the slaty variety they are tabular.

It is generally translucent on the edges, sometimes opaque.

It is in general rather softer than granular foliated limestone.

It is brittle, and easily frangible.

Its streak is generally greyish-white.

Specific

Specific gravity, Splintery, 2.600, 2.720, *Brisson.*—Opalescent Shell Marble, 2.673, *Leonhard.*—2.675, *Werner.*

Chemical Characters.

It effervesces with acids, and the greater part is dissolved; and burns to quicklime, without falling to pieces.

Constituent Parts.

Rudersdorf.		Bluish-grey Limestone.		Limestone from Sweden.		Limestone from Ettersberg *.	
Lime,	53.00	Lime,	49.50	Lime,	49.25	Lime,	33.41
Carbon. acid,	42.50	Carbon. acid,	40.00	Carbon. acid,	35.00	Carbon. acid,	42.00
Silica,	1.12	Silica,	5.25	Silica,	8.75	Silica,	10.25
Alumina,	1.00	Alumina,	2.75	Alumina,	2.50	Magnesia,	9.43
Iron,	0.75	Iron,	1.37	Iron,	2.75	Iron,	2.25
Water,	1.03	Water,	1.13	Loss,	1.75	Manganese,	1.25
						Loss,	1.41
	100		100		100		100
Simon, Gehlen's Jour. iv. s. 426.		*Simon*, Ib.		*Simon*, Ib.		*Bucholz*, Ib.	

Geognostic Situation.

This mineral occurs in vast abundance in nature, principally in secondary formations, along with sandstone, gypsum, and coal; and in small quantity in primitive mountains. The variegated varieties, which are frequently traversed by veins of calcareous-spar, occur principally in districts composed of grey-wacke and clay-slate. It is distinctly stratified, and the strata vary in thickness, from a few inches to many fathoms, and are from a few fathoms to many miles in extent. The strata generally incline to horizontal

* Some of the limestones in Fifeshire agree in composition with that of Ettersberg.

lumachella or shell marble, found at Bleiberg in Carinthia, the shells and fragments of shells, which belong to the nautilus tribe, are set in a brown-coloured basis, and reflect many beautiful and brilliant pearly inclining to metallic colours, principally the fire-red, green, and blue tints. It is named *opalescent* or *fire marble*. Another lumachella marble from Astracan, contains, in a reddish-brown basis, pearly shells of nautili, that reflect a very brilliant gold-yellow colour. In some compact marbles, the surface presents a beautiful arborescent appearance, and these are named *arborescent* or *dendritic marbles*. Such are those of Papenheim in Bavaria.

The *Florentine Marble*, or *Ruin Marble*, as it is sometimes called, is a compact limestone. It occurs on the Po and the Arno, and is worked into various articles at Florence. It is said to occur in balls. It presents angular figures of a yellowish-brown, on a base of a lighter tint, and which passes to greyish-white. Seen at a distance, slabs of this stone resemble drawings done in bister. "One is amused (says Brard) to observe in it kinds of ruins; there it is a Gothic castle half destroyed, here it presents ruined walls; in another place old bastions; and, what still adds to the illusion, is, that, in these sorts of natural paintings, there exists a kind of aërial perspective, which is very sensibly perceived. The lower part, or what forms the first plane, has a warm and bold tone; the second follows it, and weakens it as it increases the distance; the third becomes still fainter; while the upper part, agreeing with the first, presents in the distance a whitish zone, which terminates the horizon, then blends itself more and more as it rises, and at length reaches the top, where it forms sometimes as it were clouds. But approach close to it, all vanishes

nishes immediately, and these pretended figures, which, at a distance, seemed so well drawn, are converted into irregular marks, which present nothing to the eye." To the same compact limestone may be referred the variety called *Cottam Marble*, from being found at Cottam, near Bristol. It resembles in many respects the Landscape Marble.

In different parts of Scotland, compact limestone is cut and polished as marble: this was the case in the parish of Cummertrees in Dumfriesshire,—in Cambuslang parish, in Lanarkshire,—in Fifeshire, &c. In England, many compact limestones are cut and polished as marbles; such are the limestones of Derbyshire, Yorkshire, Devonshire, Somersetshire, and Dorsetshire. It is sometimes used as a building stone; and, in want of better materials, for paving streets, and making highways. When, by exposure to a high temperature, it is deprived of its carbonic acid, and converted into quicklime, it is used for mortar; also by the soap-maker, for rendering his alkalies caustic; by the tanner, for cleansing hides, or freeing them from hair, muscular substance, and fat; by the farmer, in the improvement of particular kinds of soil; and by the metallurgist, in the smelting of such ores as are difficultly fusible, owing to an intermixture of silica and alumina.

Second Kind.

Blue Vesuvian Limestone.

Blauer Vesuvischer Kalkstein, *Klaproth.*

Blauer Vesuvischer Kalkstein, *Klaproth,* Beit. b. v. s. 92. *Id. Lenz,* b. ii. s. 737.

External Characters.

Its colour is dark bluish-grey, partly veined with white.

Externally it appears as if it had been rolled; and the surface is uneven.

The fracture fine earthy, passing into splintery.

It is opaque.

It affords a white streak.

It is semi-hard in a low degree.

It is rather heavy.

Constituent Parts.

Lime, - -	58.00
Carbonic Acid, -	28.50
Water, which is somewhat ammoniacal, -	11.00
Magnesia, -	0.50
Oxide of Iron,	0.25
Carbon, - -	0.25
Silica,	1.25
	99.75

Klaproth, Beit. b. v. s. 96.

From

From this analysis, it appears, that the vesuvian limestone differs remarkably in composition from common compact limestone. In common compact limestone, 100 parts of lime are combined with at least 80 parts of carbonic acid; whereas in the vesuvian limestone, 100 parts of limestone are not combined with more than 50 parts of carbonic acid. Secondly, In common limestone, independent of the water which adheres to it accidentally, as far as we know, there is no water of composition; but in the vesuvian limestone, there are 11 parts of water of composition.

Geographic Situation.

This remarkable limestone is found in loose masses amongst unaltered ejected minerals in the neighbourhood of Vesuvius.

Observations.

It is known to some collectors under the name *Compact Blue Lava* of Vesuvius; and is employed by artists in their mosaic work, to represent the sky.

Third Kind.

Roestone, or Oolite *.

Roogenstein, *Werner.*

Hammites, *Plin.* Hist. Nat. xxxvii. 10. s. 60.?—F. E. Bruckmanni, Specimen physicum sistens, Histor. Nat. Oolithe, 1721.

---

* *Roestone*, so named on account of its resemblance in form to the roe of fishes.

1721.—Stalactites oolithus, var. *b. c. d. Wall.* t. ii. p. 384.—Roogenstein, *Wid.* s. 511.—Oviform Limestone, *Kirw.* vol. i. p. 91.—Roogenstein, *Estner,* b. ii. s. 928. *Id. Emm.* b. i. s. 442.—Tufo oolitico, *Nap.* p. 353.—L'Oolite, *Broch.* t. i. p. 529.—Chaux carbonatée compacte globuliforme, *Haüy,* t. ii. p. 171.—Roogenstein, *Reuss,* b. ii. 2. s. 270. *Id. Lud.* b. i. s. 148. *Id. Suck.* 1r th. s. 591. *Id. Bert.* s. 89. *Id. Mohs,* b. ii. s. 26. *Id. Hab.* p. 72. *Id. Leonhard,* Tabel. s. 32.—Chaux carbonatée oolithe, *Brong.* t. i. p. 203.—Chaux carbonatée globuliforme, *Brard,* p. 31.—Erbsförmiger Kalkstein, *Karsten,* Tabel. s. 50.—Roestone, *Kid,* vol. i. p. 26.—Chaux carbonatée globuliforme, *Haüy,* Tabl. p. 4.—Roogenstein, *Lenz,* b. ii. s. 738. *Id. Hoff.* b. iii. s. 12.—Schaaliger Kalkstein, *Haus.* b. iii. s. 912.

External Characters.

Its colours are hair-brown, chesnut-brown, and reddish-brown, and sometimes yellowish-grey, and ash-grey.

It occurs massive, and in distinct concretions, which are round granular, the larger are composed of fine spherical granular, and sometimes of very thin concentric lamellar concretions.

Internally it is dull.

The fracture of the grains is fine splintery; but of the mass is round granular in the small, and slaty in the large.

The fragments in the large are blunt-edged.

It is opaque.

It is semi-hard, approaching to soft.

It is rather brittle, and very easily frangible.

Specific gravity, 2.6829, 2.6190, *Kopp.*—2.585, *Breithaupt.*

Chemical

Chemical Characters.

It dissolves with effervescence in acids.

Geognostic Situation.

It occurs along with red sandstone, and *lias* limestone.

Geographic Situation.

This rock, which, in England, is known under the names Bath-Stone, Ketton-Stone, Portland-Stone, and Oolite, extends, with but little interruption, from Somersetshire to the banks of the Humber in Lincolnshire. On the Continent of Europe, it occurs in Thuringia, the Netherlands, the mountains of Jura, and in other countries.

Uses.

The Oolite, or Roestone, particularly that of Bath and Portland, is very extensively employed in architecture; it can be worked with great ease, and has a light and beautiful appearance; but it is porous, and possesses no great durability, and should not be employed where there is much carved or ornamental work, for the fine chiselling is soon effaced by the action of the atmosphere. On account of the ease and sharpness with which it can be carved, it is much used by the English architects, who appear to have little regard for futurity. St Paul's is built of this stone, also Somerset-House. The Chapel of Henry VIII. affords a striking proof of the inattention of the architects to the choice of the stone. All the beautiful ornamental work of the exterior had mouldered away in the short comparative period of 300 years. It has recently been cased with a new front of Bath-Stone, in which the carving has been correctly

correctly copied: from the nature of the stone, we may predict, that its duration will not be longer than that of the original. Both Portland and Bath stone varies much in quality. In buildings constructed of this stone, we may frequently observe some of the stones black, and others white. The black stones are those which are more compact and durable, and preserve their coating of smoke; the white stones are decomposing, and presenting a fresh surface, as if they had been recently scraped *. Roestone is also used as a manure, but when burnt into quicklime, the marly varieties afford rather an indifferent mortar; but those mixed with sand a better mortar.

Observations.

1. It passes into Sandstone, Compact Limestone, and Marl.

2. Some naturalists, as Daubenton, Saussure, Spallanzani, and Gillet Lamont, conjecture, that Roestone is carbonate of lime, which has been granulated in the manner of gunpowder, by the action of water: the most plausible opinion is that which attributes the formation of this mineral to crystallization from a state of solution.

Third Subspecies.

Chalk.

Kreide, *Werner.*

Creta alba, *Wall.* t. i. p. 27.—Kreide, *Wid.* s. 492.—Chalk, *Kirw.* vol. i. p. 77.—Kreide, *Estner,* b. ii. s. 917. *Id. Emm.* b. i.

---

* Aikin.

sized fragments of a pale yellow, brick-red, deep brown, and blackish-grey.

Portuguese Marbles.

Few marbles have hitherto been discovered in Portugal, and none of them equal in beauty the finer varieties found in Spain.

Swiss Marbles.

Granular foliated limestone occurs abundantly in Switzerland, but it has not hitherto been much used as a marble.

German Marbles.

Germany abounds in marbles, and affords many varieties, remarkable either for their beauty or singularity. They are quarried in great quantity, and carried to different parts of that vast country, or are exported into the neighbouring states. The varieties are so numerous, that we cannot, in the very brief view we are now taking, pretend to notice even the more remarkable of them, but must refer, for the particular descriptions, to the economical department of this work.

Norwegian Marbles.

Norway is poor in marbles, almost the only quarry of this stone being that of Gillebeck, in the district of Christiania.

Swedish Marbles.

Sweden does not afford many kinds of marble, and none of them are eminently distinguished for the beauty of their appearance.

appearance. The principal marble is that of Fagernech, which is white, with veins of green talc.

Russian and Siberian Marbles.

The vast Empire of Russia affords a great many different kinds of marble. Georgi, in his Description of the Russian Empire, enumerates white, grey, green, blue, yellow, and red varieties; and Patrin gives the following account of the Siberian marbles. "The Uralian Mountains furnish the finest and most variegated marbles. The greater part is taken from the neighbourhood of Catharinenburg, where they are wrought, and from thence transported into Russia, particularly to Petersburgh. The late Empress caused an immense palace to be built in her capital for Orloff, her favourite, which is entirely coated with these fine marbles, both inside and outside. The Empress built the church of Isaac with the same marbles, on a vast space, near the statue of Peter the Great." Patrin found no white statuary marble in the Uralian Mountains; but in that part of the Altain Mountains which is traversed by the river Irtish, he in two places saw immense blocks of marble, perfectly white and pure, from which blocks might be hewn.

Asiatic Marbles.

At present we are very imperfectly acquainted with the marbles of Asia.

Shaw mentions a red marble from Mount Sinai: Russell, in his Natural History of Aleppo, gives an imperfect account of the marbles of Syria; and some Persian marbles are noticed by Chardin. Mr Morier, in his Journey through Persia, mentions a very beautiful marble, under the name *Marble of Tabris*, and informs us, that the tomb

of

of the celebrated poet Hafitz is constructed with it, and that the wainscotting of the principal room of the Hafi-tan, near Schiraz, is likewise of this marble. Its colours are described as light green, with veins, sometimes of red, sometimes of blue, and it has great translucency. It is cut in large slabs; for Mr Morier saw some that measured nine feet in length, and five feet in breadth. He says, that it is not procured near the city of Tabriz, or taken from a quarry, but is said to be rather a petrifaction, found in large quantities, and in immense blocks, on the borders of the Lake Shahee, near the town of Meraugheh. If it is a mere calcareous deposition, formed in the way of calcareous-alabaster or calc-sinter, it must be considered, not as marble, but a variety of that mineral.

The marbles of Hindostan, Siam, and China, are almost unknown to us. Authors speak of a quarry of white marble in the neighbourhood of Pekin; and of a similar marble in the vicinity of the capital of Siam.

African Marbles.

Beds of marble occur in the Atlas Mountains, and in those ranges that bound the shores of the Red Sea *.

American Marbles.

A good many different marbles have been discovered in the United States. The principal quarries are at Stockbridge and Lanesborough, Massachussets: in Vermont and Pennsylvania: in New-York; and in Virginia. According to Professor Hall, as mentioned by Mr Kœnig, marble has been found in many places on the west side of the Green Mountains in Vermont. A few years since, a valuable

* Vid. Murray's interesting and valuable work, "Historical Account of Discoveries and Travels in Africa," for particulars in regard to the marbles

luable quarry was opened in Middleburg, a town situated on Otter Creek, eleven miles above Vergennes. The marble is of different colours in different parts of the bed. The principal colour, however, is bluish-grey. It takes a good polish, and is in general free of admixture of any substance that might affect its polish.

Second Subspecies.

Compact Limestone.

Dichter Kalkstein, *Werner.*

This subspecies is divided into three kinds, viz. Common Compact Limestone, Blue Vesuvian Limestone, and Roestone.

First Kind.

Common Compact Limestone.

Gemeiner Dichter Kalkstein, *Werner.*

Calcareus æquabilis, *Wall.* t. i. p. 122.—Dichter Kalkstein, *Wid.* s. 494.—Compact Limestone, *Kirw.* vol. i. p. 82.—Gemeiner Dichter Kalkstein, *Emm.* b. i. s. 437.—Pietra calcarea compacta, *Nap.* p. 33.—La pierre calcaire compacte commune, *Broch.* t. i. p. 523.—Chaux carbonatée compacte, *Haüy,* t. ii. p. 166.—Gemeiner Dichter Kalkstein, *Reuss,* b. ii. 2. s. 262. *Id. Lud.* b. i. s. 146. *Id. Suck.* 1r th. s. 585. *Id. Bert.* s. 88. *Id. Mohs,* b. ii. s. 14. *Id. Hab.* s. 71. *Id. Leonhard,* Tabel. s. 32.—Chaux carbonatée compacte, *Brong.* t. i. p. 199.—Gemeiner Kalkstein, *Haus.* s. 126.—Dichter Kalkstein, *Karsten,* Tabel. s. 50.—Chaux carbonatée compacte, *Haüy,* Tabl. p. 4.—Dichter gemeiner Kalkstein, *Lenz,* b. ii. s. 732.—Gemeiner Dichter Kalkstein, *Hoff.* b. iii. s. 8.—Common Limestone, *Aikin,* p. 160.

External

External Characters.

Its most frequent colour is grey, of which the following varieties have been observed: yellowish, bluish, ash, pearl, greenish, and smoke grey; the ash-grey passes into greyish-black; the yellowish-grey into yellowish-brown, ochre-yellow, and into a colour bordering on cream-yellow. It also occurs blood-red, flesh-red, and peach-blossom-red, which latter colour is very rare.

It frequently exhibits veined, zoned, striped, clouded, and spotted coloured delineations; and sometimes also black and brown coloured arborisations.

It very rarely exhibits a beautiful play of colours, caused by intermixed portions of pearly shells.

It occurs massive, corroded, in large plates, rolled masses, and in various extraneous external shapes, of univalve, bivalve, and multivalve shells, of corals, fishes, and more rarely of vegetables, as of ferns and reeds.

Internally it is dull, seldom glimmering, which is owing to intermixed calcareous-spar.

The fracture is small and fine splintery, which sometimes passes into large and flat conchoidal, sometimes into uneven, inclining to earthy, and it occasionally inclines to straight and thick slaty.

The fragments are indeterminate angular, more or less sharp-edged, but in the slaty variety they are tabular.

It is generally translucent on the edges, sometimes opaque.

It is in general rather softer than granular foliated limestone.

It is brittle, and easily frangible.

Its streak is generally greyish-white.

Specific

Specific gravity, Splintery, 2.600, 2.720, *Brisson.*—Opalescent Shell Marble, 2.673, *Leonhard.*—2.675, *Werner.*

Chemical Characters.

It effervesces with acids, and the greater part is dissolved; and burns to quicklime, without falling to pieces.

Constituent Parts.

Rudersdorf.		Bluish-grey Limestone.		Limestone from Sweden.		Limestone from Ettersberg *.	
Lime,	53.00	Lime,	49.50	Lime,	49.25	Lime,	33.41
Carbon. acid,	42.50	Carbon. acid,	40.00	Carbon. acid,	35.00	Carbon. acid,	42.00
Silica,	1.12	Silica,	5.25	Silica,	8.75	Silica,	10.25
Alumina,	1.00	Alumina,	2.75	Alumina,	2.50	Magnesia,	9.43
Iron,	0.75	Iron,	1.37	Iron,	2.75	Iron,	2.25
Water,	1.03	Water,	1.13	Loss,	1.75	Manganese,	1.25
						Loss,	1.41
	100		100		100		100
Simon, Gehlen's Jour. iv. s. 426.		*Simon*, Ib.		*Simon*, Ib.		*Bucholz*, Ib.	

Geognostic Situation.

This mineral occurs in vast abundance in nature, principally in secondary formations, along with sandstone, gypsum, and coal; and in small quantity in primitive mountains. The variegated varieties, which are frequently traversed by veins of calcareous-spar, occur principally in districts composed of grey-wacke and clay-slate. It is distinctly stratified, and the strata vary in thickness, from a few inches to many fathoms, and are from a few fathoms to many miles in extent. The strata generally incline to horizontal

* Some of the limestones in Fifeshire agree in composition with that of Ettersberg.

rizontal; sometimes, however, they are vertical, or variously convoluted, even arranged in concentric layers, thus presenting appearances illustrative of their chemical nature. Petrifactions, both of animals and vegetables, but principally of the former, abound in compact limestone: these are of corals, shells, fishes, and sometimes of amphibious animals. On a general view, it is to be considered as rich in ores of different kinds, particularly ores of lead and zinc: thus, nearly all the rich and valuable lead-mines in England are situated in limestone.

Geographic Situation.

It abounds in the sandstone and coal formations, both in Scotland and England; and in Ireland, it is a very abundant mineral in all the districts where clay-slate and red sandstone rocks occur. On the Continent of Europe, it is a very widely and abundantly distributed mineral; and forms a striking feature in many extensive tracts of country in Asia, Africa and America, as will be particularly described in the Geognostic part of this work.

Uses.

When compact limestone joins to pure and agreeable colours, so considerable a degree of hardness that it takes a good polish, it is by artists considered as a Marble; and if it contains petrifactions mineralized, it is named *shell* or *lumachella*, and *coral* or *zoophytic marble*, according as the organic remains are testaceous or coralline*. In one particular variety of lumachella

* The name *marmor*, is derived from the Greek *μαρμαιρειν*, *to shine*, or *glitter*, and was by the ancients applied, not only to limestone, but also to stones possessing agreeable colours, and receiving a good polish, such as gypsum, jasper, serpentine, and even granite and porphyry.

lumachella or shell marble, found at Bleiberg in Carinthia, the shells and fragments of shells, which belong to the nautilus tribe, are set in a brown-coloured basis, and reflect many beautiful and brilliant pearly inclining to metallic colours, principally the fire-red, green, and blue tints. It is named *opalescent* or *fire marble*. Another lumachella marble from Astracan, contains, in a reddish-brown basis, pearly shells of nautili, that reflect a very brilliant gold-yellow colour. In some compact marbles, the surface presents a beautiful arborescent appearance, and these are named *arborescent* or *dendritic marbles*. Such are those of Papenheim in Bavaria.

The *Florentine Marble*, or *Ruin Marble*, as it is sometimes called, is a compact limestone. It occurs on the Po and the Arno, and is worked into various articles at Florence. It is said to occur in balls. It presents angular figures of a yellowish-brown, on a base of a lighter tint, and which passes to greyish-white. Seen at a distance, slabs of this stone resemble drawings done in bister. "One is amused (says Brard) to observe in it kinds of ruins: there it is a Gothic castle half destroyed, here it presents ruined walls; in another place old bastions; and, what still adds to the illusion, is, that, in these sorts of natural paintings, there exists a kind of aërial perspective, which is very sensibly perceived. The lower part, or what forms the first plane, has a warm and bold tone; the second follows it, and weakens it as it increases the distance; the third becomes still fainter; while the upper part, agreeing with the first, presents in the distance a whitish zone, which terminates the horizon, then blends itself more and more as it rises, and at length reaches the top, where it forms sometimes as it were clouds. But approach close to it, all va-

nishes immediately, and these pretended figures, which, at a distance, seemed so well drawn, are converted into irregular marks, which present nothing to the eye." To the same compact limestone may be referred the variety called *Cottam Marble*, from being found at Cottam, near Bristol. It resembles in many respects the Landscape Marble.

In different parts of Scotland, compact limestone is cut and polished as marble: this was the case in the parish of Cummertrees in Dumfriesshire,—in Cambuslang parish, in Lanarkshire,—in Fifeshire, &c. In England, many compact limestones are cut and polished as marbles; such are the limestones of Derbyshire, Yorkshire, Devonshire, Somersetshire, and Dorsetshire. It is sometimes used as a building stone; and, in want of better materials, for paving streets, and making highways. When, by exposure to a high temperature, it is deprived of its carbonic acid, and converted into quicklime, it is used for mortar; also by the soap-maker, for rendering his alkalies caustic; by the tanner, for cleansing hides, or freeing them from hair, muscular substance, and fat; by the farmer, in the improvement of particular kinds of soil; and by the metallurgist, in the smelting of such ores as are difficultly fusible, owing to an intermixture of silica and alumina.

Second

Second Kind.

Blue Vesuvian Limestone.

Blauer Vesuvischer Kalkstein, *Klaproth*.

Blauer Vesuvischer Kalkstein, *Klaproth*, Beit. b. v. s. 92. *Id.* *Lenz*, b. ii. s. 737.

External Characters.

Its colour is dark bluish-grey, partly veined with white.

Externally it appears as if it had been rolled; and the surface is uneven.

The fracture fine earthy, passing into splintery.

It is opaque.

It affords a white streak.

It is semi-hard in a low degree.

It is rather heavy.

Constituent Parts.

Lime, - -	58.00
Carbonic Acid, -	28.50
Water, which is somewhat ammoniacal, -	11.00
Magnesia, -	0.50
Oxide of Iron,	0.25
Carbon, - -	0.25
Silica,	1.25
	99.75

Klaproth, Beit. b. v. s. 96.

From

1721.—Stalactites oolithus, var. *b. c. d. Wall.* t. ii. p. 384.—Roogenstein, *Wid.* s. 511.—Oviform Limestone, *Kirw.* vol. i. p. 91.—Roogenstein, *Estner,* b. ii. s. 928. *Id. Emm.* b. i. s. 442.—Tufo oolitico, *Nap.* p. 353.—L'Oolite, *Broch.* t. i. p. 529.—Chaux carbonatée compacte globuliforme, *Haüy,* t. ii. p. 171.—Roogenstein, *Reuss,* b. ii. 2. s. 270. *Id. Lud.* b. i. s. 148. *Id. Suck.* 1r th. s. 591. *Id. Bert.* s. 89. *Id. Mohs,* b. ii. s. 26. *Id. Hab.* p. 72. *Id. Leonhard,* Tabel. s. 32.—Chaux carbonatée oolithe, *Brong.* t. i. p. 203.—Chaux carbonatée globuliforme, *Brard,* p. 31.—Erbsförmiger Kalkstein, *Karsten,* Tabel. s. 50.—Roestone, *Kid,* vol. i. p. 26.—Chaux carbonatée globuliforme, *Haüy,* Tabl. p. 4.—Roogenstein, *Lenz,* b. ii. s. 738. *Id. Hoff.* b. iii. s. 12.—Schaaliger Kalkstein, *Haus.* b. iii. s. 912.

External Characters.

Its colours are hair-brown, chesnut-brown, and reddish-brown, and sometimes yellowish-grey, and ash-grey.

It occurs massive, and in distinct concretions, which are round granular, the larger are composed of fine spherical granular, and sometimes of very thin concentric lamellar concretions.

Internally it is dull.

The fracture of the grains is fine splintery; but of the mass is round granular in the small, and slaty in the large.

The fragments in the large are blunt-edged.

It is opaque.

It is semi-hard, approaching to soft.

It is rather brittle, and very easily frangible.

Specific gravity, 2.6829, 2.6190, *Kopp.*—2.585, *Breithaupt.*

Chemical

Chemical Characters.

It dissolves with effervescence in acids.

Geognostic Situation.

It occurs along with red sandstone, and *lias* limestone.

Geographic Situation.

This rock, which, in England, is known under the names Bath-Stone, Ketton-Stone, Portland-Stone, and Oolite, extends, with but little interruption, from Somersetshire to the banks of the Humber in Lincolnshire. On the Continent of Europe, it occurs in Thuringia, the Netherlands, the mountains of Jura, and in other countries.

Uses.

The Oolite, or Roestone, particularly that of Bath and Portland, is very extensively employed in architecture; it can be worked with great ease, and has a light and beautiful appearance; but it is porous, and possesses no great durability, and should not be employed where there is much carved or ornamental work, for the fine chiselling is soon effaced by the action of the atmosphere. On account of the ease and sharpness with which it can be carved, it is much used by the English architects, who appear to have little regard for futurity. St Paul's is built of this stone, also Somerset-House. The Chapel of Henry VIII. affords a striking proof of the inattention of the architects to the choice of the stone. All the beautiful ornamental work of the exterior had mouldered away in the short comparative period of 300 years. It has recently been cased with a new front of Bath-Stone, in which the carving has been

correctly

correctly copied: from the nature of the stone, we may predict, that its duration will not be longer than that of the original. Both Portland and Bath stone varies much in quality. In buildings constructed of this stone, we may frequently observe some of the stones black, and others white. The black stones are those which are more compact and durable, and preserve their coating of smoke; the white stones are decomposing, and presenting a fresh surface, as if they had been recently scraped *. Roestone is also used as a manure, but when burnt into quicklime, the marly varieties afford rather an indifferent mortar; but those mixed with sand a better mortar.

Observations.

1. It passes into Sandstone, Compact Limestone, and Marl.

2. Some naturalists, as Daubenton, Saussure, Spallanzani, and Gillet Lamont, conjecture, that Roestone is carbonate of lime, which has been granulated in the manner of gunpowder, by the action of water: the most plausible opinion is that which attributes the formation of this mineral to crystallization from a state of solution.

Third Subspecies.

Chalk.

Kreide, *Werner.*

Creta alba, *Wall.* t. i. p. 27.—Kreide, *Wid.* s. 492.—Chalk, *Kirw.* vol. i. p. 77.—Kreide, *Estner,* b. ii. s. 917. *Id. Emm.* b. i.

* Aikin.

sized fragments of a pale yellow, brick-red, deep brown, and blackish-grey.

Portuguese Marbles.

Few marbles have hitherto been discovered in Portugal, and none of them equal in beauty the finer varieties found in Spain.

Swiss Marbles.

Granular foliated limestone occurs abundantly in Switzerland, but it has not hitherto been much used as a marble.

German Marbles.

Germany abounds in marbles, and affords many varieties, remarkable either for their beauty or singularity. They are quarried in great quantity, and carried to different parts of that vast country, or are exported into the neighbouring states. The varieties are so numerous, that we cannot, in the very brief view we are now taking, pretend to notice even the more remarkable of them, but must refer, for the particular descriptions, to the economical department of this work.

Norwegian Marbles.

Norway is poor in marbles, almost the only quarry of this stone being that of Gillebeck, in the district of Christiania.

Swedish Marbles.

Sweden does not afford many kinds of marble, and none of them are eminently distinguished for the beauty of their appearance.

appearance. The principal marble is that of Fagernech, which is white, with veins of green talc.

Russian and Siberian Marbles.

The vast Empire of Russia affords a great many different kinds of marble. Georgi, in his Description of the Russian Empire, enumerates white, grey, green, blue, yellow, and red varieties; and Patrin gives the following account of the Siberian marbles. "The Uralian Mountains furnish the finest and most variegated marbles. The greater part is taken from the neighbourhood of Catharinenburg, where they are wrought, and from thence transported into Russia, particularly to Petersburgh. The late Empress caused an immense palace to be built in her capital for Orloff, her favourite, which is entirely coated with these fine marbles, both inside and outside. The Empress built the church of Isaac with the same marbles, on a vast space, near the statue of Peter the Great." Patrin found no white statuary marble in the Uralian Mountains; but in that part of the Altain Mountains which is traversed by the river Irtish, he in two places saw immense blocks of marble, perfectly white and pure, from which blocks might be hewn.

Asiatic Marbles.

At present we are very imperfectly acquainted with the marbles of Asia.

Shaw mentions a red marble from Mount Sinai: Russell, in his Natural History of Aleppo, gives an imperfect account of the marbles of Syria; and some Persian marbles are noticed by Chardin. Mr Morier, in his Journey through Persia, mentions a very beautiful marble, under the name *Marble of Tabriz*, and informs us, that the tomb

of

of the celebrated poet Hafitz is constructed with it, and that the wainscotting of the principal room of the Hafi-ten, near Schiraz, is likewise of this marble. Its colours are described as light green, with veins, sometimes of red, sometimes of blue, and it has great translucency. It is cut in large slabs; for Mr Morier saw some that measured nine feet in length, and five feet in breadth. He says, that it is not procured near the city of Tabriz, or taken from a quarry, but is said to be rather a petrifaction, found in large quantities, and in immense blocks, on the borders of the Lake Shahee, near the town of Meraugheh. If it is a mere calcareous deposition, formed in the way of calcareous-alabaster or calc-sinter, it must be considered, not as marble, but a variety of that mineral.

The marbles of Hindostan, Siam, and China, are almost unknown to us. Authors speak of a quarry of white marble in the neighbourhood of Pekin; and of a similar marble in the vicinity of the capital of Siam.

African Marbles.

Beds of marble occur in the Atlas Mountains, and in those ranges that bound the shores of the Red Sea *.

American Marbles.

A good many different marbles have been discovered in the United States. The principal quarries are at Stockbridge and Lanesborough, Massachussets: in Vermont and Pennsylvania: in New-York; and in Virginia. According to Professor Hall, as mentioned by Mr Kœnig, marble has been found in many places on the west side of the Green Mountains in Vermont. A few years since, a va-

* Vid. Murray's interesting and va[illegible] "Historical Account of Discoveries and Travels in Africa," fo[illegible] [illegible]ard to the ma[illegible]

luable quarry was opened in Middleburg, a town situated on Otter Creek, eleven miles above Vergennes. The marble is of different colours in different parts of the bed. The principal colour, however, is bluish-grey. It takes a good polish, and is in general free of admixture of any substance that might affect its polish.

Second Subspecies.

Compact Limestone.

Dichter Kalkstein, *Werner.*

This subspecies is divided into three kinds, viz. Common Compact Limestone, Blue Vesuvian Limestone, and Roestone.

First Kind.

Common Compact Limestone.

Gemeiner Dichter Kalkstein, *Werner.*

Calcareus æquabilis, *Wall.* t. i. p. 122.—Dichter Kalkstein, *Wid.* s. 494.—Compact Limestone, *Kirw.* vol. i. p. 82.—Gemeiner Dichter Kalkstein, *Emm.* b. i. s. 437.—Pietra calcarea compacta, *Nap.* p. 33.—La pierre calcaire compacte commune, *Broch.* t. i. p. 523.—Chaux carbonatée compacte, *Haüy,* t. ii. p. 166.—Gemeiner Dichter Kalkstein, *Reuss,* b. ii. 2. s. 262. *Id. Lud.* b. i. s. 146. *Id. Suck.* 1r th. s. 585. *Id. Bert.* s. 88. *Id. Mohs,* b. ii. s. 14. *Id. Hab.* s. 71. *Id. Leonhard,* Tabel. s. 32.—Chaux carbonatée compacte, *Brong.* t. i. p. 199.—Gemeiner Kalkstein, *Haus.* s. 126.—Dichter Kalkstein, *Karsten,* Tabel. s. 50.—Chaux carbonatée compacte, *Haüy,* Tabl. p. 4.—Dichter gemeiner Kalkstein, *Lenz,* b. ii. s. 732.—Gemeiner Dichter Kalkstein, *Hoff.* b. iii. s. 8.—Common Lime-ne, *Aikin,* p. 160.

External

External Characters.

Its most frequent colour is grey, of which the following varieties have been observed: yellowish, bluish, ash, pearl, greenish, and smoke grey; the ash-grey passes into greyish-black; the yellowish-grey into yellowish-brown, ochre-yellow, and into a colour bordering on cream-yellow. It also occurs blood-red, flesh-red, and peach-blossom-red, which latter colour is very rare.

It frequently exhibits veined, zoned, striped, clouded, and spotted coloured delineations; and sometimes also black and brown coloured arborisations.

It very rarely exhibits a beautiful play of colours, caused by intermixed portions of pearly shells.

It occurs massive, corroded, in large plates, rolled masses, and in various extraneous external shapes, of univalve, bivalve, and multivalve shells, of corals, fishes, and more rarely of vegetables, as of ferns and reeds.

Internally it is dull, seldom glimmering, which is owing to intermixed calcareous-spar.

The fracture is small and fine splintery, which sometimes passes into large and flat conchoidal, sometimes into uneven, inclining to earthy, and it occasionally inclines to straight and thick slaty.

The fragments are indeterminate angular, more or less sharp-edged, but in the slaty variety they are tabular.

It is generally translucent on the edges, sometimes opaque.

It is in general rather softer than granular foliated limestone.

It is brittle, and easily frangible.

Its streak is generally greyish-white.

Specific

Specific gravity, Splintery, 2.600, 2.720, *Brisson.*—Opalescent Shell Marble, 2.673, *Leonhard.*—2.675, *Werner.*

Chemical Characters.

It effervesces with acids, and the greater part is dissolved; and burns to quicklime, without falling to pieces.

Constituent Parts.

Rudersdorf.		Bluish-grey Limestone.		Limestone from Sweden.		Limestone from Ettersberg *.	
Lime,	53.00	Lime,	49.50	Lime,	49.25	Lime,	33.41
Carbon. acid,	42.50	Carbon. acid,	40.00	Carbon. acid,	35.00	Carbon. acid,	42.00
Silica,	1.12	Silica,	5.25	Silica,	8.75	Silica,	10.25
Alumina,	1.00	Alumina,	2.75	Alumina,	2.50	Magnesia,	9.43
Iron,	0.75	Iron,	1.37	Iron,	2.75	Iron,	2.25
Water,	1.03	Water,	1.13	Loss,	1.75	Manganese,	1.25
						Loss,	1.41
	100		100		100		100
Simon, Gehlen's Jour. iv. s. 426.		*Simon*, Ib.		*Simon*, Ib.		*Buchols*, Ib.	

Geognostic Situation.

This mineral occurs in vast abundance in nature, principally in secondary formations, along with sandstone, gypsum, and coal; and in small quantity in primitive mountains. The variegated varieties, which are frequently traversed by veins of calcareous-spar, occur principally in districts composed of grey-wacke and clay-slate. It is distinctly stratified, and the strata vary in thickness, from a few inches to many fathoms, and are from a few fathoms to many miles in extent. The strata generally incline to horizontal

* Some of the limestones in Fifeshire agree in composition with that of Ettersberg.

rizontal; sometimes, however, they are vertical, or variously convoluted, even arranged in concentric layers, thus presenting appearances illustrative of their chemical nature. Petrifactions, both of animals and vegetables, but principally of the former, abound in compact limestone: these are of corals, shells, fishes, and sometimes of amphibious animals. On a general view, it is to be considered as rich in ores of different kinds, particularly ores of lead and zinc: thus, nearly all the rich and valuable lead-mines in England are situated in limestone.

Geographic Situation.

It abounds in the sandstone and coal formations, both in Scotland and England; and in Ireland, it is a very abundant mineral in all the districts where clay-slate and red sandstone rocks occur. On the Continent of Europe, it is a very widely and abundantly distributed mineral; and forms a striking feature in many extensive tracts of country in Asia, Africa and America, as will be particularly described in the Geognostic part of this work.

Uses.

When compact limestone joins to pure and agreeable colours, so considerable a degree of hardness that it takes a good polish, it is by artists considered as a Marble; and if it contains petrifactions mineralized, it is named *shell* or *lumachella*, and *coral* or *zoophytic marble*, according as the organic remains are testaceous or coralline*. In one particular variety of lumachella

* The name *marmor*, is derived from the Greek *μαρμαιρειν*, *to shine*, or *glitter*, and was by the ancients applied, not only to limestone, but also to stones possessing agreeable colours, and receiving a good polish, such as gypsum, jasper, serpentine, and even granite and porphyry.

lumachella or shell marble, found at Bleiberg in Carinthia, the shells and fragments of shells, which belong to the nautilus tribe, are set in a brown-coloured basis, and reflect many beautiful and brilliant pearly inclining to metallic colours, principally the fire-red, green, and blue tints. It is named *opalescent* or *fire marble*. Another lumachella marble from Astracan, contains, in a reddish-brown basis, pearly shells of nautili, that reflect a very brilliant gold-yellow colour. In some compact marbles, the surface presents a beautiful arborescent appearance, and these are named *arborescent* or *dendritic marbles*. Such are those of Papenheim in Bavaria.

The *Florentine Marble*, or *Ruin Marble*, as it is sometimes called, is a compact limestone. It occurs on the Po and the Arno, and is worked into various articles at Florence. It is said to occur in balls. It presents angular figures of a yellowish-brown, on a base of a lighter tint, and which passes to greyish-white. Seen at a distance, slabs of this stone resemble drawings done in bister. "One is amused (says Brard) to observe in it kinds of ruins: there it is a Gothic castle half destroyed, here it presents ruined walls; in another place old bastions; and, what still adds to the illusion, is, that, in these sorts of natural paintings, there exists a kind of aërial perspective, which is very sensibly perceived. The lower part, or what forms the first plane, has a warm and bold tone; the second follows it, and weakens it as it increases the distance; the third becomes still fainter; while the upper part, agreeing with the first, presents in the distance a whitish zone, which terminates the horizon, then blends itself more and more as it rises, and at length reaches the top, where it forms sometimes as it were clouds. But approach close to it, all va-

 nishes

nishes immediately, and these pretended figures, which, at a distance, seemed so well drawn, are converted into irregular marks, which present nothing to the eye." To the same compact limestone may be referred the variety called *Cottam Marble*, from being found at Cottam, near Bristol. It resembles in many respects the Landscape Marble.

In different parts of Scotland, compact limestone is cut and polished as marble: this was the case in the parish of Cummertrees in Dumfriesshire,—in Cambuslang parish, in Lanarkshire,—in Fifeshire, &c. In England, many compact limestones are cut and polished as marbles; such are the limestones of Derbyshire, Yorkshire, Devonshire, Somersetshire, and Dorsetshire. It is sometimes used as a building stone; and, in want of better materials, for paving streets, and making highways. When, by exposure to a high temperature, it is deprived of its carbonic acid, and converted into quicklime, it is used for mortar; also by the soap-maker, for rendering his alkalies caustic; by the tanner, for cleansing hides, or freeing them from hair, muscular substance, and fat; by the farmer, in the improvement of particular kinds of soil; and by the metallurgist, in the smelting of such ores as are difficultly fusible, owing to an intermixture of silica and alumina.

Second Kind.

Blue Vesuvian Limestone.

Blauer Vesuvischer Kalkstein, *Klaproth.*

Blauer Vesuvischer Kalkstein, *Klaproth,* Beit. b. v. s. 92. *Id. Lenz,* b. ii. s. 737.

External Characters.

Its colour is dark bluish-grey, partly veined with white.

Externally it appears as if it had been rolled; and the surface is uneven.

The fracture fine earthy, passing into splintery.

It is opaque.

It affords a white streak.

It is semi-hard in a low degree.

It is rather heavy.

Constituent Parts.

Lime, - -	58.00
Carbonic Acid, -	28.50
Water, which is somewhat ammoniacal, -	11.00
Magnesia, -	0.50
Oxide of Iron,	0.25
Carbon, - -	0.25
Silica,	1.25
	99.75

Klaproth, Beit. b. v. s. 96.

From

From this analysis, it appears, that the vesuvian limestone differs remarkably in composition from common compact limestone. In common compact limestone, 100 parts of lime are combined with at least 80 parts of carbonic acid; whereas in the vesuvian limestone, 100 parts of limestone are not combined with more than 50 parts of carbonic acid. Secondly, In common limestone, independent of the water which adheres to it accidentally, as far as we know, there is no water of composition; but in the vesuvian limestone, there are 11 parts of water of composition.

Geographic Situation.

This remarkable limestone is found in loose masses amongst unaltered ejected minerals in the neighbourhood of Vesuvius.

Observations.

It is known to some collectors under the name *Compact Blue Lava* of Vesuvius; and is employed by artists in their mosaic work, to represent the sky.

Third Kind.

Roestone, or Oolite *.

Roogenstein, *Werner.*

Hammites, *Plin.* Hist. Nat. xxxvii. 10. s. 60.?—F. E. Bruckmanni, Specimen physicum sistens, Histor. Nat. Oolithe, 1721.

---

* *Roestone*, so named on account of its resemblance in form to the roe of fishes.

1721.—Stalactites oolithus, var. *b. c. d. Wall.* t. ii. p. 384.—Roogenstein, *Wid.* s. 511.—Oviform Limestone, *Kirw.* vol. i. p. 91.—Roogenstein, *Estner,* b. ii. s. 928. *Id. Emm.* b. i. s. 442.—Tufo oolitico, *Nap.* p. 353.—L'Oolite, *Broch.* t. i. p. 529.—Chaux carbonatée compacte globuliforme, *Haüy,* t. ii. p. 171.—Roogenstein, *Reuss,* b. ii. 2. s. 270. *Id. Lud.* b. i. s. 148. *Id. Suck.* 1r th. s. 591. *Id. Bert.* s. 89. *Id. Mohs,* b. ii. s. 26. *Id. Hab.* p. 72. *Id. Leonhard,* Tabel. s. 32.—Chaux carbonatée oolithe, *Brong.* t. i. p. 203.—Chaux carbonatée globuliforme, *Brard,* p. 31.—Erbsförmiger Kalkstein, *Karsten,* Tabel. s. 50.—Roestone, *Kid,* vol. i. p. 26.—Chaux carbonatée globuliforme, *Haüy,* Tabl. p. 4.—Roogenstein, *Lenz,* b. ii. s. 738. *Id. Hoff.* b. iii. s. 12.—Schaaliger Kalkstein, *Haus.* b. iii. s. 912.

External Characters.

Its colours are hair-brown, chesnut-brown, and reddish-brown, and sometimes yellowish-grey, and ash-grey.

It occurs massive, and in distinct concretions, which are round granular, the larger are composed of fine spherical granular, and sometimes of very thin concentric lamellar concretions.

Internally it is dull.

The fracture of the grains is fine splintery; but of the mass is round granular in the small, and slaty in the large.

The fragments in the large are blunt-edged.

It is opaque.

It is semi-hard, approaching to soft.

It is rather brittle, and very easily frangible.

Specific gravity, 2.6829, 2.6190, *Kopp.*—2.585, *Breithaupt.*

Chemical

Chemical Characters.

It dissolves with effervescence in acids.

Geognostic Situation.

It occurs along with red sandstone, and *lias* limestone.

Geographic Situation.

This rock, which, in England, is known under the names Bath-Stone, Ketton-Stone, Portland-Stone, and Oolite, extends, with but little interruption, from Somersetshire to the banks of the Humber in Lincolnshire. On the Continent of Europe, it occurs in Thuringia, the Netherlands, the mountains of Jura, and in other countries.

Uses.

The Oolite, or Roestone, particularly that of Bath and Portland, is very extensively employed in architecture; it can be worked with great ease, and has a light and beautiful appearance; but it is porous, and possesses no great durability, and should not be employed where there is much carved or ornamental work, for the fine chiselling is soon effaced by the action of the atmosphere. On account of the ease and sharpness with which it can be carved, it is much used by the English architects, who appear to have little regard for futurity. St Paul's is built of this stone, also Somerset-House. The Chapel of Henry VIII. affords a striking proof of the inattention of the architects to the choice of the stone. All the beautiful ornamental work of the exterior had mouldered away in the short comparative period of 300 years. It has recently been cased with a new front of Bath-Stone, in which the carving has been correctly

correctly copied: from the nature of the stone, we may predict, that its duration will not be longer than that of the original. Both Portland and Bath stone varies much in quality. In buildings constructed of this stone, we may frequently observe some of the stones black, and others white. The black stones are those which are more compact and durable, and preserve their coating of smoke; the white stones are decomposing, and presenting a fresh surface, as if they had been recently scraped *. Roestone is also used as a manure, but when burnt into quicklime, the marly varieties afford rather an indifferent mortar; but those mixed with sand a better mortar.

Observations.

1. It passes into Sandstone, Compact Limestone, and Marl.

2. Some naturalists, as Daubenton, Saussure, Spallanzani, and Gillet Lamont, conjecture, that Roestone is carbonate of lime, which has been granulated in the manner of gunpowder, by the action of water: the most plausible opinion is that which attributes the formation of this mineral to crystallization from a state of solution.

Third Subspecies.

Chalk.

Kreide, *Werner.*

Creta alba, *Wall.* t. i. p. 27.—Kreide, *Wid.* s. 492.—Chalk, *Kirw.* vol. i. p. 77.—Kreide, *Estner,* b. ii. s. 917. *Id. Emm.* b. i.

* Aikin.

of the celebrated poet Hafitz is constructed with it, and that the wainscotting of the principal room of the Hafi-tan, near Schiraz, is likewise of this marble. Its colours are described as light green, with veins, sometimes of red, sometimes of blue, and it has great translucency. It is cut in large slabs; for Mr Morier saw some that measured nine feet in length, and five feet in breadth. He says, that it is not procured near the city of Tabriz, or taken from a quarry, but is said to be rather a petrifaction, found in large quantities, and in immense blocks, on the borders of the Lake Shahee, near the town of Meraugheh. If it is a mere calcareous deposition, formed in the way of calcareous-alabaster or calc-sinter, it must be considered, not as marble, but a variety of that mineral.

The marbles of Hindostan, Siam, and China, are almost unknown to us. Authors speak of a quarry of white marble in the neighbourhood of Pekin; and of a similar marble in the vicinity of the capital of Siam.

African Marbles.

Beds of marble occur in the Atlas Mountains, and in those ranges that bound the shores of the Red Sea *.

American Marbles.

A good many different marbles have been discovered in the United States. The principal quarries are at Stockbridge and Lanesborough, Massachussets: in Vermont and Pennsylvania: in New-York; and in Virginia. According to Professor Hall, as mentioned by Mr Kœnig, marble has been found in many places on the west side of the Green Mountains in Vermont. A few years since, a valuable

* Vid. Murray's interesting and valuable work, "Historical Account of Discoveries and Travels in Africa," for particulars in regard to the marbles

luable quarry was opened in Middleburg, a town situated on Otter Creek, eleven miles above Vergennes. The marble is of different colours in different parts of the bed. The principal colour, however, is bluish-grey. It takes a good polish, and is in general free of admixture of any substance that might affect its polish.

Second Subspecies.

Compact Limestone.

Dichter Kalkstein, *Werner.*

This subspecies is divided into three kinds, viz. Common Compact Limestone, Blue Vesuvian Limestone, and Roestone.

First Kind.

Common Compact Limestone.

Gemeiner Dichter Kalkstein, *Werner.*

Calcareus æquabilis, *Wall.* t. i. p. 122.—Dichter Kalkstein, *Wid.* s. 494.—Compact Limestone, *Kirw.* vol. i. p. 82.—Gemeiner Dichter Kalkstein, *Emm.* b. i. s. 437.—Pietra calcarea compacta, *Nap.* p. 33.—La pierre calcaire compacte commune, *Broch.* t. i. p. 523.—Chaux carbonatée compacte, *Haüy,* t. ii. p. 166.—Gemeiner Dichter Kalkstein, *Reuss,* b. ii. 2. s. 262. *Id. Lud.* b. i. s. 146. *Id. Suck.* 1r th. s. 585. *Id. Bert.* s. 88. *Id. Mohs,* b. ii. s. 14. *Id. Hab.* s. 71. *Id. Leonhard,* Tabel. s. 32.—Chaux carbonatée compacte, *Brong.* t. i. p. 199.—Gemeiner Kalkstein, *Haus.* s. 126.—Dichter Kalkstein, *Karsten,* Tabel. s. 50.—Chaux carbonatée compacte, *Haüy,* Tabl. p. 4.—Dichter gemeiner Kalkstein, *Lenz,* b. ii. s. 732.—Gemeiner Dichter Kalkstein, *Hoff.* b. iii. s. 8.—Co[illegible] stone, *Aikin,* p. 160.

External Characters.

Its most frequent colour is grey, of which the following varieties have been observed: yellowish, bluish, ash, pearl, greenish, and smoke grey; the ash-grey passes into greyish-black; the yellowish-grey into yellowish-brown, ochre-yellow, and into a colour bordering on cream-yellow. It also occurs blood-red, flesh-red, and peach-blossom-red, which latter colour is very rare.

It frequently exhibits veined, zoned, striped, clouded, and spotted coloured delineations; and sometimes also black and brown coloured arborisations.

It very rarely exhibits a beautiful play of colours, caused by intermixed portions of pearly shells.

It occurs massive, corroded, in large plates, rolled masses, and in various extraneous external shapes, of univalve, bivalve, and multivalve shells, of corals, fishes, and more rarely of vegetables, as of ferns and reeds.

Internally it is dull, seldom glimmering, which is owing to intermixed calcareous-spar.

The fracture is small and fine splintery, which sometimes passes into large and flat conchoidal, sometimes into uneven, inclining to earthy, and it occasionally inclines to straight and thick slaty.

The fragments are indeterminate angular, more or less sharp-edged, but in the slaty variety they are tabular.

It is generally translucent on the edges, sometimes opaque.

It is in general rather softer than granular foliated limestone.

It is brittle, and easily frangible.

Its streak is generally greyish-white.

Specific

Specific gravity, Splintery, 2.600, 2.720, *Brisson.*—Opalescent Shell Marble, 2.673, *Leonhard.*—2.675, *Werner.*

Chemical Characters.

It effervesces with acids, and the greater part is dissolved; and burns to quicklime, without falling to pieces.

Constituent Parts.

Rudersdorf.		Bluish-grey Limestone.		Limestone from Sweden.		Limestone from Ettersberg *.	
Lime,	53.00	Lime,	49.50	Lime,	49.25	Lime,	33.41
Carbon. acid,	42.50	Carbon. acid,	40.00	Carbon. acid,	35.00	Carbon. acid,	42.00
Silica,	1.12	Silica,	5.25	Silica,	8.75	Silica,	10.25
Alumina,	1.00	Alumina,	2.75	Alumina,	2.50	Magnesia,	9.43
Iron,	0.75	Iron,	1.37	Iron,	2.75	Iron,	2.25
Water,	1.03	Water,	1.13	Loss,	1.75	Manganese,	1.25
						Loss,	1.41
	100		100		100		100
Simon, Gehlen's Jour. iv. s. 426.		*Simon*, Ib.		*Simon*, Ib.		*Buchols*, Ib.	

Geognostic Situation.

This mineral occurs in vast abundance in nature, principally in secondary formations, along with sandstone, gypsum, and coal; and in small quantity in primitive mountains. The variegated varieties, which are frequently traversed by veins of calcareous-spar, occur principally in districts composed of grey-wacke and clay-slate. It is distinctly stratified, and the strata vary in thickness, from a few inches to many fathoms, and are from a few fathoms to many miles in extent. The strata generally incline to horizontal

* Some of the limestones in Fifeshire agree in composition with that of Ettersberg.

rizontal; sometimes, however, they are vertical, or variously convoluted, even arranged in concentric layers, thus presenting appearances illustrative of their chemical nature. Petrifactions, both of animals and vegetables, but principally of the former, abound in compact limestone: these are of corals, shells, fishes, and sometimes of amphibious animals. On a general view, it is to be considered as rich in ores of different kinds, particularly ores of lead and zinc: thus, nearly all the rich and valuable lead-mines in England are situated in limestone.

Geographic Situation.

It abounds in the sandstone and coal formations, both in Scotland and England; and in Ireland, it is a very abundant mineral in all the districts where clay-slate and red sandstone rocks occur. On the Continent of Europe, it is a very widely and abundantly distributed mineral; and forms a striking feature in many extensive tracts of country in Asia, Africa and America, as will be particularly described in the Geognostic part of this work.

Uses.

When compact limestone joins to pure and agreeable colours, so considerable a degree of hardness that it takes a good polish, it is by artists considered as a Marble; and if it contains petrifactions mineralized, it is named *shell* or *lumachella*, and *coral* or *zoophytic marble*, according as the organic remains are testaceous or coralline*. In one particular variety of lumachella

* The name *marmor*, is derived from the Greek *μαρμαίρειν*, *to shine*, or *glitter*, and was by the ancients applied, not only to limestone, but also to stones possessing agreeable colours, and receiving a good polish, such as gypsum, jasper, serpentine, and even granite and porphyry.

lumachella or shell marble, found at Bleiberg in Carinthia, the shells and fragments of shells, which belong to the nautilus tribe, are set in a brown-coloured basis, and reflect many beautiful and brilliant pearly inclining to metallic colours, principally the fire-red, green, and blue tints. It is named *opalescent* or *fire marble*. Another lumachella marble from Astracan, contains, in a reddish-brown basis, pearly shells of nautili, that reflect a very brilliant gold-yellow colour. In some compact marbles, the surface presents a beautiful arborescent appearance, and these are named *arborescent* or *dendritic marbles*. Such are those of Papenheim in Bavaria.

The *Florentine Marble*, or *Ruin Marble*, as it is sometimes called, is a compact limestone. It occurs on the Po and the Arno, and is worked into various articles at Florence. It is said to occur in balls. It presents angular figures of a yellowish-brown, on a base of a lighter tint, and which passes to greyish-white. Seen at a distance, slabs of this stone resemble drawings done in bister. "One is amused (says Brard) to observe in it kinds of ruins: there it is a Gothic castle half destroyed, here it presents ruined walls; in another place old bastions; and, what still adds to the illusion, is, that, in these sorts of natural paintings, there exists a kind of aërial perspective, which is very sensibly perceived. The lower part, or what forms the first plane, has a warm and bold tone; the second follows it, and weakens it as it increases the distance; the third becomes still fainter; while the upper part, agreeing with the first, presents in the distance a whitish zone, which terminates the horizon, then blends itself more and more as it rises, and at length reaches the top, where it forms sometimes as it were clouds. But approach close to it, all va-

nishes immediately, and these pretended figures, which, at a distance, seemed so well drawn, are converted into irregular marks, which present nothing to the eye." To the same compact limestone may be referred the variety called *Cottam Marble*, from being found at Cottam, near Bristol. It resembles in many respects the Landscape Marble.

In different parts of Scotland, compact limestone is cut and polished as marble: this was the case in the parish of Cummertrees in Dumfriesshire,—in Cambuslang parish, in Lanarkshire,—in Fifeshire, &c. In England, many compact limestones are cut and polished as marbles; such are the limestones of Derbyshire, Yorkshire, Devonshire, Somersetshire, and Dorsetshire. It is sometimes used as a building stone; and, in want of better materials, for paving streets, and making highways. When, by exposure to a high temperature, it is deprived of its carbonic acid, and converted into quicklime, it is used for mortar; also by the soap-maker, for rendering his alkalies caustic; by the tanner, for cleansing hides, or freeing them from hair, muscular substance, and fat; by the farmer, in the improvement of particular kinds of soil; and by the metallurgist, in the smelting of such ores as are difficultly fusible, owing to an intermixture of silica and alumina.

Second Kind.

Blue Vesuvian Limestone.

Blauer Vesuvischer Kalkstein, *Klaproth.*

Blauer Vesuvischer Kalkstein, *Klaproth,* Beit. b. v. s. 92. *Id.* *Lenz,* b. ii. s. 737.

External Characters.

Its colour is dark bluish-grey, partly veined with white.

Externally it appears as if it had been rolled; and the surface is uneven.

The fracture fine earthy, passing into splintery.

It is opaque.

It affords a white streak.

It is semi-hard in a low degree.

It is rather heavy.

Constituent Parts.

Lime, - -	58.00
Carbonic Acid, -	28.50
Water, which is somewhat ammoniacal, -	11.00
Magnesia, -	0.50
Oxide of Iron,	0.25
Carbon, - -	0.25
Silica,	1.25
	99.75

Klaproth, Beit. b. v. s. 96.

From

From this analysis, it appears, that the vesuvian limestone differs remarkably in composition from common compact limestone. In common compact limestone, 100 parts of lime are combined with at least 80 parts of carbonic acid; whereas in the vesuvian limestone, 100 parts of limestone are not combined with more than 50 parts of carbonic acid. Secondly, In common limestone, independent of the water which adheres to it accidentally, as far as we know, there is no water of composition; but in the vesuvian limestone, there are 11 parts of water of composition.

Geographic Situation.

This remarkable limestone is found in loose masses amongst unaltered ejected minerals in the neighbourhood of Vesuvius.

Observations.

It is known to some collectors under the name *Compact Blue Lava* of Vesuvius; and is employed by artists in their mosaic work, to represent the sky.

Third Kind.

Roestone, or Oolite *.

Roogenstein, *Werner*.

Hammites, *Plin.* Hist. Nat. xxxvii. 10. s. 60.?—F. E. Bruckmanni, Specimen physicum sistens, Histor. Nat. Oolithe, 1721.

---

* *Roestone*, so named on account of its resemblance in form to the roe of fishes.

1721.—Stalactites oolithus, var. *b. c. d. Wall.* t. ii. p. 384.—Roogenstein, *Wid.* s. 511.—Oviform Limestone, *Kirw.* vol. i. p. 91.—Roogenstein, *Estner,* b. ii. s. 928. *Id. Emm.* b. i. s. 442.—Tufo oolitico, *Nap.* p. 353.—L'Oolite, *Broch.* t. i. p. 529.—Chaux carbonatée compacte globuliforme, *Haüy,* t. ii. p. 171.—Roogenstein, *Reuss,* b. ii. 2. s. 270. *Id. Lud.* b. i. s. 148. *Id. Suck.* 1r th. s. 591. *Id. Bert.* s. 89. *Id. Mohs,* b. ii. s. 26. *Id. Hab.* p. 72. *Id. Leonhard,* Tabel. s. 32.—Chaux carbonatée oolithe, *Brong.* t. i. p. 203.—Chaux carbonatée globuliforme, *Brard,* p. 31.—Erbsförmiger Kalkstein, *Karsten,* Tabel. s. 50.—Roestone, *Kid,* vol. i. p. 26.—Chaux carbonatée globuliforme, *Haüy,* Tabl. p. 4.—Roogenstein, *Lenz,* b. ii. s. 738. *Id. Hoff.* b. iii. s. 12.—Schaaliger Kalkstein, *Haus.* b. iii. s. 912.

External Characters.

Its colours are hair-brown, chesnut-brown, and reddish-brown, and sometimes yellowish-grey, and ash-grey.

It occurs massive, and in distinct concretions, which are round granular, the larger are composed of fine spherical granular, and sometimes of very thin concentric lamellar concretions.

Internally it is dull.

The fracture of the grains is fine splintery; but of the mass is round granular in the small, and slaty in the large.

The fragments in the large are blunt-edged.

It is opaque.

It is semi-hard, approaching to soft.

It is rather brittle, and very easily frangible.

Specific gravity, 2.6829, 2.6190, *Kopp.*—2.585, *Breithaupt.*

Chemical

Chemical Characters.

It dissolves with effervescence in acids.

Geognostic Situation.

It occurs along with red sandstone, and *lias* limestone.

Geographic Situation.

This rock, which, in England, is known under the names Bath-Stone, Ketton-Stone, Portland-Stone, and Oolite, extends, with but little interruption, from Somersetshire to the banks of the Humber in Lincolnshire. On the Continent of Europe, it occurs in Thuringia, the Netherlands, the mountains of Jura, and in other countries.

Uses.

The Oolite, or Roestone, particularly that of Bath and Portland, is very extensively employed in architecture; it can be worked with great ease, and has a light and beautiful appearance; but it is porous, and possesses no great durability, and should not be employed where there is much carved or ornamental work, for the fine chiselling is soon effaced by the action of the atmosphere. On account of the ease and sharpness with which it can be carved, it is much used by the English architects, who appear to have little regard for futurity. St Paul's is built of this stone, also Somerset-House. The Chapel of Henry VIII. affords a striking proof of the inattention of the architects to the choice of the stone. All the beautiful ornamental work of the exterior had mouldered away in the short comparative period of 300 years. It has recently been cased with a new front of Bath-Stone, in which the carving has been correctly

correctly copied: from the nature of the stone, we may predict, that its duration will not be longer than that of the original. Both Portland and Bath stone varies much in quality. In buildings constructed of this stone, we may frequently observe some of the stones black, and others white. The black stones are those which are more compact and durable, and preserve their coating of smoke; the white stones are decomposing, and presenting a fresh surface, as if they had been recently scraped *. Roestone is also used as a manure, but when burnt into quicklime, the marly varieties afford rather an indifferent mortar; but those mixed with sand a better mortar.

Observations.

1. It passes into Sandstone, Compact Limestone, and Marl.

2. Some naturalists, as Daubenton, Saussure, Spallanzani, and Gillet Lamont, conjecture, that Roestone is carbonate of lime, which has been granulated in the manner of gunpowder, by the action of water: the most plausible opinion is that which attributes the formation of this mineral to crystallization from a state of solution.

Third Subspecies.

Chalk.

Kreide, *Werner.*

Creta alba, *Wall.* t. i. p. 27.—Kreide, *Wid.* s. 492.—Chalk, *Kirw.* vol. i. p. 77.—Kreide, *Estner,* b. ii. s. 917. *Id. Emm.* b. i.

---

* Aikin.

of the celebrated poet Hafitz is constructed with it, and that the wainscotting of the principal room of the Hafl-ten, near Schiraz, is likewise of this marble. Its colours are described as light green, with veins, sometimes of red, sometimes of blue, and it has great translucency. It is cut in large slabs; for Mr Morier saw some that measured nine feet in length, and five feet in breadth. He says, that it is not procured near the city of Tabriz, or taken from a quarry, but is said to be rather a petrifaction, found in large quantities, and in immense blocks, on the borders of the Lake Shahee, near the town of Meraugheh. If it is a mere calcareous deposition, formed in the way of calcareous-alabaster or calc-sinter, it must be considered, not as marble, but a variety of that mineral.

The marbles of Hindostan, Siam, and China, are almost unknown to us. Authors speak of a quarry of white marble in the neighbourhood of Pekin; and of a similar marble in the vicinity of the capital of Siam.

African Marbles.

Beds of marble occur in the Atlas Mountains, and in those ranges that bound the shores of the Red Sea *.

American Marbles.

A good many different marbles have been discovered in the United States. The principal quarries are at Stockbridge and Lanesborough, Massachussets: in Vermont and Pennsylvania: in New-York; and in Virginia. According to Professor Hall, as mentioned by Mr Kœnig, marble has been found in many places on the west side of the Green Mountains in Vermont. A few years since, a valuable

* Vid. Murray's interesting and valuable work, "Historical Account of Discoveries and Travels in Africa," for particulars in regard to the marbles

luable quarry was opened in Middleburg, a town situated on Otter Creek, eleven miles above Vergennes. The marble is of different colours in different parts of the bed. The principal colour, however, is bluish-grey. It takes a good polish, and is in general free of admixture of any substance that might affect its polish.

Second Subspecies.

Compact Limestone.

Dichter Kalkstein, *Werner.*

This subspecies is divided into three kinds, viz. Common Compact Limestone, Blue Vesuvian Limestone, and Roestone.

First Kind.

Common Compact Limestone.

Gemeiner Dichter Kalkstein, *Werner.*

Calcareus æquabilis, *Wall.* t. i. p. 122.—Dichter Kalkstein, *Wid.* s. 494.—Compact Limestone, *Kirw.* vol. i. p. 82.—Gemeiner Dichter Kalkstein, *Emm.* b. i. s. 437.—Pietra calcarea compacta, *Nap.* p. 33.—La pierre calcaire compacte commune, *Broch.* t. i. p. 523.—Chaux carbonatée compacte, *Haüy,* t. ii. p. 166.—Gemeiner Dichter Kalkstein, *Reuss,* b. ii. 2. s. 262. *Id. Lud.* b. i. s. 146. *Id. Suck.* 1r th. s. 585. *Id. Bert.* s. 88. *Id. Mohs,* b. ii. s. 14. *Id. Hab.* s. 71. *Id. Leonhard,* Tabel. s. 32.—Chaux carbonatée compacte, *Brong.* t. i. p. 199.—Gemeiner Kalkstein, *Haus.* s. 126.—Dichter Kalkstein, *Karsten,* Tabel. s. 50.—Chaux carbonatée compacte, *Haüy,* Tabl. p. 4.—Dichter gemeiner Kalkstein, *Lenz,* b. ii. s. 732.—Gemeiner Dichter Kalkstein, *Hoff.* b. iii. s. 8.—Common Limestone, *Aikin,* p. 160.

External

External Characters.

Its most frequent colour is grey, of which the following varieties have been observed: yellowish, bluish, ash, pearl, greenish, and smoke grey; the ash-grey passes into greyish-black; the yellowish-grey into yellowish-brown, ochre-yellow, and into a colour bordering on cream-yellow. It also occurs blood-red, flesh-red, and peach-blossom-red, which latter colour is very rare.

It frequently exhibits veined, zoned, striped, clouded, and spotted coloured delineations; and sometimes also black and brown coloured arborisations.

It very rarely exhibits a beautiful play of colours, caused by intermixed portions of pearly shells.

It occurs massive, corroded, in large plates, rolled masses, and in various extraneous external shapes, of univalve, bivalve, and multivalve shells, of corals, fishes, and more rarely of vegetables, as of ferns and reeds.

Internally it is dull, seldom glimmering, which is owing to intermixed calcareous-spar.

The fracture is small and fine splintery, which sometimes passes into large and flat conchoidal, sometimes into uneven, inclining to earthy, and it occasionally inclines to straight and thick slaty.

The fragments are indeterminate angular, more or less sharp-edged, but in the slaty variety they are tabular.

It is generally translucent on the edges, sometimes opaque.

It is in general rather softer than granular foliated limestone.

It is brittle, and easily frangible.

Its streak is generally greyish-white.

Specific

Specific gravity, Splintery, 2.600, 2.720, *Brisson.*—Opalescent Shell Marble, 2.673, *Leonhard.*—2.675, *Werner.*

Chemical Characters.

It effervesces with acids, and the greater part is dissolved; and burns to quicklime, without falling to pieces.

Constituent Parts.

Rudersdorf.		Bluish-grey Limestone.		Limestone from Sweden.		Limestone from Ettersberg *.	
Lime,	53.00	Lime,	49.50	Lime,	49.25	Lime,	33.41
Carbon. acid,	42.50	Carbon. acid,	40.00	Carbon. acid,	35.00	Carbon. acid,	42.00
Silica,	1.12	Silica,	5.25	Silica,	8.75	Silica,	10.25
Alumina,	1.00	Alumina,	2.75	Alumina,	2.50	Magnesia,	9.43
Iron,	0.75	Iron,	1.37	Iron,	2.75	Iron,	2.25
Water,	1.03	Water,	1.13	Loss,	1.75	Manganese,	1.25
						Loss,	1.41
	100		100		100		100
Simon, Geblen's Jour. iv. s. 426.		*Simon*, Ib.		*Simon*, Ib.		*Buchols*, Ib.	

Geognostic Situation.

This mineral occurs in vast abundance in nature, principally in secondary formations, along with sandstone, gypsum, and coal; and in small quantity in primitive mountains. The variegated varieties, which are frequently traversed by veins of calcareous-spar, occur principally in districts composed of grey-wacke and clay-slate. It is distinctly stratified, and the strata vary in thickness, from a few inches to many fathoms, and are from a few fathoms to many miles in extent. The strata generally incline to horizontal

* Some of the limestones in Fifeshire agree in composition with that of Ettersberg.

rizontal; sometimes, however, they are vertical, ly convoluted, even arranged in concentric layers, senting appearances illustrative of their chemi Petrifactions, both of animals and vegetables, b pally of the former, abound in compact limest are of corals, shells, fishes, and sometimes of a animals. On a general view, it is to be consider in ores of different kinds, particularly ores of lead thus, nearly all the rich and valuable lead-mine land are situated in limestone.

Geographic Situation.

It abounds in the sandstone and coal formation Scotland and England; and in Ireland, it is a v dant mineral in all the districts where clay-slate sandstone rocks occur. On the Continent of Eu a very widely and abundantly distributed mineral; a striking feature in many extensive tracts of c Asia, Africa and America, as will be particularly in the Geognostic part of this work.

Uses.

When compact limestone joins to pure and agr lours, so considerable a degree of hardness that it tak polish, it is by artists considered as a Marble; and tains petrifactions mineralized, it is named *shell* or *lu* and *coral* or *zoophytic marble*, according as the or mains are testaceous or coralline*. In one particular lu

* The name *marmor*, is derived from the Greek μαρμαιρειν, glitter, and was by the ancients applied, not only to limestone, stones possessing agreeable colours, and receiving a good polish, sum, jasper, serpentine, and even granite and porphyry.

lumachella or shell marble, found at Bleiberg in Carinthia, the shells and fragments of shells, which belong to the nautilus tribe, are set in a brown-coloured basis, and reflect many beautiful and brilliant pearly inclining to metallic colours, principally the fire-red, green, and blue tints. It is named *opalescent* or *fire marble*. Another lumachella marble from Astracan, contains, in a reddish-brown basis, pearly shells of nautili, that reflect a very brilliant gold-yellow colour. In some compact marbles, the surface presents a beautiful arborescent appearance, and these are named *arborescent* or *dendritic marbles*. Such are those of Papenheim in Bavaria.

The *Florentine Marble*, or *Ruin Marble*, as it is sometimes called, is a compact limestone. It occurs on the Po and the Arno, and is worked into various articles at Florence. It is said to occur in balls. It presents angular figures of a yellowish-brown, on a base of a lighter tint, and which passes to greyish-white. Seen at a distance, slabs of this stone resemble drawings done in bister. "One is amused (says Brard) to observe in it kinds of ruins: there it is a Gothic castle half destroyed, here it presents ruined walls; in another place old bastions; and, what still adds to the illusion, is, that, in these sorts of natural paintings, there exists a kind of aërial perspective, which is very sensibly perceived. The lower part, or what forms the first plane, has a warm and bold tone; the second follows it, and weakens it as it increases the distance; the third becomes still fainter; while the upper part, agreeing with the first, presents in the distance a whitish zone, which terminates the horizon, then blends itself more and more as it rises, and at length reaches the top, where it forms sometimes as it were clouds. But approach close to it, all va-

nishes immediately, and these pretended figures, which, at a distance, seemed so well drawn, are converted into irregular marks, which present nothing to the eye." To the same compact limestone may be referred the variety called *Cottam Marble*, from being found at Cottam, near Bristol. It resembles in many respects the Landscape Marble.

In different parts of Scotland, compact limestone is cut and polished as marble: this was the case in the parish of Cummertrees in Dumfriesshire,—in Cambuslang parish, in Lanarkshire,—in Fifeshire, &c. In England, many compact limestones are cut and polished as marbles; such are the limestones of Derbyshire, Yorkshire, Devonshire, Somersetshire, and Dorsetshire. It is sometimes used as a building stone; and, in want of better materials, for paving streets, and making highways. When, by exposure to a high temperature, it is deprived of its carbonic acid, and converted into quicklime, it is used for mortar; also by the soap-maker, for rendering his alkalies caustic; by the tanner, for cleansing hides, or freeing them from hair, muscular substance, and fat; by the farmer, in the improvement of particular kinds of soil; and by the metallurgist, in the smelting of such ores as are difficultly fusible, owing to an intermixture of silica and alumina.

Second

Second Kind.

Blue Vesuvian Limestone.

Blauer Vesuvischer Kalkstein, *Klaproth.*

Blauer Vesuvischer Kalkstein, *Klaproth,* Beit. b. v. s. 92. *Id.* *Lenz,* b. ii. s. 737.

External Characters.

Its colour is dark bluish-grey, partly veined with white.

Externally it appears as if it had been rolled; and the surface is uneven.

The fracture fine earthy, passing into splintery.

It is opaque.

It affords a white streak.

It is semi-hard in a low degree.

It is rather heavy.

Constituent Parts.

Lime, - -	58.00
Carbonic Acid, -	28.50
Water, which is somewhat ammoniacal, -	11.00
Magnesia, -	0.50
Oxide of Iron,	0.25
Carbon, - -	0.25
Silica,	1.25
	99.75

Klaproth, Beit. b. v. s. 96.

From

1721.—Stalactites oolithus, var. *b. c. d. Wall.* t. ii. p. 384.—Roogenstein, *Wid.* s. 511.—Oviform Limestone, *Kirw.* vol. i. p. 91.—Roogenstein, *Estner,* b. ii. s. 928. *Id. Emm.* b. i. s. 442.—Tufo oolitico, *Nap.* p. 353.—L'Oolite, *Broch.* t. i. p. 529.—Chaux carbonatée compacte globuliforme, *Haüy,* t. ii. p. 171.—Roogenstein, *Reuss,* b. ii. 2. s. 270. *Id. Lud.* b. i. s. 148. *Id. Suck.* 1r th. s. 591. *Id. Bert.* s. 89. *Id. Mohs,* b. ii. s. 26. *Id. Hab.* p. 72. *Id. Leonhard,* Tabel. s. 32.—Chaux carbonatée oolithe, *Brong.* t. i. p. 203.—Chaux carbonatée globuliforme, *Brard,* p. 31.—Erbsförmiger Kalkstein, *Karsten,* Tabel. s. 50.—Roestone, *Kid,* vol. i. p. 26.—Chaux carbonatée globuliforme, *Haüy,* Tabl. p. 4.—Roogenstein, *Lenz,* b. ii. s. 738. *Id. Hoff.* b. iii. s. 12.—Schaaliger Kalkstein, *Haus.* b. iii. s. 912.

External Characters.

Its colours are hair-brown, chesnut-brown, and reddish-brown, and sometimes yellowish-grey, and ash-grey.

It occurs massive, and in distinct concretions, which are round granular, the larger are composed of fine spherical granular, and sometimes of very thin concentric lamellar concretions.

Internally it is dull.

The fracture of the grains is fine splintery; but of the mass is round granular in the small, and slaty in the large.

The fragments in the large are blunt-edged.

It is opaque.

It is semi-hard, approaching to soft.

It is rather brittle, and very easily frangible.

Specific gravity, 2.6829, 2.6190, *Kopp.*—2.585, *Breithaupt.*

Chemical

Chemical Characters.

It dissolves with effervescence in acids.

Geognostic Situation.

It occurs along with red sandstone, and *lias* limestone.

Geographic Situation.

This rock, which, in England, is known under the names Bath-Stone, Ketton-Stone, Portland-Stone, and Oolite, extends, with but little interruption, from Somersetshire to the banks of the Humber in Lincolnshire. On the Continent of Europe, it occurs in Thuringia, the Netherlands, the mountains of Jura, and in other countries.

Uses.

The Oolite, or Roestone, particularly that of Bath and Portland, is very extensively employed in architecture; it can be worked with great ease, and has a light and beautiful appearance; but it is porous, and possesses no great durability, and should not be employed where there is much carved or ornamental work, for the fine chiselling is soon effaced by the action of the atmosphere. On account of the ease and sharpness with which it can be carved, it is much used by the English architects, who appear to have little regard for futurity. St Paul's is built of this stone, also Somerset-House. The Chapel of Henry VIII. affords a striking proof of the inattention of the architects to the choice of the stone. All the beautiful ornamental work of the exterior had mouldered away in the short comparative period of 300 years. It has recently been cased with a new front of Bath-Stone, in which the carving has been correctly

correctly copied: from the nature of the stone, we may predict, that its duration will not be longer than that of the original. Both Portland and Bath stone varies much in quality. In buildings constructed of this stone, we may frequently observe some of the stones black, and others white. The black stones are those which are more compact and durable, and preserve their coating of smoke; the white stones are decomposing, and presenting a fresh surface, as if they had been recently scraped *. Roestone is also used as a manure, but when burnt into quicklime, the marly varieties afford rather an indifferent mortar; but those mixed with sand a better mortar.

Observations.

1. It passes into Sandstone, Compact Limestone, and Marl.

2. Some naturalists, as Daubenton, Saussure, Spallanzani, and Gillet Lamont, conjecture, that Roestone is carbonate of lime, which has been granulated in the manner of gunpowder, by the action of water: the most plausible opinion is that which attributes the formation of this mineral to crystallization from a state of solution.

Third Subspecies.

Chalk.

Kreide, *Werner.*

Creta alba, *Wall.* t. i. p. 27.—Kreide, *Wid.* s. 492.—Chalk, *Kirw.* vol. i. p. 77.—Kreide, *Estner,* b. ii. s. 917. *Id. Emm.* b. i.

* Aikin.

of the celebrated poet Hafitz is c
that the wainscotting of the principal
near Schiraz, is likewise of this ma
described as light green, with veins, a
times of blue, and it has great tran
large slabs; for Mr Morier saw som
feet in length, and five feet in breadt
not procured near the city of Tabriz,
ry, but is said to be rather a petrifi
quantities, and in immense blocks,
Lake Shahee, near the town of M
mere calcareous deposition, formed
ous-alabaster or calc-sinter, it must
marble, but a variety of that mineral.

The marbles of Hindostan, Siam,
unknown to us. Authors speak
marble in the neighbourhood of Pe
marble in the vicinity of the capital c

African Marble.

Beds of marble occur in the Atl
those ranges that bound the shores of

American Marble

A good many different marbles ha
the United States. The principal
bridge and Lanesborough, Massachu
Pennsylvania: in New-York; and i
ing to Professor Hall, as mentioned b
has been found in many places on
Green Mountains in Vermont. A

---

* Vid. Murray's interesting and valuable
of Discoveries and Travels in Africa," for particu

luable quarry was opened in Middleburg, a town situated on Otter Creek, eleven miles above Vergennes. The marble is of different colours in different parts of the bed. The principal colour, however, is bluish-grey. It takes a good polish, and is in general free of admixture of any substance that might affect its polish.

Second Subspecies.

Compact Limestone.

Dichter Kalkstein, *Werner.*

This subspecies is divided into three kinds, viz. Common Compact Limestone, Blue Vesuvian Limestone, and Roestone.

First Kind.

Common Compact Limestone.

Gemeiner Dichter Kalkstein, *Werner.*

Calcareus æquabilis, *Wall.* t. i. p. 122.—Dichter Kalkstein, *Wid.* s. 494.—Compact Limestone, *Kirw.* vol. i. p. 82.—Gemeiner Dichter Kalkstein, *Emm.* b. i. s. 437.—Pietra calcarea compacta, *Nap.* p. 33.—La pierre calcaire compacte commune, *Broch.* t. i. p. 523.—Chaux carbonatée compacte, *Haüy,* t. ii. p. 166.—Gemeiner Dichter Kalkstein, *Reuss,* b. ii. 2. s. 262. *Id. Lud.* b. i. s. 146. *Id. Suck.* 1r th. s. 585. *Id. Bert.* s. 88. *Id. Mohs,* b. ii. s. 14. *Id. Hab.* s. 71. *Id. Leonhard,* Tabel. s. 32.—Chaux carbonatée compacte, *Brong.* t. i. p. 199.—Gemeiner Kalkstein, *Haus.* s. 126.—Dichter Kalkstein, *Karsten,* Tabel. s. 50.—Chaux carbonatée compacte, *Haüy,* Tabl. p. 4.—Dichter gemeiner Kalkstein, *Lenz,* b. ii. s. 732.—Gemeiner Dichter Kalkstein, *Hoff.* b. iii. s. 8.—Common Limestone, *Aikin,* p. 160.

External

Specific gravity, Splintery, 2.600, 2.720, *Brisson.*—Opalescent Shell Marble, 2.673, *Leonhard.*—2.675, *Werner.*

Chemical Characters.

It effervesces with acids, and the greater part is dissolved; and burns to quicklime, without falling to pieces.

Constituent Parts.

Rudersdorf.		Bluish-grey Limestone.		Limestone from Sweden.		Limestone from Ettersberg *.	
Lime,	53.00	Lime,	49.50	Lime,	49.25	Lime,	33.41
Carbon. acid,	42.50	Carbon. acid,	40.00	Carbon. acid,	35.00	Carbon. acid,	42.00
Silica,	1.12	Silica,	5.25	Silica,	8.75	Silica,	10.25
Alumina,	1.00	Alumina,	2.75	Alumina,	2.50	Magnesia,	9.43
Iron,	0.75	Iron,	1.37	Iron,	2.75	Iron,	2.25
Water,	1.03	Water,	1.13	Loss,	1.75	Manganese,	1.25
						Loss,	1.41
	100		100		100		100
Simon, Gehlen's Jour. iv. s. 426.		*Simon*, Ib.		*Simon*, Ib.		*Buchols*, Ib.	

Geognostic Situation.

This mineral occurs in vast abundance in nature, principally in secondary formations, along with sandstone, gypsum, and coal; and in small quantity in primitive mountains. The variegated varieties, which are frequently traversed by veins of calcareous-spar, occur principally in districts composed of grey-wacke and clay-slate. It is distinctly stratified, and the strata vary in thickness, from a few inches to many fathoms, and are from a few fathoms to many miles in extent. The strata generally incline to horizontal

* Some of the limestones in Fifeshire agree in composition with that of Ettersberg.

rizontal; sometimes, however, they are vertical, or variously convoluted, even arranged in concentric layers, thus presenting appearances illustrative of their chemical nature. Petrifactions, both of animals and vegetables, but principally of the former, abound in compact limestone: these are of corals, shells, fishes, and sometimes of amphibious animals. On a general view, it is to be considered as rich in ores of different kinds, particularly ores of lead and zinc: thus, nearly all the rich and valuable lead-mines in England are situated in limestone.

Geographic Situation.

It abounds in the sandstone and coal formations, both in Scotland and England; and in Ireland, it is a very abundant mineral in all the districts where clay-slate and red sandstone rocks occur. On the Continent of Europe, it is a very widely and abundantly distributed mineral; and forms a striking feature in many extensive tracts of country in Asia, Africa and America, as will be particularly described in the Geognostic part of this work.

Uses.

When compact limestone joins to pure and agreeable colours, so considerable a degree of hardness that it takes a good polish, it is by artists considered as a Marble; and if it contains petrifactions mineralized, it is named *shell* or *lumachella*, and *coral* or *zoophytic marble*, according as the organic remains are testaceous or coralline *. In one particular variety of lumachella

* The name *marmor*, is derived from the Greek *μαρμαιρειν*, *to shine*, or *glitter*, and was by the ancients applied, not only to limestone, but also to stones possessing agreeable colours, and receiving a good polish, such as gypsum, jasper, serpentine, and even granite and porphyry.

lumachella or shell marble, found at Bleiberg in Carinthia, the shells and fragments of shells, which belong to the nautilus tribe, are set in a brown-coloured basis, and reflect many beautiful and brilliant pearly inclining to metallic colours, principally the fire-red, green, and blue tints. It is named *opalescent* or *fire marble*. Another lumachella marble from Astracan, contains, in a reddish-brown basis, pearly shells of nautili, that reflect a very brilliant gold-yellow colour. In some compact marbles, the surface presents a beautiful arborescent appearance, and these are named *arborescent* or *dendritic marbles*. Such are those of Papenheim in Bavaria.

The *Florentine Marble*, or *Ruin Marble*, as it is sometimes called, is a compact limestone. It occurs on the Po and the Arno, and is worked into various articles at Florence. It is said to occur in balls. It presents angular figures of a yellowish-brown, on a base of a lighter tint, and which passes to greyish-white. Seen at a distance, slabs of this stone resemble drawings done in bister. "One is amused (says Brard) to observe in it kinds of ruins: there it is a Gothic castle half destroyed, here it presents ruined walls; in another place old bastions; and, what still adds to the illusion, is, that, in these sorts of natural paintings, there exists a kind of aërial perspective, which is very sensibly perceived. The lower part, or what forms the first plane, has a warm and bold tone; the second follows it, and weakens it as it increases the distance; the third becomes still fainter; while the upper part, agreeing with the first, presents in the distance a whitish zone, which terminates the horizon, then blends itself more and more as it rises, and at length reaches the top, where it forms sometimes as it were clouds. But approach close to it, all vanishes

nishes immediately, and these pretended figures, which, at a distance, seemed so well drawn, are converted into irregular marks, which present nothing to the eye." To the same compact limestone may be referred the variety called *Cottam Marble*, from being found at Cottam, near Bristol. It resembles in many respects the Landscape Marble.

In different parts of Scotland, compact limestone is cut and polished as marble: this was the case in the parish of Cummertrees in Dumfriesshire,—in Cambuslang parish, in Lanarkshire,—in Fifeshire, &c. In England, many compact limestones are cut and polished as marbles; such are the limestones of Derbyshire, Yorkshire, Devonshire, Somersetshire, and Dorsetshire. It is sometimes used as a building stone; and, in want of better materials, for paving streets, and making highways. When, by exposure to a high temperature, it is deprived of its carbonic acid, and converted into quicklime, it is used for mortar; also by the soap-maker, for rendering his alkalies caustic; by the tanner, for cleansing hides, or freeing them from hair, muscular substance, and fat; by the farmer, in the improvement of particular kinds of soil; and by the metallurgist, in the smelting of such ores as are difficultly fusible, owing to an intermixture of silica and alumina.

Second

Second Kind.

Blue Vesuvian Limestone.

Blauer Vesuvischer Kalkstein, *Klaproth.*

Blauer Vesuvischer Kalkstein, *Klaproth,* Beit. b. v. s. 92. *Id. Lenz,* b. ii. s. 787.

External Characters.

Its colour is dark bluish-grey, partly veined with white.

Externally it appears as if it had been rolled; and the surface is uneven.

The fracture fine earthy, passing into splintery.

It is opaque.

It affords a white streak.

It is semi-hard in a low degree.

It is rather heavy.

Constituent Parts.

Lime, - -	58.00
Carbonic Acid, -	28.50
Water, which is somewhat ammoniacal, -	11.00
Magnesia, -	0.50
Oxide of Iron,	0.25
Carbon, - -	0.25
Silica,	1.25
	99.75

Klaproth, Beit. b. v. s. 96.

From

From this analysis, it appears, that the vesuvian limestone differs remarkably in composition from common compact limestone. In common compact limestone, 100 parts of lime are combined with at least 80 parts of carbonic acid; whereas in the vesuvian limestone, 100 parts of limestone are not combined with more than 50 parts of carbonic acid. Secondly, In common limestone, independent of the water which adheres to it accidentally, as far as we know, there is no water of composition; but in the vesuvian limestone, there are 11 parts of water of composition.

Geographic Situation.

This remarkable limestone is found in loose masses amongst unaltered ejected minerals in the neighbourhood of Vesuvius.

Observations.

It is known to some collectors under the name *Compact Blue Lava* of Vesuvius; and is employed by artists in their mosaic work, to represent the sky.

Third Kind.

Roestone, or Oolite*.

Roogenstein, *Werner*.

Hammites, *Plin.* Hist. Nat. xxxvii. 10. s. 60.?—F. E. Bruckmanni, Specimen physicum sistens, Histor. Nat. Oolithe, 1721.

---

* *Roestone*, so named on account of its resemblance in form to the roe of fishes.

1721.—Stalactites oolithus, var. *b. c. d. Wall.* t. ii. p. 384.—Roogenstein, *Wid.* s. 511.—Oviform Limestone, *Kirw.* vol. i. p. 91.—Roogenstein, *Estner,* b. ii. s. 928. *Id. Emm.* b. i. s. 442.—Tufo oolitico, *Nap.* p. 353.—L'Oolite, *Broch.* t. i. p. 529.—Chaux carbonatée compacte globuliforme, *Haüy,* t. ii. p. 171.—Roogenstein, *Reuss,* b. ii. 2. s. 270. *Id. Lud.* b. i. s. 148. *Id. Suck.* 1r th. s. 591. *Id. Bert.* s. 89. *Id. Mohs,* b. ii. s. 26. *Id. Hab.* p. 72. *Id. Leonhard,* Tabel. s. 32.—Chaux carbonatée oolithe, *Brong.* t. i. p. 203.—Chaux carbonatée globuliforme, *Brard,* p. 31.—Erbsförmiger Kalkstein, *Karsten,* Tabel. s. 50.—Roestone, *Kid,* vol. i. p. 26.—Chaux carbonatée globuliforme, *Haüy,* Tabl. p. 4.—Roogenstein, *Lenz,* b. ii. s. 738. *Id. Hoff.* b. iii. s. 12.—Schaaliger Kalkstein, *Haus.* b. iii. s. 912.

External Characters.

Its colours are hair-brown, chesnut-brown, and reddish-brown, and sometimes yellowish-grey, and ash-grey.

It occurs massive, and in distinct concretions, which are round granular, the larger are composed of fine spherical granular, and sometimes of very thin concentric lamellar concretions.

Internally it is dull.

The fracture of the grains is fine splintery; but of the mass is round granular in the small, and slaty in the large.

The fragments in the large are blunt-edged.

It is opaque.

It is semi-hard, approaching to soft.

It is rather brittle, and very easily frangible.

Specific gravity, 2.6829, 2.6190, *Kopp.*—2.585, *Breithaupt.*

Chemical

Chemical Characters.

It dissolves with effervescence in acids.

Geognostic Situation.

It occurs along with red sandstone, and *lias* limestone.

Geographic Situation.

This rock, which, in England, is known under the names Bath-Stone, Ketton-Stone, Portland-Stone, and Oolite, extends, with but little interruption, from Somersetshire to the banks of the Humber in Lincolnshire. On the Continent of Europe, it occurs in Thuringia, the Netherlands, the mountains of Jura, and in other countries.

Uses.

The Oolite, or Rcestone, particularly that of Bath and Portland, is very extensively employed in architecture; it can be worked with great ease, and has a light and beautiful appearance; but it is porous, and possesses no great durability, and should not be employed where there is much carved or ornamental work, for the fine chiselling is soon effaced by the action of the atmosphere. On account of the ease and sharpness with which it can be carved, it is much used by the English architects, who appear to have little regard for futurity. St Paul's is built of this stone, also Somerset-House. The Chapel of Henry VIII. affords a striking proof of the inattention of the architects to the choice of the stone. All the beautiful ornamental work of the exterior had mouldered away in the short comparative period of 300 years. It has recently been cased with a new front of Bath-Stone, in which the carving has been correctly

correctly copied: from the nature of the stone, we may predict, that its duration will not be longer than that of the original. Both Portland and Bath stone varies much in quality. In buildings constructed of this stone, we may frequently observe some of the stones black, and others white. The black stones are those which are more compact and durable, and preserve their coating of smoke; the white stones are decomposing, and presenting a fresh surface, as if they had been recently scraped *. Roestone is also used as a manure, but when burnt into quicklime, the marly varieties afford rather an indifferent mortar; but those mixed with sand a better mortar.

Observations.

1. It passes into Sandstone, Compact Limestone, and Marl.

2. Some naturalists, as Daubenton, Saussure, Spallanzani, and Gillet Lamont, conjecture, that Roestone is carbonate of lime, which has been granulated in the manner of gunpowder, by the action of water: the most plausible opinion is that which attributes the formation of this mineral to crystallization from a state of solution.

Third Subspecies.

Chalk.

Kreide, *Werner.*

Creta alba, *Wall.* t. i. p. 27.—Kreide, *Wid.* s. 492.—Chalk, *Kirw.* vol. i. p. 77.—Kreide, *Estner,* b. ii. s. 917. *Id. Emm.* b. i.

* Aikin.

b. i. s. 433.—Creta commune, *Nap.* p. 381.—**La Craie,** ***Broch.*** t. i. p. 521.—Craie, *Haüy,* t. ii. p. 166.—**Kreide,** ***Reuss,*** b. ii. 2. s. 259. *Id. Lud.* b. i. s. 145. *Id. Suck.* 1r th. s. 583. *Id. Bert.* s. 87. *Id. Mohs,* b. ii. s. 9. *Id. Hab.* s. 70. *Id. Leonhard,* Tabel. s. 32.—**Chaux carbonatée crayeuse,** ***Brong.*** t. i. p. 208. *Id. Brard,* p. 29.—Kreide, *Karsten,* Tabel. s. 50. *Id. Haus.* s. 126.—Chalk, *Kid,* vol. i. p. 18.—**Kreide,** ***Lenz,*** b. ii. s. 728. *Id. Oken,* b. i. s. 410. *Id. Hoff.* b. iii. s. 4.—Chalk, *Aikin,* p. 160.

External Characters.

Its colour is yellowish-white, which sometimes passes to greyish-white and snow-white. It is sometimes marked with yellowish-grey.

It occurs massive, disseminated, in crusts, and in extraneous external shapes.

It is dull.

The fracture is coarse and fine earthy.

The fragments are blunt-edged.

It is opaque.

It writes and soils very much.

It is soft, and sometimes very soft.

It is rather sectile, and easily frangible.

It adheres slightly to the tongue.

It feels very meagre, and rather rough.

Specific gravity, 2.252, *Muschenbroeck.*—2.315, *Kirwan.*—2.657, *Watson.*—2.226, *Breithaupt.*

Chemical Characters.

It effervesces strongly with acids.

Constituent

Constituent Parts.

Lime,	56.5
Carbonic Acid,	43.0
Water,	0.5

Buchols, in Gehlen's Journal, b. iv. s. 416.

Chalk from Gallicia.

Lime,	47.00
Carbonic Acid,	33.00
Silica,	7.00
Alumina,	2.00
Magnesia,	8.00
Iron,	0.05

Hacquet.

Lime,	53
Carbonic Acid,	42
Alumina,	2
Water,	3

Kirwan, Min. vol. i. p. 77.

Geognostic Situation.

It constitutes one of the newer secondary or flœtz formations; is usually found in low situations, and frequently on sea coasts. It is stratified, and the strata in general are horizontal. It often contains flint, which is disposed either in interrupted beds in the chalk, or in globular, tuberose, or tabular masses imbedded in it. It abounds in organic remains, and these are principally of animals of the lower orders, such as echinites, belemnites, terebratulites, pinnites, &c. These petrifactions are either in the state of carbonate, or are converted into flint, which latter is by far the most frequent. It cannot be considered as a metalliferous formation, as it contains nothing but small imbedded portions of iron-pyrites. Two principal kinds of chalk occur in chalk districts: the one is named *Hard*, the other *Soft Chalk;* the hard chalk always occurs undermost, is considerably harder than the other, and rarely contains petrifactions or flint; the soft chalk, on the contrary, rests upon the other, is softer, and abounds in flint and petrifactions.

Geographic

Geographic Situation.

It abounds in the south-eastern parts of England,—extends through several provinces in France,—occupies great tracts of country in Poland and Russia,—is met with on the shores of the Baltic,—and in the islands of Zeeland and Rugen.

Uses.

The uses of this mineral are various. The more compact kinds are employed as building-stones, when they are used either in a rough state, or are sawn into blocks of the requisite size and shape: it is burnt into quicklime, and used for mortar in different countries; thus, nearly all the houses in London are cemented with chalk-mortar *: it is also employed in great quantities in the polishing of glass and metals, and whitening the roofs of rooms, in the state of *whiting* †; in constructing moulds to cast metal in; by carpenters and others as a material to mark with. When perfectly purified, and mixed with vegetable colours, it forms a kind of pastil colour: thus, with litmus, turmeric, saffron, and sap-green, it forms durable colours, but vegetable colours that contain an acid, become blue when mixed with it. The *Vienna white* known to artists is perfectly purified chalk. It is used by starch-makers and chemists to dry precipitates

* According to Smeaton, it makes as good lime as the best limestone or marble.

† In the preparation of whiting, chalk is pounded, and diffused through water, and the finer part of the sediment is then dried; by this means, the siliceous particles are separated, which, by their hardness, would scratch the surface of metallic and other substances, in the polishing of which whiting is used.—*Aikin's Chem. Dictionary.*

precipitates on, for which it is peculiarly qualified, on account of the remarkable facility with which it absorbs water. With isinglas, or white of eggs, it forms a valuable lute or cement. In the gilding of wood, it is necessary, before laying on the gold, to cover it with a succession of coats of a mixture of whiting and size. The mineral is also used as a filtering-stone; and in a purified state, it is employed as a remedy to correct acidity in the stomach, and the morbid states which arise from this.

Observations.

1. The principal characters of chalk are its colours, fracture, soiling, and low specific gravity. These, and its easy frangibility, distinguish it from *White Clay-stone;* and its meagre feel distinguishes it from the *Clays.*

2. It is conjectured that the name *Creta*, is derived from the island of Candia, (Creta of the ancients), where this mineral is said to occur. Ancient writers seem to use the word *creta* in different senses, as appears from the following observations: "The word *creta*, though applied by Wallerius and others to chalk, is generally used by the early naturalists to express clay: "Proderit sabulosis locis *cretam* ingerere; cretosis ac nimium densis, *sabulum* *;" where, as *sabulum* certainly means sand, it is nearly evident, from the reciprocal use of the substances mentioned, compared with the opposite properties of sand and clay, that *creta* signifies the latter. "Lateres non sunt e sabuloso, neque arenoso, multoque minus calculoso ducendi solo; sed e *cretoso* †." Again, it may be observed, with respect

* Columella, p. 73.

† Plin. Nat. Hist. ed. Brot. vol. vi. p. 174.

spect to the following line,

'Hinc humilem Myconem, cretosaque rura Cimoli *,'

that the Cimolian earth is described in various passages of Pliny, &c. under characters peculiar to clay.

There are two passages in which *creta* seems to be applicable to chalk: one in Horace,

——— 'creta an carbone notandi †.'

The other in Pliny: "Alia *creta* argentaria appellatur nitorem argento reddens ‡;" this being a common use of chalk at the present day.—*Kid's Mineralogy*, vol. i. p. 18, 19.

Fourth Subspecies.

Agaric Mineral, or Rock Milk.

Berg-Milch, *Werner.*

Agaricus mineralis, *Wall.* t. i. p. 30.—Bergmilch, *Wid.* s. 490. —Agaric mineral, *Kirw.* vol. i. p. 76.—Bergmilch, *Estner,* b. ii. s. 914. *Id. Emm.* b. i. s. 430.—Agaric Mineral, *Nap.* p. 333. *Id. Lam.* p. 331.—Lait de Montagne, ou l'Agaric Mineral, *Broch.* t. i. p. 519.—Chaux carbonatée spongieuse, *Haüy,* t. ii. p. 167.—Bergmilch, *Reuss,* b. ii. 2. s. 257. *Id. Lud.* b. i. s. 145. *Id. Suck.* 1r th. s. 582. *Id. Bert.* s. 87. *Id. Mohs,* b. ii. s. 8. *Id. Leonhard,* Tabel. s. 32.—Chaux carbonatée spongieuse, *Brong.* t. i. p. 210.—Montmilch, *Haus.* s. 127.—Bergmilch, *Karst.* Tabel. s. 50.—Agaric Mineral, *Kid.*

* Ovid, Metam. lib. vii.

† Horat. Sat. iii. lib. 2.

‡ Plin. Hist. Nat. ed. Brot. vol. vi. p. 184.

Kid, vol. i. p. 38.—Bergmilch, *Lenz*, b. ii. s. 727. *Id. Oken*, b. i. s. 411. *Id. Hoff.* b. iii. s. 2.

External Characters.

Its colours are snow-white, greyish-white, and yellowish-white.

It occurs frequently in crusts, also in loosely cohering tuberose pieces.

It is dull.

It is composed of fine dusty particles.

It soils strongly.

It feels meagre.

It adheres slightly to the tongue.

It is very light, almost supernatant.

Chemical Characters.

It effervesces with acids, and is completely dissolved in them.

Constituent Parts.

It is a pure Carbonate of Lime.

Geognostic and Geographic Situations.

It is found on the north side of Oxford, between the Isis and the Cherwell, and near Chipping-Norton, also in Oxfordshire *; and in the fissures of caves of limestone mountains in Switzerland, Austria, Salzburg, and other countries.

Uses.

* Kid's Mineralogy, vol. i. p. 39.

Uses.

In Switzerland, where it occurs abundantly, it is used for whitening houses.

Observations.

1. It is formed by water passing over and through limestone rocks, and afterwards depositing in holes, fissures, and on faces of rocks, the calcareous earth it had dissolved in its course.

2. It is named *Agaric Mineral*, from its sometimes adhering to rocks with the resemblance of a fungus or agaric: the name *Rock Milk* given to it by some mineralogists, is from its white appearance when oozing from the clefts of rocks; and the name *Lac Lunæ* is sometimes given to it, from the milky-like appearance it presents in a cave in Phrygia; this cave, according to the tradition of the neighbourhood, having been formerly frequented by Diana *.

Fifth Subspecies.

Fibrous Limestone.

Fasriger Kalkstein, *Werner*.

This subspecies is divided into two kinds, viz. Common Fibrous Limestone, or Satin-Spar, and Fibrous Calc-Sinter.

First

* Kid's Mineralogy, vol. i. p. 39.

First Kind.

Common Fibrous Limestone, or Satin-Spar.

Gemeiner fasriger Kalkstein, *Werner.*

Pierre calcaire fibreuse, *Broch.* t. i. p. 549.—Gemeiner fasriger Kalkstein, *Reuss,* b. ii. 2. s. 304. *Id. Mohs,* b. ii. s. 85. *Id. Leonhard,* Tabel s. 34.—Chaux carbonatée fibreuse, *Brong.* t. i. p. 218.—Fasriger Kalkstein, *Karsten,* Tabl. s. 50. —Satin-spar, *Kid,* vol. i. p. 49.—Chaux carbonatée fibreuse-conjointe, *Haüy,* Tabl. p. 4.—Gemeiner fasriger Kalkstein, *Lenz,* b. ii. s. 750. *Id. Hoff.* b. iii. s. 81.—Fibrous Carbonate of Lime, *Aikin,* p. 159.

External Characters.

Its colours are greyish, reddish, and yellowish white.

It occurs massive; also in distinct concretions, which are coarse and fine fibrous, and either straight or curved.

Its lustre is glistening or shining, and pearly.

The fragments are splintery.

It is feebly translucent.

It is as hard as calcareous-spar.

It is easily frangible.

Specific gravity 2.70, *Pepys.*

Constituent Parts.

Carbonate of Lime,	95.75	Lime, -	50.8
Carbonate of Manganese,	4.25	Carbonic Acid,	47.6
			98.4
	Holme.		*Pepys,* in Kid's Min. vol. i. p. 49.

Stromeyer says that fibrous limestone contains some *per cents.* of gypsum.

Geognostic and Geographic Situations.

It occurs in thin layers in clay-slate at Aldstone Moore in Cumberland: in layers and veins in the middle district of Scotland, as in Fifeshire. On the Continent, at Potschappel, near Dresden; and at Schneeberg, also in Saxony.

Uses.

It is sometimes cut into necklaces, crosses, and other ornamental articles.

Second Kind.

Fibrous Calc-Sinter *.

Fasriger Kalksinter, *Werner*.

Sintricher fasriger Kalkstein, *Reuss*, b. ii. 2. s. 306.—Kalksinter, *Lud.* b. i. s. 150. *Id. Suck.* 1r th. s. 618. *Id. Bert.* s. 93. *Id. Mohs*, b. ii. s. 86. *Id. Hab.* s. 78.—Fasriger Kalksinter, *Leonhard*, Tabel. s. 34.—Sintriger Kalkstein, *Karsten*, Tabel. s. 50.—Chaux carbonatée concretionnée, *Haüy*, Tabl. p. 4.—Sintricher Kalkstein, *Lenz*, b. ii. s. 751.—Fasriger Kalksinter, *Hoff.* b. iii. s. 32.

External

* This is the Alabaster of the ancients, and is by the moderns named *Calcareous Alabaster*, to distinguish it from another mineral, gypsum, which they name *Gypseous Alabaster*.

External Characters.

The principal colour is white, of which the following are the varieties, viz. yellowish, greenish, greyish, reddish, and snow white; from yellowish-white it passes into wine, wax, and honey yellow, and into a kind of reddish-brown, passing into clove-brown; from greyish-white it passes into yellowish and pearl grey, and from this latter into reddish, seldomer into flesh and peach-blossom red, and brownish-red; lastly, it passes from greenish-white into asparagus, siskin, mountain, verdigris green, and sky-blue.

The peach-blossom colour is owing to cobalt; the flesh-red to manganese; the verdigris-green to copper; the siskin-green to nickel, and the brown to iron.

It is sometimes concentrically and reniformly striped, or it is spotted or clouded.

It occurs massive, stalactitic, globular, tubular, claviform, fruticose, curtain-shaped, cock's-comb-shaped, coralloidal, reniform, and tuberose; also in distinct concretions, which are fibrous, and these are straight, seldom curved, and sometimes scopiform or stellular; also in reniform curved lamellar concretions, and seldom in large and coarse angulo-granular concretions; very rarely we observe the longish external shapes, as the stalactitic, terminated by a three-sided pyramidal crystallization.

The surface is generally rough, and seldom fine drusy.

Internally it is glimmering, which passes on the one side into dull, on the other into glistening; and the lustre is pearly.

The fracture is fine splintery.

The fragments are splintery, or wedge-shaped.

It is translucent, or only translucent on the edges.

It is nearly as hard as calcareous-spar.

It is rather brittle, and easily frangible.

Specific gravity 2.658,–2.735.

Constituent Parts.

Lime, - - -	56.0
Carbonic Acid, -	43.0
Water, - - -	1.0
	100.0

Bucholz, in Gehlen's Journal, b. iv. s. 425.

Geognostic and Geographic Situations.

It is found encrusting the roofs, walls, and floors of caves, particularly those situated in limestone rocks. It is formed from water holding carbonate of lime in solution. Nothing is more common than the presence of carbonic acid in water; and when a superabundance of this acid is present, the acid is capable of holding in solution a portion of carbonate of lime; but when the solution comes to be agitated, or exposed to the atmosphere, or to a change of temperature, the carbonic acid makes its escape, and thus deprives the water of its solvent power. Water thus impregnated with carbonate of lime, oozes slowly through rocks of any kind, until it reaches the walls and roofs of caves: there some time elapses before a drop of sufficient size to fall by its own weight is formed, and in this interval some of the calcareous particles are separated from the water, owing to the escape of the carbonic acid, and adhere to the roof. In this manner, successive particles are separated, and attached to each other, until a *stalactite* is formed.

formed. If the percolation of the water containing calcareous particles be too rapid to allow time for the formation of a stalactite, the earthy matter is deposited from it after it has fallen from the roof upon the floor of the cave; and in this case, the deposition is called a *stalagmite*. In some cases, the separation of the calcareous matter takes place both at the roof and on the floor of the cave; and in the course of time, the substance of each deposition increasing, they both meet, and form pillars, often of great magnitude, and that appear destined to support the roof of the cave. Water charged with calcareous earth also oozes through the walls of these caves, and deposites in them a crust of calc-sinter, of various forms; so that in this manner the whole comes to be encrusted with calcareous matter; and if the infiltration continues, the cave in the process of time is entirely filled up.

Caves of this kind occur in almost every country. Maccallister's Cave, in the island of Skye, and those in the limestone hills of Derbyshire, are the most striking appearances of this kind hitherto observed in Scotland and England. But the most celebrated stalactitic cave, is that of Antiparos in the Archipelago, which has been particularly described by Tournefort. Similar caves occur in Germany, France, Switzerland, Spain, in the United States of America, and other countries.

Italy, which is so rich in fine marble, is not less so in beautiful calc-sinter or calcareous alabaster: the territory of Volterra in Tuscany, alone, furnishes no fewer than twenty different varieties. Sicily is also abundant in calc-sinter; and of these, the rose-coloured variety of Trapani is much admired.

Spain is, next to Italy, the most productive country of calcareous

calcareous alabaster. The environs of Granada and Malaga are particularly remarkable for the beautiful varieties of this mineral which they afford.

Persia also abounds in highly prized varieties of calcareous alabaster.

Uses.

Calc-sinter or Calcareous Alabaster, is used for the same purposes as marble, and is cut into tables, columns, vases, drapery for marble figures, and sometimes also into statues. It was also used by the ancients in the manufacture of their unguentary vases. A vessel of this kind is mentioned in the 26th chapter of Matthew's Gospel, where it is said, "There came unto him a woman, having an *alabaster* box of precious ointment." The most beautiful calcareous alabasters, those used by the ancients, are conjectured to have been brought from the mountains of the Thebaid, situated between the Nile and the Red Sea, near the city of Alabastron. In the National Museum in Paris, there is a colossal figure of an Egyptian deity, cut in this rare kind of alabaster. Many different varieties of this mineral are described by authors: the following are enumerated by Brard.

I. *Alabaster of One Colour.*

1. *Antique white Calcareous Alabaster.*—This variety is very rare: it is now only found amongst the ruins of ancient monuments, and particularly at Ortée, not far from Rome; but we are ignorant of the place from whence the ancients procured it.

2. *Yellowish-*

* A fine *sarcophagus* of white alabaster has been lately dug up in Egypt by that skilful and enterprising explorer *Belzoni*.

2. *Yellowish-white, inclining to rose, or Oriental Alabaster.*—The Egyptian statue already mentioned, is made of this beautiful variety of alabaster. It is supposed that the Egyptians procured it from Upper Egypt; but the same variety is found at present in the vicinity of Alicant and Valencia in Spain, and of Trapani in Sicily.

3. *Alabaster of Sienna.*—Its colour is honey-yellow, and it is nearly transparent. A similar variety is found in the island of Malta, of which statues of considerable size are made.

II. *Striped Alabaster,—Onyx Marble of the Ancients.*

The ancients procured these alabasters from the mountains of Arabia, and also from several districts in Germany.

1. *Striped Alabaster from Malaga.*—Two leagues from Malaga, there is a cave filled with wax-yellow alabaster, which, when cut perpendicularly, appears agreeably striped with two different yellow colours; but when cut in another direction, it only presents large irregular spots. The Palace of Madrid is ornamented with this alabaster.

2. *Alabaster from Montreal in Sicily.*—This variety is marked with bright red and yellow stripes.

3. *Alabaster from Caputo in Sicily.*—It is marked with yellow and white stripes.

4. *Alabaster from Mount Pellegrino.*—The stripes are narrow, and of two colours; the one yellow, the other deep black.

5. *Maltese Alabaster.*—Several varieties of alabaster are quarried in Malta: amongst others, one exhibits wax-yellow and white stripes, and another black, brown, and white stripes.

III. *Spotted*

III. *Spotted Alabaster.*

These spots are often produced by the manner in which the stone is cut. There are two very rich columns of this variety, in what used to be called the Hall of the Emperor, in the National Museum in Paris. They were discovered in the year 1780, amidst the ruins of the ancient city of Gabii, four leagues from Rome.

Observations.

1. Some varieties of calc-sinter are so porous, as to allow water to percolate through them, and these are used as filtering-stones.

2. At the springs of St Philippi in Tuscany, moulds of different kinds are suspended on the walls of the basins into which the calcareous water of these springs falls: after a certain period, these become covered with a very solid incrustation of white and very fine calc-sinter, which is easily separated from them, and is found to present excellent impressions of the moulds, whatever they are. It is said that vases, and even statues, are formed in the above manner from calcareous springs, near Guanca-Velica in Peru.

3. Some of the great caves in limestone countries are formed by masses of limestone irregularly heaped on each other, and connected together by means of calc-sinter.

4. Authors differ as to the derivation of the word *Alabaster*. It does not appear to have orignated from *albus*, as some pretend, because the white varieties were rare, and it was the yellow kinds that were most highly prized by the ancients: others are of opinion, and it is the most plausible one, that it is derived from the Greek word αλαβαστρον, which is

is by some derived from α *neg.* and λαμβανειν or λαβειν, *to hold*, because the vessels made of this mineral were without handles, and very smooth, and were therefore difficult to lay hold of. Vessels used for holding ointment or perfume were made of this stone, and were named *Alabastron*. Afterwards, the name alabastron was applied to ointment vessels made of other substances. Thus, in Theocritus, Idyll. xv. lin. 114. we have χρυσεῖ αλαβαστρα, *golden alabastra*. Raphelius remarks, on Matthew xxvi. 7., that Herodotus, among the presents sent by Cambyses to the King of Æthiopia, mentions μυρου αλαβαστρον; and Cicero, Academ. lib. ii. speaks of "alabastrum unguenti plenum." Matth. xxvi. 7.; Mark xiv. 3.; Luke vii. 37.

Sixth Subspecies.

Tufaceous Limestone, or Calc-Tuff*.

Kalk-Tuff, *Werner*.

Tuff Kalkstein, *Reuss*, b. ii. 2. s. 314.—Kalk-tuff, *Lud.* b. i. s. 157. *Id. Suck.* 1r th. s. 623. *Id. Mohs*, b. ii. s. 207. *Id. Leonhard*, Tabel. s. 34.—Tuffartiger Kalkstein, *Karsten*, Tabel. s. 50.—Calcareous Tufa, *Kid*, vol. i. p. 24.—Chaux carbonatée concretionnée incrustante, *Haüy*, Tabl. p. 4.—Kalchtuff, *Lenz*, b. ii. s. 755. *Id. Hoff.* b. iii. s. 40.—Tuffa, *Aikin*, p. 161.

External

* The term *tufa*, appears to be derived from the verb τύφω, which, in its original signification, is appropriate to volcanic productions, especially to such as are of a spongy or porous texture.—*Kid.*

calcareous springs in this country, as in those at Starly Burn in Fifeshire, and other places; and on the Continent of Europe it is also a frequent mineral.

Uses.

The hardest kinds are used for building-stones, and are also burnt into quicklime. It is sometimes also used as a filtering-stone.

Observations.

The substance called *Osteocolla* or *Beinbruch* by the older mineralogists, is a conglomerate of bones and calc-tuff.

Seventh Subspecies.

Pisiform Limestone, or Pea-stone.

Erbsenstein, *Werner*.

La pierre de Pois, ou la Pisolite, *Broch.* t. ii. p. 555.—Erbsenstein, *Lud.* b. i. s. 151. *Id. Suck.* 1r th. s. 621. *Id. Bert.* s. 93. *Id. Mohs*, b. ii. s. 93. *Id. Hab.* s. 79.—Dichter Kalksinter, *Leonhard*, Tabel. s. 34.—Chaux carbonatée concretionnée, Pisolithe, *Brong.* t. i. p. 213.—Erbsförmiger Kalkstein, *Karsten*, Tabel. s. 50.—Pisolithus, or Pea-stone, *Kid*, vol. i. p. 27.—Chaux carbonatée concretionnée testacée, incrustante, &c. but excluding globuliforme, *Haüy*, Tabl. p. 4. —Erbsenstein, *Lenz*, b. ii. s. 754. *Id. Hoff.* b. iii. s. 36.—Pea-stone, *Aikin*, p. 160.

External

:, is the almost constant occurrence of particles of and in the centre of these globular concretions. In tre instances, the centre of the concretions is empty. ral resembling peastone, occurs at the Baths of St ri in Tuscany; also at Perscheesberg in Silesia; Hungary.

Uses.

sometimes cut into plates, for ornamental purposes.

Eighth Subspecies.

Slate-Spar.

Schieferspath, *Werner.*

spath, *Wid.* s. 510.—Argentine, *Kirw.* vol. i. p. 105.—to-spatho, *Nap.* p. 355.—Shifferspath, *Lam.* t. i. p. 385. Spath schisteux, ou le Schieferspath, *Broch.* t. i. p. 558. hieferspath, *Reuss,* b. ii. s. 50. *Id. Lud.* b. i. s. 152. *Id.* 1r th. s. 626. *Id. Bert.* s. 95. *Id. Mohs,* b. ii. s. 3. *Tab.* s. 81. *Id. Leonhard,* Tabel. s. 34.—Chaux carbo- nacré argentine, *Brong.* t. i. p. 232.—Verhærteter it, *Karsten,* Tabel. s. 50.—Chaux carbonatée nacré pri-re, *Haüy,* Tabl. p. 6.—Schieferspath, *Lenz,* b. ii. s. 761. *Hoff.* b. iii. s. 46.

External Characters.

olours are greenish-white, reddish-white, yellowish-greyish-white, and snow-white.

ccurs massive, also in distinct concretions, which are ly curved lamellar, and sometimes coarse and large ar.

The

The lustre is intermediate between shining and glistening, and is pearly.

The fragments are either indeterminate angular and blunt-edged, or are tabular.

It is feebly translucent, or only translucent on the edges.

It is soft.

It is intermediate between sectile and brittle.

It is easily frangible.

It feels rather greasy.

Specific gravity, 2.647, *Kirwan.*—2.474, *Blumenbach.*—2.6300, *La Metherie.*—2.611, *Breithaupt.*

Chemical Characters.

It effervesces very violently with acids; but is infusible before the blowpipe.

Constituent Parts.

From Bremagriia.		From Kongsberg.	
Lime, -	55.00	Lime, -	56.00
Carbonic Acid,	41.66	Carbonic Acid,	39.33
Oxide of Manganese,	3.00	Silica, -	1.66
	———	Oxide of Iron,	1.00
	Bucholz.	Water, -	2.00
			———
			Suersen.

Geognostic Situation.

It occurs in primitive limestone, along with calcareous spar, brown-spar, fluor-spar, and galena; in metalliferous beds, associated with magnetic ironstone, galena, and blende; and in veins, along with tinstone.

Geographic

Geographic Situation.

It occurs in Glen Tilt, Perthshire; and in Assynt in Sutherland, in marble: in Cornwall; and near Granard in Ireland *. On the Continent, it is found along with tinstone, in the Saxon Erzgebirge; along with octahedrite, in a vein at St Christophe in Dauphiny; also in Norway, in metalliferous beds, and in limestone.

Observations.

This mineral is characterised by its colour, lustre, curved lamellar concretions, its degree of hardness and translucency. It is distinguished from *Aphrite*, by its translucency, hardness, compactness, and greater weight.

Ninth Subspecies.

Aphrite.

Schaumerde, *Werner.*

This subspecies is divided into three kinds, viz. Scaly Aphrite, Slaty Aphrite, and Sparry Aphrite.

First

* Greenough.

Scaly Aphrite.

Schaumerde, *Werner.*

Zerreiblicher Aphrit, *Karsten.*

Shaumerde, *Emm.* b. i. s. 484.—Silvery Chalk, *Kirw.* vol. i. p. 78.—L'Ecume de Terre, *Broch.* t. i. p. 557.—Ecume de Terre des Allemands, *Haüy,* t. iv. p. 360.—Schaumerde, *Reuss,* b. ii. 2. s. 317. *Id. Lud.* b. i. s. 152. *Id. Suck.* 1r th. s. 625. *Id. Bert.* s. 95. *Id. Mohs,* b. ii. s. 6. *Id. Hab.* s. 80. *Id. Leonhard,* Tabel. s. 34.—Chaux carbonatée nacré talqueuse, *Brong.* t. i. p. 252.—Zerreiblicher Aphrite, *Karsten,* Tabel. s. 50. *Id. Haus.* s. 126.—Chaux carbonatée nacré lamellaire, *Haüy,* Tabl. p. 6.—Schaumkalch, *Lenz,* b. ii. s. 757.—Erdige Schaumerde, *Oken,* b. i. s. 394.—Schaumkalk, *Hoff.* b. iii s. 46.

External Characters.

Its colours are snow, yellowish, and reddish white, sometimes passing into silver-white.

It occurs either friable or compact.

The friable varieties are composed of glistening or glimmering particles, in which the lustre is pearly. The particles are fine scaly, and feel fine, but not greasy. They are either loose, or loosely cohering.

The compact varieties are massive, disseminated, or in granular concretions, with a shining or nearly splendent lustre, which is pearly, sometimes inclining to semi-metallic.

The fragments are indeterminate angular, and blunt-edged in the great, but tabular in the small.

It

It is opaque.
It soils slightly.
It is very soft, passing into friable.
It is sectile, and uncommonly easily frangible.
It feels very fine, but not greasy.

Chemical Characters.

It effervesces most violently with acids.

Constituent Parts.

Lime, - - -	51.5
Carbonic Acid, -	39.0
Silica, - - -	5.715
Oxide of Iron, - -	3.285
Water, - - -	1.0
	100.5 *Bucholz.*

Geognostic Situation.

It occurs in nests, disseminated, or in small veins, in the flœtz or secondary limestone, and gypsum.

Geographic Situation.

It is found in Thuringia and Hessia.

Observations.

It is characterised by its colour, lustre, low degree of hardness, inconsiderable coherence, and lightness; its want of greasy feel, distinguishes it readily from *Nacrite*, with which it has been confounded.

Second Kind.

Slaty Aphrite.

Schaumschiefer, *Friesleben.*

Schaumschiefer, *Friesleben*, Geognostiche Beiträge, b. ii. s. 232.

External Characters.

Its colours are snow-white, passing into yellowish, reddish, and silver white.

It occurs massive, seldom coarsely disseminated.

It is strongly glimmering, sometimes approaching to glistening, even to shining; and the lustre is pearly, which sometimes passes into semi-metallic.

It is slaty in the great, but undulating curved foliated in the small.

It splits very easily into extremely thin tabular fragments.

It is opaque, or very feebly translucent in the thinnest folia.

It soils pretty strongly.

It feels soft, and rather silky.

It is flexible in thin plates.

Chemical Characters.

It falls into pieces with a crackling noise, when put into water. When touched with an acid, it effervesces with great violence, and is entirely dissolved in it.

Geognostic

Geognostic and Geographic Situations.

It occurs massive, imbedded, and in veins, in the first flœtz limestone, in Thuringia and Hessia.

Observations.

1. The straight slaty variety passes into Slate-Spar, and into Scaly Aphrite.

2. Meinecke, and other old observers, described this mineral as a variety of Common Talc. It was first accurately examined and described by Friesleben.

Third Kind.

Sparry Aphrite.

Schaumspath, *Friesleben.*

Schaumspath, *Friesleben*, Geognostiche Beiträge, b. ii. s. 234.

External Characters.

Its colours are snow, yellowish, and greyish white.

It seldom occurs massive, generally disseminated; sometimes in flaky crusts, in veins, or imbedded in large crystals of selenite.

It is shining, sometimes inclining to splendent, sometimes to glistening; and the lustre is pearly, which inclines to vitreous in the splendent varieties.

The fracture is foliated, sometimes straight, sometimes curved, and the folia have a single distinct cleavage.

It is opaque; feebly translucent in thin pieces.

It occurs in large and small granular distinct concretions.

It soils slightly, with glimmering dusty particles.

It is soft.

It is sectile.

Chemical Characters.

The same as in the other kinds.

Geognostic Situation.

It occurs in flœtz or secondary limestone and gypsum. According to Friesleben, it appears to be geognostically allied to selenite; and although it differs from that mineral in colour, transparency, lustre, sectility, feel, and effervescence with acids, yet it passes into it, and also into slaty aphrite, sometimes by simple gradations, sometimes by intermixture of the two minerals; and large lenticular crystals of selenite occur, which are pure at the edges, become gradually more opaque towards the centre, and in the centre are pure sparry aphrite.

Geographic Situation.

It occurs in Thuringia.

Observations.

It was first described and named by Friesleben, in his "Geognostical Contributions."

Tenth Subspecies.

Lucullite.

This subspecies is divided into three kinds, viz. Compact Lucullite, Prismatic Lucullite, and Foliated Lucullite.

First

First Kind.

Compact Lucullite.

Dichter Lucullan, *John.*

Lapis suillus, *Wall.* t. i. p. 148.—Swinestone, *Kirw.* vol. i. p. 89. —Stinkstein, *Wid.* s. 521. *Id. Estner,* b. ii. s. 1023. *Id. Emm.* b. i. s. 487.—Pierre calcaire puante, ou Pierre puante, *Lam.* t. ii. p. 58.—Chaux carbonatée fetide, *Haüy,* t. ii. p. 188. —La pierre puante, *Broch.* t. i. p. 567.—Gemeiner Stinkstein, *Reuss,* b. ii. 2. s. 335. *Id. Lud.* b. i. s. 155. *Id. Suck.* 1r th. s. 638. *Id. Bert.* s. 111. *Id. Mohs,* b. ii. s. 126.—Gemeiner Stinkstein, *Leonhard,* Tabel. s. 36.—Chaux carbonatée fetide, *Brong.* t. i. p. 236.—Gemeiner Stinkstein, *Haus.* s. 128. *Id. Karsten,* Tabel. s. 50.—Swinestone, *Kid,* vol. i. p. 29.—Chaux carbonatée fetide, *Haüy,* Tabl. p. 6.; et Chaux carbonatée bituminifere, *Id.* p. 6.—Gemeiner Stinkstein, *Lenz,* b. ii. s. 767.—Dichter Stinkstein, *Oken,* b. i. s. 407. —Swinestone, *Aikin,* p. 162.

This kind is divided into Common Compact Lucullite or Black Marble, and Stinkstone.

a. Common Compact Lucullite, or Black Marble.

Dichter Lucullan; Schwarzer Marmor, *John,* Chemisches Laboratorium, t. ii. s. 227. *Id. Lenz,* b. ii. s. 765.

External Characters.

Its colour is greyish-black.

It occurs massive.

Internally it is strongly glimmering, inclining to glistening.

The

The fracture is fine-grained uneven, and large conchoidal.

The fragments are indeterminate angular, and rather sharp-edged.

It is opaque.

It is semi-hard.

It yields a dark ash-grey coloured streak.

It is brittle, and easily frangible.

Specific gravity 3.000, *John.*

When two pieces are rubbed against each other, a fetid urinous odour is exhaled, the intensity of which is increased when we at the same time breathe on them.

Chemical Characters.

It is infusible without addition. When exposed to a high temperature in an open crucible, it burns white. With sulphuric acid, it forms a black-coloured mass: it dissolves in nitrous and muriatic acids, but leaves an insoluble black-coloured substance. During the solution and escape of the carbonic acid, a smell resembling that of sulphuretted hydrogen is evolved.

Constituent Parts.

Lime, - - - - - -	53.38
Carbonic Acid, - - - -	41.50
Black Oxide of Carbon, - -	0.75
Magnesia, and Oxide of Manganese,	0.12
Oxide of Iron, - - - - -	0.25
Silica, - - - - - -	1.13
Sulphur, - - - - - -	0.25
Potash, combinations of Muriatic and Sulphuric Acids, and Water, -	2.62
	100.00

John, Chem, Laborat. b. ii. s. 240.

Geognostic

Geognostic Situation.

The geognostic relations of this mineral are still but little known: it is said to occur in beds in primitive and older secondary rocks.

Geographic Situation.

Hills of this mineral occur in the district of Assynt in Sutherland. It is the *Assynt* or *Sutherland Marble* of artists *. Varieties of it are met with at Ashford, Matlock, and Monsaldale, in Derbyshire: at Kilkenny; at Crayleath, in the county of Down; at Kilcrump, in the county of Waterford; at Churchtown, in the county of Cork; and in the county of Galway, in Ireland. The black marbles of Dinan and Namur, in the Netherlands, are of the same nature. Faujas St Fond is said to have discovered the old quarries of this mineral worked by the ancients, two leagues from Spa, not far from Aix-la-Chapelle.

Uses.

The finer varieties of this mineral have been highly prized, and used as marble from a very remote period. It was so much admired and esteemed by the Consul Lucullus, that he gave it his own name. Pliny observes: " Post hunc Lepidum ferme quadriennio L. Lucullus Consul fuit, qui nomen (ut apparet ex re) *Luculleo Marmori* dedit, admodum delectatus illo, primusque Romam invexit, atrum alioqui, cum cætera maculis aut coloribus commendentur. Nascitur autem in Nili insula, solumque horum marmorum

* Geological Transactions, vol. ii. p. 408, 409, 410.

marmorum ab amatore nomen accepit *." It is said that Marcus Scaurus ornamented his palace with columns thirty-eight feet high of lucullite; and Ferber describes busts and pedestals of it in the Capitol, and at Albani. The mausoleum of Frederick-William, father of Frederick the Great, at Potsdam, is of black marble. The Chinese cut it into bars, and use it along with other minerals in the construction of their musical instrument named *king*. The *Paragone* mentioned by Ferber as a variety of black marble, is said to be basalt. Under the title *Nero antico*, the Italians include all the fine antique lucullites, which are now very rare, and are only to be met with in polished slabs or pieces.

The finest varieties of lucullite met with in trade in this island, are the black marbles of Sutherlandshire, Kilkenny, and Galway.

Observations.

1. It is distinguished from other *Marbles* and *Limestones* by its deep black colour, the strong fetid urinous smell it emits when rubbed, and higher specific gravity.

2. It has been confounded by Boetius de Boot, and Agricola, with several other minerals, as Obsidian, Basalt, and Lydian-stone.

3. It was first described as a particular mineral by Dr John, in the Memoirs of the Society of the Friends of Natural History in Berlin, and afterwards in his work entitled Chemical Laboratory.

b. Stinkstone.

* Plin. Hist. Nat.

b. Stinkstone, or Swinestone.

Stinkstein, *Werner*.

External Characters.

Its colours are yellowish and greyish white, smoke-grey, ash-grey, bluish-grey and brownish-grey, pitch-black, and cream-yellow, which passes into wood-brown, hair-brown, yellowish-brown, liver-brown, and blackish-brown.

It is sometimes dendritic on the surface, or clouded with greyish-black.

It occurs massive, disseminated, also in distinct concretions, which are small granular, and concentric lamellar.

Internally it is dull or glimmering.

The fracture is sometimes small splintery, sometimes imperfect conchoidal, and fine-grained uneven, which passes into earthy, or straight slaty.

The fragments are indeterminate angular, or slaty.

It is opaque, but the cream-yellow varieties are translucent on the edges.

It is semi-hard.

It affords a greyish-white coloured streak; and when rubbed, emits a fetid urinous odour.

It is brittle, and easily frangible.

Specific gravity, 2.750, slaty variety from Bottendorf.—2.677, 2.698, *Breithaupt*.

Chemical Characters.

Nearly the same as in the preceding kind.

Constituent

Constituent Parts.

	From Bottendorf.
Carbonate of Lime,	148—149.00
Silica, -	7.00
Alumina,	5.25
Oxide of Iron, - -	2.50
Oxide of Manganese, - -	1.00
Oxide of Carbon, and a little Bitumen,	0.50
Lime *, - -	1.00
Sulphur, Alkali, Salt, Water,	3.75
	170.00

John, Chem. Laborat. t. ii. s. 242.

Geognostic Situation.

This mineral occurs in beds, in secondary limestone, and occasionally alternates with the secondary gypsum, and with beds of clay. In some places, the strata are quite straight, in others have a zig-zag direction, or are more or less deeply waved, and they are occasionally disposed in a concentric manner, like the concentric lamellar concretions of greenstone. Some strata contain angular pieces of stink-stone, which at first sight might be taken for fragments; and even whole beds occur, which are composed throughout of angular portions, either connected together by means of clay, or immediately joined without any basis. These various appearances do not seem to have been occasioned by any mechanical force acting upon the strata after their formation, but are rather to be viewed as original varieties of structure,

* I have copied the above analysis from Dr John's work; yet I do not see how it is possible that 1 part of Lime could be discovered along with 149 of Carbonate of Lime.

structure, which have taken place during the formation of the strata.

It has been also met with in beds in shell limestone, and in the coal formation.

Geographic Situation.

It occurs in the vicinity of North Berwick in East Lothian, resting on red sandstone; and in the parish of Kirbean in Galloway. On the Continent, it is a frequent rock in Thuringia and Mansfield.

Uses.

In ancient times, it was used as a medicine in veterinary practice: at present, it is principally employed as a limestone, and when burnt, affords an excellent lime, both for mortar and manure. In some districts, as in Thuringia, it is used as a paving-stone and also cut into troughs, steps for stairs, door-posts, and other similar purposes.

Observations.

1. The names *Stinkstone* and *Swinestone* given to this mineral, are from the disagreeable odour it emits when pounded or rubbed.

2. Its fetid urinous odour distinguishes it from *Compact Limestone.*

3. It readily decays, and during its decomposition loses its colour, becomes friable, and gives out an unrespirable gas. Hence it is an unwelcome mineral in mines.

Second

Second Kind.

Prismatic Lucullite.

Stänglichter Lucullan, *John.*

Madreporit, *Klaproth,* Beit. b. iii. s. 272. *Id. Haüy,* t. iv. p. 378. *Id. Lucas,* p. 200.—Madreporstein, *Leonhard,* Tabel. s. 36.—Chaux carbonatée Madreporite, *Brong.* t. i. p. 229. —Madreporite, *Brard,* p. 413.—Stänglicher Anthraconit, *Haus.* s. 128.—Madreporstein, *Karst.* Tabel. s. 50.—Chaux carbonatée bacillaire-fasciculée gris-noiratre, *Haüy,* Tabl. p. 3.—Stänglichter Lucullan, *John,* Chem. Laborat. b. i. s. 243. *Id. Lenz,* b. ii. s. 770.—Stängliger Stinkstein, *Oken,* b. i. s. 407.

External Characters.

Its colours are greyish-black, pitch-black, smoke-grey, and hair-brown.

It occurs massive, in balls, also in distinct concretions, which are stellular and scopiform prismatic.

The external surface is sometimes delicately longitudinally streaked.

Externally it is sometimes dull, sometimes glistening: internally it is shining and splendent, and the lustre is intermediate between vitreous and resinous.

It has a threefold cleavage, and the folia are sometimes curved.

The fragments are indeterminate angular, sometimes inclining to rhomboidal.

It is translucent on the edges, or opaque.

It is semi-hard.

It

It affords a grey-coloured streak.

It is brittle, and easily frangible.

When rubbed, it emits a strongly fetid urinous smell.

Specific gravity, 2.653, 2.688, 2.703, *John*.

Chemical Characters.

When pounded and boiled in water, it gives out a hepatic odour, which continues but for a short time. The filtrated water possesses weak alkaline properties, and contains a small quantity of a muriatic and sulphuric salt. It does not appear to be affected by pure alkalies. It dissolves with effervescence in nitrous and muriatic acids, and leaves behind a coal-black or brownish-coloured residuum.

Constituent Parts.

From Stavern in Norway.		From Greenland.		From Garphytta, in Nericke in Sweden.	
Carbonic Acid,	41.50	Carbonic Acid,	41.53	Carbonic Acid,	41.75
Lime, -	53.37	Lime, -	53.00	Lime, -	54.00
Oxide of Manganese, -	0.75	Oxide of Manganese, -	1.00	Oxide of Manganese, -	0.50
Oxide of Iron,	1.25	Oxide of Iron,	0.75	Oxide of Iron,	0.75
Oxide of Carbon,	1.25	Oxide of Carbon,	1.00	Brown Oxide of Carbon, -	0.75
Sulphur, -	0.25	Sulphur, -	0.50	Sulphur, Alkali, Alkaline Muriate and Sulphate, Water,	2.25
Alumina, -	1.25	Alumina, -	0.75		
Silica, -	1.25	Silica, Alkali, Alkaline Muriate, Water, -	1.47		
Alkali, Alkaline Muriate, Water, Magnesia, Zirconia,	2.13				
	100.00		100.00		100.00
John, Chem. Laborat. b. ii. s. 246.		*John*, ib. s. 248.		*John*, ib. s. 250.	

Geognostic and Geographic Situations.

It occurs in balls, varying from the size of a pea to two feet in diameter, in brown dolomite, at Building Hill near Sunderland. At Stavern in Norway, it appears to occur in transition rocks: in alum-slate at Garphytta in Nericke: in Greenland: and in the Russbachthal in Salzburg.

Observations.

This mineral, which was first discovered by Von Moll, in the Russbachthal in Salzburg, was named by him *Madreporite*, on account of the resemblance of its prismatic concretions to certain lithophytes. It was first described by Schroll, and analysed by Heim *. According to Heim, it contains, Lime, 63.250, Alumina, 10.125, Silica, 12.500, Oxide of iron, 10.988. The same result is said to have been obtained in the School of Mines in Paris †; but both differ so much from the analysis of Klaproth, that we do not hesitate in considering them as erroneous. The publication of Klaproth's chemical examination, induced Von Moll to name it *Anthraconite*, on account of the carbon which it contains ‡; and Dr John, from its intimate connection, both mineralogical and chemical, with Common Lucullite and Stinkstone, arranges it in the system along with these minerals.

Third

* Von Moll's Jahrbücher der Berg und Hüttenkunde, Ister Band, s. 291.—304.

† Haüy, Traité de Mineralogie, t. iv. p. 378, 379.

‡ Ephem. der Berg und Hütt. 2. b. ii. lief s. 305.

Third Kind.

Foliated or Sparry Lucullite.

Späthiger Lucullan, *John.*

Späthiger Lucullan, *John*, Chem. Laborat. b. ii. s. 250. *Id.* *Lenz*, b. ii. s. 772.

External Characters.

Its colours are yellowish, greyish, and greenish-white; also bluish-grey, and greyish and velvet black.

It occurs massive, disseminated, in small granular concretions, and crystallized in acute six-sided pyramids.

Internally it alternates from glimmering to shining.

The fragments are generally rhomboidal.

It is translucent, or translucent on the edges.

It is semi-hard, approaching to soft.

It is brittle, and easily frangible.

When rubbed, it emits an urinous smell.

Specific gravity 2.650, *John.*

Chemical Characters.

They agree with those of the preceding subspecies: in its solution in acids, there remains a minute black-coloured residuum.

Constituent

Constituent Parts.

From Moscau.	From Garphytta. Translucent variety.	From Garphytta. Black variety.
Carbonate of Lime, 96.50	Carbonate of Lime, 99.1	Carbonate of Lime, 95.0
Carbonate of Manganese, Magnesia and Iron, 1.50	Carbonate of Manganese, Magnesia, and Iron, 0.9	Carbonate of Manganese, Magnesia and Iron, 1.5
Oxide of Iron, 1.00	A trace of Carbon, and of an odorous substance.	Mixture of Alum-slate, and of Iron-pyrites, as constituent parts, 3.5
Lime, Alumina, Carbon, Silica, and Water, 1.00		
100.00	100.0	100.0
John, Chem. Laborat. b. iii. s. 90.	*Hisinger* and *Berzelius*, in Afhandlingar i Fysik Kemi och Mineralogi.	*Hisinger* and *Berzelius*, ib.

Geognostic and Geographic Situations.

It occurs in veins, and also in small cotemporaneous masses, in a bed of limestone in clay-slate, at Andreasberg in the Hartz: in veins of silver-ore in hornblende-slate at Kongsberg in Norway: also in transition alum-slate in larger and smaller elliptical masses, the centre of which is of iron-pyrites, and the periphery sparry lucullite, at Andrarum in Schonen, Garphytta in Nericke, and Christiania in Norway.

Eleventh Subspecies.

Marl.

Mergel, *Werner.*

This subspecies is divided into two kinds, viz. Earthy Marl and Compact Marl.

First

First Kind.

Earthy Marl.

Mergel Erde, *Werner.*

Mergel Erde, *Wid.* s. 523.—Earthy Marle, *Kirw.* vol. i. p. 94. —Mergelerde, *Estner,* b. ii. s. 1027. *Id. Emm.* b. i. s. 491.—Marna terrosa, *Nap.* p. 360.—La Marne terreuse, *Broch.* t. i. p. 569.—Erdiger Mergel, *Reuss,* b. ii. 2. s. 339.—Mergelerde, *Lud.* b. i. s. 156. *Id. Suck.* 1r th. s. 643. *Id. Bert.* s. 114. *Id. Mohs,* b. ii. s. 129.—Erdiger Mergel, *Hab.* s. 73. *Id. Leonhard,* Tabel. s. 36. *Id. Karsten,* Tabel. s. 50.—Mergel Erde, *Haus.* s. 127. *Id. Lenz,* b. ii. s. 777.—Erd Mergel, *Oken,* b. i. s. 406.—Mergelerde, *Hoff.* b. iii. s. 67.

External Characters.

Its colours are yellowish-grey, and seldom pale smoke-grey. These are the colours it exhibits when dry: when moist, and in its original repository, its colours are pale blackish-brown or brownish-black *.

It consists of fine dusty particles, which are either loose or feebly cohering.

It is dull.

The particles feel fine, or rather rough and meagre.

It soils slightly.

It is light.

* Some of the varieties have generally a brown colour, and emit an urinous smell; these are by some authors considered as Earthy Stinkstone.

Chemical Characters.

It effervesces strongly with acids. It emits a strong urinous smell when first dug up; but after exposure to the air it loses this quality.

Constituent Parts.

It is said to be composed of Lime, Alumina, Silica, and Bitumen.—*Friesleben.*

Geognostic and Geographic Situations.

It occurs in beds in the flœtz or secondary limestone and gypsum formations, along with stinkstone, in Thuringia and Mansfeld.

Observations.

1. The grey colours, dusty particles, meagre feel, and lightness, characterise this mineral. It passes into Compact Marl, when its particles become more coherent, and also into Stinkstone, and Black Clay.

2. It is sometimes mixed with mica and calcareous spar, also with iron-ochre, and seldomer with pure clay, quartz, sand, gypsum, and aphrite.

3. Masses of stinkstone and limestone, of various sizes and shapes, occur in the beds of marl-earth, which at first sight might be confounded with fragments, although, when closely examined, they prove to be of cotemporaneous formation with it.

4. It is described by some writers under the name Asche, Flozasche, Aschengebirge, and Erdiger Stink-kalk.

Second Kind.

Compact or Indurated Marl.

Verhärteter Mergel, *Werner*.

Marga, *Plin.* Hist. Nat. xvii. 4. s. 6. (in part).—Verhärteter Mergel, *Wid.* s. 524.—Indurated Marl, *Kirw.* vol. i. p. 95.—Verhärteter Mergel, *Emm.* b. i. s. 493.—Marna indurita, *Nap.* p. 361.—La Marne endurcie, *Broch.* t. i. p. 571.—Argile calcifere, ou Marne, *Haüy*, t. iv. p. 445.—Verhärteter Mergel, *Reuss*, b. ii. 2. s. 341. *Id. Lud.* b. i. s. 156. *Id. Suck.* 1r th. s. 642. *Id. Bert.* s. 115. *Id. Mohs*, b. ii. s. 130. *Id. Hab.* s. 73. *Id. Leonhard*, Tabel. s. 36. *Id. Karsten*, Tabel. s. 50. *Id. Haus.* s. 127. *Id. Lenz*, b. ii. s. 779.—Stein Mergel, *Oken*, b. i. s. 407.—Verhärteter Mergel, *Hoff.* b. iii. s. 69.—Marl, *Aikin*, p. 163.

External Characters.

Its colours are yellowish-grey, smoke-grey, and muddy bluish-grey. It is sometimes spotted reddish and brownish in the rents, and marked with dendritic delineations.

It occurs massive, in blunt angular pieces, vesicular, in flattened balls; and frequently contains petrifactions of fishes and crabs, also of gryphites, belemnites, chamites, pectinites, ammonites, terebratulites, ostracites, musculites, and mytulites.

It is dull both externally and internally, and only glimmering when intermixed with foreign parts.

The fracture is generally earthy, which approaches sometimes to splintery, sometimes to conchoidal; in the great inclines to thick and straight slaty.

the chalk in different parts of France, as in the vicinity of Paris.

Uses.

Several different kinds of compact marl occur in nature: these are calcareous marl, in which the calcareous earth predominates; clay marl, in which the aluminous earth is in considerable quantity; and ferruginous marl, in which the mass contains a considerable intermixture of oxide of iron. This latter kind occurs in spheroidal concretions, called *septaria* or *ludi Helmontii*, that vary from a few inches to a foot and a half in diameter. When broken in a longitudinal direction, we observe the interior of the mass intersected by a number of fissures, by which it is divided into more or less regular prisms, of from three to six or more sides, the fissures being sometimes empty, but oftener filled up with another substance, which is generally calcareous spar. From these septaria are manufactured that excellent material for building under water, known by the name of *Parker's Cement* *. The calcareous and aluminous marl are used for improving particular kinds of land; also for mortar; in some kinds of pottery; and in the smelting of particular ores of iron.

Observations.

1. Its meagre feel and inferior sectility distinguish it from *Clays*; its colour, inferior hardness, and greater sectility distinguish

---

* These marly septaria abound in the Isle of Shepey, in the Medway, and often contain in their interior globular portions of heavy-spar, having diverging fibrous concretions. Similar septaria occur in Derbyshire, and in the county of Durham, in which latter district, the internal fissures are filled with quartz.

distinguish it from *Claystone*, and these characters, along with more considerable specific gravity, distinguish it from *Tripoli*.

2. The *Leutrite* of Lenz and Sartorius appears to be a marly sandstone.

3. The *Tutenmergel* or *Nagelkalk* is a variety of marl inclining to compact limestone, disposed in broken conical lamellar concretions, in which the surfaces are transversely streaked. It is found at Gorarp in Sweden.

4. Pliny, Vitruvius, and Varro, describe this mineral under the name *Marga*, and say it was used for improving the soil.

Twelfth Subspecies.

Bituminous Marl-Slate.

Bituminöser Mergelschiefer, *Werner.*

Bituminöser Mergelschiefer, *Wid.* s. 526.—Bituminous Marlite, *Kirw.* vol. i. p. 103.—Bituminöser Mergelschiefer, *Estner*, b. ii. s. 1035. *Id. Emm.* b. i. s. 498.—Schisto marno bituminoso, *Nap.* p. 363.—Le Schiste marneuse bitumineux, *Broch.* t. i. p. 574.—Bituminöser Mergelschiefer, *Reuss*, b. ii. 2. s. 376. *Id. Lud.* b. i. s. 157. *Id. Suck.* 1r th. s. 646. *Id. Bert.* s. 116. *Id. Mohs*, b. ii. s. 132. *Id. Hab.* s. 74 *Id. Leonhard*, Tabel. s. 36. *Id. Karsten*, Tabel. s. 50. *Id. Haus.* s. 127. *Id. Lenz*, b. ii. s. 786. *Id. Oken*, b. i. s. 405. *Id. Hoff.* b. iii. s. 72.

External Characters.

Its colour is intermediate between greyish-black and brownish-black.

It

It occurs massive, and frequently contains impressions of fishes and plants.

Its lustre is glimmering, glistening, or shining, and resinous.

Its fracture is straight, or curved slaty.

The fragments are slaty in the large, but indeterminate and rather sharp angular in the small.

It is opaque.

It is shining and resinous in the streak.

It is soft, and feels meagre.

It is rather sectile, and easily frangible.

Specific gravity 2.631, 2.690, *Breithaupt.*

Constituent Parts.

It is said to be a Carbonate of Lime united with Alumina, Iron, and Bitumen.

Geognostic Situation.

It occurs in secondary or flœtz limestone. It frequently contains cupreous minerals, particularly copper-pyrites, copper-glance, variegated copper-ore, and more rarely, native copper, copper green, and blue copper. It contains abundance of petrified fishes, and these are said to be most numerous in those situations where the strata are basin-shaped. Many attempts have been made to ascertain the genera and species of these animals, but hitherto with but little success. It would appear, that the greater number resemble fresh-water species, and a few the marine species. It also contains fossil remains of lizards, shells, corals, and of cryptogamous fresh-water plants.

Geographic

Geographic Situation.

Europe.—It abounds in the Hartzgebirge; also in Magdeburg and Thuringia. It is a frequent mineral in Upper and Lower Saxony: it occurs also in Franconia, Bohemia, Bavaria, Silesia, Suabia, Hesse-Cassel, and Switzerland.

America.—It is said to occur in the Cordilleras of South America.

4. Prismatic Limestone, or Arragonite.

Prismatischer Kalk-Haloide, *Mohs.*

Arragon, *Werner.*

This species is divided into two subspecies, viz. Common Arragonite, and Coralloidal Arragonite.

First Subspecies.

Common Arragonite.

Gemeiner Arragon, *Werner.*

Arragon-Spar, *Kirw.* vol. i. p. 87.—Arragon, *Estner.* b. ii. s. 1039. *Id. Emm.* b. v. s. 357.—L'Arragonite, *Broch.* t. i. p. 576. *Id. Haüy,* t. iv. p. 337.—Excentrischer Kalkstein, *Reuss,* b. ii. 2. s. 300. *Id. Lud.* b. i. s. 158.—Excentrischer Kalkstein, *Suck.* 1r th. s. 615.—Arragon, *Bert.* s. 97. *Id. Mohs,* b. ii. s. 98.—Arragonite, *Lucas,* p. 192.—Exzentrischer Kalkstein, *Leonhard,* Tabel. s. 34.—Chaux carbonatée Arragonite, *Brong.* t. i. p. 221.—Arragonite, *Brard,* p. 403.

p. 403.—Gemeiner Arragonit, *Haus.* s. 127.—Arragon, *Karsten*, Tabel. s. 50.—Arragon-Spar, *Kid*, vol. i. p. 55.—Arragonite, *Haüy*, Tabl. p. 6.—Arragonit, *Haus.* Handb. b. iii. s. 970. *Id. Hoff.* b. iii. s. 77. *Id. Aikin*, p. 161.

External Characters.

Its colours are greyish-white, greenish-white, and yellowish-white; from greenish-white it passes into greenish-grey, mountain and verdigris green; and from yellowish-white into yellowish-grey. It is sometimes violet-blue. In some crystals, green and blue colours occur together, and sometimes also grey.

It occurs massive; and in distinct concretions, which are thick, thin, and very thin prismatic, and sometimes scopiformly diverging. It is frequently crystallized.

The primitive form is di-prismatic, or is an oblique four-sided prism, bevelled on the extremities; the lateral edges are 115° 56′, and 64° 4′; and the bevelling edge 109° 28′. The following are some of its secondary figures:

1. Irregular six-sided prism, frequently with four lateral edges of about 116°, and two of 128°; or with three lateral edges of 128°, two of 116°, and one of 104°. These are formed by the grouping of several oblique four-sided prisms, bevelled on the extremities. Sometimes this prism is so flat, that it appears like a table.
2. Six-sided table.

   When, on the contrary, the long six-sided prism becomes acicular, there is formed
3. Long, and generally acicular double six-sided pyramids.

The crystals are middle-sized and small; they are generally

Fragments of calcareous-spar, when placed in a similar situation, undergo no alteration.

It is completely soluble, with effervescence, in the nitric and muriatic acids.

Constituent Parts.

Lime,	58.5	Lime,	56.327	Lime,	54.5	Lime,	55.5
Carbon. acid,	41.5	Carb. acid,	43.045	Carbon. acid,	41.5	Carbon. acid,	43.7
		Water,	0.628	Water,	3.5	Water,	0.8
				Loss, -	0.5		
	100.0				100.0		100.0
Fourcroy & Vauquelin, Annal. du Mus. t. iv. p. 405.		*Biot & Thenard*, N. Bull. des Sciences, de la Soc. Ph. t. i. n. 32.		*Buchols*, Gehlen's Jour. b. iii. s. 80.		*Holme*, Annals of Phil. vol. i. p. 334.	

From Molina in Arragon.		From Bastanes.	
Carbonate of Lime,	94.5757	Carbonate of Lime,	94.8249
Carbonate of Strontian,	3.9662	Carbonate of Strontian,	4.0836
Hydrate of Iron, -	0.7060	Protoxide of Manganese, with a trace of Iron,	0.0939
Water of Crystallization,	0.3000	Water of Crystallization,	0.9831
	99.5489		99.6855

Stromeyer, in Gilbert's Annalen der Physik, xiv. 217. October 1813.

Geognostic and Geographic Situations.

Europe.—It occurs along with galena in the lead mines of Leadhills, and in secondary trap-rocks in different parts of Scotland. It is one of the many interesting minerals met with in the secondary trap-rocks of the island of Iceland, and in the trap-rocks of the Department of the Puy de Dome, of Caupenne near Dax, and at Bastanes in Bearn, all in France. The trap-rocks of Bohemia, of the Breisgau, and of Lower Italy, occasionally contain beautiful specimens

specimens of it. At Schwatz in the Tyrol, it is associated with copper-green, grey copper-ore, ochry brown iron-ore, iron-pyrites, quartz, and calcareous-spar; in Spain it occurs imbedded in gypsum, along with reddish-brown quartz crystals; near Iglo in Hungary, it is accompanied with calcareous-spar, ochry-brown iron-ore, and copper-green; near to Schemnitz in Hungary, its accompanying minerals are brown-spar, brittle silver-ore, and galena; at Salfeldt, the principal mineral with which it is grouped is compact brown iron-ore; at Leogang in Salzburg, it is superimposed on brown iron-ore, along with blue copper, and pyramidal calcareous-spar; and it is found in compact limestone at Wolfstein in the Upper Palatinate.

America.—It is found in the trap-rocks in Kannioak in North Greenland, and in the Haasen Island, also in North Greenland. Specimens of it have been met with at Guanaxuato in Mexico, but not in Peru.

Asia.—It occurs in the trap-rocks of Van Diemen's Land, and in the neighbouring islands.

Africa.—It is enumerated amongst the simple minerals contained in the trap or lava rocks of the isle of Bourbon.

Second Subspecies.

Coralloidal Arragonite.

Arragonite coralloide, *Haüy*, Tabl. p. 7.—Fasriger Kalksinter, *Hoff.* b. iii. s. 32.

External Characters.

Its most frequent colours are varieties of white.

It

It occurs massive, reniform, tuberose, coralloidal, imperfect globular; in distinct concretions, which are fibrous, generally straight, seldom curved, and stellular and scopiform; sometimes also in reniform curved lamellar, and large angulo-granular concretions.

The lustre glimmering, or glistening and pearly.

The fracture is fine splintery.

The fragments are wedge-shaped and splintery.

It is translucent, or translucent on the edges.

In other characters, agrees with the preceding subspecies.

Geognostic and Geographic Situations.

It is found in Dufton Fell in Cumberland, also in the iron mines of Stiria and Carinthia, and also at Saint Marié aux Mines.

Observations.

1. The preceding descriptions include nearly all the varieties of arragonite hitherto described by authors. To these Mohs adds the two following:

a.

1. *Colour.* Flesh-red and pearl-grey.

Form. Massive.

Distinct Concretions. These are fibrous, straight, stellular and scopiform, and are collected into others which are large angulo-granular.

Fracture—Is not visible.

Fragments. Splintery and wedge-shaped.

Observations.—Several of the varieties of the fibrous brown-spar of Hoffmann belong to this kind of arragonite.

b.

b.

2. *Colour.* White, and sometimes tile-red.

Form. Massive.

Distinct concretions. These are prismatic, sometimes straight, sometimes curved, and always parallel.

Fracture—Not visible.

Fragments. Splintery.

Observations.—This kind of arragonite is described by Haüy under the name Arragonite fibreux, and part of the common fibrous limestone of Hoffmann also belongs here.

2. All the varieties of arragonite are distinguished from *Calcareous-Spar*, by superior hardness, and specific gravity.

3. Arragonite has received different names at different periods: the common prismatic varieties have been named *Arragonian Apatite*, *Arragonian Calc-Spar*, and *Hard Calcareous-Spar*; and the pyramidal varieties have been described under the names *Iglit*, or *Igloit*.

4. The Coralloidal Arragonite has been described under the name *Flos Ferri*.

Genus II.—APATITE.

This Genus contains one Species, viz. Rhomboidal Apatite.

1. Rhomboidal

1. Rhomboidal Apatite.

Rhomboidrischer Flus Haloide, *Mohs.*

Apatit, *Werner.*

THIS Species is divided into three Subspecies, viz. Foliated Apatite, Conchoidal Apatite, and Lamellar Apatite.

First Subspecies.

Foliated Apatite.

Geminer Apatite, *Werner.*

Phosphorite, *Kirw.* vol. i. p. 128.—Gemeiner Apatit, *Wid.* s. 528.—Phosphorit, *Estner,* b. ii. s. 1049. *Id. Emm.* b. i. s. 502.—Fosforite lamellare, *Nap.* p. 367.—Apatit, *Lameth.* t. ii. p. 85.—L'Apatite, *Broch.* t. i. p. 580.—Chaux phosphatée, *Haüy,* t. ii. p. 234.—Gemeiner & Blättricher Apatit, *Reuss,* b. ii. 2. s. 355. & 362.—Apatit, *Lud.* b. i. s. 159. *Id. Suck.* 1r th. s. 655. *Id. Bert.* s. 99. *Id. Mohs,* b. ii. s. 139. —Chaux phosphatée, *Lucas,* p. 11.—Apatit, *Leonhard,* Tabel. s. 38.—Chaux phosphatée Apatite, *Brong.* t. i. p. 240.—Chaux phosphatée, *Brard,* p. 44.—Blättriger Apatit, *Haus.* s. 123.—Apatit, *Karsten,* Tabel. s. 52.—Crystallised Phosphate of Lime, *Kid,* vol. i. p. 80.—Chaux phosphatée, *Haüy,* Tabl. p. 7.—Gemeiner Apatit, *Lenz,* b. ii. s. 804.—Blättriger Apatit, *Oken,* b. i. s. 397.—Apatit, *Hoff.* b. iii. s. 84.—Blättricher Apatit, *Haus.* Handb. b. iii. s. 869.—Apatit, *Aikin,* p. 172.

External Characters.

It most frequent colours are snow-white, yellowish-white, reddish-white, and greenish-white; from greenish-white it passes

passes into mountain-green, celandine-green. leek-gr emerald-green, and olive-green. It occurs also hyaci red, flesh-red, rose-red, and pearl-grey, from which it ses into violet-blue, lavender-blue, and seldom into ind blue. Sometimes it is pale wine-yellow, and yellow brown. Frequently several of these colours occur in same piece.

It sometimes occurs massive and disseminated, also distinct concretions, which are large and small angulo-nular, and sometimes thin and straight lamellar; gene ly crystallized.

The primitive form is a di-rhomboid, in which the an are 131° 49′, and 109° 28′. The following are the se dary figures:

I. Prism.

Low equiangular six-sided prism *.

a. Perfect, fig. 148. Pl. 8.

b. With truncated lateral edges †, fig. 149. Pl

c. With truncated lateral and terminal edg fig. 150. Pl. 8.

d. With truncated terminal edges ‖, fig. 151. P

e. With truncated terminal edges and angl fig. 152. Pl. 8.

f. With bevelled terminal edges and trunc angles.

g. With rounded edges, so that the prism app cylindrical.

h. I

---

* Chaux phosphatée primitive, Haüy.

† Chaux phosphatée peridodecaedre, Haüy.

‡ Chaux phosphatée emarginée, Haüy.

‖ Chaux phosphatée annulaire, Haüy.

§ Chaux phosphatée unibinaire, Haüy.

h. Flatly acuminated on one extremity with six planes, which are set on the lateral planes. In this figure, the apex of the acumination, all the angles, and the alternate lateral edges, are slightly truncated.

i. Acuminated on both extremities with six planes, the apices, lateral edges, and angles, occasionally truncated.

k. The preceding acumination again very flatly acuminated with six planes, which are set on the planes of the first acumination. The apices of the acuminations truncated.

II. Table:

a. Equiangular six-sided table, in which the edges and angles are sometimes truncated.

b. Eight-sided table, in which four of the terminal edges are truncated.

The crystals are small, very small, and middle-sized; and occur sometimes single, sometimes many irregularly superimposed on each other.

The lateral planes are seldom smooth, generally longitudinally streaked; the truncating and acuminating planes are smooth.

Externally it is splendent or shining; internally glistening, and the lustre is resinous.

It has a fourfold cleavage, in which three of the cleavages are parallel with the lateral planes of the prism, and one (the most perfect), with the terminal planes of the prism.

The fracture is intermediate between uneven and imperfect conchoidal.

 The

The fragments are indeterminate angular, and rather sharp-edged.

It is generally translucent; seldom nearly transparent, when it refracts single.

It is harder than fluor-spar, but not so hard as felspar.

It is brittle, and easily frangible.

Specific gravity, 3.179, *Lowry*.—3.248, *Breithaupt*.

Physical Characters.

It becomes electric by heating, and also by being rubbed with woollen cloth.

Chemical Characters.

When thrown on glowing coals, it emits a pale grass-green phosphoric light. It dissolves very slowly in the nitric acid, and without effervescence. It gradually loses its colour, when heated before the blowpipe, but its lustre and transparency are heightened. It is infusible without addition.

Constituent Parts.

Lime, - -	55
Phosphoric Acid, and trace of Manganese,	45
	100

Klaproth, Bergm. Journ. 1788. b. i. s. 269.

Geognostic Situation.

It occurs in tinstone veins, and also imbedded in talc.

Geographic

Geographic Situation.

Europe.—It occurs in yellow foliated talc, and, along with fluor-spar, in the mine called Stena-Gwyn, in St Stephens, in Cornwall, also at St Michael's Mount, Godolphin-bal in Breage, also in Cornwall: at Schlackenwald in Bohemia, in tinstone veins, along with tungsten, wolfram, topaz, and fluor-spar; at Ehrenfriedersdorf in Saxony, along with tinstone, copper, and arsenical pyrites, fluor-spar, steatite, lithomarge, talc, and quartz; imbedded in lepidolite near Rosena in Moravia; in a mixture of quartz and felspar at Nantes, Four-au-Diable, in department of the Lower Loire; at Arendal in Norway, along with magnetic ironstone, garnet, hornblende, and limestone; in veins, on St Gothard in Switzerland; and in Estremadura in Spain, in small tables, along with lamellar apatite or phosphorite.

America.—It occurs in grains or hexahedral prisms in granite, near Baltimore in Maryland; in granite and gneiss, along with beryl, garnet, and schorl, at Germantown in Pennsylvania; in iron-pyrites at St Anthony's Nose, in the Hudson in New York; in granite, at Milford-hills, near New-Haven in Connecticut; and at Topsham in Maine, in granite.

3. The same figure, truncated on the lateral edges of the prism *, fig. 154. Pl. 8.

The crystals are middle-sized, small, and very small; sometimes longitudinally streaked, and sometimes traversed by cross rents.

Externally the crystals are splendent, and vitreous: internally shining, and resinous.

It has an imperfect cleavage.

The fracture is small and imperfect conchoidal.

The fragments are rather blunt-edged.

It alternates from transparent to translucent.

In other characters agrees with the foliated apatite.

Specific gravity, 3.200, from Utö, *Klaproth*.—3.190, from Zillerthal, *Klaproth*.

Chemical Characters.

Some varieties of this subspecies do not phosphoresce when exposed to heat.

Constituent Parts.

Apatite from Utö.	
Phosphate of Lime,	92.00
Carbonate of Lime,	6.00
Silica, -	1.00
Loss in heating,	0.50
Manganese a trace.	——
	99.50

Klaproth, Beit. b. v. s. 181.

From Zillerthal.	
Lime, -	53.75
Phosphoric Acid,	46.25
	——
	100

Klaproth, Beit. b. iv. s. 197.

Geognostic

* Chaux phosphatée didodecaedre, Haüy.

Geognostic and Geographic Situations.

Europe.—It occurs imbedded in gneiss near Kincardine in Ross-shire; also in beds of magnetic ironstone, along with sphene, calcareous-spar, hornblende, quartz, and augite, at Arendal in Norway; imbedded in green-talc, in the Zillerthal in Salzburg; in granite, near Nantes, and in basalt, at Mount Ferrier, in France; and in a porous ironshot limestone, near Cape de Gate, in Murcia in Spain.

America.—Imbedded in granite at Baltimore *; in gneiss at Germantown; and in mica-slate in West Greenland †.

Observations.

1. This mineral was at one time described as a kind of Schorl; afterwards as a variety of Beryl, on account of colours and figure; and some authors have arranged it with Fluor-spar, and others along with Chrysolite. Werner ascertained that it was a distinct species from any of those just enumerated, and named it *Apatite*, from the Greek word ἀπατάω, *to deceive*, on account of its having been confounded with so many other minerals. It was Klaproth who first analysed it.

2. The conchoidal subspecies has been considered by some authors as a mere variety of the common apatite; whilst others have raised it to the rank of a species: thus, many of the French mineralogists arrange it with the varieties of common apatite; while some German mineralogists describe the asparagus-green varieties under the name *Asparagus-stone*, and certain green and blue varieties under

* Gilmor, in Bruce's Mineralogical Journal, p. 228.

† Giesecké.

der the name *Moroxite*. The name moroxite given to this mineral by Karsten, is borrowed from the Morochites of Pliny, concerning which, that author says, " Est gemma, per se porracea viridisque, trita autem candicans."—*Histor. Natur.* l. xxxvii.

3. Apatite is distinguished from Beryl, Schorl, and Chrysolite, by its inferior hardness: its greater hardness and non-effervescence with acids, distinguish it from Calcareous-spar.

Third Subspecies.

Phosphorite.

Phosphorit, *Werner.*

This Subspecies is divided into two Kinds, viz. Common Phosphorite and Earthy Phosphorite.

First Kind.

Common Phosphorite.

Gemeiner Phosphorit, *Karsten.*

Gemeiner Phosphorit, *Haus.* s. 123. *Id. Karsten,* Tabel. s. 52. —Chaux phosphatée terreuse, *Haüy,* Tabl. p. 8.—Gemeiner Phosphorit, *Lenz,* b. ii. s. 801.—Dichter Phosphorit, *Haus.* Handb. b. iii. s. 872.—Phosphorit, *Hoff.* b. iii. s. 92.—Massive Apatite, *Aikin,* p. 172.

External Characters.

Its colour is yellowish-white, sometimes approaching to

greyish-white. It is occasionally spotted pale ochre-yellow and yellowish-brown.

It occurs massive, and in distinct concretions, which are thin and curved lamellar.

The surface is uneven and drusy.

It is dull or glistening.

Its cleavage is imperfect curved, and generally floriform.

The fracture is uneven.

The fragments are indeterminate angular, and rather blunt-edged.

It is opaque, or feebly translucent on the edges.

It is soft, rather brittle, and very easily frangible.

Chemical Characters.

It becomes white before the blowpipe, and, according to Proust, melts with difficulty into a white-coloured glass. When rubbed in an iron mortar, it emits a green-coloured phosphoric light; and the same effect is produced when it is pounded and thrown on glowing coals.

Constituent Parts.

Lime, - - - - - -	59.0
Phosphoric Acid, - - -	34.0
Silica, - - - - - -	2.0
Fluoric Acid, - - - - -	2.5
Muriatic Acid, - - - - -	0.5
Carbonic Acid, - - - - -	1.0
Oxide of Iron, - - - - -	1.0
	100.0

Pelletier, Journal des Mines, N. 166.

Geognostic

Geognostic and Geographic Situations.

It occurs in crusts, and crystallized, along with apatite and quartz, at Schlackenwald in Bohemia, but most abundantly near Leigrosan, in the province of Estremadura in Spain, where it is sometimes associated with apatite, and frequently with amethyst, and forms whole beds, that alternate with limestone and quartz.

Second Kind.

Earthy Phosphorite.

Erdiger Phosphorit, *Karsten.*

Erdiger Phosphorit, *Haus.* s. 123. *Id. Karsten,* Tabel. s. 52.—Chaux phosphatée pulverulente, *Haüy,* Tabl. p. 8.—Erdiger Phosphorit, *Lenz,* b. ii. s. 802. *Id. Haus.* Handb. b. iii. s. 873.

External Characters.

Its colours are greyish-white, greenish-white, and pale greenish-grey.

It consists of dull dusty particles, which are partly loose, partly cohering, and which soil slightly, and feel meagre and rough.

Chemical Characters.

It phosphoresces when laid on glowing coals.

Constituent

Constituent Parts.

Earthy Phosphorite from Marmarosch.	
Lime, - -	47.00
Phosphoric Acid, -	32.25
Fluoric Acid,	2.50
Silica, -	0.50
Oxide of Iron,	0.75
Water, - -	1.00
Mixture of Quartz and Loam,	11.50
	95.50

Klaproth, Beit. b. iv. s. 373.

Geognostic and Geographic Situations.

It occurs in a vein, in the district of Marmarosch in Hungary.

Observations.

1. It was for some time described in systems of mineralogy as a variety of Fluor-spar.

2. The celebrated Prussian chemist John, has published the description and analysis of a mineral under the name *Ratofkite*, from Ratofka, near Werea in Russia, which appears to be nearly allied to this subspecies, and may possibly prove to be a new species, intermediate between fluor and apatite. The following is the description and analysis of it, as given by John, in the second continuation of his work entitled, "Chemische Untersuchungen," &c.

"Its colour is lavender-blue; and it is composed of loose dusty dull particles, that soil slightly, and are not particularly heavy. It is contained in aphrite. *Constituent Parts:* Fluate of Lime, 49.50. Phosphate of Lime, 20.00.

20.00. Muriatic Acid, 2.00. Phosphate of Iron, 3.75. Water, 10.00."

Genus III. FLUOR.

This Genus contains but one species, viz. Octahedral Fluor.

1. Octahedral Fluor.

Flus, *Werner*.

Octaedrisches Flus Haloide, *Mohs*.

It is divided into three subspecies, Compact Fluor, Foliated Fluor, and Earthy Fluor.

First Subspecies.

Compact Fluor.

Dichter Flus, *Werner*.

Fluor solidus, *Wall.* t. i. p. 542. ?—Dichter Fluss, *Wid.* s. 542. —Compact Fluor, *Kirw.* vol. i. p. 127.—Dichter Fluss, *Estner*, b. ii. s. 1067. *Id. Emm.* b. i. s. 516.—Fluorite compatta, *Nap.* p. 374.—Le Fluor compacte, *Brock.* t. i. p. 594.—Dichter Fluss, *Reuss*, b. ii. 2. s. 379. *Id. Lud.* b. i. s. 161. *Id. Suck.* 1 th. s. 663. *Id. Bert.* s. 103. *Id. Mohs*, b. ii. s. 150. *Id. Hab.* s. 89.—Chaux fluatée massive compacte, *Lucas*, p. 247.—Dichter Fluss, *Leonhard*, Tabel. s. 38.—Chaux fluatée,

fluatée compacte, *Brong.* t. i. p. 245.—Dichter Fluss, *Hau.* s. 124. *Id. Karsten,* Tabel. s. 52.—Chaux fluatée compacte, *Haüy,* Tabl. p. 9.—Dichter Fluss, *Lenz,* b. ii. s. 823. *Id. Oken,* b. i. s. 899.—Dichter Flus, *Hoff.* b. iii. s. 95.—Compact Fluor, *Aikin,* p. 175.

External Characters.

Its colours are greenish-grey and greenish-white; from greenish-grey it passes into pearl-grey, and into a colour intermediate between flesh-red and brownish-red; from greenish-white it passes into dark mountain-green, and even into greenish-black. These colours are either simple, or in flamed or spotted delineations.

It occurs massive.

Externally and internally it is dull, or feebly glimmering.

The fracture is even, which passes on the one side into small splintery, on the other into flat conchoidal.

The fragments are rather sharp-edged.

It is more or less translucent.

It is harder than calcareous-spar, but not so hard as apatite.

It is brittle, and easily frangible.

Specific gravity 3.150, 3.191, *Breithaupt.*

Chemical Characters.

The chemical characters are the same as those of the following subspecies.

Geognostic and Geographic Situations.

It is found in veins, associated with fluor-spar, at Stolberg in the Hartz: the veins traverse rocks of grey-wacke, and

and besides fluor-spar, contain heavy-spar and copper-pyrites. It has also been met with at Stripasen in Norbergs-Bergslag in Sweden.

Second Subspecies.

Foliated Fluor.

Flus-Spath, *Werner.*

Fluor spathosus; Fluor granularis, et Fluor cristallisatus, *Wall.* t. i. p. 180. 183.—Spath fusible ou vitreux, *Romé de Lisle,* t. ii. p. 1.—Chaux fluorée, *De Born,* t. i. p. 355.—Fluss-spath, *Wid.* s. 558.—Foliated or Sparry Fluor, *Kirw.* vol. i. p. 127. —Fluss-spath, *Estner,* b. ii. s. 1070. *Id. Emm.* b. i. s. 519.—Fluorite lamellare, *Nap.* p. 375.—Fluor, *Lam.* t. i. p. 78.—Chaux fluatée cristallisée, *Haüy,* t. ii. p. 247.—Le Spath Fluor, *Broch.* t. i. p. 595.—Spathiger Fluss, *Reuss,* b. ii. 2. s. 381. *Id. Lud.* b. i. s. 162. *Id. Suck.* 1r th. s. 664. *Id. Bert.* s. 103. *Id. Mohs,* b. ii. s. 151. *Id. Hab.* s. 83.—Chaux fluatée, *Lucas,* p. 12.—Spathiger Fluss, *Leonhard,* Tabel. s. 38.—Chaux fluatée spathique, *Brong.* t. i. p. 243.—Chaux fluatée, *Brard,* p. 47.—Gemeiner, stänglicher, schaaliger, & körniger Fluss-spath, *Haus.* s. 123, 124.—Spathiger Fluss, *Karsten,* Tabel. s. 52.—Fluat of Lime, or Fluor-spar, *Kid,* vol. i. p. 73.—Chaux fluatée, *Haüy,* Tabl. p. 9.—Fluss-spath, *Lenz,* b. ii. s. 824. *Id. Hoff.* b. iii. s. 96.—Crystallised Fluor, *Aikin,* p. 175.

External Characters.

Its most common colours are white, yellow, green and blue, seldomer red, grey and brown, and least frequently

black.

black. The white varieties are reddish, yellowish, greenish, and greyish; the latter passes into smoke-grey, the reddish-white into rose-red; from this into pearl-grey, violet-blue, smalt-blue, and sky-blue; from this latter into verdigris-green, celandine-green, mountain-green, emerald-green, grass-green, asparagus-green, and oil-green; further into wax, wine, honey yellow, and yellowish-brown. The violet-blue sometimes passes into bluish-black.

The colours are of all degrees of intensity, and sometimes pieces occur spotted or striped. Green cubes occur with blue angles, &c. Some colours, as sky-blue, fade by keeping, particularly in warm places.

It occurs massive, disseminated, also in distinct concretions, which are large, coarse, small, and fine granular, sometimes straight prismatic, which are traversed by others that are thick and fortification-wise curved lamellar. The striped colour delineation is in the direction of these concretions. It occurs crystallized, in the following figures:

1. Cube, which is the most frequent crystallization, and is also the fundamental figure of the species *, fig. 155. Pl. 8.
2. Cube, truncated on all the edges †, fig. 156. Pl. 8. When these truncating planes increase so much as to cause the faces of the cube to disappear, there is formed
3. The rhomboidal or garnet dodecahedron ‡, fig. 157. Pl. 8.

4. Cube,

* Chaux fluatée cubique, Haüy.

† Chaux fluatée cubo-dodecaedre, Haüy.

‡ Chaux fluatée dodecaedre, Haüy.

4. Cube, with truncated angles *, fig. 158. Pl. 8.
When these truncating planes increase, so as to cause the faces of the cube to disappear, there is formed an
5. Octahedron, or regular double four-sided pyramid †, fig. 159. Pl. 8. This figure is sometimes truncated on the edges, as fig. 160. Pl. 8., or on the edges and angles at the same time; and varieties of it occur, in which the planes or faces are cylindrically convex ‡.
6. Cube, with bevelled edges ||, fig. 161. Pl. 8.
When the bevelling planes enlarge so much, as to cause the original faces of the cube to disappear, a tessular crystal, with 24 triangular planes, is formed, fig. 162. Pl. 8.
7. Cube, in which all the angles are acuminated with three planes, which are set on the lateral planes.
8. Cube, in which all the angles are acuminated with six planes, which are set on the lateral planes.

The cubes vary from very large to very small; the other crystals are only small and middle sized.

The crystals are generally placed on one another, and form druses; but are seldom single.

The surface is smooth and splendent, or drusy and rough,

* Chaux fluatée cubo-octaedre, Haüy.

† Chaux fluatée primitive, Haüy.

‡ Chaux fluatée spheroidale, Haüy.

|| Chaux fluatée bordée, Haüy.

rough, as in the rhomboidal dodecahedron, and some octahedrons.

Internally the lustre is specular-splendent, or shining, and is vitreous.

It has a fourfold equiangular cleavage, which is parallel with the planes of an octahedron.

The fragments are octahedral or tetrahedral.

It alternates from translucent to transparent, and refracts single.

It is harder than calcareous-spar, but not so hard as apatite.

It is brittle, and easily frangible.

Specific gravity, 3.0943, 3.1911, *Haüy*.—3.148, *Gellert*.—3.092, *Brisson*.—3.156, 3.184, *Muschenbroeck*.—3.138, 3.228, *Karsten*.

Chemical Characters.

Before the blowpipe it generally decrepitates, gradually loses its colour and transparency, and melts, without addition, into a greyish-white glass. When two fragments are rubbed against each other, they become luminous in the dark. When gently heated, or laid on glowing coal, it phosphoresces, (particularly the sky-blue, violet-blue, and green varieties,) partly with a blue, partly with a green light. When brought to a red-heat, it is deprived of its phosphorescent property. The violet-blue variety from Nertschinsky, named *Chlorophane*, when placed on glowing coals, does not decrepitate, but soon throws out a beautiful verdigris-green and apple-green light, which gradually disappears as the mineral cools, but may be again excited, if it is heated; and this may be repeated a dozen of times, provided

provided the heat is not too high. When the chlorophane is exposed to a red-heat, its phosphorescent property is entirely destroyed. Pallas mentions a pale violet-blue variety spotted with green, from Catharinenburg, which is so highly phosphorescent, that when held in the hand for some time, it throws out a pale whitish light; when placed in boiling water, a green light; and exposed to a higher temperature, a bright blue light. When sulphuric acid is added to heated fluor-spar, in the state of powder, a white penetrating vapour (the fluoric acid) is evolved, which has the property of corroding glass *.

Constituent Parts.

	Northumberland.	Gersdorf.	Gersdorf.
Lime,	67.34	67.75	65.0
Fluoric Acid,	32.66	32.25	35.0
	100.00	100.00	100.0
	Thomson, in Wern. Mem. vol. i. p. 11.	*Klaproth*, Beit. b. iv. s. 365.	*Richter*, Uber die Neueren Gegenst. v. Chem. b. iv. s. 25.

Geognostic Situation.

It occurs principally in veins, that traverse primitive, transition, and sometimes secondary rocks; also in beds, associated

* M. Th. de Grotthaus describes a fluor under the name *Pyro-Emerald* and *Chlorophane*: it is the violet fluor of Nertschinsky. After being exposed some time to the sun's rays, it preserves its phosphorescence for weeks. Its affinity for light is so great, that it absorbs that of a candle or the electric spark, which it gives out in the dark

Second Subspecies.

Conchoidal Apatite or Asparagus-Stone.

Muschlicher Apatit, *Hausmann.*

Spargelstein, *Werner.*

Chrysolith ordinaire, ou proprement dite, *De Lisle,* t. ii. p. 271. —Chrysolithe, *De Born,* t. . p. 68. 2. E. *a.* 3.—Spargelstein, *Emm.* b. iii. s. 359.—La Pierre d'Asperge, *Broch.* t. i. p. 586. —Muschlicher Apatit, *Reuss,* b. ii. 2. s. 358.—Chaux phosphatée crysolithe, *Brong.* t. i. p. 240.—Muschlicher Apatit, *Haus.* s. 123. *Id. Lenz,* b. ii. s. 808. *Id. Oken,* b. i. s. 397. *Id. Haus.* Handb. b. iii. s. 870.—Spargelstein, *Hoff.* b. ii. s. 89.

External Characters.

Its colours are mountain-green, leek-green, pistachio-green, asparagus-green, olive-green, and siskin-green, which passes into wine-yellow, bordering on orange-yellow. It also occurs sky-blue, greenish and yellow grey, and clove-brown.

It sometimes occurs massive and disseminated, also in distinct concretions, which are large granular; but most frequently crystallized, and in the following figures:

1. Equilateral, longish, six-sided prism, acuminated with six planes, which are set on the lateral planes *, fig. 158. Pl. 8.
2. Sometimes the acumination ends in a line †.

3. Th

* Chaux phosphatée pyramidée, Haüy.

† Chaux phosphatée cuneiforme, Haüy.

3. The same figure, truncated on the lateral edges of the prism *, fig. 154. Pl. 8.

The crystals are middle-sized, small, and very small; sometimes longitudinally streaked, and sometimes traversed by cross rents.

Externally the crystals are splendent, and vitreous: internally shining, and resinous.

It has an imperfect cleavage.

The fracture is small and imperfect conchoidal.

The fragments are rather blunt-edged.

It alternates from transparent to translucent.

In other characters agrees with the foliated apatite.

Specific gravity, 3.200, from Utö, *Klaproth*.—3.190, from Zillerthal, *Klaproth*.

Chemical Characters.

Some varieties of this subspecies do not phosphoresce when exposed to heat.

Constituent Parts.

Apatite from Utö.	
Phosphate of Lime,	92.00
Carbonate of Lime,	6.00
Silica, -	1.00
Loss in heating,	0.50
Manganese a trace.	——
	99.50

Klaproth, Beit. b. v. s. 181.

From Zillerthal.	
Lime, -	53.75
Phosphoric Acid,	46.25
	——
	100

Klaproth, Beit. b. iv. s. 197.

Geognostic

* Chaux phosphatée didodecaedre, Haüy.

Geognostic and Geographic Situations.

Europe.—It occurs imbedded in gneiss near Kincardine in Ross-shire; also in beds of magnetic ironstone, along with sphene, calcareous-spar, hornblende, quartz, and augite, at Arendal in Norway; imbedded in green-talc, in the Zillerthal in Salzburg; in granite, near Nantes, and in basalt, at Mount Ferrier, in France; and in a porous ironshot limestone, near Cape de Gate, in Murcia in Spain.

America.—Imbedded in granite at Baltimore *; in gneiss at Germantown; and in mica-slate in West Greenland †.

Observations.

1. This mineral was at one time described as a kind of Schorl; afterwards as a variety of Beryl, on account of colours and figure; and some authors have arranged it with Fluor-spar, and others along with Chrysolite. Werner ascertained that it was a distinct species from any of those just enumerated, and named it *Apatite*, from the Greek word απατάω, *to deceive*, on account of its having been confounded with so many other minerals. It was Klaproth who first analysed it.

2. The conchoidal subspecies has been considered by some authors as a mere variety of the common apatite; whilst others have raised it to the rank of a species: thus, many of the French mineralogists arrange it with the varieties of common apatite; while some German mineralogists describe the asparagus-green varieties under the name *Asparagus-stone*, and certain green and blue varieties under

* Gilmor, in Bruce's Mineralogical Journal, p. 228.

† Giesecké.

der the name *Moroxite*. The name moroxite given to this mineral by Karsten, is borrowed from the Morochites of Pliny, concerning which, that author says, " Est gemma, per se porracea viridisque, trita autem candicans."—*Histor. Natur.* l. xxxvii.

3. Apatite is distinguished from Beryl, Schorl, and Chrysolite, by its inferior hardness: its greater hardness and non-effervescence with acids, distinguish it from Calcareous-spar.

Third Subspecies.

Phosphorite.

Phosphorit, *Werner*.

This Subspecies is divided into two Kinds, viz. Common Phosphorite and Earthy Phosphorite.

First Kind.

Common Phosphorite.

Gemeiner Phosphorit, *Karsten*.

Gemeiner Phosphorit, *Haus.* s. 123. *Id. Karsten*, Tabel. s. 52. —Chaux phosphatée terreuse, *Haüy*, Tabl. p. 8.—Gemeiner Phosphorit, *Lenz*, b. ii. s. 801.—Dichter Phosphorit, *Haus.* Handb. b. iii. s. 872.—Phosphorit, *Hoff.* b. iii. s. 92.—Massive Apatite, *Aikin*, p. 172.

External Characters.

Its colour is yellowish-white, sometimes approaching to greyish-

greyish-white. It is occasionally spotted pale ochre-yellow and yellowish-brown.

It occurs massive, and in distinct concretions, which are thin and curved lamellar.

The surface is uneven and drusy.

It is dull or glistening.

Its cleavage is imperfect curved, and generally floriform.

The fracture is uneven.

The fragments are indeterminate angular, and rather blunt-edged.

It is opaque, or feebly translucent on the edges.

It is soft, rather brittle, and very easily frangible.

Chemical Characters.

It becomes white before the blowpipe, and, according to Proust, melts with difficulty into a white-coloured glass. When rubbed in an iron mortar, it emits a green-coloured phosphoric light; and the same effect is produced when it is pounded and thrown on glowing coals.

Constituent Parts.

Lime, - - - - - -	59.0
Phosphoric Acid, - - -	34.0
Silica, - - - - - -	2.0
Fluoric Acid, - - - - -	2.5
Muriatic Acid, - - - - -	0.5
Carbonic Acid, - - - - -	1.0
Oxide of Iron, - - - - -	1.0
	100.0

Pelletier, Journal des Mines, N. 166.

Geognostic

Geognostic and Geographic Situations.

It occurs in crusts, and crystallized, along with apatite and quartz, at Schlackenwald in Bohemia, but most abundantly near Leigrosan, in the province of Estremadura in Spain, where it is sometimes associated with apatite, and frequently with amethyst, and forms whole beds, that alternate with limestone and quartz.

Second Kind.

Earthy Phosphorite.

Erdiger Phosphorit, *Karsten.*

Erdiger Phosphorit, *Haus.* s. 123. *Id. Karsten,* Tabel. s. 52.—Chaux phosphatée pulverulente, *Haüy,* Tabl. p. 8.—Erdiger Phosphorit, *Lenz,* b. ii. s. 802. *Id. Haus.* Handb. b. iii. s. 873.

External Characters.

Its colours are greyish-white, greenish-white, and pale greenish-grey.

It consists of dull dusty particles, which are partly loose, partly cohering, and which soil slightly, and feel meagre and rough.

Chemical Characters.

It phosphoresces when laid on glowing coals.

Constituent

Constituent Parts.

Earthy Phosphorite from Marmarosch.	
Lime, - -	47.00
Phosphoric Acid, -	32.25
Fluoric Acid,	2.50
Silica, -	0.50
Oxide of Iron,	0.75
Water, - -	1.00
Mixture of Quartz and Loam,	11.50
	95.50

Klaproth, Beit. b. iv. s. 373.

Geognostic and Geographic Situations.

It occurs in a vein, in the district of Marmarosch in Hungary.

Observations.

1. It was for some time described in systems of mineralogy as a variety of Fluor-spar.

2. The celebrated Prussian chemist John, has published the description and analysis of a mineral under the name *Ratofkite*, from Ratofka, near Werea in Russia, which appears to be nearly allied to this subspecies, and may possibly prove to be a new species, intermediate between fluor and apatite. The following is the description and analysis of it, as given by John, in the second continuation of his work entitled, " Chemische Untersuchungen," &c.

" Its colour is lavender-blue; and it is composed of loose dusty dull particles, that soil slightly, and are not particularly heavy. It is contained in aphrite. *Constituent Parts:* Fluate of Lime, 49.50. Phosphate of Lime,

20.00.

20.00. Muriatic Acid, 2.00. Phosphate of Iron, 3.75. Water, 10.00."

Genus III. FLUOR.

This Genus contains but one species, viz. Octahedral Fluor.

1. Octahedral Fluor.

Flus, *Werner.*

Octaedrisches Flus Haloide, *Mohs.*

It is divided into three subspecies, Compact Fluor, Foliated Fluor, and Earthy Fluor.

First Subspecies.

Compact Fluor.

Dichter Flus, *Werner.*

Fluor solidus, *Wall.* t. i. p. 542.?—Dichter Fluss, *Wid.* s. 542.—Compact Fluor, *Kirw.* vol. i. p. 127.—Dichter Fluss, *Estner,* b. ii. s. 1067. *Id. Emm.* b. i. s. 516.—Fluorite compatta, *Nap.* p. 374.—Le Fluor compacte, *Brock.* t. i. p. 594.—Dichter Fluss, *Reuss,* b. ii. 2. s. 379. *Id. Lud.* b. i. s. 161. *Id. Suck.* 1 th. s. 663. *Id. Bert.* s. 103. *Id. Mohs,* b. ii. s. 150. *Id. Hab.* s. 89.—Chaux fluatée massive compacte, *Lucas,* p. 247.—Dichter Fluss, *Leonhard,* Tabel. s. 38.—Chaux fluatée,

lith, *Karsten*, Tabel. s. 48. *Id. Haus.* s. 121.—Alumine fluatée alkaline, *Haüy*, Tabl. p. 22.—Kryolith, *Lenz*, b. ii. s. 943. *Id. Haus.* Handb. b. iii. s. 846. *Id. Hoff.* b. iii. s. 204. *Id. Aikin*, p. 174.

External Characters.

Its colours are pale greyish-white, snow-white, yellowish-brown and yellowish-red.

It occurs massive, disseminated, and in straight and thick lamellar distinct concretions.

It is shining, inclining to glistening, and the lustre is vitreous, inclining to pearly.

It has a fourfold cleavage, in which the folia are parallel with an equiangular four-sided pyramid; sometimes the folia are parallel with the diagonals of a rectangular four-sided prism, or with the terminal planes.

The fracture is uneven.

The fragments are cubical or tabular.

It is translucent.

It is harder than gypsum, and sometimes as hard as calcareous-spar.

It is brittle, and easily frangible.

Specific gravity, 2.949, *Haüy*.—2.953, *Karsten*.—2.9, 3.0, *Mohs*.

Chemical Characters.

It becomes more translucent in water, but does not dissolve in it. It melts before it reaches a red heat, and when simply exposed to the flame of a candle. Before the blowpipe, it at first runs into a very liquid fusion, then hardens, and at length assumes the appearance of a slag.

Constituent

Constituent Parts.

Alumina, - -	24.0	21.0
Soda, - - -	36.0	32.0
Fluoric Acid, and Water,	40.0	47.0
	100.0	100.0
	Klaproth, Beit. b. iii. s. 214.	*Vauquelin*, Haüy, Traité, t. ii. p. 400.

Geognostic and Geographic Situations.

This curious and rare mineral has been hitherto found only in West Greenland, and but in one place of that dreary and remote region, viz. the Fiord or arm of the sea named Arksut, situated about thirty leagues from the colony of Juliana Hope. It occurs in two thin layers in gneiss: one of these contains the greyish and snow white cryolite, and is not intermixed with other minerals; the other is wholly composed of the yellowish-brown coloured variety, mixed with galena, iron-pyrites, sparry iron, quartz, and felspar. They are situated very near each other: the first is washed at high water by the tide, and a considerable portion of it is exposed, the superincumbent gneiss being removed. It varies from one foot to two feet and a half in thickness *.

Observations.

1. As this mineral, when exposed to a very low heat, melts almost like ice, it was named *Cryolith*, from κρυος, *ice*, and λιθος, *stone*.

2. It

* Allan and Giesecké, in Thomson's Annals, vol. ii. p. 389.

2. It has been confounded with Heavy-spar, from which it is distinguished, by inferior specific gravity, and its easy fusibility before the blowpipe: it might also be mistaken for some varieties of Gypsum, but is distinguished from these by superior specific gravity, and its not exfoliating when exposed to the blowpipe.

GENUS VI.—GYPSUM *.

Gyps Haloide, *Mohs.*

THIS Genus contains two species, viz. 1. Prismatic Gypsum, 2. Axifrangible Gypsum.

1. Prismatic

---

* *Gypsum* is from the Greek word Γυψος. The following explanation of the term γυψος, shows that it was applied by the ancients to an earthy substance that had been exposed to the action of fire: Γύψος αἰονὲς γύψος τίς οὕτω· ἡ ἑψηθεῖσα γῆ (a): in which it corresponds with the gypsum of the moderns. The ancient naturalists sometimes seem to apply the term to sulphate of lime, the gypsum of the present day, and sometimes to a calcined carbonate of lime, or quicklime, which they called *calx*. In the following passage, it is applied to a sulphate of lime; "Cognata calci res gypsum est. Qui coquitur lapis non dissimilis alabastritæ esse debet: omnia autem optimum fieri compertum est e *lapide speculari*, squamamve talem habente (b):" the term *lapis specularis* applying very closely to our selenite, which is a sulphate of lime. "Gypsoma dicto statim utendum est, quoniam *celerrime* coit (c):" the word *celerrime* being more applicable to the comparatively

(a) Vid. Etymolog. Magn.

(b) Plin. Hist. Nat. lib. xxxvi.

(c) Plin. Hist. Nat. lib. xxxvi.

1. Prismatic Gypsum or Anhydrite.

Prismatisches Gyps-Haloide, *Mohs*.

Muriacit, *Werner*.

It is divided into five subspecies, viz. Sparry Anhydrite, Scaly Anhydrite, Fibrous Anhydrite, Conchoidal Anhydrite, Compact Anhydrite. * Vulpinite. * Glauberite.

First Subspecies.

Sparry Anhydrite or Cube-Spar.

Wurfelspath, *Werner*.

Spathiger Karstenit, *Haus.* s. 124.—Spathiger Muriacit, *Karsten,* Tabel. s. 52.—Chaux sulphatée Anhydre laminaire, *Haüy,* Tabl. p. 10.—Wurfelspath, *Lenz,* b. ii. s. 946.—Wurflicher Muriacit, *Hoff.* b. iii. s. 124.

External Characters.

Its chief colour is white, which passes on the one side into blue, and on the other into red: the colours form the following

---

ratively rapid consolidation of calcined gypsum, than to that of common mortar. There is a passage in Theophrastus, in which a ship is said to have been set on fire, in consequence of the moistening of its cargo, which consisted of gypsum and wearing-apparel: in this case, there can be little doubt, that the substance called gypsum could not have been of the same nature with the gypsum of the present day, which in no instance, perhaps, contains such a proportion of carbonate of lime, as, when calcined, would be sufficient to produce this effect.—*Kid's Min.* vol. i. p. 69, 70, 71.

following series; a colour intermediate between brick-red and flesh-red, reddish-white, yellowish-white, greyish-white, pearl-grey, which sometimes inclines to rose-red, and also pale violet-blue. Sometimes it is slightly iridescent.

It occurs massive; also in distinct concretions, which are thin and straight lamellar, collected into others which are large granular. It is sometimes crystallized. The primitive figure is an oblique prism, in which the angles are 108° 8′ and 79° 56′. The following are some of the secondary forms:

1. Rectangular four-sided prism: it is sometimes so low as to appear as a four-sided table.
2. Broad six-sided prism.
3. Eight-sided prism.
4. Broad rectangular four-sided prism, acuminated on the extremities with four planes, which are set on the lateral edges, and the apex of the acumination deeply truncated.

Externally it is shining or splendent, and pearly: internally splendent and pearly.

It has a threefold cleavage, and the cleavages are perpendicular to each other; two of them are parallel with the lateral planes of the primitive prism, the third parallel with the terminal planes.

The fragments are cubical.

The fracture is conchoidal.

It alternates from transparent to strongly translucent, and refracts double.

It scratches calcareous-spar, but does not affect fluor-spar.

It is brittle, and very easily frangible.

Specific

Specific gravity, 2.957, *Bournon.*—2.964, *Klaproth.*—2.7, 3.0, *Mohs.*

Chemical Characters.

When exposed to the blowpipe, it does not exfoliate, and melt like gypsum, but becomes glazed over with a white friable enamel.

Constituent Parts.

	From Bern.	From Tyrol.
Lime, - -	40	41.75
Sulphuric Acid, -	60	55.00
Muriate of Soda, -		1.00
	100	97.75
	Haüy, Traité, t. iv. p. 349.	*Klaproth*, Beit. b. iv. s. 235.

Geognostic and Geographic Situations.

It is sometimes met with in the gypsum of Nottinghamshire *. In the salt-mines of Hall in the Tyrol; in those of Bex in Switzerland; in quartz, along with talc, sulphur, and iron-pyrites, in the mine of Pesay, also in Switzerland; Lauterberg in the Hartz; Tiede, near Brunswick; and in the large copper-mine of Fahlun in Sweden.

Second

* Greenough.

Second Subspecies.

Scaly Anhydrite.

Anhydrit, *Werner*.

Schupiger Muriacit, *Karsten*.

Schuppiger Karstenit, *Haus.* s. 124.—Schuppiger Muriacit, *Karsten,* Tabel. s. 52.—Chaux sulphatée Anhydre lamellaire, *Haüy,* Tabl. p. 10.—Schuppiger Anhydrit, *Lenz,* b. i. s. 849.

External Characters.

Its colours are snow, greyish, and milk white, which latter passes into smalt-blue, and rarely into grey.

It occurs massive, and in small granular concretions.

The lustre is splendent and pearly.

The cleavage is imperfect and curved.

The fragments are not particularly blunt-edged.

It is translucent on the edges.

It is easily frangible.

Specific gravity 2.957, *Haüy.*

Constituent Parts.

Lime, - - - - - -	41.75
Sulphuric Acid, - - -	55.00
Muriate of Soda, - - -	1.00
	97.75

Klaproth, Beit. b. iv. s. 235.

Geognostic and Geographic Situations.

It is found in the salt-mines of Hall in the Tyrol, 5088 feet above the level of the sea.

Third

Third Subspecies.

Fibrous Anhydrite.

Fasriger Muriacit, *Werner.*

Fasriger Muriacit, *Karsten,* Tabel. s. 52.—Fasriger Karstenit, *Haus.* Handb. b. iii. s. 883.—Fasriger Muriacit, *Hoff.* b. iii. s. 136.—Fasriger Anhydrit, *Lenz,* b. ii. s. 724.

External Characters.

Its colours are brick-red, and pale blood-red; also indigo-blue, Berlin-blue, smalt-blue, and smoke-grey.

It occurs massive, and in coarse fibrous concretions, which are straight or curved, and sometimes stellular.

Internally it is glimmering and glistening, and pearly.

The fragments are long splintery.

It is translucent on the edges, or feebly translucent.

It is rather easily frangible.

Specific gravity 3.002, *Breithaupt.*

Geographic Situation.

It is found in the salt-mines of Berchtesgaden, and at Ischel in Upper Austria, at Hall in the Tyrol, Salz on the Neckar, Carinthia, and Tiede near Brunswick.

Uses.

The blue varieties are sometimes cut and polished for ornamental purposes.

Fourth Subspecies.

Convoluted Anhydrite.

Gekröstein, *Werner.*

Chaux sulphatée Anhydre concretionnée, *Haüy.*—Gekröstein, *Hoff.* b. iii. s. 131.

External Characters.

Its colour is dark milk-white.

It occurs massive; also in distinct concretions, which are thick lamellar, and intestinally convoluted or contorted, and these are again composed of others which are thin prismatic.

Internally it is glistening or glimmering, and the lustre is pearly.

The fracture is small and fine splintery.

The fragments are indeterminate angular, and rather sharp-edged.

It is translucent on the edges, or translucent.

Same hardness as other subspecies.

Specific gravity 2.850, *Klaproth*.

Constituent Parts.

Lime, - - -	42.00
Sulphuric Acid, - -	56.50
Muriate of Soda, - -	0.25
	98.75

Klaproth, Beit. b. iv. s. 236.

Geognostic and Geographic Situations.

It occurs in the salt-mines of Bochnia, and at Wieliczka in Poland.

Observations.

It was first described as a variety of compact heavy-spar and is by many named *Pierre de Tripes*, from its convoluted concretions.

Fifth Subspecies.

Compact Anhydrite.

Dichter Muriacit, *Werner.*

Dichter Muriacit, *Karsten,* Tabel. s. 52.—Dichter Karstenit, *Haus.* s. 124.—Chaux sulphatée Anhydre compacte, *Haüy,* Tabl.—Dichter Anhydrit, *Lenz,* b. ii. s 847.—Dichter Muriaticit, *Hoff.* b. iii. s. 133.

External Characters.

Its colours are bluish-grey, greyish-white, and colours intermediate between ash and smoke grey, and between ash-grey and tile-red. Sometimes with spotted delineations.

It occurs massive; also in granular distinct concretions.

It is feebly glimmering, or dull.

The fracture is small splintery, passing into even and flat conchoidal.

The fragments are more or less sharp-edged.

It alternates from translucent to translucent on the edges.

Same hardness as other subspecies.

Specific gravity, 2.850, *Klaproth.*—2906, *Rose.*—2.969, *Breithaupt.*

Constituent Parts.

Lime, -	41.48	Lime, -	42.00
Sulphuric Acid,	56.28	Sulphuric Acid,	56.50
Water, - -	0.75	Muriate of Soda,	0.25
		Loss, - -	1.28
	100.00		100.00
	Rose, in Karsten's Tabellen.		*Klaproth,* Beit. b. iv. s. 233.

 Geognostic

Geognostic and Geographic Situations.

It occurs in beds in the salt-mines of Austria and Salzburg; and also in secondary gypsum, on the eastern foot of the Hartz mountains.

* Vulpinite.

Vulpinite, *De La Metherie*, Tableaux.—Chaux sulphatée quartzifere, *Haüy*, t. iv. p. 355.—Pierre de Vulpino, dans le Bergamasce, *Fleuriau de Belleveu*.—Chaux anhydro-sulphatée quartzifere, *Haüy*, Tabl. p. 11.—Vulpinit, *Lenz*, b. ii. s. 851.

External Characters.

Its colour is greyish-white, and veined with bluish-grey.

It occurs massive.

Internally it is splendent.

The fracture is foliated, and is said to exhibit a threefold slightly oblique cleavage.

The fragments are rhomboidal.

It occurs in granular distinct concretions.

It is translucent on the edges.

It is soft.

It is brittle.

It is easily frangible.

Specific gravity 2.878, *Haüy*.

Chemical Characters.

It melts easily before the blowpipe into a white opaque enamel; and becomes feebly phosphorescent when thrown on glowing coals.

Constituent

Constituent Parts.

Sulphate of Lime, -	92.0
Silica, - - -	8.0
	100.0

Vauquelin, in Bulletin des Sciences de la Société Philomatique, N. 9.; Journal de Physique, t. xlvii. p. 101.; Journal des Mines, N. xxxiv.

Geognostic and Geographic Situations.

It occurs along with granular foliated limestone, and is sometimes associated with quartz, and occasionally with sulphur. It is found at Vulpino in Italy.

Uses.

1. It takes a very fine polish, and is employed by the statuaries of Bergamo and Milan for making slabs, chimney-pieces, &c. It is known to artists by the name *Marmo bardiglio di Bergamo.*

2. It was first particularly noticed by Fleuriau.

* Glauberite.

Glauberite, *Brongniart.*

Glauberite, *Brong.* Journal des Mines, t. xxiii. p. 5. *Id. Haüy*, Tabl. p. 23. *Id. Lenz*, b. ii. s. 950. *Id. Aikin*, p. 139.

External Characters.

Its colours are greyish-white, and wine-yellow.

It

It occurs crystallized, in very low oblique four-sided prisms, the lateral edges of which are 104° 28′ and 75° 32′, and in which the terminal planes are set on obliquely.

The crystals occur singly, or in groups.

The lateral planes are transversely streaked; the terminal planes are smooth.

It is shining.

The fracture parallel with the terminal planes and edges is foliated; in other directions it is conchoidal.

It is softer than calcareous-spar.

It is transparent.

It is brittle.

Specific gravity 2.700.

Chemical Characters.

It decrepitates before the blowpipe, and melts into a white enamel. In water it becomes opaque, and is partly soluble.

Constituent Parts.

Dry Sulphate of Lime,	49.0
Dry Sulphate of Soda,	51.0
	100.0

Brongniart, J. des Mines, t. xxiii. p. 17.

Geognostic and Geographic Situations.

It is found imbedded in rock-salt at Villaruba, near Ocana in New Castile, in Spain.

Observations.

It was brought from Spain to Paris by M. Dumeril, and first analysed and described by Brongniart.

2. Axifrangible

2. Axifrangible Gypsum.

Axentheilendes Gyps-Haloide, *Mohs.*

This species contains six subspecies, viz. Sparry Gypsum or Selenite, Foliated Granular Gypsum, Compact Gypsum, Fibrous Gypsum, Scaly Foliated Gypsum, and Earthy Gypsum. * Montmartrite.

First Subspecies.

Sparry Gypsum or Selenite.

Fraueneis, *Werner.*

Lapis specularis, *Plin.* Hist. Nat. xxxvi. 22. 145.—Gypsum selenites, *Wall.* t. i. p. 165.—Selenite, *Romé de Lisle,* t. i. p. 441. —Fraueneis, *Wern.* Cronst. s. 53.—Broad foliated Gypsum, *Kirw.* vol. i. p. 123.—Fraueneis, *Emm.* b. i. s. 540.—Chaux sulphatée cristallisée, *Haüy,* t. ii. p. 266.—La Selenite, *Broch.* t. i. p. 609.—Spathiger Gyps, *Reuss,* b. ii. 2. s. 406.—Fraueneis, *Lud.* b. i. s. 164. *Id. Suck.* 1r th. s. 675.—Grossblättriger Gyps, *Bert.* s. 109.—Fraueneis, *Mohs,* b. ii. s. 183. *Id. Hab.* s. 84.—Spathiger Gyps, *Leonhard,* Tabel. s. 37.—Chaux sulphatée Selenite, *Brong.* t. i. p. 171.—Spathiger Gyps, *Karsten,* Tabel. s. 52. *Id. Haus.* s. 124.—Selenite, *Kid,* vol. i. p. 66.—Gyps-spath, *Lenz,* b. ii. s. 840.—Durchschtiger Gyps, *Oken,* b. i. s. 400.—Spathiger Gyps, *Haus.* Handb. b. iii. s. 887.—Fraueneis, *Hoff.* b. iii. s. 117.—Selenite, *Aikin,* p. 167.

External Characters.

Its colours are smoke-grey, greyish-white, snow-white, greenish-

crystals are pushed into each other in the direction of their length, a

4. Quadruple crystal is formed.

The crystals occur of all degrees of magnitude, and are often long and acicular.

The lateral planes of the prism are sometimes smooth, sometimes longitudinally streaked, and shining; the convex terminal faces, and the lens, are rough and dull.

Internally the lustre is splendent and pearly.

The cleavage is three-fold; the most distinct cleavage is perpendicular to the axis of the prism; the other two are parallel with the lateral planes of the primitive prism. The cleavages are generally straight, and sometimes curved.

The fragments are rhomboidal, in which two of the sides are smooth and splendent, and four are streaked and shining.

It alternates from semi-transparent to transparent, and in the latter case is observed to refract double.

It yields readily to the nail; scratches talc, but not calcareous-spar.

It is sectile.

It is very easily frangible.

In thin pieces it is flexible, but not elastic.

Specific gravity, 2.322, *Muschenbröck*—2.3065, 2.3117, 2.3846, *Kopp.*—2.2, 2.4, *Mohs.*

Chemical Characters.

It exfoliates before the blowpipe, and, if the flame is directed towards the edge of the folia, it melts into a white enamel, which, after a time, falls into a white powder.

Constituent

Lime, -	33.9
Sulphuric Acid,	43.9
Water, -	21.0
Loss,	2.1
	100.0

Bucholz, in Gehlen's Journ. b. v. s. 158.

Geognostic Situation.

It occurs principally in the secondary or flœtz gypsum formation, in thin layers: less frequently in rock-salt; more rarely as a constituent part of metalliferous veins; but in considerable quantity in that deposite known in the south of England under the name Blue or London Clay. Crystals of this substance are daily forming in gypsum hills, in old mines, and in mining heaps.

Geographic Situation.

It is not unfrequent in the blue clay in the south of England, as at Shotover Hill, near Oxford; Newhaven, Sussex; Isle of Shepey in the Medway; and at Alston in Cumberland. It occurs in the secondary gypsum around Paris; in veins of copper-pyrites and grey copper-ore, at Herrengrund, near Newsohl; in a vein of galena or lead-glance at Teschen, in Bohemia; and all over the Continent of Europe, and in other quarters of the globe where foliated granular gypsum occurs.

Uses.

At a very early period, before the discovery of glass, selenite was used for windows; and we are told, that in the time

time of Seneca, it was imported into Rome from Spain, Cyprus, Cappadocia, and even from Africa. It continued to be used for this purpose until the middle ages; for Albinus informs us, that in his time, the windows of the dome of Merseburg were of this mineral. The first greenhouses, those invented by Tiberius, were covered with selenite. According to Pliny, bee-hives were incased in selenite, in order that the bees might be seen at work. It is used for the finest kind of stucco, and the most delicate pastil-colours. When burnt, and perfectly dry, it is used for cleansing and polishing precious stones, work in gold and silver, and also pearls. It was formerly much used by Roman Catholics for *frosting* the images of the Virgin Mary: hence the names *Glacies Mariæ* or *Frauen-glas* given to it. It has also been named *Lapis specularis*, and *Gypsum speculare* and *glaciale*, from its resemblance to glass or ice.

Second Subspecies.

Foliated Granular Gypsum.

Blættriger Gyps, *Werner.*

Gypsum lamellare, *Wall.* t. i. p. 165.—Blættriger Gyps, *Wid.* s. 548.—Granularly Foliated Gypsum, *Kirw.* vol. i. p. 123.—Blættriger Gyps, *Estner,* b. ii. s. 1109. *Id. Emm.* b. i. s. 532.—Gesso lamellare, *Nap.* p. 381.—Le Gyps lamelleux, *Broch.* t. i. p. 606.—Körniger Gyps, *Reuss,* b. ii. 2. s. 400.—Blættriger Gyps, *Lud.* b. i. s. 163. *Id. Suck.* 1r th. s. 673. *Id. Bert.* s. 107. *Id. Mohs,* b. ii. s. 180.—Körnigblättriger Gyps, *Hab.* s. 85.—Körniger Gyps, *Leonhard,* Tabel. s. 37. *Id. Karst.* Tabel. s. 52.—Schupiger Gyps, *Haus.* s. 125.—Körniger

ger & Klein blättriger Gyps, *Lenz*, b. ii. s. 838. *Id. Hof.* b. iii. s. 110.—Massive Gypsum, with a granularly lamellar structure, *Aikin*, p. 168.

External Characters.

Its most common colours are white, grey, and red; seldomer yellow, brown, and black. The white colours are, snow, greyish, yellowish, and reddish white; from reddish-white, it passes into flesh-red, blood-red, and brick-red; the greyish-white passes into ash-grey, and smoke-grey, and greyish-black; and the yellowish-grey passes into wax-yellow. It seldom occurs of a hair-brown colour, and this only when it is intermixed with stinkstone. The colours sometimes occur in spotted, striped, and veined delineations.

It occurs massive, also in distinct concretions, which are large, coarse, small, and fine angulo-granular, seldom prismatic, and these are broad, narrow, short, wedge-shaped, scopiform or stellular. It is sometimes crystallized, in small conical lenzes, in which the surface is rough.

The lustre passes from shining through glistening to glimmering, and is pearly.

It has the same cleavage as selenite.

The fragments are very blunt-edged.

It is translucent.

It is very soft.

It is sectile, and very easily frangible.

Specific gravity, 2.2741, 2.3108, *Brisson.*

Constituent

Constituent Parts.

Lime, -	32
Sulphuric Acid,	30
Water, -	38
	98 *Kirwan.*

A reddish-white variety, found near Lüneburg, according to Hausmann, afforded, besides Sulphate of Lime, 4 parts of Muriate of Lime.—*Haus.* Nord. Deutsch. Beit. st. ii. s. 98.

Geognostic Situation.

It occurs in beds in primitive rocks, as gneiss and mica-slate: in a similar repository in transition clay-slate; but most abundantly in beds in the rocks of the secondary or flœtz class. In these rocks it is associated with selenite, compact gypsum, fibrous gypsum, rock-salt, stinkstone, and limestone.

Geographic Situation.

Europe.—It occurs in Cheshire and Derbyshire; at the Segeberg, near Kiel; and at Lüneburg, where it contains crystals of boracite, and sometimes of quartz. It is associated with flœtz rocks in Thuringia, Mansfeldt, Silesia, Suabia, Bavaria, Austria, Switzerland, the Tyrol, Poland, Spain, and France. At Airolo, in the St Gothard group, it occurs in beds in gneiss; at St Meul in the Valais, it alternates with hornblende-slate; and with mica-slate on Mount Cenis: in Salzburg, it is associated with transition limestone, and clay-slate, and there it sometimes contains sparry

sparry gypsum or selenite, and also grey copper-ore, copper-pyrites, iron-pyrites, galena, and cinnabar.

Asia.—It is found in Persia, Caramania, and in different parts of Siberia; both in flœtz and primitive mountains. Pallas mentions his having met with granular gypsum, along with mica, serpentine, and felspar, in Siberia.

America.—It is found near Athapuscau Lake, where rock-salt occurs; also in the United States of America; in Nova Scotia; and at the foot of the Andes, in South America.

Uses.

The foliated and compact subspecies of gypsum, when pure, and capable of receiving a good polish, are by artists named simply *Alabaster*, or, to distinguish them from calc-sinter, or what is called calcareous alabaster, *Gypseous Alabaster*. The finest white varieties of granular gypsum, are selected by artists for statues and busts: the variegated kinds are cut into pillars, and various ornaments for the interior of halls and houses; and the most beautiful variegated sorts are cut into vases, columns, plates, and other kinds of table furniture. Those varieties that contain imbedded portions of selenite, when cut across, exhibit a beautiful iridescent appearance, and are named *Gypseous Opal*. In Derbyshire, and also in Italy, the very fine granular varieties, are cut into large vases, columns, watch-cases, plates, and other similar articles. If a lamp is placed in a vase of snow-white translucent gypsum, a soft and pleasing light is diffused from it through the apartment. It is said the ancients being acquainted with this property, used gypsum in place of glass, in order that the light in their temples might be pale and mysterious, and in harmony with the place. The *phengites* of the ancients would appear to have been foliated gypsum. According to Pliny, it was employed instead of glass in windows, on account of its translucency; and in the Temple of Fortune, and the gilded palace of Nero, the chambers were lined with the finest and most highly prized kinds of phengites. Domitian, who, towards the close of his life became suspicious and distrustful of all around him, had the portico in which he used to walk lined with phengites, in order that he might see what was passing behind him. Both subspecies are used in agriculture. Much difference of opinion has prevailed among agriculturists with respect to the uses of gypsum. It is said to have been very advantageously employed in America; and also in the county of Kent: but it has failed in most of the other counties of England, though tried in various ways, and for different crops. When peat-ashes contain a considerable portion of gypsum, they may be advantageously employed as a top-dressing for cultivated grasses, on such soils as contain little or no sulphate of lime. The pure white varieties of granular gypsum are used as ingredients in the composition of earthen-ware and porcelain; and the glaze or enamel with which porcelain is covered, has the purest gypsum, or even selenite, as one of its ingredients. Its most important use is in the preparation of *stucco:* for this purpose, the gypsum is first exposed to a heat sufficient to drive off its water of crystallization, then finely ground, mixed with a small portion of fine sand and quicklime, and, lastly, a determinate proportion of water is added, which occasions the compound first to swell, and then to contract and harden. Stucco is sometimes used for lining walls and roofs of apartments, in place of common plaster;

ter; and occasionally for covering the floors of summer-houses or churches. The finest kind of stucco is used for casts of figures, and statues of various kinds. Artificial coloured marbles are also made of stucco, which are used for covering pillars, walls, altars, pavements of churches, or halls.

Observations.

1. Its inferior weight and hardness distinguish it from *Dolomite* and *Granular Foliated Limestone*.

2. It has frequently a porphyritic structure; the imbedded minerals are generally crystals of quartz; sometimes crystals of Arragonite, as in certain Spanish varieties, and rarely crystals of Boracite.

Third Subspecies.

Compact Gypsum.

Dichter Gyps, *Werner.*

Gypsum alabastrum, et Gypsum æquabile, *Wall.* t. i. p. 161, 162.—Dichter Gyps, *Wid.* s. 544.—Compact Gypsum, *Kirw.* vol. i. p. 121.—Dichter Gyps, *Estner,* b. ii. s. 1098. *Id. Emm.* b. i. s. 529.—Gesso compatto alabastro, *Nap.* p. 384.—Alabastrite, *Lam.* t. ii. p. 76.—Chaux sulphatée compacte, *Haüy,* t. ii. p. 266.—Le Gypse compacte, *Broch.* t. i. p. 602.—Dichter Gyps, *Reuss,* b. ii. 2. s. 393. *Id. Lud.* b. i. s. 163. *Id. Suck.* 1r th. s. 670. *Id. Bert.* s. 105. *Id. Mohs,* b. ii. s. 179. *Id. Hab.* s. 86.—Chaux sulphatée compacte, *Lucas,* p. 13.—Dichter Gyps, *Leonhard,* Tabel. s. 37.—Chaux sulphatée Gypse compacte, *Brong.* t. i. p. 174.—Chaux sulphatée compacte,

pacte, *Brard*, p. 52.—Dichter Selenit, *Haus.* s. 125.—Dichter Gyps, *Karsten*, Tabel. s. 52.—Chaux sulphatée compacte, *Haüy*, Tabl. p. 10.—Dichter Gyps, *Lenz*, b. ii. s. 834. *Id. Hoff.* b. iii. s. 108.—Massive Gypsum, *Aikin*, p. 168.

External Characters.

Its colours are snow, yellowish, reddish, and greyish white; smoke, yellowish, ash, bluish and greenish grey, pale sky-blue, and violet-blue; a colour intermediate between brownish and brick red, seldom flesh-red; sometimes honey-yellow. Frequently several colours occur in the same piece, and these are either spotted, flamed, striped, or veined.

It occurs massive.

It is generally dull, seldom feebly glimmering.

The fracture is fine splintery, passing on the one side into even, on the other into fine-grained uneven.

The fragments are indeterminate angular, and blunt-edged.

It is translucent on the edges.

It is very soft; sectile; and easily frangible.

Specific gravity, 2.1679, before absorption of water; 2.2052, after absorption of water, *Brisson.*—2.288, *Kirwan.*—2.287, *Breithaupt.*

Chemical Characters.

All the different varieties of gypsum, when exposed to heat, are deprived of their water of crystallization, become opaque, fall into a powder, which, when mixed with water, speedily hardens on exposure to the air. They are difficultly fusible before the blowpipe, without addition, and

Fourth Subspecies.

Fibrous Gypsum.

Fasriger Gyps, *Werner.*

Gypsum striatum, *Wall.* t. i. p. 167.—Fasriger Gyps, *Wid.* s. 546. —Fibrous Gypsum, *Kirw.* vol. i. p. 122.—Fasriger Gyps, *Estner,* b. i. s. 1105. *Id. Emm.* b. i. s. 536.—Gesso fibroso, *Nap.* p. 386.—Chaux sulphatée fibreuse, *Haüy,* t. ii. p. 266. —La Gypse fibreuse, *Broch.* t. i. p. 604.—Fasriger Gyps, *Reuss,* b. ii. 2. s. 396. *Id. Suck.* 1r th. s. 678. *Id. Bert.* s. 106. *Id. Mohs,* b. ii. s. 182.—Chaux sulphatée fibreuse, *Lucas,* p. 13.—Fasriger Gyps, *Leonhard,* Tabel. s. 37.—Chaux sulphatée Gypse fibreuse, *Brong.* t. i. p. 174.—Chaux sulphatée fibreuse, *Brard,* p. 52.—Fasriger Gyps, *Karsten,* Tabel. s. 52. *Id. Haus.* s. 125.—Fibrous Gypsum, *Kid,* vol. i. p. 65.—Chaux sulphatée fibreuse-conjointe, *Haüy,* Tabl. p. 10.—Fasriger Gyps, *Lenz,* b. ii. s. 844. *Id. Hoff.* b. iii. s. 115.—Fibrous Gypsum, *Aikin,* p. 168.

External Characters.

Its principal colours are white, grey, and red; of white, it possesses the following varieties, viz. yellowish, greyish, snow, and reddish white; from reddish-white it passes into brick-red, flesh-red, and brownish-red; the yellowish-white passes into yellowish-grey, wine-yellow, and honey-yellow. Sometimes several colours occur together in the same specimen.

It occurs massive, and dentiform; also in fibrous distinct concretions, which are parallel, generally straight, and sometimes curved.

Its

Its lustre passes from glistening, through shining to splendent, and is pearly.

The fragments are long splintery.

It is translucent.

It is very soft, sectile; and easily frangible.

Constituent Parts.

Lime,	33.00
Sulphuric Acid,	44.13
Water,	21.00
	98.13

Bucholz, N. Allg. Journ. d. Chem. b. v. H. ii. s. 160.

Geognostic Situation.

It occurs along with the other subspecies of this species.

Geographic Situation.

It occurs in red sandstone near Moffat; in red clay, on the banks of the Whitadder in Berwickshire; in Dunbartonshire; also in Cumberland, Yorkshire, Cheshire, Worcestershire, Derbyshire, Somersetshire, and Devonshire *. On the Continent of Europe, it is met with in Thuringia, Mansfeldt, Bavaria, Salzburg, Switzerland, France, &c.

Uses.

When cut *en cabachon*, and polished, it reflects a light not unlike that of the cat's-eye, and is sometimes sold as that

* Greenough.

that stone. It is also cut into necklaces, ear-pendants, and crosses; and in this form, it is also sold for a harder mineral, the Fibrous Limestone, or even imposed on the ignorant for that variety of felspar named Moonstone.

Observations.

It might be confounded with Fibrous Limestone, and Asbestus, but is readily distinguished from these minerals by its inferior hardness, and the alteration it undergoes at a low red heat.

Fifth Subspecies.

Scaly Foliated Gypsum.

Schaumgyps, *Werner.*

Schaumgyps, *Hoff.* b. iii. s. 106.—Chaux sulphatée niviforme, *Haüy.*

External Characters.

Its colours are yellowish-white and snow-white.

It occurs massive and disseminated; also in distinct concretions, which are small and scaly granular.

Internally it is glistening and pearly.

The fracture is small scaly foliated.

The fragments are indeterminate angular and blunt-edged.

It is opaque, or translucent on the edges.

It is very soft, passing into friable.

It is sectile, and easily frangible.

Geognostic

Geognostic and Geographic Situations.

It occurs along with selenite and compact gypsum at Montmartre, near Paris, in that formation of gypsum named by Werner the third or yellow flœtz gypsum formation.

Observations.

This subspecies is characterized by its scaly foliated aspect.

Sixth Subspecies.

Earthy Gypsum.

Gyps-erde, *Werner.*

Gypsum terrestre farinaceum ; Farina fossilis, *Wall.* t. i. p. 36. —Gypserde, *Wid.* s. 543.—Farinaceous Gypsum, *Kirw.* vol. i. p. 120.—Gypserde, *Estner,* b. ii. s. 1095. *Id. Emm.* b. i. s. 527.—Gesso terroso, *Nap.* p. 379.—Le Gypse terreux, *Broch.* t. i. p. 601.—Chaux sulphatée terreuse, *Haüy,* t. ii. p. 278.—Erdiger Gyps, *Reuss,* b. ii. 2. s. 391. *Id. Lud.* b. i. s. 163. *Id. Suck.* 1r th. s. 669. *Id. Bert.* s. 105. *Id. Mohs,* b. ii. s. 178.—Chaux sulphatée terreuse, *Lucas,* p. 13.—Erdiger Gyps, *Leonhard,* Tabel. s. 37.—Chaux sulphatée, Gypse terreux, *Brong.* t. i. p. 174.—Chaux sulphatée terreuse, *Brard,* p. 52.—Erdiger Selenit, *Haus.* s. 125.—Erdiger Gyps, *Karsten,* Tabel. s. 52.—Farinaceous Gypsum, *Kid,* vol. i. p. 65. —Chaux sulphatée terreuse, *Haüy,* Tabl. p. 10.—Gypserde, *Lenz,* b. ii. s. 833. *Id. Hoff.* b. iii. s. 107.—Earthy Gypsum, *Aikin,* p. 168.

External Characters.

Its colour is yellowish-white, which passes into yellowish-grey, and sometimes inclines to snow-white.

It

It is composed of fine scaly or dusty particles, which are more or less cohering.

It is feebly glimmering.

It feels meagre, and rather fine.

It soils slightly.

It is light.

Geognostic Situation.

It is found immediately under the soil, in beds several feet thick, resting on gypsum, and also in nests or cotemporaneous masses imbedded in it. It is conjectured to have been formed in some instances by the decay of previously existing gypsum beds; in others, it appears to be an original deposite, of cotemporaneous formation with the solid kinds of gypsum.

Geographic Situation.

It is found in Saxony, Switzerland, Salzburg, and Norway.

Use.

In some districts, it is used as a manure.

Observations.

1. It is distinguished from *Agaric Mineral* by its colour, scaly particles, and its soiling feebly; and from *Earthy Heavy-spar*, by its inferior specific gravity.

2. It is the *Himmels-mehl* and *Gyps-guhr* of some authors.

* Montmartrite.

Lightning Source UK Ltd.
Milton Keynes UK
UKHW011124020119
334816UK00016B/1697/P